STRATEGIC MINERALS

VOLUME II

STRATEGIC MINERALS

VOLUME II

MAJOR MINERAL-CONSUMING REGIONS OF THE WORLD

Issues and Strategies

W. C. J. van Rensburg
with Paul Anaejionu

University of Texas at Austin

PRENTICE-HALL, INC., Englewood Cliffs, New Jersey 07632

Library of Congress Cataloging-in-Publication Data

VAN RENSBURG, W. C. J.
    Strategic minerals.

    (Prentice-Hall international series in world
resources, energy, and minerals)

    Includes bibliographies and indexes.
    Contents: v. 1. Major exporting regions of the world,
issues and strategies—v. 2. Major mineral-consuming
regions of the world, issues and strategies.
    1. Mineral industries—Government policy—Case
studies.   2. Metal trade—Government policy—Case
studies.   3. Strategic materials—Government policy—
Case studies.   I.   Title.
HD9506.A2V35 1986   333.8'5   85-6569
ISBN 0-13-851387-2 (v. 1)
ISBN 0-13-851411-9 (v. 2)

Cover design: Edsal Enterprises
Manufacturing buyer: Rhett Conklin

PRENTICE-HALL INTERNATIONAL SERIES IN WORLD RESOURCES,
    ENERGY, AND MINERALS

W. C. J. van Rensburg, Series Editor

Printed in the United States of America

10 9 8 7 6 5 4 3 2 1

ISBN 0-13-851411-9 01

Prentice-Hall International (UK) Limited, *London*
Prentice-Hall of Australia Pty. Limited, *Sydney*
Prentice-Hall Canada Inc., *Toronto*
Prentice-Hall Hispanoamericana, S.A., *Mexico*
Prentice-Hall of India Private Limited, *New Delhi*
Prentice-Hall of Japan, Inc., *Tokyo*
Prentice-Hall of Southeast Asia Pte. Ltd., *Singapore*
Editora Prentice-Hall do Brasil, Ltda., *Rio de Janeiro*
Whitehall Books Limited, *Wellington, New Zealand*

Contents

7-7-87 Midwest 96.03

Preface

As the first term of the Reagan administration came to an end, it was clear that its concern about the availability and security of strategic minerals was continuing. The efforts of Congress and industry representatives to create in the White House a council on minerals and materials found expression in S.373, a congressionally approved piece of legislation intended to fulfill this objective. At the same time, the Secretary of the Interior's Strategic Materials and Minerals Advisory Committee was considering options for reducing America's vulnerability to supply disruptions. However, in the meantime, Congress and the Executive Branch continue to keep America's public lands closed to mineral exploration and development. It is possible, over the next few years, that an additional 80 million acres may be closed to mining.

The National Materials and Minerals Policy Research and Development Act of 1980 was intended to supplant the traditional ad hoc decision-making and policy coordination process used to develop and implement national minerals and materials policy with a more formal mechanism. Although the Reagan administration has taken some steps to implement this act, a recent report by the General Accounting Office concluded that there are still serious deficiencies in U.S. minerals policy formulation.

Over the past 30 years the United States has become much more dependent on foreign sources of minerals and energy. With the granting of independence to numerous developing countries in Africa, Asia, and Latin America, the political control of the West over the mineral resources of the Third World has largely evaporated. The decline of the American fleet, together with the massive buildup of the Soviet fleet, means that the supply of minerals from these countries to the West has become far less secure.

As a result of the oil crises of 1974 and 1980, energy costs have increased substantially. This, together with growing opposition to mining and mineral processing on environmental grounds, has made it much more attractive for the United States to import minerals in processed forms. As a result,

we have seen the virtual demise of the U.S. mineral processing industry.

During the Truman and Eisenhower administration, a substantial strategic minerals stockpile was developed. However, for the next 20 years there were no significant stockpile purchases, and massive amounts of stockpiled materials were disposed of. The Reagan administration's efforts to restore the stockpile were largely defeated by massive federal deficits. As a result, much of the material in the stockpile has deteriorated, no longer complies with modern specifications, and/or is stockpiled in the wrong form or in insufficient quantities. The deficiencies of the stockpile are far more serious than is generally appreciated because in the event of a major global conflict the United States would almost certainly have to share these materials with Western Europe and Japan, which have totally inadequate domestic stockpiles.

Western Europe has continued to rely on international trade for most of its energy and mineral supplies. Special relations with former colonies have been cultivated in pursuit of more secure supplies. However, the two energy crises, the invasion of Shaba Province in Zaire, and the Falkland Islands War have raised concern about continued reliability and security of supplies, and have spurred efforts in a number of countries to establish at least modest strategic mineral stockpiles. Western Europe remains highly vulnerable to the disruption of its energy and mineral supplies—a situation that is clearly understood in the USSR.

After nearly three decades of spectacular economic growth, with concomitant increases in energy and mineral consumption, Japan faced the shock of the two energy crises and the realization of its great vulnerability to the disruption of raw material supplies. With its advanced technology, Japan appears to have taken a conscious decision to make a structural change in its economy and to move toward the production and export of high-technology products and away from mineral processing. This may reduce Japan's reliance on imports of unprocessed minerals, and like the United States, Japan will increasingly import minerals in processed forms. Whether this will reduce Japan's vulnerability to supply disruptions remains to be seen.

Since World War II, the USSR has become a superpower in industrial and military might. Soviet mineral production has soared, and a policy of energy and mineral self-sufficiency, regardless of cost, has been followed. There is a difference of opinion regarding the present energy and mineral position of the USSR. Two CIA reports have predicted that Soviet oil production would soon peak. Others have pointed to increasing Soviet imports of chromium, manganese, copper, lead, and bauxite. Their conclusion was that the USSR was approaching the depletion of several of her energy and mineral resources, and that, like the West, she would soon become dependent on imports of these commodities. Lacking sufficient hard-currency resources to acquire these commodities by normal international trade, the Soviets may be forced to appropriate them by force. Other observers scoffed at this, and maintained that the Soviets have no serious problems in meeting virtually all of their energy and mineral requirements from domestic sources. It now appears that the Soviets may have exhausted some of their energy and mineral resources in the western (European) part of the country, and that a massive move to the vast, inhospitable areas of the East is imminent. Producing energy and minerals in the Asian part of the USSR will require huge amounts of capital, vast developments in infrastructure, and massive movements of population. It will take time.

The countries covered in this volume—the United States, Western Europe, Japan, and the USSR—consume a disproportionate share of the world's energy and minerals.

ACKNOWLEDGMENTS

A study of this type necessarily draws on the work of hundreds of other scholars. It is not possible to mention them all by name. However, they are recognized in the bibliographies and many lists of references in these two volumes.

We wrote to numerous individuals, government departments, and mining companies in dozens of countries in our quest for completeness. A very high percentage of them responded generously. We wish to thank all of them.

A special vote of thanks is due to the commodity and area specialists of the U.S. Bureau of Mines, who responded patiently and competently to our numerous enquiries. The Australian Department of Trade and Resources helped substantially with the Australian section of this study. The Canadian Consulate in Dallas, and the Centre for Resource Studies at Queens University, provided much useful information about Canada.

We would like to acknowledge the contributions of all the other co-supervisors of the theses on which these two volumes are based. A particular vote of thanks is due to Dr. James McKie.

Other people who made important contributions are Dr. Margot Wojcieckowski, Dr. Susan Bambrick, and our editor, Barbara Zeiders. The artwork was done by Barbara Hartman, and the computer program for the compilation of the index to Volume II was written by Ann Dolce.

W. C. J. van Rensburg
Paul Anaejionu

Executive Summary

THE UNITED STATES

The rapid economic and industrial development of the United States during the nineteenth and twentieth centuries was greatly facilitated by an abundance of natural resources, including energy and mineral resources. The United States attained the highest material standard of living the world has ever witnessed, and this economy was, and still is, based on very high per capita consumption of energy and minerals. With a population of over 230 million, the aggregate consumption of minerals over the past 50 years has been enormous. As a result, many of the richest mineral deposits have been depleted, and in future most domestic mining will probably be from smaller, lower-grade, complex, or remote deposits. Having lost its technological lead in mining and metallurgical technology, suffering from high labor costs, and having imposed stringent and costly pollution control standards, the United States is finding it increasingly difficult to compete with foreign mineral producers. This situation has been exacerbated by the recent trend for governments in developing countries to become directly involved with mineral extraction and processing.

The situation varies from one mineral commodity to another. For minerals such as molybdenum, phosphates, and lithium, the United States is still a formidable competitor on world markets. For others, such as copper and zinc, the decline in the U.S. industries has not been the result of the depletion of domestic reserves, but of other social, technological, economic, and environmental factors. For minerals such as chromium, manganese, and the platinum group, the United States simply never had sufficient domestic reserves.

Apart from becoming more dependent on mineral imports, the nature of U.S. import dependence has also changed during the past two decades. Twenty years ago the United States imported ores and concentrates and processed them in domestic metallurgical plants. Now minerals are increasingly imported in processed or refined forms. Until shortly after World War II, the United States and its allies had political control over most of the world's mineral reserves, as well as over the most important sea-lanes, so imports were secure. The former British, French, Portuguese, and Belgian empires no longer exist. Instead, their mineral reserves are controlled by nationalistic, independent, and largely nonaligned Third World governments. The Soviet fleet has become a major factor on the high seas.

There are differences of opinion about the significance of this changing mineral supply position of the United States. Liberals view it with equanimity, believing that mineral supply is simply a matter of price, and that the United States will therefore always be able to meet all its mineral requirements—at a price. Conservatives are much more concerned, and believe that mineral and energy import dependence could easily deteriorate into supply vulnerability. They see signs that the USSR is already exploiting the situation in a resource war.

The United States has limited options for reducing the risk of supply disruptions. The national strategic stockpile probably offers the most economic and rapid means of ensuring adequate supplies during emergencies. Greater domestic production is possible for some minerals, but would require a more sympathetic investment climate and greater access to federal lands. Substitution of scarcer materials by more abundant ones could be achieved in some instances. Increased recycling may be feasible for a few metals. Technological advances may allow the conservation of a few minerals of concern. Complete self-sufficiency of mineral supplies is not a technically or economically feasible option for the United States.

An additional concern for the United States is the fact that our allies in Western Europe and Japan are far more import dependent than the United States and that they have not stockpiled minerals to nearly the same extent as has the United States. Should the United States share scarce supplies of strategic minerals with them during an emergency?

The changing patterns of U.S. production, consumption, and imports are examined for a number of strategic nonfuel and energy minerals for the period from 1960 to the present. It is concluded that, with few exceptions, the security of U.S. supplies has declined during this period. During the past decade the U.S. mineral processing industry has deteriorated rather rapidly. This, in most instances, has reduced the number of foreign countries capable of exporting a particular commodity in processed forms. The decline in the processing industries will, in turn, have an adverse effect on U.S. manufacturing industries, and this will affect the balance of payments and employment.

Although the U.S. public became aware of U.S. dependence on imported oil during the crises of 1973 and 1979, they remain blissfully unaware of this country's far greater dependence on imported strategic minerals.

JAPAN

Japan is a resource-poor country whose industries require large quantities of imported strategic minerals and energy supplies. As a result, energy and mineral trade plays an essential role in the Japanese economy.

The mineral industry of Japan, like that of other industrialized countries, has had to adapt to the changing conditions in international energy and mineral resource markets. To remain competitive, Japanese mineral companies employ various strategies, including resource conservation, emphasis on the production of higher-value goods, business diversification, and export of minerals-related technology.

Japanese strategies to secure resource supplies include the development of domestic resources; the development of diverse foreign sources through equity investments, loans, technological cooperation, and technology transfer; and long-term purchase contracts. Other strategies that have increased in importance are investments in overseas processing capacity, stockpiling, deep-seabed mining, and a switch from energy-intensive industries to knowledge- and service-intensive industries.

Japan's trade restrictions and surpluses, and its level of defense spending in the late 1970s and early 1980s, threaten the security of its mineral supply. Good trade relations and a strong sea-lane defense are necessary to guarantee Japan's resource supplies.

There are significant differences between the mineral policy of Japan and that of the United States. The Japanese policy emphasizes diversification of foreign supply sources, whereas the proposed U.S. policy emphasizes increased domestic production and stockpiling. In addition, there is far more cooperation between government and industry in Japan than in the United States.

WESTERN EUROPE

Despite the existence of the European Economic Community, Western Europe remains a region characterized by diverse national minerals and energy policies. The policies of some nations are probably not in the interest of the region as a whole. Western Europe has a history of competition between the various countries for available energy and mineral supplies.

The policies of the various countries range from the free-market approach of West Germany to the much more socialistic policies of France, with the United Kingdom falling somewhere in between. While Western Europe as a whole is highly dependent on imported sources of energy and strategic minerals, the level and nature of import dependence varies considerably from one country to another. Most European countries are highly dependent on oil imported from the Middle East and North Africa. Attempts have been made to reduce this dependence by heavier reliance on domestic and imported steam coal, on nuclear energy (notably in France), and on natural gas imported in increasing amounts from the USSR. Norway, on the other hand, enjoys long-term self-sufficiency in domestic oil supplies. The United Kingdom has a temporary self-sufficiency for oil, and the Netherlands for natural gas. Several Western European countries have extensive mineral processing industries. In the case of the United Kingdom, France, and Belgium, these were based on mineral supplies from their former colonies.

In the past few years, several European countries have initiated strategic mineral stockpiles. France is probably most advanced in this regard. West Germany, on the other hand, has virtually abandoned similar attempts, fearing that stockpiled materials would quickly fall into the hands of Warsaw Pact forces in the event of a war.

Like Japan, Western Europe has been heavily dependent on imported sources of strategic minerals for so long that they have had time to invest in overseas mines and processing plants and thus diversify their sources of supply. However, most countries in the area are heavily dependent on southern Africa for a number of important commodities. Growing unrest in that subcontinent is causing concern about continued availability and security of supplies.

THE SOVIET UNION

The USSR has emerged as one of the world's major minerals producers. Its reserves of the main industrial and strategic minerals are substantial, enabling the Soviets to attain a very high degree of self-sufficiency. Minerals development has historically had an important place in the rapid industrial expansion of the USSR.

The most all-encompassing element affecting the Soviet minerals industry is the shift of the sources of supply from the European to the Asian areas of the nation. Newer deposits are increasingly remote, delineated by Arctic climates, sparse population, adverse terrain, and poor supporting infrastructure. Deficiencies in the Soviet planned economic system further impede their efforts to develop an efficient minerals sector. Problems include economic stagnation, labor shortages, inadequate bureaucratic coordination, low quality of industrial output, few incentives to promote innovation, and excessive waste. These geographic and economic constraints promise to greatly escalate production costs.

The USSR is a major producer of all primary energy minerals. Oil is vital for domestic economic growth, for export to the West to earn hard currency, and for export to Eastern Europe to help stabilize trade deficits there. World concern over a postulated need for the USSR to become a net importer of oil by 1985 appears unfounded, due to the potential for production increments from offshore development, enhanced recovery, or future discoveries. Natural gas remains the fuel of the future. Western European participation in the development of the huge Siberian gas fields ensures rapid increases in production. Soviet energy policy revolves around the substitution of gas, coal, and nuclear power for oil. Coal will fire power plants east of the Urals, while nuclear energy is the projected fuel in Soviet Europe.

The USSR is a major producer of virtually all important ferrous, nonferrous, precious, and nonmetallic minerals. Soviet mineral policy is predicated on self-sufficiency, this being based on ideological and political objectives. Unique aspects of mineral policy include the advanced level of recovery from Arctic areas, the secondary importance of prices, and the use of trade as a political rather than an economic instrument. Current trade goals embrace the acquisition of hard currency and the increasing integration of Eastern European economies with that of the USSR.

Ore-grade declines, changing patterns of trade in minerals, increasing domestic consumption, and economic problems indicate that the potential for future mineral import independence is uncertain. However, the existence of vast undeveloped Asian mineralized regions provides the Soviets with the option of renewing their commitment to this policy. Until new mines and infrastructure are in place in the Asian USSR, however, some importation may be necessary to meet requirements. This does not signify an abandonment of the strategy of self-sufficiency.

The possibility that the Soviet presence in southern Africa is a tactic of denying the West access to vital minerals has been dismissed by many authors. They overlook the potential for preemptive buying. It is not feasible at this stage to conclusively determine whether Soviet actions are motivated primarily by resource considerations.

Introduction

Although mineral consumption in the more advanced developing countries has grown significantly in recent years, a major proportion of total world production of most minerals is still consumed in the industrialized nations of Western Europe, the United States, Japan, and the USSR. Hence these countries have a considerable effect on world mineral markets.

The mineral positions of these developed nations differ considerably. Western Europe and Japan have long since depleted most of their domestic mineral resources, and are heavily dependent on imports for most of their mineral requirements. Having had to contend with import dependence for many decades, these nations have developed considerable skills in acquiring minerals from abroad. Japan, the most import dependent of them all, has developed a range of strategies to ensure adequate and secure supplies of strategic minerals. These are examined in considerable detail, and compared to the strategies employed by Western Europe and the United States. Western European nations, many of which had previously relied on colonial empires as their primary source of raw materials, now have to depend on international trade for the bulk of their mineral supplies.

The United States, once the world leader in mineral production and exports, still has a large domestic mining industry but demand has far outstripped the capacity of this industry. Since World War II, the United States has not only become much more dependent on mineral imports, but the nature of that mineral import dependence has changed dramatically. Where once America imported mainly ores and concentrates, and processed them in domestic metallurgical plants, most imports are now in the form of processed minerals and refined metals. The reasons for this trend are examined, and although it is noted that the causes vary from one commodity to another, it is concluded that higher energy costs, increasingly stringent environmental regulations, depletion of high-grade domestic deposits, government interference, and high labor costs are the main culprits.

The USSR, with its huge land area, diverse geology, and relatively new industrial economy, is far more self-sufficient in energy and strategic mineral resources than the other countries covered in this study. The USSR also places far more emphasis on mineral and energy self-sufficiency, and follows a policy aiming at achieving this as an integral part of its national strategy. Recently, some Western observers have concluded that the USSR is being forced to change from a policy of self-sufficiency at all costs to one of outward access to minerals because of the depletion of its domestic deposits. The validity of this observation is examined, and it is concluded that, while it may hold true for a few commodities, in general the Soviet position is characterized by a need to develop new deposits in the remote Asian reaches of the country, by a lack of infrastructure, capital, and technology, and by a need to earn more hard currency.

While the USSR buys time to develop its Asian energy and mineral resources, an outward access policy serves the dual purpose of obtaining mineral supplies more economically and of denying these resources to the West. Whether this strategy constitutes a "resource war" with the Western countries, aimed at depriving them of the oil resources of the Middle East and the strategic mineral resources of central and southern Africa, remains a highly contentious issue. Certainly, the USSR regards the dependence of the United States and its allies on imported energy and strategic minerals as their "weak link."

1

THE UNITED STATES

CHAPTER I-1

Introduction

The United States is dependent on foreign minerals for more than 50% of half of the 40 minerals considered to be essential to the economy and to national defense. Many of those sources are regarded as economically and politically unstable. The presence of the USSR in and around central Africa—source of some of the most vital strategic minerals—is part of an overall objective that the Soviets have pursued for more than three decades. That objective is to develop an oil and strategic minerals denial strategy either through physical disruption, market manipulation, or political and military domination of producer or neighboring states.

The United States now imports from southern Africa 53% of its manganese ore and 38% of its ferromanganese, 40% of its chromite and 71% of its ferrochromium, 53% of its platinum-group metals, 55% of its vanadium, and 55% of its cobalt. The concern is that this import dependence on strategic minerals, which do not occur in economic concentrations within the United States, may turn into an import vulnerability.

One of the major tasks of U.S. diplomacy should be to keep access to these strategic minerals open and available to the United States. This is a formidable task in view of the American public's strong opposition to the South African government's policy of apartheid, and their almost total ignorance about the strategic minerals problem.

Although the United States has never been completely self-sufficient in strategic minerals, the position has changed dramatically since World War I. Before World War I, the United States was the world's major producer of oil and of many other important minerals. However, rapidly changing technology during and after that war vastly expanded the range of commodities vital to industry and national defense. After World War I, C. K. Leith and others expressed concern about the growing U.S. dependence on imported minerals. They called for the development of a national minerals policy, designed to reduce America's vulnerability to mineral supply disruptions. One of the elements of such a policy would be a national strategic stockpile.

President Franklin D. Roosevelt opposed the concept of a strategic minerals stockpile. This turned out to be a costly mistake, and the United States obtained adequate supplies of strategic minerals only with great difficulty during World War II. Had the United States and the British Empire not had political control over a very large share of the world's mineral reserves, and substantial control of vital sea-lanes, the situation could have been much worse. As it was, the United States had to obtain some percentage of all minerals except iron ore, coal, and salt from abroad. A large number of minerals were obtained only from foreign sources.

After World War II, the Truman and Eisenhower administrations recognized the importance of strategic minerals, and developed a respectable strategic stockpile. However, for the next 20 years the stockpile suffered gross neglect, and it has only received some additional material during the Reagan administration.

In recent years, there has been a polarization of views regarding strategic minerals along liberal-conservative lines. Conservatives point to statements made by Leonid Breshnev to the effect that it is the Soviet intention to deprive the West of its two main treasure troves: the oil reserves of the Middle East, and the strategic mineral reserves of central and southern Africa. They point to the much greater mineral import dependence of Western Europe and Japan, and conclude that denial of southern Africa's mineral resources would leave these nations hostages to the USSR, thereby drastically changing the world balance of power. The conservatives see this as a low-cost, low-risk option for the Soviets, who do regard the energy and mineral import dependence of the Western countries as their Achilles' heel. Conservatives conclude that the West is already engaged in a low-level "resource war" with the Soviet bloc. As supporting evidence they point to the Soviet and Soviet surrogate presence in Afghanistan, Angola, Zimbabwe, and Mozambique.

Liberals strongly dispute these views. They regard the Soviet presence in these countries as evidence that the USSR is merely pursuing targets of opportunity, and that the presence of extensive mineral reserves is incidental. They regard the resource war thesis as implausible, pointing out that the total value of Western mineral imports is far less than that of oil imports, and conclude that the USSR would not be prepared to commit large numbers of its ground forces to southern Africa.

Most of the liberal proponents of a mineral policy that relies almost exclusively on the market mechanism are economists who see mineral supply solely in terms of a response of supply to price. They do not appreciate that economic mineral deposits represent geological freaks, that there are long lead times between exploration for and production of new mineral supplies, or that the concern of the United States and its allies is with the *availability* rather than the *cost* of mineral supplies.

In this book a number of key nonfuel and energy minerals are analyzed over the period 1960–1983. Changes in the level and nature of domestic production and import dependence are examined. The changing nature of America's mineral supply situation between World War I and the present is illustrated. It is concluded that U.S. import dependence has grown considerably during that time, and that the nature of our import dependence has changed significantly. Where once America imported mainly ores and concentrates, and jealously protected domestic mineral-processing industries, this nation must now increasingly import minerals in processed forms.

Finally, strategies for reducing strategic mineral supply disruptions are examined. Liberals and conservatives alike agree that the strategic stockpile is the quickest and cheapest option available. The history and present status of the stockpile are examined in considerable detail. A nation as large as the United States should have considerable potential for additional mineral discoveries. The potential for increased domestic mineral production is analyzed, and it is concluded that one of the major constraints in this regard is the growing inaccessibility of federal lands to mineral exploration and development. While increased domestic production of certain minerals is possible, a policy of complete self-sufficiency, such as that practiced by the USSR, would be unrealistic for the United States. It would, in fact, constitute a self-imposed embargo.

Continued dependence on foreign sources of supply for many minerals is therefore inevitable. The potential for diversification of foreign supply sources is examined. Other options for reducing our vulnerability to mineral supply disruptions are also examined. Substitution, recycling, and conservation are foremost among these. It is concluded that these options may yield positive results if they are combined with an expanded national minerals research and development program.

A major task for the future is to educate the American public about the role of strategic minerals in the economy, and in maintaining national security. They need to understand that increasingly stringent environmental regulations, and the withdrawal of federal lands from mineral exploration, have already exacted a heavy toll on the domestic mining and mineral-processing industries.

The Mineral Industries of the United States

GEOLOGY OVERVIEW

Given its status as a major mineral producer, the United States would logically have a varied geologic setting on which to develop a minerals industry. Such is the case. In fact, the geology of the United States is extremely complex. This is to be expected of a nation almost 9.4 million square kilometers in area[1] which runs the entire width of a continent. However, for the purpose of this study, an overview of U.S. geology will suffice. Figure I-2-1 shows the major tectonic regions of the United States, and a brief discussion follows.

Cordilleran Foldbelt

The Cordilleran Foldbelt in the western United States is an integral portion of the cordillera extending from Alaska to South America. This foldbelt resulted from the collision of the Pacific Plate with the westward-moving American Plate during the late Mesozoic and early Tertiary Ages. The more recent periods of geologic activity developed a region of extremely complex geology. Five major units can be identified.

The first major unit is the terrestrial volcanics of Tertiary and Quaternary Age. There are several areas in the cordillera with this rock type. Perhaps the most famous area is the Cascade Mountain range, which runs from Washington to northern California. The Cascades are the western edge of a larger area of terrestrial volcanics which extends east through northern Nevada and southern Idaho into northwestern Wyoming. Several other areas are located in south-central Colorado, southwest New Mexico, and west Texas. Mercury, lead, zinc, and copper are found here.

The second major unit is the terrestrial basin fill of the late Tertiary and Quaternary Ages. This is found in the Central Valley of California and also in an area extending from the Nevada-Utah border through southern Arizona and New Mexico, and then into Colorado and eastern Wyoming. Major minerals found in these miogeosynclinal deposits include copper, molybdenum, and potash.

The third unit is the coastal marine deposits found in southern California and parts of Oregon and Washington. These, too, are of Tertiary Age.

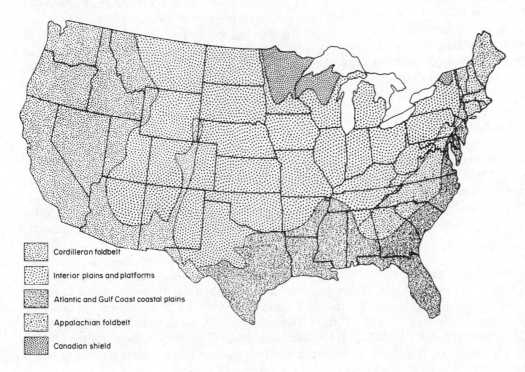

Cordilleran foldbelt

Interior plains and platforms

Atlantic and Gulf Coast coastal plains

Appalachian foldbelt

Canadian shield

FIGURE I-2-1 Major tectonic regions of the United States.

Minerals found here include asbestos, mercury, and petroleum.

The fourth unit is made up of eugeosynclinal deposits. The deposits are found in much of northern California and parts of northern Nevada and Utah. They are also found just to the west of the terrestrial volcanics in western Oregon and Washington. Mafic and ultramafic rocks are found throughout this region, and are especially prevalent in northern California and southwestern Oregon. Some of the minerals found here include nickel, cobalt, and chromium.

The fifth and final unit is comprised of granite plutons of the Mesozoic Age. These deposits are found along the California-Nevada border, in southern Idaho, and in northeastern Washington. Silver, gold, lead, and zinc are major minerals found here.

Interior Plains and Platforms

Much of the interior part of the continental United States consists of Paleozoic and younger sedimentary rocks. These rocks thinly cover the Precambrian basement rocks of the southern Canadian Shield. The Interior Region stretches from the Great Plains in the west through the interior lowlands and then east to the Appalachian foldbelt. No major folding took place in this region, but there are many basins, ranges, and domes. Minerals found in this region include coal, oil, lead, and zinc.

Atlantic and Gulf Coast Coastal Plains

This region stretches in a broad arc from southeast Texas up to the Atlantic coastal states of Delaware and New Jersey. These are marine sedimentary rocks; only these rocks overlay Paleozoic Age basement rocks. Minerals found here include petroleum, sulfur, salt, uranium, and phosphates.

Canadian Shield

The northern part of the continental United States has two major areas where the Canadian Shield's Precambrian rocks are exposed. The western outcropping is around Lake Superior (i.e., in Minnesota, Wisconsin, and upper Michigan). To the east, in upper New York, are folded Precambrian rocks. Schists, greisses, and granites predominate in this area. Major minerals in this area include iron, copper, and zinc.

Overall Reserves

As the discussion above indicates, the United States is very well endowed with mineral deposits. Although the United States lacks certain minerals, such as manganese and chromium, it is still very rich mineralogically. This is illustrated by Table I-2-1, which shows the latest estimates of the U.S.

share of the world reserves of certain minerals. As is apparent from the table, the United States is particularly well endowed with such minerals as copper, lead, molybdenum, silver, zinc, and coal. These and other minerals have formed the basis of of a strong mineral mining and processing sector in the United States.

ECONOMIC OVERVIEW OF THE INDUSTRY

With the exception of intellectual ability, practically everything of value in an economy is derived from either minerals or agricultural products. As such, the mining and processing of minerals form the foundation for economic growth in the United States and all other industrial nations. This is quickly seen when one considers, for example, the rapid growth of the U.S. economy following the discovery of iron and coal deposits in the nineteenth century.

A further illustration of this concerns the monetary contributions mineral mining and processing make to the gross national product (GNP) of the United States. In 1981, the direct output of nonfuel mineral mining was worth some $25 billion. Processing those minerals added $251 billion to the value of those minerals, for a total value of $276 billion.[6] This made up about 9.4% of the U.S. GNP of $2938 billion in 1981.[7] Obviously, the share of U.S. GNP held by the minerals industry is even higher; the figures above do not include production and processing of fuels such as coal and petroleum.

The driving force behind such a large domestic mineral industry is domestic consumption. Once an ore is mined and processed, it is then fabricated into a usable consumer good. For example, iron ore becomes steel, which in turn helps build an automobile. Using 1981 figures again, the per capita consumption of nonfuel minerals in 1981 was over 18,000 lb.[8] The bulk of this consumption level was made up of basic construction materials such as stone, sand, gravel, and cement. These are readily available. Over 1000 lb of that total was in the form of iron and steel. This basic industry, as well as many high-technology and defense industries, depends on some materials not widely found in the United States. Examples include fluorspar, manganese, columbium, and chromium.

To meet the demands of domestic (and foreign) consumers, the U.S. mining industry has had to exploit a large number of domestic (and foreign) mineral deposits. Table I-2-2 shows the number of mines operating in 1980 to recover a number of important minerals. The table's total count of 631 operating mines gives an indication of the need to

TABLE I-2-1   U.S. Share of World Reserves, Selected Minerals

Mineral	Date of estimate	Unit[a]	World reserves	U.S. reserves	U.S. share (%)
Asbestos	1981	C	123,100.0	3,650.0	3.0
Bauxite	1981	D	23,400.0	40.0	0.2
Cadmium	1981	C	680.0	110.0	16.2
Chromite	1981	D	3,541.0		
Cobalt	1981	C	3,665.0		
Columbium	1981	C	7,940.0		
Copper	1981	C	550,800.0	101,000.0	18.3
Fluorspar	1981	C	303,000.0	14,500.0	4.8
Gold	1981	B	32,254.2	1,399.7	4.3
Ilmenite	1981	C	394,100.2	29,000.0	7.4
Iron Ore	1981	D	93,600.0	3,630.0	3.9
Lead	1981	C	156,700.0	42,000.0	26.8
Manganese	1981	D	1,835.0		
Molybdenum	1981	C	9,480.0	4,130.0	43.6
Nickel	1981	C	82,030.0	200.0	0.2
Phosphate	1981	D	70,920.0	8,500.0	12.0
Platinum group	1981	B	36,778.0	31.0	0.1
Potash	1981	D	9,080.0	300.0	3.3
Rutile	1981	C	28,900.0	2,000.0	6.9
Silver	1981	B	230,675.0	46,950.0	20.4
Tantalum	1981	B	65,910.0		
Tin	1981	C	9,715.0	40.0	0.4
Tungsten	1981	C	2,634.6	124.7	4.7
Vanadium	1981	C	15,935.0	105.0	0.7
Zinc	1981	C	241,020.0	48,000.0	19.9
Coal	1982	A		482,954.0	
Natural gas	1983	E	3,199.0	198.0	6.2
Petroleum	1983	F	609.3	27.3	4.1
Uranium	1981	C	2,590.2 (free world)	708.0	27.3

[a]A, million tons; B, metric tons; C, 1000 metric tons; D, million metric tons; E, trillion cubic feet; F, billion barrels.

Sources: Refs. 2, 3, 4, and 5.

operate many mines to satisfy an existing level of demand. Moreover, it shows that mines of many different sizes are operated.

Table I-2-3 shows that the mines are different not only in terms of size, but also in terms of location. This is to be expected in a nation as large and as geologically diverse as the United States. Given the incentive to operate profitably, a mining firm will operate at a site most likely to operate at a cost low in relation to expected income. Thus a mine might be located in only one site even though the mined mineral is found in other locations.

MAJOR NONFUEL MINERAL INDUSTRIES

Although the United States produces some 82 nonfuel minerals,[9] this study concentrates only on 25 major metals, nonmetals, and industrial minerals.

Aluminum

Aluminum is the most abundant metallic element in the earth's crust, and it is experiencing a rapid growth in demand. Its strength, light weight, resistance to corrosion, and other properties have made aluminum such a popular metal that it now is used more than any other metal except iron. Consumption by end-use sector in the United States was as follows for 1983: 37% packaging, 19% transportation, 17% building, 9% electrical, 8% consumer durables, and 10% other uses.[9]

Aluminum metal is produced by refining bauxite and then smelting alumina by processes in use for roughly a century. Bauxite is the preferred ore for use in aluminum production because bauxite, a hydrated aluminum oxide, is broken down with far less energy than is required for aluminum silicates. The United States has very small reserves of bauxite,

TABLE I-2-2   Domestic Mines by Size, Selected Minerals, 1980 (Tons)

Commodity	Number of mines	Output volume 1-1000	1000-10,000	10,000-100,000	100,000-1,000,000	1-10 million	10 million plus
Bauxite	10		1	4	5		
Copper	39		1	5	7	17	9
Gold							
Lode	44	20	10	5	6	3	
Placer	36	8	10	12	6		
Iron ore	35		2	4	8	14	7
Lead	33	15	6	3	2	7	
Platinum	1				1		
Silver	37	20	10		7		
Titanium (ilmenite)	5				1	4	
Tungsten	28	26			2		
Uranium	265	43	73	103	44	2	
Zinc	20	1	1	4	13	1	
Other[a]	18	4	4	5	3	2	2
Asbestos	4		1		2	1	
Fluorspar	5		2	2	1		
Phosphate	44		4	2	11	15	12
Potash	7					7	
Total	631						

[a]Antimony, beryllium, manganiferous ore, mercury, molybdenum, nickel, rare earth, tin, and vanadium.
Source: Ref. 6.

TABLE I-2-3   Principal Producing States of Selected Minerals, 1981

Mineral	Leading Producers (in order of decreasing output)	Other producers
Asbestos	California, Vermont, Arizona	
Bauxite	Arkansas, Alabama, Georgia	
Copper (mine)	Arizona, Utah, New Mexico, Montana	10 other states
Fluorspar	Illinois, Nevada, Texas	
Gold (mine)	Nevada, South Dakota, Utah, Arizona	10 others
Iron ore	Minnesota, Michigan, California, Wyoming	9 others
Lead (mine)	Missouri, Idaho, Colorado, Utah	10 others
Manganiferous ores	Minnesota, South Carolina, New Mexico	
Molybdenum	Colorado, Arizona, Utah, New Mexico	California
Nickel	Oregon	
Phosphate	Florida, Idaho, North Carolina, Tennessee	Alabama, Montana, Utah
Platinum-group metals	Alaska	
Potash	New Mexico, California, Utah	
Silver (mine)	Idaho, Arizona, Nevada, Colorado	12 others
Tin	Alaska, Colorado	
Titanium concentrates	New Jersey, New York, Florida	
Tungsten concentrates	California, Colorado, Nevada, Montana	5 others
Vanadium	Colorado, Utah, Idaho, Arkansas	Arizona, New Mexico
Zinc (mine)	Tennessee, Missouri, New York	12 others

Source: Ref. 6.

as was shown in Table I-2-1. If domestic bauxite reserves were to be the sole source of aluminum ore, the domestic aluminum industry would quickly shut down. As a result, the United States is forced to import bauxite to supplement the output of domestic mines. The six mines operating in the

United States in 1983 were located in Arkansas, Alabama, and Georgia. Arkansas mines accounted for over 70% of total domestic bauxite output.[9]

The United States has a large base of nonbauxite alumina resources, and at times there have been serious efforts to utilize them. Murray[10] has identified five major resources, and all are plentiful in the United States. The most promising source is high-alumina clays, mainly kaolinite, with a content of over 30% Al_2O_3. Most of these clays are found in a belt of lenticular sedimentary bodies running from Georgia to South Carolina, but some clays are also found in Arkansas and Oregon. A second source, alunite, a hydrous sulfate of aluminum, is found in deposits throughout the western United States. Murray estimated inferred reserves of 2.5 billion metric tons of alumina in these deposits.[10]

The other resources identified by Murray are: (1) dawsonite, found in oil shales in Colorado; (2) anorthosite, found throughout the western United States; and (3) nepheline syenite, deposits of alkali feldspars and nepheline in Arkansas and the western United States. All five resources have shown very high alumina extraction costs in pilot plants, and it is unlikely that these resources will be utilized with further increases in energy costs. The lower shipping costs accrued by using these domestic nonbauxitic resources would perhaps help negate the higher processing costs.

In 1981, there were nine alumina and 32 primary aluminum producers operating in the United States. Relevant data on them is presented in Table I-2-4. One interesting point about the location of the alumina refineries is that they are all on or near the U.S. Gulf coast. This was to take advantage of both low energy costs and low shipping costs due to proximity to nearby bauxite exporters. Total refinery capacity was 18.8% of world capacity.[8]

Figure I-2-2 shows U.S. production, imports, and consumption of bauxite from 1960 to 1983. Domestic bauxite production was relatively steady until the recession in the early 1980s, but always was far short of demand. Figure I-2-2 also shows that bauxite imports generally increased in line with consumption, but again fell after 1980 due to the recession. Major sources of bauxite imports have been Jamaica, Guinea, and Surinam. From 1979 to 1982, those three nations accounted for 81% of U.S. bauxite imports.[9]

Table I-2-5 shows U.S. trade in bauxite and alumina from 1970 to 1983. It is readily apparent that bauxite imports decreased and alumina imports increased over this time period. This transition took place for three major reasons. First, imports

TABLE I-2-4   U.S. Alumina Refineries and Aluminum Smelters, 1981 (Thousand Tons)

(a) Alumina refineries

Firm	Location	Annual capacity
Reynolds	Corpus Christi, Tex.	1400
Kaiser	Baton Rouge, La.	955
Kaiser	Gramercy, La.	770
Alcoa	Point Comfort, Tex.	1325
Martin Marietta	St. Croix, Virgin Islands	635
Alcoa	Mobile, Ala.	800
Ormet	Burnside, La.	545
Reynolds	Hurricane Creek, Ark.	650
Alcoa	Bauxite, Ark.	340
Total		7420

(b) Aluminum smelters

State	Number of smelters	1978 Capacity
Indiana	1	290
Washington	7	1211
Oregon	2	220
Alabama	2	318
Tennessee	2	360
Louisiana	2	296
Montana	1	180
Maryland	1	176
North Carolina	1	125
Missouri	1	140
West Virginia	1	163
Kentucky	2	300
New York	2	341
Texas	4	624
Arkansas	2	183
Ohio	1	260
Total		5197[a]

[a]Annual smelter capacity in 1981 was 5,467,000 tons.
Source: Ref. 11.

of alumina saved U.S. producers the costs of refining bauxite. Second, importing alumina allowed U.S. producers to cut shipping costs by one-half. Third, by staying with certain exporting nations, U.S. producers had to import more alumina. This is because many bauxite-producing nations increased domestic alumina production to increase the economic contribution of bauxite production. Refining bauxite into alumina increased the value of the exported product by roughly 700%.[12] This shift toward more domestic bauxite processing began about 20 years ago in such countries as Australia and Jamaica. From 1979 to 1982, Australia, Jamaica, and Surinam provided 98% of the alumina imported by the United States.[9]

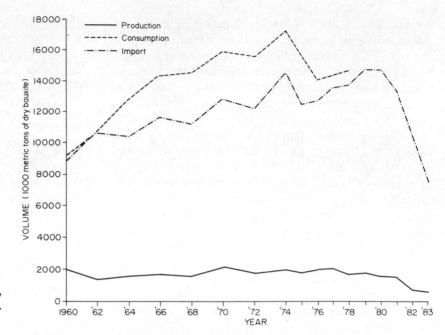

FIGURE I-2-2 U.S. production, imports, and consumption of bauxite, 1960–1983. (From Refs. 6, 9, and 11.)

TABLE I-2-5 U.S. Imports and Exports of Bauxite and Alumina, 1970–1983 (Thousand Metric Tons)

	Bauxite		Alumina	
Year	Imports	Exports	Imports	Exports
1983	7,600	70	3,900	650
1982	10,400	65	3,183	567
1981	13,300	52	3,978	737
1980	14,700	33	4,358	1,138
1979	14,800	15	3,837	849
1978	14,500	13	3,967	878
1976	13,500	15	3,288	1,050
1974	16,000	16	3,290	927
1972	12,168	29	2,585	879
1970	12,822	3	2,339	1,100

Sources: Refs. 6, 9, and 11.

American production of aluminum metal was valued at $5.7 billion in 1983 and was provided by 27 different reduction plants. Production was 3,640,000 tons of metal, a level far short of the total industry capacity of 5.5 million tons of metal. Overall, U.S. capacity utilization was 66.2% for 1983, while the worldwide figure was 76% (14,900,000 tons produced; 19,600,000 tons capacity).[9]

Table I-2-6 shows that the U.S. share of total world aluminum production has fallen from 1960 to 1983. In 1960, the United States produced 40.4% of the world's aluminum; by 1983, the share held by the United States was down to 24.4%. Part of this relative shift in location of production is due to the introduction of new reduction facilities around the world. However, some of it is due to

temporary and permanent plant shutdowns in the United States caused by high operating costs. Spector noted that rising energy costs have made certain smelters obsolescent in the United States, Japan, and other nations.[13] In fact, Japanese energy costs are so high that some aluminum production capacity has been dismantled.

The prospect for holding down increases in energy costs in the United States looks dim. The best example of this problem is found in the Pacific Northwest. Some 35% of U.S. aluminum smelting capacity is located there because of extensive hydropower facilities. However, the cost of hydropower has risen dramatically for customers of the Bonneville Power Authority. In 1980, smelters purchased electricity at 9 mills/kWh.[12] By November 1983, the same smelters were charged 26.8 mills/kWh.[9]

Faced with these power increases, U.S. aluminum producers can follow any of three options. First, they can increase the secondary aluminum production. This is already done to a great extent. In 1979, roughly 25% of the total U.S. supply of aluminum came from the recovery of purchased scrap.[12] A second option is to shut down capacity. This, too, is being done at times. The third option, that of importing aluminum metal, is also being followed, and to an increasing extent.

Figure I-2-3 shows U.S. production, trade, and consumption of aluminum from 1960 to 1983. It shows that even though domestic production and consumption generally have dropped in the last few years, imports have increased. In fact, the United States went from being a net exporter in 1980 to a net importer from 1981 to 1983.[9] From 1979 to 1982, the major sources of U.S. imports of

TABLE I-2-6   U.S. and World Aluminum Production, 1960–1983[a]

Year	World output	U.S. output	U.S. share of world output (%)
1960	4,985	2,014	40.4
1962	5,595	2,118	37.8
1964	6,531	2,553	39.1
1966	7,661	2,968	39.3
1968	8,839	3,255	36.8
1970	10,630	3,976	37.4
1972	12,133	4,122	33.9
1974	14,528	4,903	33.7
1976	13,787	4,251	30.8
1978	15,510	4,804	30.9
1980	16,940	5,130	30.3
1982	14,626	3,609	24.7
1983e[b]	14,900	3,640	24.4

[a]Production figures are for 1000 tons of primary aluminum.
[b]e, estimated.
Sources: Refs. 6 and 9.

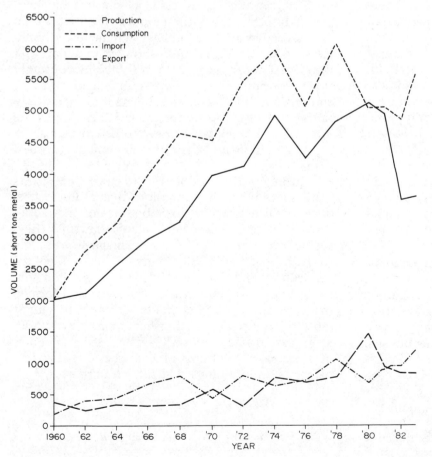

FIGURE I-2-3 U.S. production, imports, exports, and consumption of aluminum, 1960–1983. (From Refs. 6 and 9.)

aluminum have been Canada (63%), Ghana (14%), Venezuela (6%), and Japan (5%).[9] It is important to note that the first three sources still have cheap hydropower available.

The U.S. Bureau of Mines has estimated that U.S. aluminum consumption will increase by roughly 6.5% per year through 1990. On the other hand, bauxite and alumina demand is expected to fall somewhat.[9] Should these estimates be accurate, the United States will see a further increase in the imports of metal and a slowdown in domestic refining and smelting capacity expansions. Evidence already points to the accuracy of this estimate. After looking at Bonneville Power Administration quotes for electricity, Alumax, Inc. withdrew its plans to build a smelter in Washington. Now Alumax is working on an agreement to build a smelter jointly with Pechiney Ugine Kuhlman at Becancour, Quebec. Alumax said the attractive rates offered by Hydro-Quebec were a major incentive for this decision.[14]

Asbestos

The principal variety of asbestos found in the United States (and most other nations) is chrysotile. Chrysotile is a hydrous magnesium silicate found in serpentine. Major U.S. deposits are in California and Vermont, where the asbestos is found as irregular cross-fiber veins or shear-fiber zones in the host rock. Minor deposits are also found in Arizona in the form of horizontal asbestos-bearing serpentines in thin limestone beds. As shown in Table I-2-1, the U.S. reserves of asbestos are fairly small in comparison with world reserves.

Asbestos mines in the United States have historically met only a small fraction of total domestic demand. As a result, a great deal of asbestos has been imported, particularly from the vast, rich chrysotile deposits in Quebec. In 1983, only three mines produced asbestos in the United States. They are described briefly in Table I-2-7.

A fourth mine owned by Jaquays Mining Corp. operated in Gila County, Arizona until the early 1980s, but has since closed. This closure reduced the domestic asbestos production capacity to something less than the 100,000 metric tons per year capacity in 1979–1980.[11]

Figure I-2-4 shows U.S. production, imports, and consumption of asbestos from 1960 to 1983. It is readily apparent that domestic production of asbestos has historically met only a small portion of domestic demand. The U.S. historically was a major consumer of annual world asbestos output. In 1960, the United States consumed some 29.1% of total world production. By 1976, U.S. consumption had increased in an absolute sense, but had fallen in relation to total world output—only 18.4%.[6] The shortfall of U.S. production in relation to both domestic demand and world production is shown further by Table I-2-8.

Canada has historically provided over 90% of total U.S. asbestos imports.[6] From 1979 to 1982, Canada provided 97% of U.S. imports, while small amounts of amosite and crocidolite accounted for the rest of imports.[9]

There is little concern over the adequacy of potential sources of asbestos. Rather, the concern in the United States is that of the environmental and health hazards associated with the use of asbestos. Although there is some debate as to whether chrysotile is as dangerous as crocidolite, it is clear that environmental and health problems have been a major factor behind the slump in consumption in the United States. As Figure I-2-4 shows, U.S. consumption and imports began falling dramatically in the mid-1970s, long before the two most recent economic recessions.

With time, many asbestos uses have been eliminated. In other cases, the amount of asbestos used in a particular application has been decreased. End-use breakdowns for 1983 show that major uses were friction products (21% of total U.S. consumption), flooring (21%), asbestos-cement pipe (15%), and roofing (3%).[9] This should change dramatically in the very near future, for the U.S. Environmental Protection Agency is planning to propose a ban or a gradual phase-out of almost all remaining uses of asbestos.[15]

Although this action might draw complaints from some users, it is possible that asbestos producers will approve. This is because of the monumental number of health-related lawsuits filed against producers. Claims and damage awards were so high that, in 1982, Manville Corp. filed for bankruptcy and then, in 1983, the same firm sold its Canadian asbestos mines. Those mines formed the basis for what once made Manville the largest asbestos company in the noncommunist world. Things are different now. The proposed ban on use of asbestos will not hurt Manville because the company has been finding substitutes for asbestos.[15]

TABLE I-2-7   U.S. Asbestos Mines, 1983

Company	State and county	Mine name	Mining method
Calaveras Asbestos Corp.	California, Calaveras	Copperopolis	Open-pit
Union Carbide Corp.	California, San Benito	Santa Rita	Plowed, air-dried
Vermont Asbestos Group	Vermont, Orleans	Lowell	Open-pit

Sources: Refs. 6, 9, and 11.

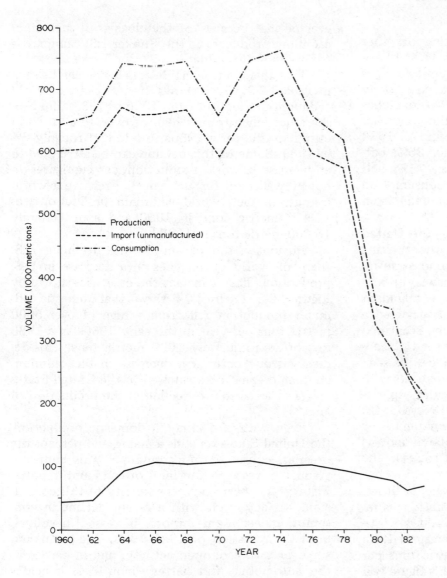

FIGURE I-2-4 U.S. production, imports, and consumption of asbestos, 1960–1983. (From Refs. 6 and 9.)

TABLE I-2-8 U.S. and World Asbestos Production, 1960–1983 (Thousand Metric Tons)

Year	World output	U.S. output	U.S. share of world output %
1960	2113	41	1.9
1962	2771	48	1.7
1964	2767	92	3.3
1966	3047	114	3.7
1968	3007	109	3.6
1970	3489	114	3.3
1972	3776	119	3.2
1974	4162	102	2.5
1976	5086	105	2.1
1978	5154	93	1.8
1980	4818	80	1.7
1981	4981	76	1.5
1982	4200	64	1.5
1983	4400	70	1.6

Sources: Refs. 6, 9, and 11.

Cadmium

Long used for coating and plating of other metals, cadmium has lately found new and very exotic applications. The nickel-cadmium battery is the primary application in which demand is likely to increase, but further research into photovoltaics could spur demand for cadmium sulfide and cadmium telluride.[9] End uses for cadmium in 1983 were as follows: coating and plating, 35%; batteries, 25%; pigments, 20%; plastics, 15%; and alloys, 5%.[9] It is expected that U.S. consumption of cadmium will increase by 1.8% annually from 1981 to 1990.[9]

Domestic cadmium production in the United States is entirely as a by-product of the smelting of zinc concentrates. Cadmium is found as an independent mineral in the sulfides hawleyite and greenockite. However, neither sulfide is abundant enough in the United States to be considered a potential ore mineral.[11] However, when recovered as a zinc by-product, the reserve base for U.S. cadmium production is quite large. This can be seen by referring to Table I-2-1. The zinc deposits with high cadmium contents are (1) irregular replacement deposits in Utah, Colorado, and Nevada; (2) contact metamorphic rocks in California and New Mexico; (3) veins in Idaho and Montana; and (4) stratiform deposits in carbonate host rocks in the Mississippi Valley.[11]

Since cadmium is produced at zinc facilities, the discussion of U.S. production operations is included in the section on zinc. However, it was estimated that domestic production capacity in 1985 would be on the order of 3300 metric tons per year.[11] There is little likelihood that this figure will

ever be met because of the closings of such zinc/cadmium producers as the Bunker Hill Company's facility in Kellogg, Idaho.

That this is a reasonable conclusion can be seen in Table I-2-9, which shows the U.S. share of world cadmium production from 1960 to 1983. The U.S. share has fallen from a high of over 40% of world output in the early 1960s to a low of roughly 6% in 1983. Some of this decline can be attributed to an increase in world production, yet the figures do not fully account for such a decline relative to world output. In fact, world cadmium production was 11,474 metric tons in 1960 and approximately 16,000 metric tons in 1983.[6]

Obviously, the reason for the declining U.S. share of world output lies in a decrease in U.S. production. This is in fact the case, as is shown by Figure I-2-5. Figure I-2-5 shows that domestic cadmium production fell from a high of over 5000 metric tons per year in the early 1960s to a 1983 low of approximately 1000 metric tons. This decrease is due both to a decrease in the cadmium content of zinc ores smelted in the United States and also an absolute decline in the production of zinc.[6]

Even with a decline in domestic production, the United States remains a major, and perhaps the premier, consumer of cadmium.[11] This implies a gradual increase in the level of cadmium imports. Figure I-2-5 bears out the validity of this conclusion. In fact, even with a recent decline in consumption, cadmium imports increased in volume until the recession of 1982–1983. The recent upsurge in U.S. economic activity, and in particular the automobile and battery industries, indicates

TABLE I-2-9   U.S. Share of World Smelter Production of Cadmium, 1960–1983 (Metric Tons Cadmium Content)

Year	World output	U.S. output	U.S. share (%)
1960	11,474	4,737	41.3
1962	12,289	5,051	41.1
1964	12,672	4,743	37.4
1966	13,019	4,744	36.4
1968	15,013	4,830	32.2
1970	16,617	4,292	25.8
1972	16,662	3,759	22.6
1974	17,267	3,023	17.5
1976	16,773	2,047	12.2
1978	16,765	1,653	9.9
1980	17,720	1,578	8.9
1981	17,400	1,580	9.1
1982	16,140	1,007	6.2
1983	16,000	1,000	6.3

Sources: Refs. 6 and 9.

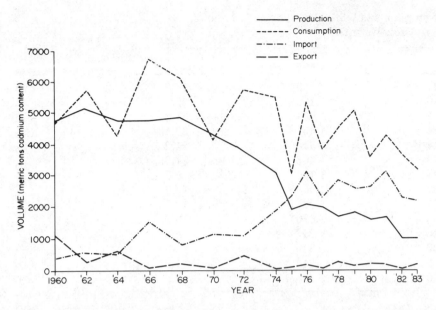

FIGURE I-2-5   U.S. production, imports, exports, and consumption of cadmium, 1960–1983. (From Refs. 6 and 9.)

that cadmium consumption will pick up. However, domestic cadmium production will probably continue to slump due to the problems of the domestic zinc industry. These problems are discussed later in this chapter.

Chromium

Chromium is used in metallurgical, chemical, and refractory applications; as such, it is one of the most important raw materials of industry. Two particularly important applications are the use of chromium to produce stainless steel and to plate various metals. Its use in superalloys also makes it a key strategic metal for defense-related applications. The end-use breakdown for 1983 was: transportation, 20%; construction, 19%; machinery, 18%; chemical, 12%; and others, 31%.[9]

As was seen in Table I-2-1, the United States had no reserves of chromium. There are deposits of chromium, especially in Montana, Oregon, California, and Alaska, yet all suffer from small deposit sizes, low ore grades, and long distances from demand centers. Thus they are not economically recoverable at today's prices. The U.S. Bureau of Mines estimated that total demonstrated chromite resources amount to 5.6 million metric tons of Cr_2O_3.[16] This is insignificant compared to the reserves and resources of Zimbabwe and South Africa.

The U.S. resources occur as both primary and secondary deposits. The major primary deposits are the widespread stratiform deposits of the Stillwater Complex in south central Montana. These deposits are typically high-iron chromite deposits. Other primary deposits are the high-chromium or high-aluminum podiform deposits found in Alaska and northern California. Secondary deposits occur as

both placers in the beach sands of Oregon and laterites along the coasts of California and Oregon.[16]

There has been no domestic production of chromite since the early 1960s. Production rarely exceeded 1000 metric tons per year except during both world wars and the stockpiling years of the 1950s. The stockpile buildup used domestic chromite made available only because of production subsidies offered through the Defense Production Act. Domestic production ended upon the expiration of the Defense Production Act subsidies in 1961.[11]

In 1979, six companies produced ferrochrome, chromium alloys, and chromium metal for the steel and alloy industries. Seven firms produced chromium products for refractory uses, and three produced chromium products for chemical uses. By 1983, two of the metallurgical chromium producers had dropped out of the business, ostensibly because of high energy costs.[9] It is possible that more of these businesses will close in the near future.

The chromium industry has exhibited two trends common to many other minerals in demand in the United States. First, U.S. consumption has fallen in relation to total world consumption and production. This is illustrated in Table I-2-10, which shows that U.S. chromite consumption as a portion of world production fell from a high of 31.3% in 1964 to a 1983 low of 2.5%. Part of this is due to a fall in demand caused by the recent recession. The second trend also helps to account for the drop in chromite consumption. Chromite imports and consumption have dropped while ferrochrome imports and consumption increased. This is illustrated by Figure I-2-6.

Figure I-2-6 also illustrates just how important

TABLE I-2-10 U.S. Consumption and World Production of Chromite, 1960–1983 (Thousand Tons Gross Weight)

Year	World output	U.S. consumption	U.S. share (%)
1960	4,885	1,220	24.9
1962	4,840	1,131	23.4
1964	4,632	1,451	31.3
1966	4,974	1,461	29.4
1968	5,444	1,316	24.2
1970	6,672	1,403	21.0
1972	6,725	1,140	16.9
1974	8,187	1,450	17.7
1976	9,372	1,006	10.7
1978	9,920	1,010	10.2
1980	10,725	977	9.1
1981	9,900	889	8.9
1982	10,907	545	4.9
1983e[a]	11,200	285	2.5

[a]e, estimated.

Sources: Refs. 6 and 9.

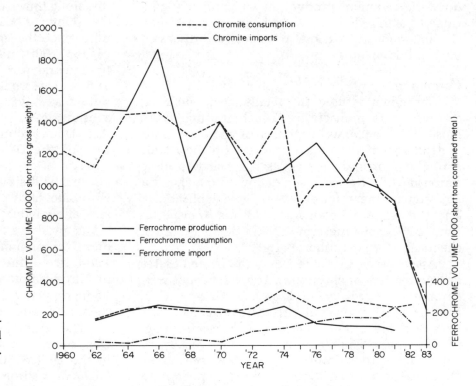

FIGURE I-2-6 U.S. production, imports, and consumption of chromite and ferrochrome, 1960–1983. (From Refs. 6, 9, 11, and 16.)

imports are in meeting domestic demand for chromium. The United States has historically depended on imports for over 90% of its chromium supply in any given year.[9] Import sources are relatively constrained, very few countries produce chromite, and only a few more produce ferrochrome. For example, from 1979 to 1982, South Africa alone accounted for 48% and 44% of U.S. imports of chromite and ferrochrome, respectively.[9] Given the strategic value of chromium, many feel that the United States should try to become less import dependent.

One option would be to resume domestic production of chromite. However, this would most likely necessitate a federal government subsidy few would find acceptable. A recent study by the U.S. Bureau of Mines concluded that production of metallurgical- and chemical-grade chromite would require prices of $237 and $188 per metric ton, respectively.[16] However, the 1983 year end price for metallurgical-grade chromite was $110 per metric ton; for chemical-grade chromite, it was $52 per metric ton.[9] It is very unlikely that the federal government will consider paying such massive subsidies at this time.

A second option for reduction of import dependence would be to increase the recovery and reuse of spent chromium-containing products and waste. This is being done to a small extent. Some 20% of 1983 demand was met by the recycling of chromium in stainless steel scrap.[9] Moreover, other studies by the U.S. Bureau of Mines indicated that some 73,000 metric tons of chromium is lost to domestic industry every year.[6] A fair amount of that waste and scrap could be recycled.

However, demand is certain to rebound from the 1983 low. If it rises at the 2.2% per year level predicted by the U.S. Bureau of Mines, there seems little chance that the United States will significantly reduce its dependence on a few sources of chromite and ferrochrome.

Cobalt

Long used as a coloring additive, cobalt has lately developed into an important and even essential element in many applications. It is used extensively in the chemical industry as a catalyst, drier, and so on. Moreover, it is of essential importance in the alloying industries. Cobalt imparts such desirable qualities as heat resistance, magnetic properties, high strength, and wear resistance to alloys. As such, cobalt is of great use to the aerospace and electrical product industries. In 1983, cobalt use in the United States was segmented among the following end-use sectors: superalloys (especially gas turbine engines), 35%; magnets, 15%; driers, 16%; catalysts, 10%; cutting tool bits, 5%; and other applications, 19%.[9]

The United States currently has no reserves of cobalt; this is shown in Table I-2-1. However, the United States does have large resources of cobalt. In fact, the U.S. Bureau of Mines estimated that the United States may have the third largest resource base of cobalt worldwide, after Zaire and Cuba.[17] Unfortunately, this does not mean that the United States can quickly mine its domestic resources. Almost all cobalt is recovered as a by-product of mining of more readily available minerals such as copper and nickel. Cobalt is rarely mined as a primary product due to its low ore grade.[11]

Domestic mining of cobalt once took place in three different locations. The Blackbird district of Idaho was one of the few mines that produced cobalt as a primary product, but the mine closed in June 1959. Cobalt was recovered as a by-product of copper and nickel mines in Frederickton, Missouri, and as a by-product of iron ore mining in Cornwall, Pennsylvania. These two mines closed in 1961 and 1971, respectively.[11] There has been no domestic mine production of cobalt in the United States since then. There is currently one domestic cobalt refinery: the Braithwaite, Louisiana, facility operated by AMAX Nickel, Inc. This refinery recovers cobalt from imported matte, most of which comes from South Africa.[6]

Figure I-2-7 shows the level of U.S. cobalt imports and consumption from 1960 to 1983. Although the levels initially seem quite erratic, there is an obvious, though gradual, increase in consumption until the late 1970s. At that point, the recession, cutback in both airline travel and industry expansion plans, and other factors tended to drop the consumption level. However, 1983 consumption levels increased by some 16% over 1982 levels.[9]

Even when the United States did have mine production of cobalt, domestic output usually fell short of domestic demand. Thus the United States has had to import most of its cobalt needs. Historically, the primary source of U.S. imports has been Zaire (once known as the Belgian Congo). This is not surprising because Zaire is the largest cobalt producer and exporter in the world. Zaire's direct share of U.S. imports ranged between 40 and 80% from 1960 to the late 1970s.[6] From 1979 to 1982, Zaire provided 37% of U.S. imports. These figures are somewhat misleading, though, for they do not include U.S. imports from Belgium which originated in Zaire. From 1979 to 1982 alone, the Belgian-based imports added another 8% to the share of U.S. imports held by Zaire.[9]

Other major sources of cobalt imports have been Zambia and Canada. Directly and indirectly, these two producers accounted for another 28% of the U.S. imports from 1979 to 1982.[9] Although the Canadian supplies are thought to be reasonably secure, the strong dependence on southern African suppliers has many Americans worried. The threat of supply cutoffs and/or price hikes has generated great concern among those who appreciate the vital uses of cobalt.

It has been estimated that domestic cobalt consumption will increase by 2% a year from 1981 to 1990.[9] Much of the increase will be in rare earth magnets, alloys and catalysts.[6] This could lead to even greater import dependence unless steps are taken in the United States to increase domestic production.

The most obvious response would be to resume

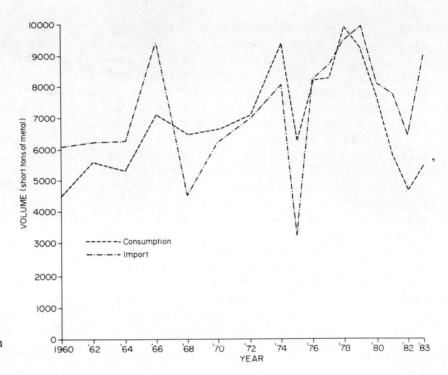

FIGURE I-2-7 U.S. Imports and consumption of cobalt, 1960–1983. (From Refs. 6 and 9.)

domestic mine production of cobalt. There are currently three possible sites. Two of them are the formerly operated mining districts in Idaho (Blackbird district), and southeast Missouri. The most promising site could be the Gasquet Mountain Ni-Co-Cr laterite deposit close to the Oregon-California border. Preliminary studies indicate a possible annual output of 2.2 million pounds of cobalt alone.[18] This would amount to roughly one-fifth of the (admittedly low) 1983 U.S. consumption level.

However, all three projects are faced with serious obstacles. In all cases, the cost of recovery is expected to run at least $20 a pound. This does not compare favorably with Zaire's cost of $3 to $6 a pound.[17] Use of production subsidies through the Defense Production Act has been widely called for, but the political acceptability of such an action seems dubious at best.[17,18] This is particularly true since the spot-market price for cobalt dropped to about $6 a pound.[9]

Barring domestic mine production, a second option would be to reduce overall cobalt consumption. This is being attempted through both recycling and substitution programs. In 1983, some 6% of reported domestic consumption was met by recycling cobalt-bearing scrap. This could increase somewhat.[9] A variety of attempts have been made at finding cobalt substitutes, but these have not been very successful. For strategic applications such as gas turbine engine parts, no acceptable substitute has yet been found (although ceramics seem promising for some uses).[11] In less critical uses

such as magnets or driers, substitutes are available, but the end result is almost always higher cost and/or reduced performance. In fact, some acceptable substitutes (nickel or chrome, for example) would only increase the U.S. level of import dependence for the substitutes.

The final option is not a new one: stockpiling. However, the desire to maintain an adequate stockpile is apparently a newly rediscovered idea. The Truman and Eisenhower administrations built up a stockpile of over 100 million pounds of cobalt to ensure that the U.S. would not run short even in a protracted emergency. However, changing economic and political realities led to the initiation of stockpile sales in the early 1960s. These sales took place at an annual level of about 6 million pounds a year.[19] Sales continued until 1977, and by then the stockpile level was just over 40 million pounds.[6] Recently, there has been a push by the Reagan administration to build up the cobalt stockpile, but there have been problems.

One problem is that much of the existing cobalt stockpile is, according to the American Society for Metals, inadequate for most critical application.[20] Thus, about 40 million pounds of stockpiled cobalt must either be replaced or upgraded. A second problem is that stockpile purchases could once again influence the cobalt market. Initial purchases encouraged world mine development, but the stockpile sales of the 1960s and 1970s reduced the rate of mine development. These alternating expansions and contractions of supply capacity have already resulted in significant price hikes, market panics, and, lately, excess capacity.[19] A third problem, perhaps most important in the near term, is that there seems to be little incentive to allocate money for stockpile purchases when the federal deficit is so high. In any event, the United States will have difficulty dealing with its cobalt import dependence.

Columbium and Tantalum

Although they have slightly different uses, columbium and tantalum have several things in common. First, both minerals have not been produced in the United States since 1959. Second, the United States is between 90 and 100% import dependent. Ferrocolumbium is used in the steel industry, while columbium metal and alloys have important aerospace applications. End uses in 1983 for columbium were: construction, 41%; transportation, 23%; oil and gas industry, 17%; machinery, 13%; other, 6%. Tantalum's properties make it particularly important to the electronics industries. End uses for tantalum in 1983 were: electronic components, 66%; machinery, 20%; transport, 11%; and other, 3%.[9]

Table I-2-1 showed that the United States has no reserves of either columbium or tantalum. The United States does have some resources of columbium in pyrochlore in carbonatite deposits. Similarly, some tantalum resources are found in tiny deposits in parts of the United States. However, in both cases, the ores cannot be economically recovered at today's prices.[9]

Since there is no domestic mine production of either mineral, U.S. consumption needs must be met through imports. Here, too, the minerals show similarities: in this case, in forms and supply sources. Columbium is imported in the form of concentrates, tin slags, and other crude forms as well as ferrocolumbium. Tantalum is imported as concentrates, tin slags, and scrap. Nigeria and, lately, Brazil have been the dominant sources of U.S. columbium imports.[6] From 1979 to 1982, Brazil supplied 75% of U.S. imports. Brazil, Canada, and Thailand supplied 87% of U.S. imports for that period.[9] The United States has consistently been 100% dependent on imports.[9]

Brazil, Canada, and Thailand have also been the dominant supply sources for U.S. tantalum imports. From 1979 to 1982, these three countries provided 61% of U.S. imports. In this case, though, Thailand was the primary supplier, with 42% of U.S. imports to its credit.[9] In the early 1970s, Canada was the primary supply source, while the Belgian Congo held that status in the 1960s.[6] Although U.S. import dependence for tantalum is not the 100% level it is for columbium, it has almost always been at 90% or better since 1960.[9]

Tables I-2-11 and I-2-12 show the levels of U.S. imports and consumption of columbium and tantalum from 1972 to 1983. Table I-2-11 is significant because it shows that U.S. columbium imports have exhibited a trend common to other minerals: U.S. imports are being processed to an increasing extent before they leave the exporter's shores. In this case, ferrocolumbium imports have generally increased while imports for columbium-bearing slags and concentrates have fallen.

Table I-2-12 also shows that the United States

TABLE I-2-11   U.S. Imports and Consumption of Columbium, 1972–1983 (Thousand Pounds Contained Columbium)

Year	Concentrate, slags	Ferrocolumbium	Consumption
1972	2105	1530	6383
1974	1010	3276	7193
1975	992	1947	4231
1976	2497	2221	6003
1977	2433	2676	4777
1978	2418	4159	6585
1979	2827	5515	7120
1980	3810	5918	7534
1981	1892	6068	8118
1982	1230	3128	6600
1983e[a]	1300	2000	6000

[a]e, estimated.

Sources: Refs. 6 and 9.

TABLE I-2-12   U.S. Imports, Exports, and Consumption
of Tantalum, 1972–1983 (Thousand Pounds Contained
Tantalum)

Year	Imports, all forms	Consumption	Exports
1972	1157	5878	
1974	1730	1998	400
1975	933	265	412
1976	1310	1328	432
1977	2058	1476	539
1978	1409	1114	607
1979	1914	1439	721
1980	2280	1187	706
1981	1580	1262	222
1982	1087	1060	340
1983e[a]	1000	970	220

[a]e, estimated.

Sources: Refs. 6 and 9.

has been a significant exporter of tantalum, often
in finished products such as electronics. That the
level of exports decreased faster than the import
levels during the last recession should come as no
surprise to those familiar with the problems of the
U.S. minerals industry in particular and the U.S.
economy in general.

Moreover, it is not likely that this trend will
be reversed in the cases of columbium and tantalum. Although their consumption levels are low
compared with that of other minerals, their applications are primarily in up-and-coming industries.
Thus the U.S. Bureau of Mines has forecasted
annual rates of demand growth from 1981 to 1990
of 5% and 3% for columbium and tantalum, respectively.[9] Since the startup of U.S. mine production
of either mineral is unlikely, it can be safely assumed that the United States will continue to be
roughly 100% import dependent for both minerals
in the near future.

Copper

Copper has a history of use longer than that of
practically any other metal or nonmetal mineral.
Moreover, its uses are very widespread in both industrial and consumer applications. In 1983, U.S.
consumption of copper was distributed among the
following end-use sectors: building construction,
32%; electrical and electrical components, 26%;
industrial machinery and equipment, 19%: transportation, 10%; and consumer products, 13%.[9] It
is expected that demand will grow at an annual
rate of 1.1% until 1990.[9] This should continue the
status of the United States as the major copper
consumer in the world.

Fortunately, the United States has a copper
reserve level commensurate with its demand for the
metal. As was shown in Table I-2-1, the United
States possesses about 18% of world reserves of
copper; U.S. resources are even larger. Most of the
U.S. copper ores are sulfides found in porphyry
deposits. This has meant that a great deal of U.S.
production is from surface mines followed by flo-
tation systems to concentrate the ore.[6] Average ore
grade of U.S. production has steadily fallen in the
last 50 plus years—from 2.11% in 1933 to 0.51% in
1981.[6] This has been overcome by advances in
large-scale mining, with its resultant economies of
scale.

Thirty-five major copper mines operated in the
United States in 1983. Table I-2-2 shows that most
of the mines operating in the early 1980s were very
large, often processing a million or more tons of
ore per year. This is to be expected from porphyry
mining operations. Arizona has been the premier
producing state for quite some time. In 1983, 66%
of total U.S. mine production of copper came from
Arizona. Utah accounted for 17%, and 10 other
states produced the rest.[9]

Domestic production capacity also involved a
number of processing facilities in 1983. There were
15 primary copper smelters, 43 secondary smelters,
13 electrolytic refineries, and nine fire refineries.[9]
Most of these facilities are located in the western
United States close to operating mines.

The overall magnitude of U.S. operations can
easily be pinpointed. For example, mine production was valued at $1.8 billion in 1983, and that
was a year of extremely depressed demand and
production.[9] Moreover, the copper mines and mills
in the United States had directly employed as
many as 36,000 workers in the 1970s.[9] This figure
has recently been cut in half due to industry cutbacks, but copper is still a very important mineral
in the U.S. economy.

Table I-2-13 shows the U.S. share of world mine
production of copper from 1960 to 1983. It is
readily apparent that the United States is still a significant producer of copper, but the United States
no longer is the dominant producer it once was.
From 1960 on, the U.S. share of world mine production has dropped from a high of roughly 26% to
a low, in 1983, of 13.2%. This can be attributed to
many factors: (1) declining production in the
United States due to falling ore grades, strikes, and
so on; (2) the introduction of new producers in the
world market; and (3) the refusal of some producers
to curtail production even when market conditions
indicate such an action. These factors are addressed
below.

TABLE I-2-13   U.S. Share of World Copper Production, 1960–1983
(Thousand Metric Tons Copper Content)

| Year | Mine production | | |
	World production	U.S. production	U.S. share (%)
1960	4218	979	23.2
1962	4617	1114	24.1
1964	4799	1131	23.6
1966	5251	1296	24.7
1968	5116	1093	21.4
1970	6016	1559	25.9
1972	6641	1510	22.7
1974	7314	1448	19.8
1976	7451	1457	19.6
1978	7489	1358	18.1
1980	7630	1181	15.5
1981	7800	1538	19.7
1982	8040	1140	14.2
1983	7960	1050	13.2

Sources: Refs. 6, 9, and 21.

Figure I-2-8 illustrates the production, imports, and consumption of copper by the United States from 1960 to 1983. Several conclusions can immediately be drawn from the graph. First, U.S. consumption of copper appears to be fairly price responsive. Consumption fell after both oil price hikes during the 1970s, fell at the beginning of the latest U.S. recession, and then increased in 1983, due at least partly to low copper prices. Second, domestic mine output seems to be stabilizing, per-

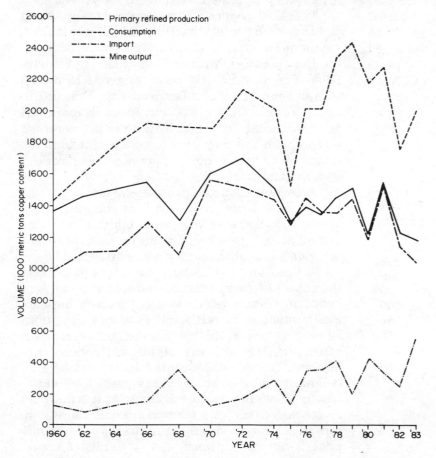

FIGURE I-2-8   U.S. production, imports, and consumption of copper, 1960–1983. (From Refs. 6 and 9.)

haps even decreasing. This was suggested by Table I-2-13, but it remains to be seen if this will hold true for the intermediate- to long-term future.

A third conclusion to be drawn from Figure I-2-8 is that domestic production of refined metal has been falling. Primary refined metal production hit a high of 1.7 million metric tons in 1972 and has steadily fallen since then. Primary refined metal production hit a low of 1.18 million metric tons in 1983.[9] In conjunction with the drop in U.S. production of primary refined metal, there has been a steady increase in imports of refined metal. This is shown by both Figure I-2-8 and Table I-2-14, which shows the level of U.S. imports and exports of refined copper from 1960 to 1983.

Table I-2-14 makes it quite clear that the United States has, in the last 23 years, gone from a net exporter to a net importer of refined metal. Many factors underlie this switch, and these are discussed below. However, one particularly important factor merits attention at this point. A look at the sources of U.S. copper imports over the last 25 years shows that Chile and Canada have consistently been the major sources of U.S. supply. However, both nations, particularly Chile, have changed the composition of their exports to the United States from predominantly blister copper to largely refined copper. For example, in 1962, roughly 95% of Chile's 204,000 metric tons of copper exports to the United States was blister copper. This changed by 1974, when just over half of Chile's exports were in refined form. By 1978, over 76% of Chile's

TABLE I-2-14   U.S. Imports and Exports
of Refined Copper, 1960–1983
(Thousand Metric Tons Copper Content)

| Year | Refined metal | |
	Imports	Exports
1960	129	403
1962	89	306
1964	126	286
1966	148	248
1968	363	219
1970	119	200
1972	174	166
1974	285	115
1976	346	102
1978	415	92
1980	427	14
1981	331	24
1982	258	31
1983	550	100

Sources: Refs. 6 and 9.

copper exports to the United States were in refined form.[6]

It would be unfair to blame this turn of events entirely on Chile and other copper-exporting nations. It is important to note that a shift away from domestic mine production and processing of domestic and imported concentrates means the domestic industry is having serious problems. That the U.S. copper industry, and, in turn, the U.S. economy, is suffering is relatively easy to show.

A great many minerals are recovered together with copper in mining and processing stages. Widely quoted figures from the U.S. Bureau of Mines show this point quite well.[21] During the concentrating stage of copper ores, 32% of the molybdenum, 100% of the rhenium, and roughly 1% each of the lead and zinc produced annually in the United States is recovered. Over 6% of the sulfuric acid and all of the arsenic produced annually in the United States is recovered during the smelting of copper concentrates. Finally, the refining of blister copper yields 37% of the gold, 32% of the silver, and all the platinum, palladium, tellurium, and selenium produced domestically.[21] Although the percentages might vary somewhat from year to year, it is easy to see just how important the copper industry is to the U.S. economy in general.

The U.S. copper industry seems to be taking punishment from all directions. The first problem, mentioned above, is that ore grades are falling in the United States. Moreover, as of late, prices have fallen for most of the other minerals recovered with copper. These two factors decrease the profitability of a mining operation. In some cases, the mine is forced to close. This situation contrasts sharply with the very high ore grades found in the mines of certain other copper-producing nations such as Zambia.[11]

A second industry problem involves the high energy and labor costs of domestic producers. These, too, decrease the profitability of a mining or processing operation. Details on these problems are widely available, so they are not described here.

A third and particularly important problem is the impact of environmental and safety regulations. Pollution control regulations particularly hurt domestic smelters by restricting the range of options allowed to meet control standards. The result is often that U.S. exports of ore and concentrates increase.[21] One estimate places the cost of U.S. pollution controls at about 15 cents a pound.[22] Given today's depressed market price, it is not hard to see why such an added cost would be of concern. This is particularly true because the copper price is set internationally. As a result, U.S. producers cannot hope to pass the cost of pollution controls on to consumers. As long as cheaper copper exists in the market, for whatever reason, U.S. producers will be pricing themselves out of the market.[21]

A fourth problem of the U.S. copper industry is that demand is being threatened domestically with the advent of copper substitutes. Plastics are finding widespread acceptance in piping, to give just one example. Another one is the substitution of fiber optics for copper communications wire. The magnitude of this substitution is great. For example, a New Jersey-based unit of the American Telephone and Telegraph Co. has plans to install 1000 miles of fiber-optic cables. This is expected to be the equivalent of 3.5 million to 4 million pounds of copper.[23]

A fifth problem of particular concern to domestic copper smelters is the change in the global trade of copper concentrates which took place over the last 25 years. Japan adopted a copper supply strategy which emphasized the domestic smelting of imported concentrates. This was to increase the value added gain to Japan, to enhance Japan's security of supply, and to provide a needed source of sulfur for sulfuric acid production.[21] The success of this policy is easily seen: Japan imported no concentrate in 1950, and yet by 1980 Japan imported 835,000 metric tons of concentrate.[24] That amounted to some 70% of the world trade in concentrates.[25]

Japan accomplished this by establishing an indirect subsidy which allowed Japanese smelters to outbid other smelters for available concentrates.[25] This resulted in many U.S. custom smelters losing their foreign concentrate sources. The situation has been exacerbated in two ways. First, since 1980, South Korea, Taiwan, and Brazil adopted even more aggressive versions of the Japanese effort.[25] Second, Japan, in order to acquire new concentrate sources, has recently contracted to buy American concentrates even when U.S. smelters could easily handle the extra supply.[24]

The end result is that U.S. smelters face more competition than ever before to acquire stable concentrate supplies. Thus it is likely that more U.S. smelters will be priced out of the market and will have to shut down. Many industry observers are particularly upset that Japanese smelters bought U.S. concentrates. They feel that such a callous disregard for the long-term impact of these policies on U.S. production capacity and economic strength is totally uncalled for.[24] The true extent of the damages will be felt only when a supply crunch hits and foreign producers supply their markets first.

With little or no concentrates available for U.S. smelters, they would have to close down.

To make matters worse, there is a sixth problem felt by U.S. copper producers. Developing nations that depend on copper as a major earner of foreign exchange, such as Chile, have been ignoring market signals lately. They have expanded production, capacity, and exports even when doing so would mean they earned a lower return per unit. They apparently can afford to do so because of loans from the World Bank and similar organizations.[26] The result is that supply is not responding to global demand conditions. Thus higher-cost producers that still obey market forces have had to cut back production to levels far beyond what should have been called for.

The two major losers here have been the copper industries of the United States and Canada. Mines have been shut down, thousands are out of work, and the industry is being severely weakened. In early 1984, 11 U.S. copper companies asked the U.S. International Trade Commission to restrict copper imports to save the domestic industry.[27] Unless something is done, either through a change in policies by Chile and other nations or a policy change in the United States, it is very likely that permanent losses will be sustained by the U.S. copper industry.

Fluorspar

Fluorspar, also known as fluorite (CaF_2), is the major source of fluorine, an element with a variety of important industrial applications. Fluorspar is used as a flux in steelmaking; as the source of hydrofluoric acid used in the aluminum, chemical, and uranium industries; and in various ceramics and glassmaking operations.[11] Fluorspar consumption in the United States in 1983 fell into the following broad categories: hydrofluoric acid, 57%; steelmaking, 41%; glassmaking and other, 2%.[9]

Table I-2-1 shows that the United States has about 5% of current world reserves of fluorspar. Domestic reserves are normally found as stratiform replacements in carbonate rocks, with over 50% of the reserves located in Illinois and Kentucky. This region is also the site of the vast bulk of domestic fluorspar resources. Other U.S. reserves take the form of stockworks found scattered around the western United States and more stratiform replacements in Colorado and Utah.[11]

The number of domestic fluorspar producers has dropped from 13 in 1977 to only five. Three producers (Ozark-Mahoning, Inverness Mining, and Hastie Mining) operate in southern Illinois. They produce over 90% of domestic output. One firm,

J. Irving Crowell, Jr., operates out of Beatty, Nevada, and a final company, D&F Minerals, is located in far west Texas, near Alpine. Total U.S. production capacity in 1980 was about 120,000 tons per year. Capacity is expected to drop to 100,000 tons per year by 1985.[11]

Additional domestic production comes in the form of fluorosilicic acid recovered from phosphoric acid facilities. Eleven such facilities operated in the late 1960s through early 1970s. Of that total, six facilities were located in Florida near phosphate mining operations. The other five facilities were in the southeastern United States.[11]

Domestic production has fallen short of domestic demand since the early 1950s. Figure I-2-9, which shows U.S. production, imports, and consumption of fluorspar, illustrates several important points. First, the level of domestic production of fluorspar has steadily dropped since 1960. Second, the gap between consumption and demand increased dramatically from 1960 to 1972. Since 1972, this gap has decreased until the 1982–1983 gap between production and consumption is about that of the 1960 gap. However, the 1982–1983 figures are lower. This could perhaps be attributed to dropping fluorspar use caused by a slump in the steel industry, and also by increased usage of hydrofluoric acid.

Mexico has historically been the dominant source of U.S. fluorspar imports. From 1960 to the mid-1970s, Mexico provided roughly 75% of annual U.S. imports.[6] This figure has since dropped somewhat. From 1979 to 1982, Mexico provided 58% and South Africa provided another 19% of U.S. fluorspar imports.[9] Mexico's falling share of the U.S. fluorspar import market can largely be attributed to a rise in South African exports to the United States.

Domestic consumption is expected to increase at an annual rate of 2.3% from 1981 to 1990.[9] Even with declining production, the United States should have little trouble in finding adequate

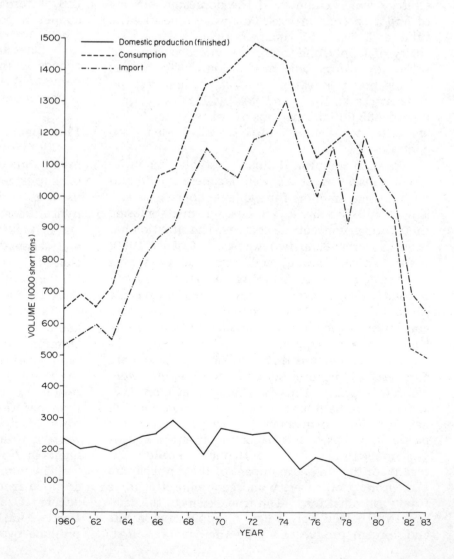

FIGURE I-2-9 U.S. production, imports, and consumption of fluorspar, 1960–1983. (From Refs. 6 and 9.)

import sources of fluorspar. Thus it is almost certain that the United States will continue to be 80 to 90% dependent on imported fluorspar.

Gold

Long considered a precious metal, gold has a history of use which revolves around jewelry and decorative items. However, gold is becoming increasingly important in industrial, and especially electronic, applications. A breakdown of major gold uses in the United States in 1983 shows that jewelry and the arts accounted for 55% of total domestic gold consumption. Industrial (electronic) uses took 33% of the gold, dental applications took 11%, and the production of bars consumed less than 1%.[9]

Table I-2-1 shows that the United States holds about 4% of current world gold reserves. About three-fourths of the U.S. reserves and resources are in the form of lode deposits.[11] The second largest resource type in the United States is thought to be deposits of disseminated gold. Finally, large amounts of gold occur in placers in Alaska and California. All told, most of the resources are in the western United States, while smaller amounts are found in Alabama, Virginia, North and South Carolina, and Georgia.[11]

Production of gold in 1983 came from a variety of sources. Some 20% of the gold was recovered as a by-product of base metal processing. It is expected that this amount will once again increase if domestic base metal mines increase their output. The remainder of gold production was distributed between some 30 placer mines (mainly in Alaska), 160 lode mines (most in the western states), and several hundred small lode/placer mines scattered across the United States.[9] California has been the largest cumulative producer in the United States, but the two major producers in recent years have been Nevada and South Dakota.[11] Utah and Alaska contribute smaller amounts.

Figure I-2-10 shows U.S. production, imports, exports, and consumption of gold from 1960 to 1983. The one steady line is that of production, which, until the 1980s, showed a slow but steady decrease in production. This has been going on since the early 1940s, when annual domestic mine production approached 5 million troy ounces.[6] Domestic consumption has dropped dramatically after price hikes in 1973–1974 and again in 1979.

Total U.S. import reliance for gold has varied dramatically, usually as a result of the level of U.S. exports and consumption. For example, U.S. exports peaked at 16.5 million troy ounces in 1979.[9]

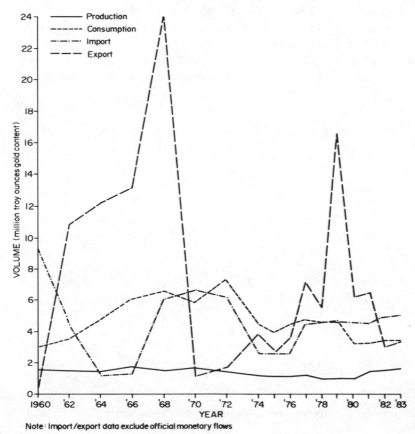

Note: Import/export data exclude official monetary flows

FIGURE I-2-10 U.S. production, imports, exports, and consumption of gold, 1960–1983. (From Refs. 6 and 9.)

In that year, according to the definition of import reliance used by the U.S. Bureau of Mines, the United States was 50% import dependent. The following year the United States was only 18% import dependent. This was due to a fall of 1.5 million troy ounces in consumption and a drop of 10 million troy ounces in gold exports.[11] From 1979 to 1982, Canada provided 59% of U.S. imports, South Africa (through Switzerland), 13%, and the USSR provided 11% of U.S. imports.

From a 1981 base, the U.S. Bureau of Mines expects U.S. annual consumption of gold to increase by 2.0% until 1990.[11] Even if the mine production stays constant at about 1.0 to 1.5 million troy ounces per year, recycling of gold scrap should provide an annual supply of about 3 million troy ounces.[11] However, it is possible that domestic gold production will continue to increase annually as it has since 1981.

This is because of the increase in exploration activity spurred on by the high price of gold. In fact, recent work has fostered the development of entirely new exploration concepts and, in turn, gold deposit discoveries. A prime example of this is the discovery of gold in Algoma-type iron formations in south-central Wyoming.[28] The relationship between iron and gold occurrences is just now becoming recognized as a valid exploration tool.

In sum, there is a strong possibility that domestic gold production will continue to increase. Thus, even with increasing consumption levels, it is possible that U.S. import dependence could decrease somewhat.

Iron Ore, Iron, and Steel

As the primary source of iron, the most widely used metal today, iron ore has a long history of use. Domestic use of iron ore to produce iron dates back to at least 1608.[11] The industry has grown extensively since then and has of late had serious problems.

Reserves of iron ore held by the United States make up roughly 4% of the world total, as shown by Table I-2-1. This is a small amount for a nation which has historically been one of the largest producers and consumers of iron ore worldwide. Total known resources in the United States amount to about 9% of the world resources.[11] The major U.S. deposits are sedimentary Algoma-type banded-iron deposits found in the region of Lake Superior. Other U.S. deposits include oolitic ironstones of Paleozoic age found in the southeast United States, and also both magmatic segregations and contact-metamorphic deposits.

Current operations in the United States consist of 18 open-pit mines and one underground mine, 15 concentrating plants, and 12 pelletizing plants. Minnesota, with 70% and Michigan, with 21%, accounted for the vast bulk of 1983 iron ore production. Average ore grade was 64% iron. Capacity of the 12 pelletizing centers was 85 million long tons per year in 1983.[9]

Figure I-2-11 shows U.S. production and imports of iron ore from 1930 to 1983 as well as U.S. consumption from 1960 to 1983. U.S. imports of iron ore began to increase after World War II, when U.S. firms began developing foreign deposits to

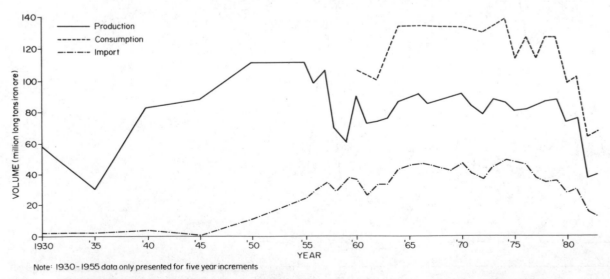

Note: 1930-1955 data only presented for five year increments

FIGURE I-2-11 U.S. iron ore production, imports, and consumption, 1930–1983. (From Refs. 6 and 9.)

off-set a depletion of domestic high-grade ores.[11] Domestic production peaked in 1953, remained reasonably stable from 1960 to 1979, and then plummetted when the U.S. steel industry began to experience severe problems. Even after permanently closing 13 mines by year end 1982, domestic ore producers operated at less than 50% of existing capacity throughout all of 1983.[9] Table I-2-15 shows how U.S. production has dropped as a share of world output.

The two major supply sources of iron ore to the United States have been Venezuela and Canada. In both cases these exporting facilities were developed by U.S. interests. The share of U.S. imports has switched over time; Venezuela was the dominant U.S. source in the 1950s and early 1960s, but Canada soon took the leading spot. From 1979 to 1982, 67% of the U.S. iron ore imports came from Canada, 15% came from Venezuela, and 8% each came from Brazil and Liberia.[9]

Since imports closely follow production and consumption, it is not surprising that U.S. iron ore imports have dropped by almost two-thirds since 1979. This is shown in Figure I-2-11. Since domestic demand for iron ore is expected to fall at an annual rate of 0.3% until 1990,[11] it is likely that iron ore imports will continue to fall.

The reason for this decline in demand lies with the problems of the U.S. steel industry. The data in Table I-2-16 show the extent of the problem: the U.S. share of world production of pig iron and steel has fallen from 23.3% and 26.0%, respectively, in 1960 to 10.3% and 12.0%, respectively, in 1983. This drop in relative share took place even

during the early 1970s, when U.S. production was increasing. Once the latest recession hit the U.S. steel industry, the bottom fell out. This is shown in Figure I-2-12.

The year 1982 was the nadir of the latest industry slump, but even with the 1983 increases in iron and steel production, the industry used only 55% of its capacity.[9] Steel companies suffered huge losses, facilities closed, and many workers lost jobs. Employment in blast furnaces and steel mills fell from 523,000 in 1974 to 295,000 in 1983. Not surprisingly, employment at iron ore mining and processing facilities also fell. In this case, employment fell from a steady level near 21,000 in the early to mid-1970s to a 1983 low of 7300.[9]

Several factors are behind the slump in the iron ore, iron, and steel industries. The most obvious one is the recession-induced drop in consumption. However, it is now expected that demand for iron and steel will increase 1.5% annually until 1990.[9] The bulk of the demand increase will be in sheet steel used in automobiles and consumer goods, while capital-good demands for iron and steel is expected to decline. The expectations of demand growth could be premature, though. High inflation, federal deficits, a continued strong U.S. dollar, or an end to voluntary automobile import quotas could easily change the picture.

A second factor behind the recent slump might simply be attributed to poor planning and bad management. Many examples come to mind. For instance, many U.S. steel producers failed to replace outmoded blast furnaces with cheaper and more efficient electric arc furnaces. Many foreign

TABLE I-2-15   U.S. Share of World Iron Ore Production, 1960–1983 (Million Metric Tons)

Year	World output	U.S. output	U.S. share (%)
1960	514	89	17.3
1962	499	72	14.4
1964	572	85	14.9
1966	628	90	14.3
1968	668	86	12.9
1970	754	90	11.9
1972	765	75	9.8
1974	881	84	9.5
1976	869	80	9.2
1978	841	82	9.8
1980	874	70	8.0
1981	853	73	8.6
1982	791	35	4.4
1983e[a]	762	38	4.9

[a]e, estimated.

Sources: Refs. 6 and 9.

TABLE I-2-16   U.S. Share of World Pig Iron and Steel Production, 1960–1983 (Million Tons)

	Pig iron			Steel		
Year	World output	U.S. output	U.S. share (%)	World output	U.S. output	U.S. share (%)
1960	285.3	66.5	23.3	381.6	99.3	26.0
1962	291.8	65.6	22.5	396.3	98.3	24.8
1964	350.0	85.5	24.4	483.0	127.1	26.3
1966	382.0	91.3	23.9	525.0	134.1	25.5
1968	418.1	88.8	21.2	584.0	131.5	22.5
1970	471.0	91.3	19.4	655.4	131.5	20.1
1972	500.0	88.9	17.8	693.0	133.2	19.2
1974	565.0	95.5	16.9	780.0	145.7	18.7
1976	541.2	86.8	16.0	742.1	128.6	17.3
1978	558.4	87.7	15.7	783.4	137.0	17.5
1980	560.0	69.0	12.3	780.0	111.9	14.3
1981	552.0	74.0	13.4	768.0	120.0	15.6
1982	500.0	43.3	8.7	708.0	74.6	10.5
1983	495.0	51.0	10.3	708.0	85.0	12.0

Sources: Refs. 6, 9, and 11.

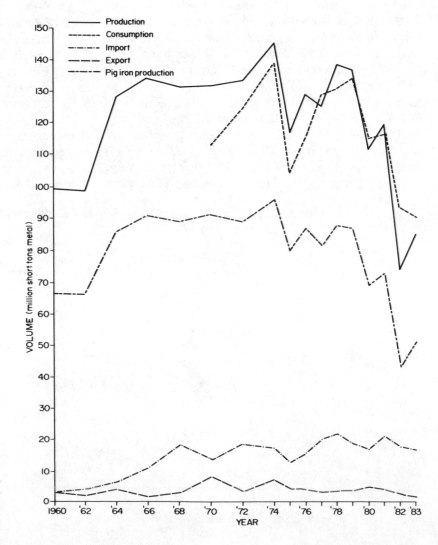

FIGURE I-2-12 U.S. production, imports, exports, and consumption of steel, and production of pig iron, 1960–1983. (From Refs. 6, 9, and 11.)

nations made the switch, though. So, as of late, Japan, South Korea, and other producers have been able to beat U.S. prices. Another example involves poor planning for iron ore capacity. Many U.S. steel producers assumed that demand would continue indefinitely and also ignored the ability of electric arc furnaces to utilize vast amounts of scrap. Thus U.S. iron ore mining and pelletizing capacity is perhaps 30% higher than could ever be needed.[29] U.S. capacity has dropped by one-third from its high of over 120 million metric tons of pellets, but only after industry suffered extensive losses.

The mismanagement that took place in U.S. steel industries partly explains the drop in U.S. pellet, iron, and steel production and the increase in imports of steel. It does not completely explain the shift, though. A good deal of this shift has taken place because of unfair foreign competition. There is by now a long list of foreign producers thought to be subsidizing their steel exports to the United States. In several cases, the U.S. Department of Commerce and International Trade Commission have ruled that foreign producers have dumped their steel in the United States and have injured the domestic industry. One recent example was the ruling against Argentina, Mexico and Brazil.[30]

Although subsidized foreign competition has hurt the U.S. industry, it is wrong to place the blame for the U.S. industry's problem entirely on foreign competitors. Fortunately, it appears that steps are now slowly being taken to correct the source of some of the problems. Although some

solutions, mergers for example, are controversial, at least the industry is now responding to a bad situation.

Lead

Another metal used historically for many centuries is lead. Early uses in pewter, glass, and the like have shifted to a more industrialized use breakdown. In 1983, lead used in batteries, gasoline additives, and other transport-related uses accounted for 75% of all domestic consumption. Construction applications such as paint took 20% of consumption, and the remainder was used in glass and ceramics.[9] The United States has historically been the primary consumer of lead in the world. In 1979, 33% of world consumption took place in the United States.[11]

As shown in Table I-2-1, the United States has about 27% of the world's reserves of lead. This makes the United States the holder of the largest reserves in the world.[11] Major deposits in the United States occur in three types. The most important, stratabound deposits in limestone and carbonate, are generally found in southeast Missouri and the region centered on the borders between Missouri, Kansas, and Oklahoma. A second type is the replacement deposit such as hydrothermal deposits in carbonate rocks. These occur in Utah, Nevada, and Colorado. The last deposit type is vein deposits, which are common in parts of Idaho. Lead commonly occurs with zinc, silver, antimony, and other minerals in these three deposit types.

Domestic mine production is centered in Missouri, as is shown in Figure I-2-13. In 1983, Missouri

Lead

Zinc

FIGURE I-2-13 U.S. lead and zinc mining regions, 1981. (From Refs. 6 and 21.)

mines produced 91% of total domestic output.[9] Table I-2-17 shows the location and production capacities of domestic lead mines. High ore grades and large, good formations made the Missouri mines low-cost producers, so Missouri dominates the mining scene.[21]

The situation is also true for the location and capacities of primary lead smelters and refineries. That is, most capacity is in Missouri. This is shown by Table I-2-18. Six primary facilities operated in Missouri, Texas, Montana, and Nebraska in 1983, and 40 secondary plants also operated.[9] Primary smelters and refineries operated at 86% of capacity in 1983, but secondary smelters operated at only 43% of capacity. As a result, some 42,000 tons of secondary capacity was permanently closed in 1983.[9]

Figure I-2-14 illustrates U.S. mine production, imports, and consumption of lead from 1960 to 1983. Several interesting points can be drawn from the figure. The first is that reported consumption vastly exceeds the sum of domestic mine produc-

tion and lead imports. This is because a large part of domestic consumption is met by recycling old scrap. In 1983 alone, 55% of consumption was met this way.[9] A second interesting point is that imports of metal have not increased in conjunction

TABLE I-2-18   U.S. Primary Lead Smelter and Refinery Capacities, 1983

Company	Location	Capacity (tons/year)
St. Joe Lead	Herculaneum, Mo.	230,000
Amax/Homestake	Buick, Mo.	140,000
ASARCO	Glover, Mo.	110,000
	East Helena, Mont.	84,000[a]
	El Paso, Tex.	92,000[a]
	Omaha, Nebr.	176,000[b]
Total		832,000

[a]Smelter only.
[b]Refinery only.

Sources: Refs. 6 and 21.

TABLE I-2-17   U.S. Lead and Zinc Mine Capacities, Mid-1981

State	Mine number	Mine name	Capacity (tons/year) Lead	Capacity (tons/year) Zinc
Missouri	1	Fletcher		
	2	Viburnum	244,000	24,000
	3	Brushy Creek		
	4	Indian Creek		
	5	Buick	176,000	44,000
	6	Milliken	83,000	8,000
	7	Magmont	84,000	11,000
Idaho	8	Star	17,000	17,000
	9	Sunshine	21,000	
Utah	10	Park City	13,000	
	11	Ontario		15,000
Colorado	12	Leadville	9,000	14,000
New Jersey	13	Sterling		36,000
New York	14	Balmat		66,000
	15	Pierrepont		
Pennsylvania	16	Friedensville		29,000
Virginia	17	Austinville Ivanhoe		18,000
Tennessee	18	Copperhill		5,000
	19	Elmwood		26,000
	20	Idol		10,000
	21	Immel		
	22	New Market		66,000
	23	Young		
	24	Jefferson City		17,000
	25	Zinc Mine Works		20,000
Total			647,000	426,000

Sources: Refs. 6 and 21.

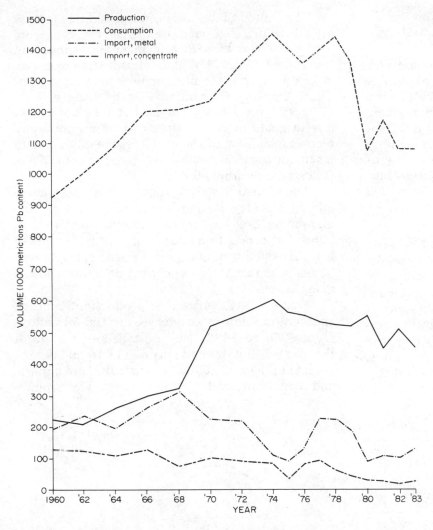

FIGURE I-2-14 U.S. mine production, imports, and consumption of lead, 1960–1983. (From Refs. 6 and 9.)

with a decrease in concentrate imports, as has happened with copper and other minerals.

A third reason is that U.S. mine production has not plummetted recently, even in the face of falling demand. Production has decreased some, but not significantly. This, too, is probably due to the buffer allowed by extensive recycling of domestic waste. Because of this, the United States has maintained its position as a major mine and refinery producer of lead. Table I-2-19, which shows U.S. and world mine and refinery production from 1960 to 1983, makes it clear that the U.S. share of world production has not slipped much since 1960.

The United States has not become highly import dependent with lead as it has with other minerals. In fact, the United States has at times been a net exporter of lead.[9] When the United States does import lead, the source is usually Canada or another country from North, Central, or South America. From 1979 to 1982, Peru, Honduras, and Canada provided 40%, 33%, and 11%, respectively, of U.S. imports of concentrate. Metal and scrap imports

TABLE I-2-19 U.S. and World Lead Production, 1960–1983 (Thousand Metric Tons)

Year	World production		U.S. production	
	Mine	Primary refined	Mine	Primary refined
1960	2385	2322	224	347
1962	2503	2408	215	341
1964	2530	2526	259	408
1966	2840	2710	297	400
1968	3007	2948	326	424
1970	3379	3285	519	605
1972	3448	3376	561	617
1974	3476	3499	602	611
1976	3303	3370	553	693
1978	3445	3469	530	566
1980	3520	3225	550	548
1981	3400	3216	446	498
1982	3450	3205	512	517
1983e[a]	3350	3159	450	510

[a] e, estimated.

Sources: Refs. 6 and 9.

for the same time period came mainly from Canada (46%), Mexico (34%), Australia (8%), and Peru (7%).[9]

Although the production side of the industry looks good, questions remain about demand. The use of lead as an additive in paint and gasoline has been largely eliminated, and the U.S. Environmental Protection Agency apparently will soon call for a total ban on these lead uses. Unless battery-powered electric vehicles, and various forms of lead-using electrical devices and systems come into demand, it is likely that the U.S. lead industry will have to cut back capacity over a period of time.[9]

Manganese

Without manganese, the production of steel would be extremely difficult, if not impossible. Manganese strips sulfur and oxygen from steel, imparts important qualities such as strength and hardenability, and is in general an indispensable part of the steelmaking process. It is also used in the production of cast iron and aluminum. However, the importance of manganese to steel production is such that use of manganese worldwide directly follows steel production.[11]

The United States has no reserves of manganese ore containing 35% or more manganese. Some ore containing less than 35% manganese is found in the United States, but it is classified as iron ore, after the mineral the manganese is associated with. There are some 74 million tons of manganese resources in the United States, but they are all too low in grade or too complex to be economically recoverable. Most of the resources are in Minnesota (as oxides and carbonate deposits) and in Maine (silicates and carbonates).[11]

The United States did have a small manganese mining industry at one time. Production of 35% manganese ore was up to 229,000 short tons in 1959, but production ended in 1972.[6] Since there is little recycling of manganese possible, the United States is almost 100% dependent on imports for its manganese needs.

Figure I-2-15 shows U.S. production, imports, and consumption of manganese ore and ferroalloys from 1960 to 1983. The most striking feature of the graph is the dramatic drop in all trend lines beginning in the mid-1970s. In particular, ore imports and consumption fell dramatically. The replace-

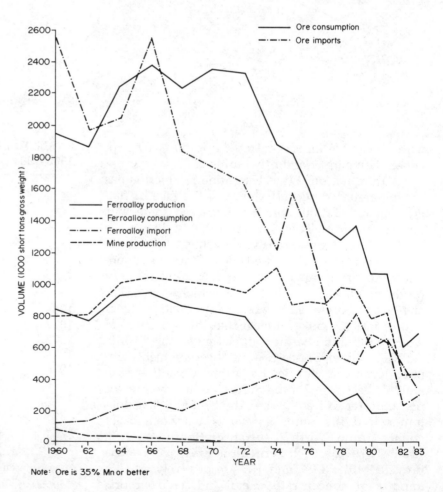

FIGURE I-2-15 U.S. production, imports, and consumption of manganese ores and ferroalloys, 1960-1983. (From Refs. 6 and 9.)

Note: Ore is 35% Mn or better

ment was an increase in the level of ferromanganese consumption. However, domestic production of ferroalloys has been falling since 1966 due to plant shutdowns. As a result, ferromanganese imports have significantly increased.

This shift has generated a great deal of concern, especially because sources of manganese imports are fairly limited. They are also increasingly distant. In the early 1960s, Brazil was the dominant supplier of U.S. manganese ore imports, while France was the major supplier of ferromanganese imports.[6] This began to change by the 1970s. From 1979 to 1982, South Africa was the leading source of U.S. imports of both manganese ore and ferromanganese. The second most important suppliers were Gabon for ore and France for ferroalloys.[9]

The increasing reliance on a small number of distant suppliers has lead to great concern about possible supply disruptions. The fact that U.S. ferromanganese production capacity is dropping only exacerbates the situation, for ore is generally available from a larger number of sources than is ferromanganese.[11] The likelihood of very limited domestic growth in demand (due to a slump and dislocation in the steel industry) is little consolation. Thus it would seem imperative that the United States ensure that an adequate stockpile is available. This has not yet been the case.[11] This issue is discussed in more detail in Chapter I-3.

Molybdenum

Molybdenum is a metallic element used primarily in metallurgical applications; it improves strength, hardenability, toughness, and resistance to wear and corrosion in steel. Additional uses in chemical applications, and an overall lack of adequate substitutes for most uses, make molybdenum a very important mineral. Some 75% of the 1983 U.S. consumption of molybdenum was for the iron and steel industries, while 15% was for chemical uses.[9]

If there is a mineral with which the United States is well endowed, it is molybdenum. Table I-2-1 shows that the United States possesses almost 44% of the world's reserves. Deposits normally are of three types: (1) porphyry or disseminated deposits, (2) contact metamorphic zones, and (3) quartz veins. All three types are hydrothermal deposits in which metals precipitate from high-temperature solutions. Due to the relationship between these deposits and the pressures and temperatures involved in mountain formations, it is not surprising that most of the world's reserves of molybdenum are found in the western cordillera of North and South America. Canada, the United States, and Chile together possess over 85% of world molybdenum reserves.[11]

In 1983, three firms mined molybdenum as a primary product. This was from porphyry molybdenum mines in Idaho, New Mexico, and Nevada. Eleven other firms recovered molybdenite as a by-product of copper, tungsten, and/or uranium mines in Arizona, California, New Mexico, and Utah. Some of the molybdenite recovered in 1983 was converted to molybdic oxide for use in making metal powders. This work was done by seven facilities.[9] Most of the molybdenum mined in 1983 came as a by-product, for low molybdenum prices forced the closure of porphyry molybdenum mines until late in the year. Utilized mine capacity was 15% in 1983.[9]

The obvious production slump suggested by the foregoing numbers is illustrated in Figure I-2-16, which shows U.S. production, consumption, and exports of molybdenum from 1960 to 1983. Concentrate production and exports of both concentrate and oxides fell dramatically from the levels of the late 1970s. Consumption fell at a lower rate, and much demand was met by simply trying to draw down increasingly large private stocks.

In fact, private stocks reached unusually high levels. Total private stocks in 1979 amounted to over 27 million pounds of contained molybdenum, yet by 1982 stocks exceeded 89 million pounds.[9] That this took place even though U.S. mine production began to fall in 1980 means that the problem was not just with U.S. producers. The key to this problem is that the United States is a major exporter of molybdenum; as such, it is subject to market influences.

However, as Figure I-2-16 shows, U.S. exports began a steep decline beginning in 1979. Obviously, other molybdenum producers and exporters did not cut back on output in accordance with market dictates. Once again, as was seen with copper, the culprit appears to be Chile. Table I-2-20 shows that U.S. production averaged between 60 and 80% of world output from 1960 to 1981. By 1983, the U.S. share was only 25.0%. The United States cut back its production from 83 million pounds of contained molybdenum in 1982 to 30 million in 1983, a decrease of about 276%.[9] Canada reduced its production by almost 200%. However, Chile, the other of the three largest producers, cut back only by 38% (from 44 million to 32 million pounds). Peru actually increased output.[9]

The data going back a few years only strengthens the case against Chile. From 1978 to 1981,

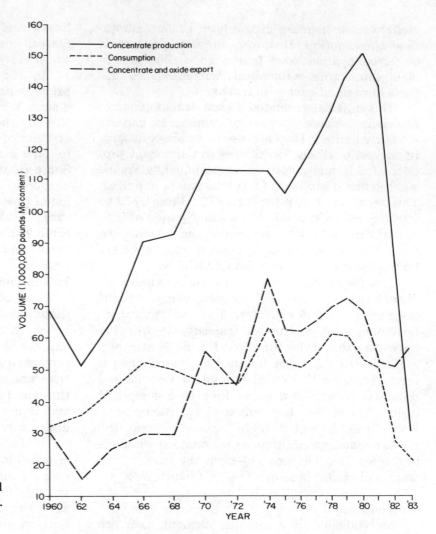

FIGURE I-2-16 U.S. production, exports, and consumption of molybdenum, 1960–1983. (From Refs. 6 and 9.)

TABLE I-2-20 U.S. Share of World Molybdenum Production, 1960–1983 (Thousand Pounds Molybdenum Content)

Year	World mine output	U.S. mine output	U.S. share (%)
1960	89,100	68,237	76.6
1962	75,100	51,244	68.2
1964	77,829	65,605	84.3
1966	124,967	90,532	72.4
1968	144,771	93,447	64.5
1970	177,982	111,352	62.6
1972	178,423	112,138	62.8
1974	189,274	112,011	59.2
1976	195,473	113,233	57.9
1978	220,922	131,843	59.7
1980	239,190	150,686	62.9
1981	223,800	139,900	62.5
1982	199,000	83,000	41.7
1983e[a]	120,000	30,000	25.0

[a]e, estimated.

Sources: Refs. 6 and 9.

Chile's production and exports of both copper and molybdenum increased.[6] Comparable figures for the United States show a general decline. Then, when market conditions really called for cutbacks, Chile continued to maintain output.

As long as a world oversupply situation exists, and as long as some producers refuse, for whatever reason, to adjust output according to market conditions, U.S. producers can expect to take it on the chin. The prospects for this changing in the near future look very remote.

Nickel

Nickel's ability to impart strength and corrosion resistance to metals and nonmetals, as well as its catalytic properties, make it a very important strategic mineral. The strategic usefulness of nickel is shown in that nickel demand rates during wars and other crises are well above the general rate of economic growth.[11] In 1983, domestic nickel consumption fell into the following general classes: stainless and alloy steels, 41%; nonferrous alloys, 32%; electroplating, 20%; and other uses, 7%.[9] The usefulness of nickel is seen in that, even with about 40% of its current uses in a depressed steel industry, nickel is expected to have an annual demand increase of 2.1% from 1981 to 1990.[9]

Table I-2-1 shows that the United States has only 0.2% of world reserves of nickel. All U.S. reserves are in the form of high-iron laterites near Riddle, Oregon. Garnierite and other minerals with an ore grade of 0.8 to 1.3% nickel are recovered from altered peridotites in the area. The United States has a fairly large nickel resource base, most of which is in the form of laterites in California, Oregon, and Washington. Some nickel resources are also found in sulfides in Minnesota's Duluth Gabbro. Lateritic resources have ore grades similar to that of the Riddle ore body, while the Duluth Gabbro deposit averages 0.2% nickel.[11]

The only domestic mine producer of nickel is the Hanna Mining Company facility in Riddle. Since nickel is recovered as a by-product of the iron ore mining operations in the area, the facility is able to overcome the higher processing costs associated with laterites. The mine has an on-site ferronickel smelter with a capacity, in 1979, of 13,000 tons per year of contained nickel (in the form of ferronickel).[11]

One other U.S. facility recovers nickel from imported copper-nickel matte. This is the AMAX Nickel Company refinery at Port Nickel (Braithwaite), Louisiana. Rated capacity in 1979 was 40,000 tons per year of nickel briquettes and powder.[11] Seven countries, including Canada, Botswana, and Australia, supply the matte.[6]

Table I-2-21 shows how U.S. nickel consumption compares with world mine production from 1960 to 1983. It is obvious that the United States consumes one-sixth to one-fourth of annual mine production of nickel. Figure I-2-17 shows that domestic production of nickel has consistently fallen far below the level of domestic consumption. Thus most consumption is met by imports of nickel.

Canada has historically been the dominant supplier of U.S. nickel imports, which is not surprising given Canada's proximity and Canada's status as the major western nickel producer and exporter. Canada's share of U.S. imports has dropped from over 90% in the 1960s to 51% from 1979 to 1982.[6] As of late, Australia and Botswana have provided almost a quarter of U.S. nickel imports.[9]

American dependence on nickel imports has created several political-economic conflicts. An early one was over the use of government subsidies to encourage increased levels of domestic mine production of nickel. A more recent conflict involves U.S. nickel imports from Soviet bloc nations. Exisiting U.S. laws banned nickel imports from Cuba, but it was discovered that Cuban nickel was funneled through the USSR to the United States. This was already on top of what some called alarmingly high Soviet nickel exports to the United States. Although the price was low, it was apparently subsidized by the Soviets. As a result, the U.S. Treasury Department in December 1983 banned all

TABLE I-2-21 U.S. Consumption versus World Mine Production of Nickel, 1960–1983
(Thousand Tons Nickel Content)

Year	U.S. consumption	World production
1960	108.2	359.0
1962	118.7	401.0
1964	146.9	408.9
1966	187.8	440.1
1968	159.3	547.9
1970	155.7	694.1
1972	159.3	673.8
1974	208.4	870.7
1976	162.9	883.9
1978	180.7	731.4
1980	156.3	850.4
1981	144.9	716.0
1982	103.9	669.8
1983e[a]	122.0	630.0

[a]e, estimated.

Sources: Refs. 6 and 9.

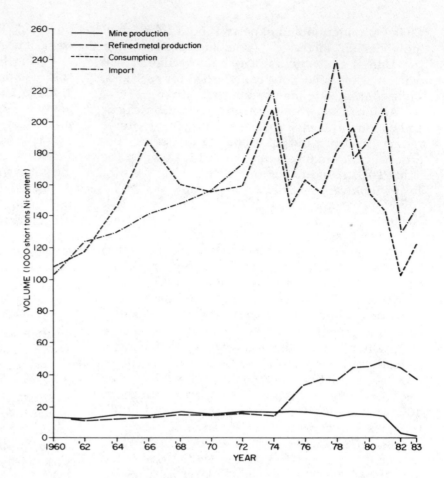

FIGURE I-2-17 U.S. production, imports, and consumption of nickel, 1960–1983. (From Refs. 6 and 9.)

direct or indirect imports of Soviet-fabricated nickel and nickel-bearing items.[31]

One promising domestic note involves the possibility that a new nickel mine might open up soon. California Nickel Company's Gasquet Mountain mine in far northern California could produce up to 18 million pounds of nickel per year from a lateritic deposit.[32] Sources indicate the mine may be a very low cost operation because of advances in recovery technologies and the value of by-products like cobalt and chromite. An actual startup date is not known due to such obstacles as an attempt to prevent mining because the deposit is in a national forest.

Phosphate

Phosphate is one of the three truly essential fertilizer minerals—without it, agriculture as we know it would simply not exist. Fortunately for the United States and its huge agricultural system, the United States has about 12% of world reserves and a much larger share of world resources.[6] This is shown by Table I-2-1. About 80% of American reserves are located in Florida and North Carolina, and smaller amounts are found in Alabama, Tennessee, Montana, Idaho, and Utah.[33]

Phosphate production in 1983 had a market value of $840 million. Twenty-two producers accounted for production. The 13 which were in Florida (12) and North Carolina (one) accounted for about 85% of total U.S. output.[9] The remainder of domestic output came from smaller operations in the five other states mentioned above.

Figure I-2-18 shows U.S. production, consumption, and exports of phosphate rock from 1960 to 1983. In addition, the figure shows world phosphate production for the same period. A quick glance reveals the dominant position the United States has in world production; that is, the United States has for many years been the primary phosphate producer. For example, in 1960, the United States produced 40.5% of total world production.[6] Comparable U.S. shares were 41% for 1964, 44.6% for 1968, 41.1% for 1972, 41.4% for 1976, 39.7% for 1980, and 32.8% for 1983.[6,9]

In fact, U.S. output of phosphate rock is so high that this is one of the few minerals for which the United States is a net exporter. As Figure I-2-18 shows, the United States typically exports one-third to one-fifth of its annual production. The bulk of U.S. exports go to Canada, Western Europe, and such Pacific Rim nations as Japan and South

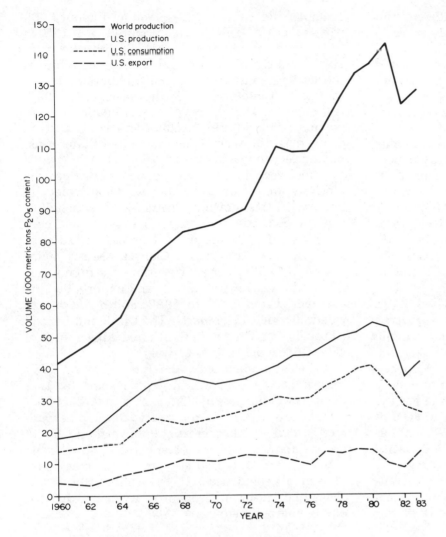

FIGURE I-2-18 World production, and U.S. production, exports, and consumption of phosphate rock, 1960–1983. (From Refs. 6 and 9.)

Korea.[11] Exports in 1982–1983 were somewhat lower than normal, largely due to increased Moroccan exports.[9]

In any event, U.S. consumption is expected to soon reverse the slump it experienced beginning in 1980. Consumption is expected to increase by 2% per year until 1990, according to the U.S. Bureau of Mines.[9]

How much U.S. production can be expected to increase in the near future will depend on two factors. The first will simply be the demand for phosphates. If U.S. demand and exports do remain strong, several proposed mines should soon open. One would be the Aurora City, North Carolina, mine owned by the North Carolina Phosphate Company. Expected production is to be 3.6 million metric tons per year by 1987.[34]

A second and perhaps far more significant factor controlling future U.S. production will be land use and environmental control laws. Much of the reserves in the United States are on federal lands, so availability is questionable. Mining in

Florida has always been controversial because of the coastal impacts. In fact, Department of Interior studies indicate pollution control measures for Florida mining operations account for almost 10% of total production costs in the 1980s.[33] Finally, the complex permit process can delay project start-ups to the point where they are no longer economical.[33]

Platinum-Group Metals

Once used primarily for jewelry and artistic purposes, the six platinum-group metals have become indispensable to a variety of industrial applications. Most domestic consumption is now for industrial purposes, and most industrial uses involve the use of the metal(s) as a catalyst. In 1983, 72% of total U.S. consumption was in automotive, electrical, and chemical applications, most of which were catalytic in nature. The remaining 28% of domestic consumption was largely for dental and jewelry applications.[9]

The platinum-group metals (PGM) are among

the scarcest of all metals worldwide, so it is not too surprising that the United States has only small reserves. In fact, Table I-2-1 shows that the United States has only 0.1% of world PGM reserves. It is little consolation that almost all other nations have the same scarcity of reserves. South Africa, the USSR, and Canada easily account for well over 99% of known reserves.[11]

U.S. reserves are largely found in Montana, Minnesota, and Alaska. Some 7 million troy ounces of reserves are in the differentiated mafic and ultramafic rocks of the Stillwater Complex in south-central Montana. Two deposits in Minnesota's Duluth Gabbro contain about 800,000 troy ounces of reserves in nickel-copper ores. Finally, PGM placers, estimated at 500,000 troy ounces, are found in the Goodnews Bay section of the Salmon River on the west coast of Alaska.[35]

Mine production of PGM in the United States is currently restricted to small amounts recovered as the by-product of copper mining and refining and as placers from the Goodnews Bay operation. The latter operation closed in 1976, reopened briefly in 1980, and then reopened again in early 1984. Moreover, the Goodnews Bay mine has the reputation of being the leading producer of PGM in the United States. Since it began operations, it has recovered some 641,000 troy ounces.[35] Capacity is currently about 10,000 troy ounces a year.

Even if U.S. mines were to operate at full output, the United States would produce only a fraction of a percent of world output. In 1983, for example, the United States produced about 8000 troy ounces of PGM; this amounted to 0.12% of world mine production.[9] This disparity becomes even more significant because the United States is the leading consumer of PGM production. This is shown in Table I-2-22 for the years 1960–1983. Although the U.S. demand levels are taking a lower portion of world mine production each year, U.S. demand is still huge. Thus the United States must rely on recycling and imports.

Since recycling is, even now, not practiced extensively, most consumption needs are met by imports. This is shown in Figure I-2-19, which illustrates U.S. production, imports, exports, and consumption of platinum-group metals from 1960 to 1983. In the early 1960s, Canada was the major source of U.S. imports.[6] With time, the dominance of South African production and exports became apparent. From 1979 to 1982, South Africa provided 56% of U.S. imports. The USSR and Canada (through its refineries in the United Kingdom) provided 16% and 11%, respectively.[9]

The aforementioned catalytic uses of PGM means that demand will grow. It is expected that demand will increase by 2.9% each year until 1990.[9] Should this occur, the pressure to develop new domestic mines will increase. Policymakers should not be fooled, though. At any price, all possible U.S. mines could produce enough to meet only 10% of expected needs.[35] The United States will always be import dependent to a great extent for platinum-group metals.

TABLE I-2-22   U.S. Consumption versus World Mine Production of Platinum-Group Metals, 1960–1983 (Thousand Troy Ounces)

Year	World production	Reported U.S. consumption	U.S. share (%)
1960	1275.0	775.2	60.8
1962	1630.0	866.5	53.2
1964	2545.8	1117.7	43.9
1966	3039.4	1675.8	55.1
1968	3393.7	1367.9	40.3
1970	4238.9	1388.5	32.8
1972	4269.9	1562.2	36.6
1974	5773.7	1981.0	34.3
1976	5978.4	1603.1	26.8
1978	6332.2	2259.6	35.7
1980	6830.0	2206.0	32.3
1981	6780.0	1921.0	28.3
1982	6500.0	1855.0	28.5
1983e[a]	6600.0	1800.0	27.3

[a]e, estimated.

Sources: Refs. 6, 9, and 11.

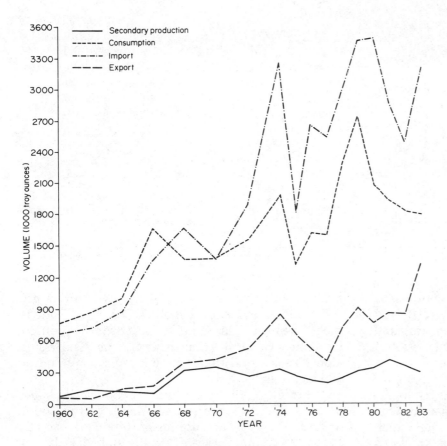

FIGURE I-2-19 U.S. production, imports, exports, and consumption of potash, 1960–1983. (From Refs. 6 and 9.)

The limited variety of possible supply sources is restricted even more by the unreliability of Soviet supplies. Given such an unstable supply picture, it would seem sensible to keep large stocks of such important metals. However, as of late 1983, government stockpiles of platinum, palladium, and iridium are well below stated goals.[9] Even with increased recovery and reuse of existing scrap, and so on, currently available stockpiles would fall short of U.S. demand requirements for a period of several years.[9]

Potash

Potash, another of the three essential fertilizer minerals, is also a mineral for which the United States must depend on foreign sources. Total U.S. reserves amount to only 3.3% of world potash reserves, as shown in Table I-2-1. Another 8 billion metric tons of potash resources are found at depths below 4000 feet, and are presently uneconomical.[9] Most U.S. potash deposits are sylvinite and are found in salt beds which formed during the evaporation of shallow areas of ancient seas.

Most deposits are extensive and tabular in form. The bulk of U.S. reserves and resources are found in Montana and North Dakota (an extension of the Williston Basin deposits of Saskatchewan) and in

Utah's Paradox Basin.[9] Smaller amounts are found in California and New Mexico. Current production occurs in New Mexico, California, and Utah. Most U.S. production is in southeast New Mexico, where conventional underground mining of bedded deposits takes place. Solution mining is often used at other sites, especially for deposits located more than 3500 feet below ground.

Figure I-2-20 illustrates U.S. production, consumption, and trade in potash from 1960 to 1983. It is obvious that a major shift in potash supply and demand took place in the early 1960s. Before 1964, the United States was a net exporter of potash, and U.S. production exceeded domestic consumption. This has since changed. Production has steadily dropped, exports have recently begun to drop, and so only increasing levels of imports can meet demand levels. By 1983, the United States was 75% dependent on imports of potash.[9]

Canada has, for the last 20 years, been the premier source of U.S. imports of potash. With the richest, largest reserves of potash in the free world, and with such close proximity to the United States, it is natural that Canadian supplies entered the United States. Canadian potash has typically provided roughly 95% of total U.S. imports, but that level dropped to 93% from 1979 to 1982.[9] The use

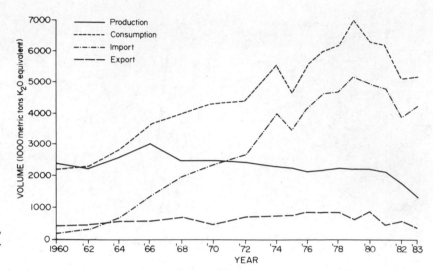

FIGURE I-2-20 U.S. production, imports, exports, and consumption of potash, 1960–1983. (From Refs. 6 and 9.)

of Canadian imports has always been acceptable to the United States because of their low cost and also because many Canadian mines were initiated by U.S. firms. Thus Canadian imports have not been perceived as a threat to U.S. producers.

However, certain other potash exporters have apparently been subsidizing their exports to the extent that even Canadian prices are undercut. One source has it that the major culprits are Israel, Spain, and the USSR. Moreover, the availability of these supplies has forced U.S. mines to close, and Canada's share of the U.S. market to fall. Should this continue, it is feared that the United States could lose some 600 million metric tons of mineable reserves.[36] This loss in domestic production capacity would only be exacerbated if the United States actually does experience a predicted 2% annual demand growth through 1990.[9] The loss of reserves amounts to 100 years of consumption at 1983 demand levels.[9]

Silver

Although this precious metal has long been coveted for its beauty and cosmetic appearance, silver has become a critical metal for a variety of industrial applications. By far the most important is its use in photography. Uses in 1983 were as follows in the United States: photography, 43%; electrical/electronic, 26%; flatware and jewelry, 15%; brazing alloys and solders, 6%; other uses, 10%.[9]

According to Table I-2-1, the United States is in good shape with respect to reserves. Over 20% of current world reserves are in the United States. Silver is found in the United States in three major deposit types. The first type consists of primary silver found as hydrothermal disseminations in veins in country rock. Lead is commonly associated with silver in this type. A second deposit finds

silver in intermediate felsic rocks (e.g., epithermal injections in andesite). Nevada's Comstock Lode is such a deposit. This third type of deposit, typified by Idaho's Coeur d'Arlene district, consists of veins along fissures and shears in Precambrian quartzites and argillites.[11]

Currently, domestic mine production takes place in 17 states, with Idaho, thanks to the Coeur d'Alene district, the dominant producer overall. Idaho, Montana, Arizona, Utah, and Nevada accounted for 87% of mine output in 1983.[9] Figure I-2-21 shows U.S. silver production, consumption, and trade from 1960 to 1983. It is obvious that U.S. mine production has been relatively steady during the time period of concern. However, U.S. industrial consumption, imports, and exports have varied dramatically due to changes in price, stockpile sales, and so on. The recent increase in production is attributed to an increase in price.

The most recent price increase has resulted in an upturn in silver exploration efforts and discoveries. This will be necessary if U.S. demand grows at the expected 2.2% per year through 1990.[9] However, it may not be reasonable to assume that U.S. mine production of silver will necessarily increase because of new discoveries. The reason for this is that most increases in reserves will probably come from base metal discoveries. Obviously, because of the uncertainties in the U.S. copper, lead, and zinc industries, it may be unreasonable to expect that U.S. silver output will increase if the base metal markets do not improve.[9]

As a result, it is likely that the United States will remain a net importer of silver. Fortunately, a variety of supply sources exist; moreover, the sources are convenient to the United States and are relatively secure. From 1979 to 1982, Canada continued its tradition of being the major source of

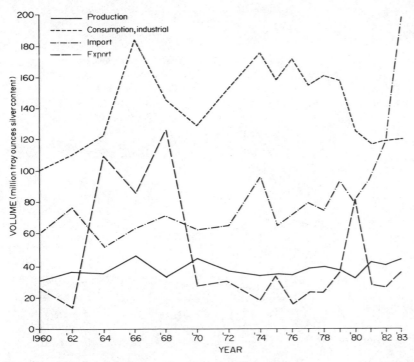

FIGURE I-2-21 U.S. production, imports, exports, and consumption of silver, 1960–1983. (From Refs. 6 and 9.)

U.S. imports. Canada accounted for 34% of U.S. imports, while Mexico had 23%, Peru 21%, the United Kingdom had 8%, and other sources contributed 10% of total U.S. imports.[9]

Tin

Tin has a history of use longer than that of almost all other metallic minerals. It was first combined with copper to form bronze about 3500 B.C.[11] Long used in containers, tin is also used in solder, but it is losing ground to aluminum in a number of applications.[9]

Table I-2-1 shows that the United States holds an extremely small share of world tin reserves; the amount was only 0.4% in 1981. Most U.S. reserves occur as lodes and placer deposits in areas with granites and their extrusive equivalents. Major U.S. deposits exist on Alaska's Seward Island, while lesser amounts occur in California, Colorado, New Mexico, South Dakota, and Texas.[11]

Domestic production in 1983 came from only two sources: as a by-product of molybdenum mining in Colorado and as a residual placer in Alaska. The only domestic primary tin smelter is the Texas City, Texas, facility, operated by the Gulf Chemical and Metallurgical Company. This facility produces tin from domestically produced concentrates as well as secondary materials and tin-bearing slags.[11] As of year-end 1981, the United States was still the world's largest producer of secondary tin.[11] This is obviously one case where the United States has reasonably good recycling and recovery facilities and systems.

Figure I-2-22 shows that domestic secondary smelter production of tin has been steady since the early 1960s except for a slow decline since the late 1970s. The trends for other important tin statistics are not as steady. First, U.S. consumption of tin has fallen steadily from a high in the mid-1960s. Second, production of primary tin has decreased since 1960; production in 1983 was only one-sixth that of 1960. Third, metal imports, although erratic, have been consistently far in excess of the level of ore imports. Moreover, ore imports have steadily fallen. As was shown earlier, this is due to the increased pre-export processing now taking place in nations that mine raw ores.

The United States has been at least 70 to 80% import dependent for tin for at least the last 20 years.[9] Major import sources have traditionally been Malaysia, Thailand, Indonesia, and Bolivia. Even though U.S. demand is expected to grow at less than 1% a year through 1990, if U.S. production also keeps slipping, it is quite likely that current levels of import dependence will at least be maintained, if not increased.[9]

Titanium

Titanium is an element with both a solid base of present applications and a promising future. Used as a metal, titanium's strength-to-weight ratio, heat and corrosion resistance, and other qualities make it of considerable benefit to many applications. Titanium metal consumption for the United States in 1983 fell into the following categories: jet and space applications, 60%; steel and alloys, 20%; and

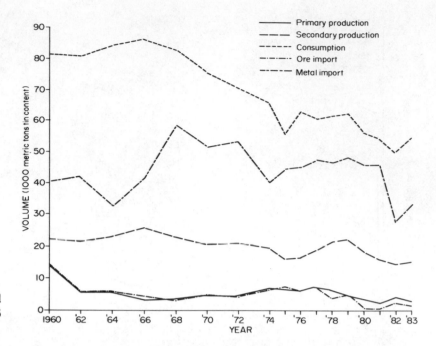

FIGURE I-2-22 U.S. production, imports, and consumption of tin, 1960–1983. (From Refs. 6 and 9.)

chemical processing, power generation, and marine applications, 20%.[9] When used in the form of titanium dioxide (TiO_2), chemical and optical applications are most important. American use of TiO_2 in 1983 was: paint pigments, 48%; paper, 27%; plastics, 13%; rubber, 3%; ceramics, 1%; and other uses, 8%.[9]

Titanium is principally derived from two minerals, ilmenite and rutile. As shown in Table I-2-1, the United States has about 7% of world reserves of both minerals. Ilmenite deposits in the United States are usually either found in titaniferous iron ores associated with ilmenite or in beach and stream placers. The former deposit type is typified by the deposits of upper New York, whereas the latter is typified by the beach deposits of Florida, New Jersey, and Georgia. The ilmenite reserves of New York and Florida make up almost 70% of total U.S. reserves.[11] Rutile deposits are also found in certain anorthosite complexes, but most U.S. reserves are in beach sand deposits in Florida, Georgia, and Tennessee. Resources of rutile are also found in Arizona, Arkansas, California, North and South Carolina, Utah, and Virginia.[11]

Ilmenite was produced by firms in both New York and Florida in late 1983. Early in that same year, a New Jersey ilmenite facility run by ASARCO was permanently closed. That one facility had accounted for 24% of total U.S. ilmenite output.[37] Almost 99% of total 1983 ilmenite production went for the production of pigments, while the remainder was used to produce synthetic rutile at a Mobile, Alabama, facility.[9]

Rutile production in 1983 was restricted to three mining operations in Florida. One recovered rutile from beach sands, while the other two recovered rutile from bulk concentrates consisting mainly of ilmenite and leucoxene. Some 82% of total 1983 production went for the production of TiO_2 pigment, while the remainder was converted to titanium tetrachloride and then titanium metal.[9]

Six firms, located primarily in Nevada, Ohio, and Oregon, accounted for all domestic production of titanium sponge in 1983. They operated at only about 40% of their total production capacity of some 33,500 tons of sponge per year.[9] Nine producers found across the nation accounted for domestic ingot production. Their total capacity as of 1981 was 50,000 tons a year.[6]

Figure I-2-23 shows U.S. production, imports, and consumption of ilmenite and rutile from 1960 to 1983. It is readily apparent that U.S. consumption has slowly increased, whereas production has fallen significantly since the early 1960s. As a result, the United States has become increasingly dependent on imports of both ilmenite and rutile. Although figures are often withheld, it is obvious that the United States imports 25 to 75% of its ilmenite needs and something like 90% or more of its rutile needs in any one year.[9] Canada is by far the primary source of U.S. ilmenite and titanium slag imports, whereas Australia accounts for almost 75% of U.S. rutile imports.[9]

Table I-2-23 shows a more detailed breakdown of U.S. imports and consumption of ilmenite and rutile from 1960 to 1983. Even a rudimentary

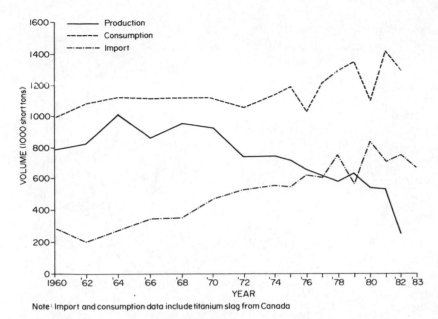

Note: Import and consumption data include titanium slag from Canada

FIGURE I-2-23 U.S. production, imports, and consumption of ilmenite and rutile, 1960–1983. (From Refs. 6 and 9.)

TABLE I-2-23 U.S. Imports and Consumption of Ilmenite[a] and Rutile, 1960–1983 (Thousand Tons)

	Imports		Consumption	
Year	Ilmenite	Rutile	Ilmenite	Rutile
1983	560	120	W[b]	270
1982	596	163	1055	239
1981	505	202	1133	285
1980	552	282	802	298
1978	458	290	1000	263
1976	340	282	917	238
1974	319	246	1110	293
1972	313	221	1051	243
1970	231	243	1099	188
1968	178	174	1102	160
1966	187	152	1095	136
1964	173	111	1109	79
1962	166	36	1083	32
1960	266	29	989	24

[a]Ilmenite figures include slag from Canada.
[b]W, withheld.
Sources: Refs. 6 and 9.

glance will indicate the increasing share held by rutile in the total levels of imports and consumption of both minerals. Ilmenite consumption has grown very slowly, whereas rutile consumption has increased to about 10 times its 1960 level. This is primarily due to the rapidly increasing use of titanium metal in aeronautical and aerospace applications beginning in the early 1960s.

This increase in the use of sponge metal is documented by Figure I-2-24, which shows U.S. production, imports, and consumption of titanium

sponge metal from 1960 to 1983. Figure I-2-24 shows several important points. First, although the United States produces almost as much sponge as it consumes, the United States does import a fair volume of sponge metal. Japan has been the primary supplier, and from 1979 to 1982, accounted for 85% of U.S. imports.[9] Second, consumption and production have generally increased since 1960, except for several significant periods of decline. Third, these periods of declining consumption and production are caused largely by the varying demand for titanium metal in commercial aircraft and military applications.[9] The graph in Figure I-2-24 shows significant falls in 1974–1976, when the first OPEC oil price hike hit the hardest, and from 1980 on, when the United States entered its worst economic slump since the 1930s. In both cases, expansion plans for commercial airlines were shelved and the same held true even for some military projects.

The U.S. Bureau of Mines forecasts an annual demand increase of about 2% for both ilmenite and rutile through 1990.[9] For sponge, demand is expected to increase 5% per year, assuming, of course, continued expansion of airline fleets.[9] These figures indicate continued import dependence for ilmenite and rutile, but things are much less certain for sponge production.

The United States currently has enough excess sponge production capacity to meet all domestic needs through the mid-1980s.[38] However, if demand does increase at about 5% per year, U.S. consumption in the year 2000 will be roughly twice what it is now. This would require an almost 100% increase in domestic production capacity, and this

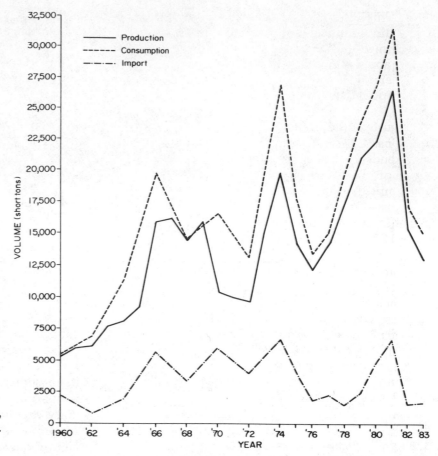

FIGURE I-2-24 U.S. production, imports, and consumption of tungsten, 1960–1983. (From Refs. 6 and 9.)

would be just to meet demand and not to help fill stockpile goals.[37]

One concern is that foreign competition is being subsidized to such an extent that U.S. producers might scale back domestic expansion plans. In fact, in late 1983, several U.S. sponge producers charged that Japan and the United Kingdom were dumping sponge in the U.S. market to help their domestic producers.[37] The U.S. International Trade Commission later investigated the charges and ultimately agreed. Thus the U.S. Commerce Department was recently given the task of assessing the extent of damages suffered by U.S. firms.

Tungsten

Tungsten imparts hardness and heat and wear resistance to other elements, so it is widely used for specific applications. In 1983, tungsten was used in the United States for the following purposes: metalworking and other machinery, 70%; transport items, 12%; lamp filaments, 9%; electrical uses, 5%; and other applications, 4%.[9]

The United States holds almost 5% of world tungsten reserves, as is shown by Table I-2-1. Most U.S. reserves are in the form of tactites, or contact metamorphic deposits, from which tungsten-bearing scheelite is recovered.[11] These deposits occur mainly

in California and, to a lesser extent, Nevada and Montana. Other important U.S. tungsten deposits are intrusive stockworks in granites. One such example is the tungsten-bearing intrusive stockworks of the Climax, Colorado, porphyry molybdenum deposit.[11]

Domestic mine production of tungsten is restricted to the western United States, where actual output varies with the state of the mineral industry. In 1981, 37 mines in the western United States produced tungsten ores and concentrates.[6] About 95% of the total came from four mines (two in California and one each in Nevada and Colorado). By 1983, the two mines in California alone accounted for 95% of that year's total, albeit reduced, production level. This was because of the shutdown of some Alaskan mines and particularly the Climax, Colorado, mine. The Climax mine, given its status as a primary molybdenum producer, had to close due to falling demand conditions; thus, the recovery of tungsten as a by-product was also stopped.[9] One California mine, the Pine Creek facility near Bishop, kept operating partly due to the revenues from such by-products as silver and gold.[6]

A great deal of domestically produced tungsten concentrate is now converted to ammonium-para-tungstate (APT) at the mine site and at other locations. This intermediate product can be used as is in certain chemical applications, but it is largely transformed into tungsten metal powder for use in the applications described above. Current APT producers have the capacity to turn out over 4.0 million pounds of tungsten per year in the form of APT.[6]

Figure I-2-25 shows U.S. production, imports, and consumption of tungsten from 1960 to 1983. The trends shown in the graph are not unusual for many of the minerals used in the United States. Consumption gradually increased until the early 1980s, when the recession caused consumption to plummet. Mine production has gradually decreased since 1970, while imports have increased. Table I-2-24 illustrates the U.S. share of world production and consumption of tungsten concentrates from 1960 to 1983. It is obvious that the relative share held by the United States has fallen considerably, especially in the last few years. This in part helps explain why U.S. mine capacity utilization was only about 20% in 1982 and 1983.[9]

The United States has depended on imports for some 40 to 60% of its consumption needs since 1960.[9] China was normally the leading source of U.S. imports, but Canada has taken that position in the last few years. From 1979 to 1982 Canada provided 20% of U.S. imports, Bolivia 18%, and China, 17%, while other countries such as the United Kingdom provided the rest.[9] The threat of economic damage due to a supply cutoff is mitigated somewhat by the abundance of supply sources. Moreover, except for uses in drilling bits and the

TABLE I-2-24   U.S. and World Production and Consumption of Tungsten Concentrates, 1960–1983 (Metric Tons Tungsten Content)

	Mine production		Consumption	
Year	U.S.	World	U.S.	World
1960	3,162	31,162	5,263	
1962	3,638	31,636	6,209	
1964	3,989	28,083	5,583	18,988
1966	4,010	29,042	8,189	22,208
1968	4,101	31,010	5,006	29,209
1970	4,223	34,264	7,573	37,901
1972	3,195	38,264	6,396	34,730
1974	3,554	36,964	7,391	37,194
1976	2,662	41,223	7,305	36,463
1978	3,129	45,408	8,528	39,195
1980	2,738	54,112	9,268	
1981	3,544	52,334	9,839	
1982	1,575	44,872	4,506	
1983e[a]	1,100	37,350	4,300	

[a]e, estimated.

Sources: Refs. 6 and 9.

like, tungsten has reasonably cost-effective replacements for most applications. Finally, tungsten is one of the few metals held in excess of goals in the government stockpile. In fact, the U.S. Government Services Administration tried to sell excesses in 1983. With all these factors in play, the threat of a cutoff seems small even if U.S. consumption does rise along with an improvement in steel and other industries.

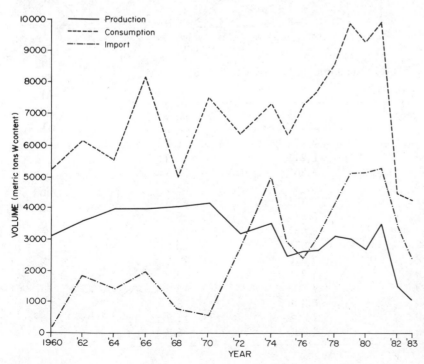

FIGURE I-2-25 U.S. production, imports, and consumption of tungsten, 1960–1983. (From Refs. 6 and 9.)

Vanadium

Vanadium's importance comes largely from its many uses in the steel industry, which consumes roughly 85% of total annual U.S. demand for vanadium.[6] Vanadium improves the ductibility, strength, and resiliency of steel, and is also the principal alloy in new high-strength low-alloy (HSLA) steels. The remainder of U.S. demand for vanadium consists of chemical applications, especially the use of vanadium as a catalyst in the production of sulfuric acid.

Table I-2-1 shows that the United States holds less than 1% of world reserves of vanadium. Most U.S. reserves are found in epigenetic uraniferous sandstones in the Colorado Plateau and in vanidiferous phosphatic shales in Idaho and Wyoming. Smaller amounts are found in alkalic igneous rocks in Arkansas. The bulk of U.S. resources occur in the shales of Idaho and Wyoming, and also in titaniferous magnetites located in Arkansas, Minnesota, Wyoming, and especially New York.[11]

Domestic production comes from a variety of sources. A major source is the Colorado Plateau sandstones, where vanadium is recovered as a byproduct of uranium mining operations. Vanadium is also recovered from ferrophosphorus slags from Idaho, petroleum residues, imported and domestic iron slags, and several other sources.[9] By far, the two largest sources of current U.S. production are the uranium and phosphorus recovery operations in Colorado, Utah, and Idaho.[11]

Figure I-2-26 shows that U.S. vanadium production ranged between 4000 and 7500 tons of contained metal from 1960 to 1976, but rapidly fell after that. The relevance of this decline in production is shown by Table I-2-25, which shows the

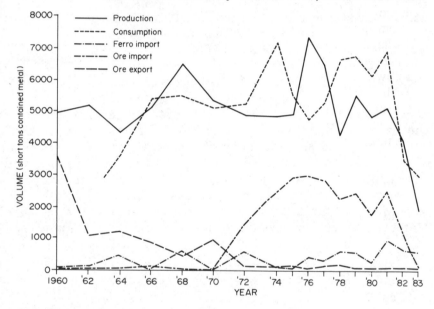

FIGURE I-2-26 U.S. production, imports, exports, and consumption of vanadium, 1960–1983. (From Refs. 6 and 9.)

TABLE I-2-25   U.S. Share of World Vanadium Production, 1960–1983 (Tons Vanadium Content)

Year	World production	U.S. production	U.S. share (%)
1960	7,236	4,971	68.7
1962	8,286	5,211	63.3
1964	8,573	4,362	50.9
1966	10,029	5,166	51.5
1968	13,331	6,483	48.6
1970	19,939	5,319	26.7
1972	20,239	4,887	24.1
1974	20,762	4,870	23.5
1976	31,209	7,316	23.4
1978	34,219	4,272	12.5
1980	39,550	4,806	12.2
1981	39,250	5,300	13.5
1982	36,500	4,098	11.2
1983	32,750	1,800	5.5

Sources: Refs. 6 and 9.

U.S. share of world vanadium production fell from 68.7% in 1960 to 5.5% in 1983. Part of this is due to the significant cuts in domestic production. A good deal of this production cutback is attributable to the dramatic fall in uranium concentrate sales after the Three Mile Island accident in 1979.[6]

Figure I-2-26 shows that U.S. consumption has dropped somewhat. This is due largely to the steel industry problems taking place since 1980. In fact, 1983 consumption was the lowest level in 20 years.[9] Prior to 1980, domestic consumption of vanadium was strong enough that, given production decreases, ore imports overtook ore exports. The United States became a net importer of ore in 1972, and has also seen a slow increase in the imports of ferrovanadium since 1975.[6] South Africa has become the dominant source of U.S. imports. From 1979 to 1982, the United States got 54% of its vanadium imports from South Africa.

From a 1981 base, it is expected that vanadium consumption will increase 3% annually until 1990.[9] To a large extent, this growth will be dependent on the strength of the U.S. steel industry. However, the lower energy costs of producing HSLA steels as well as supply problems with other alloy metals may mean that vanadium consumption could increase somewhat even if the steel industry remains, in general, stagnant.

Zinc

On the basis of amounts consumed, zinc is the fourth most heavily used metal in the United States, behind steel, aluminum, and copper. Zinc's uses are many, but the galvanizing of steel predominates. In 1983, galvanizing of steel accounted for 48% of all slab zinc uses.[9] Other important zinc applications include the production of brass, diecasting for, in particular, automobile parts, and chemical and electrical uses.[9]

Zinc is one mineral of which the United States has an abundance. Table I-2-1 shows that the United States has about 20% of total world reserves of zinc. Most zinc reserves are in the form of sphalerite,

which contains 67% zinc. Most deposits are (1) irregular breccia or replacement stratabound types in carbonate rocks, as in Tennessee and Missouri; or (2) massive sulfides in metamorphic rocks.[11] In many cases, sphalerite is found associated with galena, the principal lead mineral.

Table I-2-17 shows the production capacities of U.S. zinc mining operations, whereas Figure I-2-13 shows the location of these mines. As is readily seen, Tennessee (39%) is the largest domestic producing state.[9] Moreover, most of the zinc mined in eastern U.S. states is recovered with no by-products, whereas silver and lead are frequently recovered along with zinc in western U.S. mines.[21]

Table I-2-26 shows the annual production capacity of existing primary zinc smelters. The five smelters described in Table I-2-26 are a far cry from what once existed in the United States. In 1951, there were 18 primary zinc smelters with a total production capacity of over 1,007,000 metric tons per year.[21] The bulk of the contraction in smelting capacity took place during the late 1960s to late 1970s, when over 40% of capacity was permanently closed. These closures were due to such factors as obsolete equipment, insufficient or nonexistent concentrate feed sources, and high pollution control costs.[11] To illustrate further the decline in smelter capacity, consider that the United States was, in 1951, 95% self-sufficient in slab zinc. The corresponding figure for 1982 was 32%.[21]

This same pattern holds true for U.S. mine production of zinc. In 1950, the United States accounted for 30% of world mine production of zinc. By 1980, the United States accounted for only 8%.[21] Moreover, the United States was 67% self-sufficient in mine production of zinc in 1951; by 1982, the United States was only 21% self-sufficient.[21] This trend, for both mine and smelter production, is illustrated in Table I-2-27, which shows the U.S. share of world zinc production from 1960 to 1983.

Figure I-2-27 further illustrates the decline in domestic mine and smelter production from 1960

TABLE I-2-26 U.S. Primary Zinc Smelters, 1983

Company	Location	Smelter type	Capacity (metric tons/yr)
AMAX Zinc	Sauget, Ill.	Electrolytic	76,000
ASARCO	Corpus Christi, Tex.	Electrolytic	104,000
Jersey Miniere	Clarksville, Tenn.	Electrolytic	82,000
National Zinc	Bartlesville, Okla.	Electrolytic	51,000
St. Joe Zinc	Monaca, Pa.	Vertical retort	91,000
Total			404,000

Sources: Refs. 21, 11, and 9.

TABLE I-2-27 U.S. Share of World Zinc Production, 1960–1983 (Thousand Metric Tons Zinc Content)

Year	Mine production		Primary slab zinc	
	World	U.S.	World	U.S.
1960	3338	395.0	3025	725.3
1962	3565	458.6	3406	797.7
1964	4028	521.5	3693	865.5
1966	4483	519.4	4081	929.9
1968	4975	480.3	4628	926.1
1970	5464	484.6	4827	796.3
1972	5436	433.9	5131	574.4
1974	5781	453.5	5609	503.6
1976	5690	439.5	5362	452.6
1978	5878	302.7	5614	406.7
1980	5745	317.0		340.0
1981	5850	312.0		344.0
1982	6010	300.0		228.0
1983	6160	280.0		220.0

Sources: Refs. 6 and 9.

to 1983. Moreover, the figure shows that domestic production decreased at a faster rate than did domestic consumption. As a result, imports of zinc have become increasingly important. Figure I-2-28 shows U.S. imports and exports of zinc for the same time period. Two important points are made in the graph. First, since at least 1960, U.S. imports have vastly exceeded exports. Second, since 1972, imports of metal have exceeded imports of ores and concentrates.

To illustrate this more clearly, consider that, in 1952, the United States imported 449,000 tons of contained zinc in the form of ores and concentrates. Slab zinc imported amounted to just over 115,000 tons. In 1970, the respective figures were roughly 525,000 and 271,000 tons. In 1975, however, imports of metal amounted to 380,000 tons, and ore/concentrate imports were 145,000 tons.[21] This trend has continued since 1975.

In 1983, the United States depended on imports to meet 66% of its domestic demand.[9] From 1979 to 1982, Canada was the primary supplier of both ores and concentrates (56%) and also metal (54%).[9] Peru, Spain, Mexico, and Australia were other important zinc suppliers.

The rise in domestic consumption from 1982 to 1983 is attributed largely to the recovery in the automobile and residential construction industries.[9] However, a possibly even more important demand use is penny zinc purchases. Some have argued that these purchases were the most significant new zinc

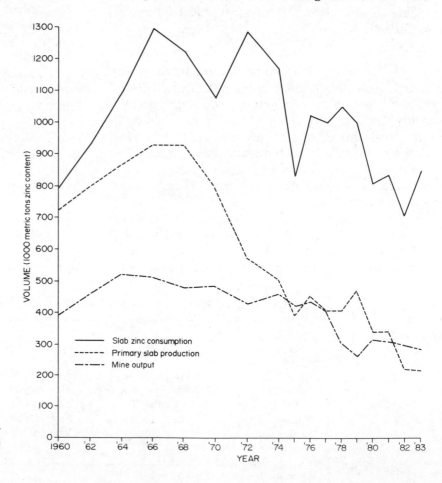

FIGURE I-2-27 U.S. production and consumption of zinc, 1960–1983. (From Refs. 6 and 9.)

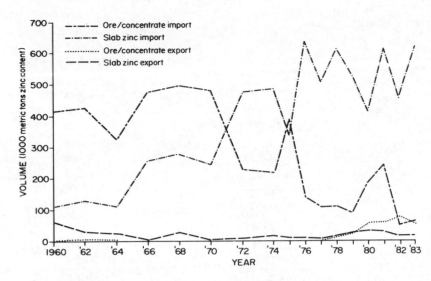

FIGURE I-2-28 U.S. imports and exports of zinc, 1960–1983. (From Refs. 6 and 9.)

use developed in the 1970s.[39] At any rate, it is expected that the United States will experience annual growth in demand of 2.7% from 1981 to 1990.[9] If this does occur, and if new mines and smelters do not come on line for whatever reason, the United States can expect to remain roughly two-thirds import dependent for a mineral of which it has a vast supply.

MAJOR FUEL MINERAL INDUSTRIES

Coal

Figure I-2-29 shows the active coal producing districts of the continental United States in the early 1980s. The figure makes it readily apparent that much of the United States is underlain with coal. In fact, the United States is estimated to have some

483 billion tons of coal reserves, as is shown in Table I-2-1. By some estimates, the United States holds roughly 25% of world coal reserves.[40]

Distribution of coal reserves is not uniform in the United States, however. In general, the eastern U.S. coals tend to be located in the Appalachian and Central (Midwestern states—Illinois, etc.) states. These coals tend to have medium- to high-Btu ranges, are usually high in sulfur, and are often found at depths necessitating the use of underground recovery techniques. Western U.S. coals are usually lower in both sulfur and Btu levels, but they also occur close enough to the surface that strip mining is feasible. Table I-2-28 gives a more detailed illustration of the distribution of coals in the United States.

Such an abundance of coal lent itself to a high

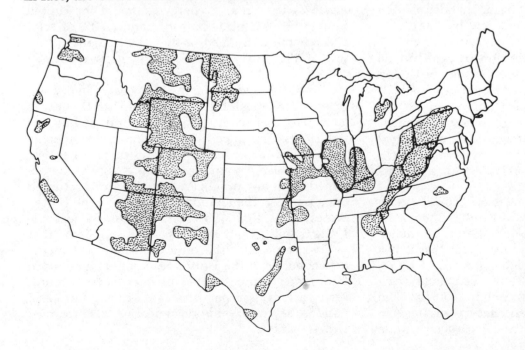

FIGURE I-2-29 Coal producing districts of the United States. (From U.S. Department of Energy, Energy Information Administration. *Coal Distribution January–September 1983.* Washington, D.C.: U.S. Government Printing Office, December 1983.)

TABLE 1-2-28   U.S. Coal Reserves, 1982 (Million Tons)

(a) By region and rank

Region	Anthracite	Bituminous	Subbituminous	Lignite	Total
Appalachia	7,204.7	102,875.1	—	1,083.0	111,162.7
Central	104.1	121,509.1	—	13,505.9	135,119.2
Western	27.8	24,740.9	181,652.4	30,251.1	236,672.1
Total	7,336.6	249,125.1	181,652.4	44,840.0	482,954.0

(b) By mining method

Region	Underground	Surface	Total
Appalachia	90,148.2	21,014.5	111,162.7
Central	93,814.8	41,304.4	135,119.2
Western	140,954.3	95,717.8	236,672.1
Total	324,917.3	158,036.6	482,954.0

(c) By state

State	Region	Reserves (billion tons)
Montana	Western	120.3
Illinois	Central	78.8
Wyoming	Western	69.7
West Virginia		39.3
Kentucky	Appalachian	33.5
Pennsylvania		30.0
Others		111.4
Total		483.0

Source: Ref. 46.

level of use in the United States. Once industrialization of the U.S. economy began, coal became the primary energy source in the nation. By the 1920s, some two-thirds of U.S. energy needs were met by coal.[41] This level of consumption of coal had its attractive features. A major one was that the United States was, and is, for all intents and purposes, self-sufficient in coal. This factor was, even by the 1920s, an issue of concern to many prominent Americans. A second advantage of the use of coal was that it employed many thousands of people in jobs directly and indirectly related to the production of coal.

Even with the advantages of coal, certain other factors contributed to a drop in the use of coal, in both absolute and relative terms. The major disadvantages of coal were (1) the fact that it is a dirty fuel; (2) its low-unit, high-place value with respect to other fuels; and (3) its unsuitability as a fuel for the burgeoning automobile industry. The development of petroleum and natural gas fields in the United States, and the ease with which these two fuels could be used, put an end to coal's dominance as the primary energy source in the United States. From a high share of almost 70% of total U.S. energy consumed in the 1920s, coal fell to a low of 17.7% of total U.S. energy consumption in 1978.[42]

Until the late 1970s, the U.S. coal industry was faced with a bleak future. This changed, though, in large part due to the two world oil price hikes of the 1970s. The price hikes convinced many people that they were wrong to consider oil as a limitless and dependable source of cheap energy to be obtained from any producer. Thus the focus returned once again to the abundance of coal in the United States. This time, however, it was not just the United States that eyed U.S. coal supplies. Other nations, especially the ones even more oil-import dependent than the United States, were anxious to find more secure supplies of energy. As a result, such nations as Japan and the Western European nations began to import increasing quantities of U.S. steam coal.

U.S. coal production jumped in response to increased domestic and foreign demand. As is seen in Table I-2-29, U.S. coal production in 1974 was 610.0 million tons. As late as 1978, U.S. production was up to only 670.2 million tons. However, the 1979 oil price hike stimulated coal production. By 1980, production hit 829.7 million tons of coal. Some estimates even predicted production would shortly reach 1 billion tons of coal a year.

U.S. coal exports increased also (Table I-2-29). Coal exports were 60.7 million tons in 1974 and only 40.7 million tons in 1978. The vast bulk of these exports consisted of metallurgical coal bound mainly for the steel industries of Japan and Western Europe. Shortly after 1978, the picture began to change. Coal exports hit a high of 112.5 million tons in 1982; of that amount, some 45.0 million tons were steam coal.[43]

This rapid increase in both production and exports made the U.S. number one in both categories of all world producers. Many in the U.S. coal industry expected that this boom would continue in both domestic and foreign markets. For example, one source argued that Pacific Rim nations, such as Japan and Taiwan, could end up importing some 200 million tons of steam coal alone by the year 2000.[44] Such export market opportunities in turn spurred a number of projects designed to improve the ability of the United States to meet export demands. These projects included the development of new mines, but the primary focus was on transportation infrastructure.

By mid-1982, many rosy export forecasts were being revised as the world economy entered its latest recession. Many U.S. firms had already committed to major expansions to meet a market that suddenly did not exist. Thus many of these firms are in serious financial straits. Other firms have backed away from similar development projects even though the projects make good long-term sense.[45]

TABLE I-2-29   U.S. Coal Production and Exports, 1974-1982 (Million Tons)

| Year | Production | Bituminous coal exports | | |
		Steam	Metallurgical	Total
1974	610.0	8.3	51.6	59.9
1976	684.9	11.6	47.8	59.4
1978	670.2	9.6	30.2	39.8
1980	829.7	26.8	63.1	89.9
1981	823.8	45.0	65.2	110.2
1982	838.1	40.7	64.6	105.2

Source: Ref. 43.

Future prospects for the U.S. coal industry are mixed. On the domestic front, concerns over transportation modes and costs, pollution controls, and the rate of growth in electricity usage will play major roles in determining how the industry does. Export markets will be influenced by the adequacy of U.S. export facilities, the competitiveness of other coal-exporting nations, and the rate at which electric utilities switch from oil to coal. Although the level of reserves in the United States would indicate good long-term prospects for the use of coal, many factors could play a serious, and possibly stultifying, role in the level of coal use in the near and intermediate future.

Petroleum

Ever since its discovery in the United States in the nineteenth century, crude oil has been a critical energy source. Because of its widespread availability, diversity of uses, and reasonably low cost, crude oil became the major source of energy used in the U.S. economy. In fact, from 1955 to 1983, crude oil and natural gas liquids annually provided from 40 to 50% of U.S. energy consumption.[42,47]

Major domestic sources of crude oil changed with time. Pennsylvania was initially the dominant producer, but Texas took the lead producer position early in the twentieth century. It has not relinquished that position since then. In fact, the top four producing states have seen only one change in many years. The development of the North Slope field pushed Alaska to the third and then second leading position as a producing state in the 1970s. Otherwise, Texas, Louisiana, California, and Oklahoma have held the top four spots.[48] Although this attests to the size of the fields in these states, it also indicates that many fields are becoming quite old. The eventual depletion of these old fields has fostered and will continue to foster exploration efforts in the remaining open areas of the United States, especially Alaska and the coastal waters.

The usefulness of oil as a fuel, especially compared to coal, was quickly seized on in the United States, particularly after the advent of the automobile. Thus U.S. consumption of crude oil quickly rose to the point where demand outstripped domestic production. The United States has long been an importer of crude oil, but the annual volume of imports never amounted to much until after World War II. By 1960, it became fairly obvious that the United States had become a major importer of oil.

Table I-2-30 shows U.S. production, imports, and consumption of crude oil from 1960 to 1983. It shows that U.S. domestic production peaked in

TABLE I-2-30   U.S. Production, Imports, and Consumption of Crude Oil, 1960–1983 (Thousand Barrels per Day)

Year	Production	Imports	Consumption
1960	7,036	1,015	8,098
1962	7,332	1,126	8,438
1964	7,614	1,199	8,833
1966	8,295	1,225	9,470
1968	9,095	1,290	10,342
1970	9,637	1,324	10,909
1972	9,477	2,222	11,757
1974	8,774	3,477	12,164
1976	8,132	5,287	13,457
1978	8,707	6,356	15,090
1980	8,597	5,263	13,841
1981	8,572	4,396	13,097
1982	8,649	3,488	12,297
1983	8,665	3,398	12,221

Sources: Ref. 48.

1970 and has fallen since then. On the other hand, domestic consumption and imports continued to rise until they peaked in the late 1970s. In 1977, oil imports accounted for 45% of total U.S. consumption, a level that many perceived to be dangerously high.[48]

A matter of additional concern was that much of the oil imported by the United States came from the Middle East, an area thought to be politically unstable. Even though, as shown in Table I-2-31, the top source of U.S. imports since 1960 was usually a Western Hemisphere country, the level of Middle East–based imports continued to increase. It eventually reached the point where then-President Carter announced in the late 1970s that the United States was prepared to go to war to protect Mideast oil supplies. Because of the prolonged war between Iran and Iraq, and the recent attacks on oil tankers, there is even today great concern over the security of oil supplies from the Middle East.

However, the U.S. import-dependence picture has changed somewhat since the late 1970s. The 1979 price hike combined with concerns over vulnerability of oil import cutoffs to foster many changes in the U.S. economy. These included:

1. Gradual downsizing of automobiles, together with improved fuel efficiency
2. Increased emphasis on conservation techniques
3. Attempts to convert fuel oil-fired boilers to natural gas and, preferably, coal
4. Some optimism regarding the future of the nuclear power industry

The most important factor behind the drop in U.S. consumption and imports of oil was simply the high price of oil and then the recsssion of the early 1980s.

One other important result of the price hikes of the 1970s deserves mention. With the attention being given to the need to reduce oil imports, and given the higher price of oil in the 1970s, many attempts were made to exploit unconventional sources of oil in the United States. The high cost of extraction seemed to suddenly be within profitable reach, so a great many attempts were made to develop various projects.

In 1980, the U.S. government established the Synthetic Fuels Corporation, a quasi-public company to promote the development of various syn-

TABLE I-2-31   Top Three Crude Oil Import Sources for the United States, 1960–1983 (Thousand Barrels per Day)

Year	No. 1 (amount)	No. 2 (amount)	No. 3 (amount)
1960	Venezuela (472)	Kuwait (130)	Canada (113)
1962	Venezuela (463)	Canada (233)	Kuwait (112)
1964	Venezuela (476)	Canada (278)	Saudi Arabia (97)
1966	Venezuela (404)	Canada (347)	Saudi Arabia (125)
1968	Canada (463)	Venezuela (344)	Libya (114)
1970	Canada (672)	Venezuela (268)	United Arab Emirates (63)
1972	Canada (856)	Venezuela (256)	Nigeria (244)
1974	Canada (791)	Nigeria (697)	Saudi Arabia (438)
1976	Saudi Arabia (1225)	Nigeria (1016)	Indonesia (539)
1978	Saudi Arabia (1142)	Nigeria (910)	Libya (638)
1980	Saudi Arabia (1254)	Nigeria (843)	Libya (549)
1981	Saudi Arabia (1112)	Nigeria (611)	Mexico (469)
1982	Mexico (645)	Saudi Arabia (530)	Nigeria (510)
1983	Mexico (796)	United Kingdom (368)	Indonesia (326)

Source: Ref. 48.

fuels projects. Some $15 billion was set aside to help fund these projects. The goal certainly seemed worthwhile. The Green River shales in Utah, Colorado, and Wyoming alone contain some 80 billion barrels of recoverable oil, according to one estimate.[42] This is just one of dozens of potential sites which, if exploited, could dramatically alter the level of U.S. dependence on oil imports.

However, shortly after the formation of the Synthetic Fuels Corporation, the economics of synfuels changed dramatically. The worldwide recession, global oil glut, and falling oil prices meant that projects once thought to be close to being economic were now doomed to failure. As a result, the Synthetic Fuels Corporation began to receive more requests for out-and-out price supports than for money to be spent on research and development. By year end 1983, the Corporation was faced with helping only a few remaining projects, and the help will amount to providing price guarantees of up to $100 a barrel.[49]

This provoked a profound reassessment of the goals of, and need for, the Synthetic Fuels Corporation. Many in the industry want help to show that the United States can depend on synthetic fuels in the future. However, other parties have argued that short-term issues such as federal deficits are more important than supporting an industry of no use when oil prices are low. It appears that the Synthetic Fuels Corporation will be dismantled in the near future.[49] How the United States will deal with short- and long-term issues of oil supply is as yet uncertain.

Natural Gas

As shown in Table I-2-1, the United States held in 1983 some 198 trillion cubic feet of natural gas reserves, or some 6.2% of the world total. Although this is a slightly better position than the U.S. share of global oil reserves (4.1%), the 1983 reserve estimate still seems fairly low for a nation that consumes as much natural gas as the United States does.

Since at least 1955, natural gas has provided from 20 to 30% of the total energy consumed annually in the United States.[42] For example, in 1975, natural gas provided 19.95 quads (28.2%) of the 70.71 quads of energy the United States consumed that year. In 1983, natural gas accounted for 17.43 quads (24.7%) of the 70.45 quads of energy consumed.[42]

Historically, the dominant domestic producers of natural gas have been Texas, Louisiana, and Oklahoma. Since 1960, for example, these three states have accounted for roughly 80% of the natural gas annually produced in the United States.[48]

Domestic production often exceeded domestic consumption until the mid-1960s.[42] After that, however, the United States was forced to import small quantities of natural gas. Canada and, to a lesser extent, Mexico have provided almost all of the natural gas imports. This may be expected, given the difficulties and costs involved in transporting natural gas over long distances, especially in liquefied form. If these costs could be reduced, it is possible that large reserves of Alaskan natural gas could be exploited. However, for now, Alaska fields have no nearby market, and proposals to expand sales of liquefied natural gas to Japan have hit snags.[50]

Since Alaskan natural gas seems unavailable, for now, to the lower 48 states, any U.S. shortfall of domestic production will likely be met by Canadian production. However, Canadian gas prices will have to drop before U.S. distributors will enter into any new contracts. Even without the existing glut of natural gas and other fuels, the present Canadian price of $4.40 per thousand cubic feet at the U.S. border is almost $2 more than the average U.S. wellhead price.[51]

The deregulation of U.S. gas prices could reduce the price disparity between Canadian and U.S. gas, but it is still quite possible that more Canadian gas could enter the United States. This is because the gas glut and production overcapacity is greater in Canada than in the United States. Barring government intervention, prices of Canadian gas could fall. This could ultimately have an adverse impact on U.S. coal producers, for coal, even at a competitive price, is not as environmentally benign as natural gas.[51] Thus, if present conditions hold true for some time, natural gas could increase its share of the total energy consumed in the United States.

Uranium

According to Table I-2-1, the United States held in 1981 some 27% of the free world's uranium reserves. The distribution of U.S. reserves, and the producing regions, is shown in Figure I-2-30. As shown in Figure I-2-30, there are three major producing regions in the United States: the Wyoming Basin, the Colorado Plateau, and the Texas Gulf Coast. These three regions contain tabular or roll-type sandstone deposits which comprise roughly 90% of the total U.S. reserve base.[52] Other reserves are in the form of igneous-metamorphic veins (Colorado), marine black shales (east-central states such as Tennessee), and phosphate deposits (Florida).[52] The one major new uranium province discovered in the last 20 years is located in south-central Virginia. This deposit alone contains over 30 million pounds of U_3O_8.[53]

Since the vast majority of U.S. uranium con-

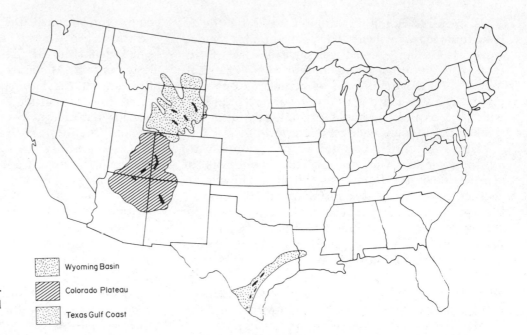

FIGURE I-2-30 Uranium producing regions of the United States. (From Ref. 52.)

Wyoming Basin

Colorado Plateau

Texas Gulf Coast

sumption is as a fuel, there has been a correlation between the development of the U.S. nuclear power industry and the status of the uranium industry. In the 1950s, when the nuclear power industry first developed, uranium production in the United States was almost nonexistent. At that time, imports supplied a good deal of the uranium needed in the United States. By the early 1960s, however, the U.S. uranium industry was in full swing, as is shown in Table I-2-32, which shows U.S. uranium production from 1950 to 1982. At that time, the United States was the world leader in uranium production.

The strength of the domestic industry was

TABLE I-2-32 U.S. Uranium Production, 1950–1982

Year	Production (tons U_3O_8)
1950	893
1955	2,784
1960	17,767
1962	17,631
1964	12,529
1966	10,589
1968	12,368
1970	12,905
1972	12,900
1974	11,569
1976	12,740
1978	18,460
1979	18,720
1980	21,840
1981	17,550
1982	13,430

Sources: Refs. 54 and 55.

shaky, though. By the mid-1960s a flood of uranium imports, and a slowdown in the expansion of the nuclear power industry, seriously threatened the U.S. uranium producers. As a result, the Atomic Energy Commission placed a ban on all imports of uranium. This act seriously hurt foreign producers, many of which were established to supply U.S. needs as a primary objective. Even so, the ban did its job in time. Domestic producers were able to hang on until the late 1960s, at which time the nuclear power industry was starting to grow. Thus domestic uranium producers once again had a market.

This increase in the development of nuclear power plants continued through most of the 1970s, so there was general optimism in the uranium industry. As the estimates for power plant development increased, so did exploration and development in the uranium industry. As it turned out, the bubble burst for two reasons. The 1979 accident at the Three Mile Island facility in Pennsylvania convinced many people that nuclear power plants were not safe. Moreover, a rash of cost overruns (in the billions of dollars) and construction delays in plants under construction drastically reduced the willingness of the public to support new and sometimes existing projects. This translated into a slowdown in orders, and even some work stoppages, the effect of which was exacerbated by the recession of the early 1980s.

The nature of the relationship between the expansion plans of the nuclear power industry and the strength of the uranium industry in the United States is shown by Tables I-2-33 and I-2-34. Table

TABLE I-2-33 Development Status of the U.S. Nuclear Power Industry, 1973-1983

(a) Industry status

Year	Licensed reactors	Number of construction permits granted	Number of construction permits pending	Units on order	Announced units	Total
1973	40	51	58	48	20	217
1983	83	53	0	2	0	138

(b) Current generating capacity

Year	Licensed reactors	Generation (million kWh)	Nuclear share of total U.S. electricity generation (%)
1974	55	113,976	6.1
1976	65	191,104	9.4
1978	72	276,403	12.5
1980	72	251,116	11.0
1981	74	272,674	11.9
1982	79	282,773	12.6
1983	83	292,051	12.6

Source: Ref. 47.

TABLE I-2-34 U.S. Uranium Exploration Expenditures, 1973-1984 (Million Dollars)

Year	Total	Exploration drilling	Development drilling	Land acquisition	Other exploration	Future estimates
1973	49.4	19.5	5.8	7.6	16.5	
1974	79.0	34.9	9.8	12.6	21.7	
1975	121.9	51.9	21.8	16.7	31.5	
1976	170.5	70.7	38.3	13.8	47.7	
1977	258.0	99.4	55.6	28.2	74.8	
1978	314.2	113.3	56.4	30.7	113.8	
1979	315.9	119.6	43.4	44.5	108.4	
1980	266.9	94.8	30.9	35.0	106.2	
1981	144.6	56.4	11.4	11.4	65.4	
1982	73.6	20.9	6.9	11.3	34.5	
1983						39.9
1984						39.9

Source: Ref. 56.

I-2-33 illustrates the development status of the nuclear power industry from 1973 to 1983, whereas Table I-2-34 shows the domestic expenditures for uranium exploration during the same time frame. It is readily apparent that exploration expenditures tend to rise and fall with the fortunes of the nuclear power industry. This is the primary factor, although other factors also play a role, most notably the availability of low-cost foreign uranium.

The future of the U.S. uranium industry is cloudy. The nuclear power industry faces great problems; among them are:

1. The lack of a consistent national energy policy

2. Regulatory impacts (construction delays) on the industry
3. Increasing costs for power plant construction
4. Decreased public support for, and belief in, the industry

All these problems will ultimately affect the domestic uranium industry. Moreover, the uranium industry is also faced with serious problems. One is an overall low ore grade, which tends to make foreign imports cost-competitive. Another concern is the environmental restrictions on mining and processing, as well as the disposal of spent fuel.

Until the issues noted above are addressed,

chances are that the U.S. uranium industry faces a depressed near- to medium-term future. This is somewhat ironic for a nation that holds such large uranium reserves, but this is not uncommon for U.S. mineral industries.

REFERENCES

1. U.S. Central Intelligence Agency, National Foreign Assessment Center. *The World Factbook—1981*. Washington, D.C: U.S. Government Printing Office, April 1981.
2. Schmidt, Helmut, and Manfred Kruszona. *Regional Distribution of Mining Production and Reserves of Mineral Commodities in the World*. Hanover, West Germany: Federal Institute for Geosciences and Natural Resources, January 1982.
3. U.S. Department of Energy, Energy Information Administration. *Weekly Coal Production: November 5, 1983*. Washington, D.C.: U.S. Government Printing Office, November 1983.
4. U.S. Department of the Interior. *Energy Resources of Federally Administered Lands*. Washington, D.C.: U.S. Government Printing Office, November 1981.
5. *Oil and Gas Journal*, 26 December 1983.
6. U.S. Department of the Interior, Bureau of Mines. *Minerals Yearbook, Centennial Edition 1981*. Washington, D.C.: U.S. Government Printing Office, 1983, and other back issues.
7. U.S. Department of Commerce, Bureau of Census. *Statistical Abstract of the United States, 1982-1983*, 103rd ed. Washington, D.C.: U.S. Government Printing Office, 1983.
8. U.S. Department of the Interior, Bureau of Mines. *The Domestic Supply of Critical Minerals*. Washington, D.C.: U.S. Government Printing Office, 1983.
9. U.S. Department of the Interior, Bureau of Mines. *Mineral Commodity Summaries 1983*. Washington, D.C.: U.S. Government Printing Office, 1984, and back issues.
10. Murray, Hadyn H. "Nonbauxite Alumina Resources." In *Cameron Volume on Unconventional Mineral Deposits*. Edited by Wayne C. Shanks III. New York: Society of Mining Engineers, 1983, pp. 111-120.
11. U.S. Department of the Interior, Bureau of Mines. *Mineral Facts and Problems, 1980 Edition*. Washington, D.C.: U.S. Government Printing Office, 1980.
12. Peterson, G. R., and S. J. Arbelbide. *Aluminum Availability—Market Economy Countries*. U.S. Bureau of Mines Information Circular 8917. Washington, D.C.: U.S. Department of the Interior, 1983.
13. Spector, Stewart R. "Price and Availability of Energy in the Aluminum Industry." *Journal of Metals*, June 1981, pp. 138-139.
14. Freeman, Alan. "Alumax Likely to Join Venture with Pechiney." *The Wall Street Journal*, 6 February 1984.
15. Pasztov, Andy, and Raymond A. Joseph. "EPA to propose, in Next Few Months, Ban or Phase-out of Remaining Asbestos Uses." *The Wall Street Journal*, 4 October 1983, p. 20.
16. Lemons, Jim F., Jr., et al. *Chromium Availability— Domestic*. U.S. Bureau of Mines Information Circular 8895. Washington, D.C.: U.S. Department of the Interior, 1982.
17. Nulty, Peter. "How to Pay a Lot for Cobalt." *Fortune*, 4 April 1983, pp. 151-155.
18. Miller, James. "Cal Nickel Targets Gasquet Mountain Strategic Minerals." *Alert Letter on the Availability of Raw Materials*, No. 52 (September 1983).
19. Miller, James. "Experts Make Cobalt Policy Recommendations." *Alert Letter on the Availability of Raw Materials*, No. 68 (February 1984).
20. "Most Stockpile Cobalt Not Good for Defense." *American Mining Congress Journal* 69, No. 22 (7 December 1983), p. 3.
21. Everest Consulting Association, Inc., and CRV Consultants, Inc. *The International Competitiveness of the U.S. Non-ferrous Smelting Industry and the Clean Air Act*. Princeton Junction, N.J., and New York, April 1982.
22. *American Mining Congress Journal* 70, No. 4 (23 February 1984), p. 2.
23. Brown, Stuart. "Fiber Optics to Replace 4M lbs. of Copper at AT&T." *American Metal Market/Metalworking News*, 23 August 1982, p. 24.
24. White, Lane. "Custom Copper Concentrates." *Engineering and Mining Journal*, May 1982, pp. 72-75.
25. Lesemann, Robert H. "U.S. Smelters Losing Access to Copper Concentrates." *American Mining Congress Journal* 69, No. 19 (19 October 1983), p. 18.
26. MacAvoy, Paul W. "A Policy That Closes U.S. Copper Mines." *The New York Times*, 19 December 1982.
27. Hughes, Kathleen A. "Eleven U.S. Copper Producers to Request Trade Commission Put Limits on Imports." *The Wall Street Journal*, 26 January 1984, p. 45.
28. "Precious Metal Possibilities in the U.S. Rocky Mountains." *Mining Magazine* 149, No. 5 (November 1983) pp. 286-287.
29. O'Boyle, Thomas F. "Steelmaker's Excess Ore Capacity Hindering the Industry's Recovery." *The Wall Street Journal*, 29 August 1983, p. 15.
30. "U.S. Tentatively Rules 3 Countries Subsidize Various Steel Exports." *The Wall Street Journal*, 8 February 1984, p. 3.
31. *Alert Letter on the Availability of Raw Materials*, No. 64 (January 1984), p. 8
32. "Cal Nickel Targets Gasquet Mountain Strategic Minerals." *Alert Letter on the Availability of Raw Materials*, No. 52 (September 1983).
33. "Phosroc: U.S. Industry Disadvantaged." *Mining Journal* 301, No. 7729 (7 October 1983), p. 257.
34. "Phosphate Production Plans at New North Carolina Mine." *Mining Magazine* 150, No. 3 (March 1984) p. 195.
35. Anstett, T. F., et al. *Platinum Availability—Market Economy Countries*. U.S. Bureau of Mines Information Circular 8897, Washington, D.C.: U.S. Department of the Interior, 1982.
36. Paul, John H. "Offshore Imports Threaten Potash Industry." *Mining Congress Journal*, 7 March 1984, p. 16.
37. "Titanium Producers at Odds." *Mining Journal* 302, No. 7744 (20 January 1984), pp. 33-35.

38. *Mining Journal* 300, No. 7712 (10 June 1983), p. 393.
39. "115th Annual Review and Outlook." *Engineering and Mining Journal* 185, No. 3 (March 1984).
40. Wilson, Carroll. *Coal—Bridge to the Future: Report of the World Coal Study*. Cambridge, Mass.: Ballinger Publishing Co., 1980.
41. Park, Charles. *Earthbound: Minerals, Energy and Man's Future*. San Francisco: Freeman, Cooper & Company, 1975.
42. U.S. Department of the Interior. *Energy Resources of Federally Administered Lands*. Washington, D.C.: U.S. Government Printing Office, November 1981.
43. U.S. Department of Energy, Energy Information Administration. *Historical Overview of U.S. Coal Exports, 1973-1982*. Washington, D.C.: U.S. Government Printing Office, November 1983.
44. Miller, Richard E. "Export Opportunities for Western Coal." *Mining Congress Journal*, February 1982.
45. U.S. Coal Exports Derailed." *Mining Journal*, 2 December 1983.
46. U.S. Department of Energy, Energy Information Administration. *Weekly Coal Production: November 5, 1983*. Washington, D.C.: U.S. Government Printing Office, November 1983.
47. U.S. Department of Energy, Energy Information Administration. *Monthly Energy Review, December 1983*. Washington, D.C.: U.S. Government Printing Office, March 1984.
48. "Forecast/Review." *Oil and Gas Journal*, 30 January 1984, and earlier annual review issues.
49. Pasztor, Andy. "U.S. Backed Synfuels Program Is Likely to End in 1984; $5 Billion Seen Returned." *The Wall Street Journal*, 19 December 1983.
50. "The Gas Glut Has Alaska and Canada Hustling." *Business Week*, 25 October 1982.
51. Bayless, Alan. "Natural-Gas Pressure Builds to a Head." *The Wall Street Journal*, 9 January 1984.
52. Mobray, Jo. *Geography of the United States Uranium Supply: Resources, Production, and Institutions*. Public Information Report 6. Austin, Tex.: University of Texas, Center for Energy Studies, 1981.
53. "Temporary Respite for Uranium?" *Mining Journal*, 4 February 1983.
54. Dupree, Walter, et al. *Energy Perspectives 2*, U.S. Department of Energy. Washington, D.C.: U.S. Government Printing Office, June 1976.
55. Technical Information Center, Institute of Gas Technology. *Energy Statistics 1st Quarter 1984*, Vol. 7, No. 1. Chicago: IGT, 1984.
56. U.S. Department of Energy, Energy Information Administration. *1982 Survey of U.S. Uranium Exploration Activity*. Washington, D.C.: U.S. Government Printing Office, August 1983.

The Changing Nature of the U.S. Mineral Supply Situation

The United States has never been self-sufficient in terms of mineral production. Rather, the United States has long been a world leader in mineral production and consumption, with a strong reliance on mineral trade to meet shortfalls in domestic needs. From 1900 to 1929, U.S. mineral production met roughly 90% of domestic demand (by value). Moreover, the United States exported more minerals than it imported during that period.[1]

During this time, an implicit minerals policy evolved in the United States. The policy emphasized a dependence on liberal mineral trade, a desire to expand capacity, and especially reduce unit costs of minerals, and a reliance on private firms to ensure an adequate supply of minerals.[2] This orientation was established and maintained even though the experiences of World War I convinced many in the United States that security of supply should be the primary goal of any U.S. minerals policy.

World War II further complicated the U.S. mineral supply situation. To meet the resource demands of war mobilization, U.S. mines were operated at full capacity. By the end of the war, many U.S. mineral deposits were exhausted, leaving the U.S. with lower ore grades in remaining deposits.[3] A choice had to be made between increasing domestic production capacity and increasing U.S. reliance on foreign supply sources for such minerals as chromium, manganese, tungsten, and others. Both options had disadvantages, and these weighed heavily on the minds of decision makers.

The option of increasing U.S. domestic production capacity was contingent on the discovery and exploitation of new deposits. Even by the 1940s, it was obvious that average ore grades of known deposits were dropping, so it appeared that development of new domestic deposits would be costly. In the cases of certain minerals, such as columbium, the United States probably had no chance of even finding domestic ores.

On the other hand, the United States could continue, and even increase, an already growing dependence on foreign supplies of certain minerals.

This option was not without its problems, as was demonstrated by the events of World War II. It was not enough that Germany's resource insecurity led to the war. In addition, during the war, German submarines repeatedly sank ships attempting to carry bauxite and other needed minerals to the United States. The United States and other Allied nations recognized the value of such "resource denial" practices, and their use by Allied forces greatly contributed to Germany's downfall.[3]

Even with widespread recognition of the problems that nations had with securing resource supplies, the United States chose to increase its reliance on foreign supply sources. This decision was made largely to keep costs down for the American consumer, and to keep profits up for the private sector. Thus, beginning in the late 1930s, the United States found that its mineral exports no longer exceeded mineral imports. By the 1940s, exports equalled imports, and shortly afterward, imports exceeded exports of minerals. By the late 1960s, mineral imports were three times that of mineral exports in terms of value.[1]

The United States carried out policies of foreign investment and reliance on imports. These policies were very successful in terms of the overriding concerns of that time. Increasing resource consumption, first in the United States and later in the rebuilding of Japan and Western Europe, led to strong economic growth. Moreover, foreign investment in mineral (and other) ventures helped keep down costs for American consumers even as it built up a global trading network for American businesses.

Because of the apparent success of the heavy U.S. investment in foreign mineral supplies, most of the voices calling for more emphasis on security of supply were not heard. However, they were not entirely silenced. For instance, the Truman and Eisenhower administrations both attempted to establish a stockpile of critical minerals so that the United States would not be caught offguard as it was at the beginning of World War II.[3]

In addition, the Cold War concerns of the 1950s

reinforced the importance of resource security. The Korean War was fought partly out of a concern over the availability of tungsten.[3] The United States and the USSR began a series of struggles over resource acquisition as one aspect of the Cold War. Attempts to control resources in various countries were often made with one eye on the possibility of denying those resources to the opposing nation. For example, the Soviets supported Iran while America cooperated with Saudi Arabia. After the USSR gained favor in Cuba, the Kennedy administration worked to retain the Belgian Congo and Indonesia.[3]

Unfortunately, many in the American government and the public failed to understand what was behind these actions. Instead, the average American simply demanded more and more goods while insisting on lower prices. Maintaining and improving material standards of living was the major goal of American society.

Thus, over the years, a variety of actions took place which effectively weakened the ability of the United States to ensure a secure supply of minerals. Short-term concerns took precedence over long-term issues, as when items were sold from the U.S. stockpiles to keep market costs down and balance the budget. Industrial, environmental, and other policies were developed without adequate consideration of their impact on the mineral industries. The result was that, slowly but surely, the United States found it increasingly difficult to meet domestic minerals demand with domestic production.

More recently, the changing nature of U.S. mineral import dependence, and Soviet actions in the resource-rich countries of southern Africa, have persuaded some observers that the West and the Soviet bloc are already engaged in a "resource war."

HISTORY OF AMERICAN MINERAL IMPORT DEPENDENCE

The oil crises of 1973 and 1979 have focused public attention on America's dependence on imported oil, although public opinion polls indicate that a significant proportion of the populace does not believe that a problem exists. The problem of U.S. dependence on imported sources of nonfuel minerals is hardly recognized. However, resources depletion and supply disruptions are not merely recent concerns. Since before World War I, competition for secure supplies of strategic metals and minerals has been an underlying element of international relations.

Although the United States has a large resource base for a number of strategic minerals, it has followed a minerals strategy aimed at the exploitation of cheap resources abroad largely because of the low grade of most domestic deposits. Access to these resources has become increasingly uncertain because of a wave of nationalism that has swept through the Third World and also left its mark on developed mineral-exporting countries such as Australia and Canada. At home, increased regulation and litigation, increasingly stringent pollution control standards, and public opposition to the environmental effects of mining and metallurgical operations have reduced domestic mineral production.

Statistical evidence illustrates that the United States is becoming more dependent on imported sources of nonfuel minerals, that some of these commodities are becoming more expensive in real terms, and that there is growing competition for the available resources from other consuming nations. The effects of this growing import dependence include large balance-of-payments deficits, the decline (until recently) of the U.S. dollar, rising unemployment, and decreasing standards of living. Less obvious at this stage are the adverse effects on national security. Yet rich and abundant mineral resources were a vital factor in the emergence of the United States as a global power after the Civil War.

During World War I, three American geologists, Charles K. Leith, George Otis Smith, and Josiah E. Spurr, emphasized the critical role of mineral resources in shaping international politics and worked to alert American leaders about the nation's resources deficiencies during the interwar period.[3] All three men advocated the conservation of mineral resources and called for policies aimed at achieving mineral resource independence. However, unlike certain of their modern counterparts, they did not oppose growth or economic development. They did foresee eventual U.S. dependence on imported minerals and the intense competition between industrial nations for overseas resources that this foreshadowed. They considered the struggle for minerals as the main reason for many international political controversies. Given the irregular distribution of key minerals around the world, Leith considered a certain amount of interference with national sovereignty inevitable.[4]

The history of America's growing mineral import dependence has been analyzed by Eckes.[3] Much of the next several pages is summarized from his book. At the beginning of this century, a popular view was that the physical abundance of America's mineral resources would ensure a bright future. However, Theodore Roosevelt and other members of the conservation movement were concerned about the depletion of nonrenewable resources. Roosevelt encouraged the National Conservation Commission to make an inventory of the nation's natural resources.

Although today the United States is in part import dependent, in 1913 it produced 64% of the world's petroleum and held first place in the production of 13 of the 30 most important mineral commodities: iron, copper, zinc, lead, silver, molybdenum, tungsten, arsenic, phosphates, salt, petroleum, natural gas, and coal. Some officials saw World War I as an opportunity to achieve even greater mineral self-sufficiency. Secretary of the Interior at that time, Franklin K. Lane, suggested that the European war would stimulate the production of domestic minerals and the discovery of new reserves by interfering with manufacturing and interrupting imports. He did not believe that U.S. mineral consumers would thereafter again turn to foreign sources of supply. There was little appreciation of the fact that technological development was changing the mix of minerals required in peace and in war, or of the fact that many of the European powers had depleted their domestic mineral resources, and that a greatly expanded scale of mineral consumption eventually promised the same fate even for the vast United States.

Military control over mineral resources influenced the outcome of World War I. The United States and the British Empire controlled a far greater percentage of world mineral resources than did Germany. Supplies of nickel, copper, and tin were particularly scarce in Germany during that war. However, World War I also demonstrated the importance of sea-lane control during times of conflict, pointed up by the success of German submarines in reducing foreign mineral supplies to the United States and Britain. Even among the victorious countries, there were mineral supply dislocations and price increases for certain metals. The war clearly illustrated that consumption of strategic metals increases sharply during conflicts.

A major controversy, which was to influence U.S. minerals policies for decades ahead, arose during World War I. Fundamentally, the controversy was whether to stimulate domestic high-cost production of minerals as an emergency measure or to rely on limited imports in addition to normal domestic production. The Department of the Interior favored domestic production and a strong, self-reliant minerals industry, whereas the War Industries Board was concerned about the exorbitant cost of using low-grade American ores, and fa-

vored some reliance on imported minerals. Another valuable lesson learned from World War I was the need to stockpile strategic minerals during peacetime.[3]

Whereas previously domestic resources of coal and iron ore were sufficient to ensure a strong military base, World War I clearly demonstrated the strategic importance of a number of other metals: chrome, manganese, nickel, and tungsten, reserves of which are unequally distributed on a global basis. With postwar technological development, this list was to expand greatly.[4] After World War I, the industrial nations searched for means to reduce their vulnerability to mineral shortages. The main concern was exclusion from overseas supplies, rather than of depletion of resources. The development of substitutes, protection of high-cost domestic resources, and the safeguarding of overseas mineral interests all received some attention.[5]

William C. Redfield[6] pointed out that the United States produced insufficient quantities of 30 materials for its domestic peacetime requirements. These included chrome, manganese, antimony, nickel, tin, tungsten, and mica. During this period U.S. mineral experts issued a call for the formulation of a national minerals policy. Elements of such a policy were to include acknowledgment of the fact that many strategic minerals are unevenly distributed and that no nation is self-sufficient in all minerals, the promotion of free trade in minerals, and stockpiling of certain minerals (such as antimony and chrome). However, there was also strong pressure for protection, and in 1922 the Fordney-McCumber Tariff was approved; duties were imposed on a number of imported ores, such as bauxite, manganese, and tungsten. The object of these tariffs was to encourage domestic production of minerals. These measures were partly successful in the case of tungsten but failed to encourage production of other minerals.

The 1920s also witnessed the first reasonably successful mineral cartels. These included the U.S.-dominated copper export association, and similar schemes for potash, nitrates, and rubber. The copper cartel collapsed as a result of new capacity in South America and Africa, the potash cartel as a result of new discoveries in New Mexico, and the nitrates monopoly as a result of the invention of the Haber nitrogen-fixation process. Herbert Hoover, Secretary of Commerce at the time, recognized the economic implications of foreign cartels and launched a campaign against them. The U.S. government employed such tactics as withdrawal of diplomatic representation, reprisals against participating governments, denials of loans, antitrust prosecutions, resource conservation, substitution, and the development of new foreign resources outside the cartel's control.[3]

There is little appreciation today of the role that raw material problems played in the outbreak of World War II. However, the dominance of the United States and the British Empire prevented minerals supplies in Germany, Italy, and Japan from being secure and contributed to these countries' turning to territorial expansion in the quest for minerals. This problem was exacerbated by the growing need for a multiplicity of mineral supplies, many of them available from only a few sources.

Hitler's public statements repeatedly included references to Germany's need to obtain the iron ore, coal, manganese, oil, and molybdenum resources of the Urals.[7] He also said: "In four years Germany must be completely independent of foreign countries so far as concerns those materials which by any means through German skill, through our chemical and machine industry or through our mining industry we can ourselves produce."[8]

Japan also lacked most vital strategic minerals from domestic sources. There were only two options for securing these supplies: rely on international trade, or secure colonies or economic dependencies in Asia. With an expanding population, Japan decided to follow the British example and build an empire.

Noting these developments in Germany, Italy, and Japan, a U.S. Planning Committee for Mineral Policy, which included C. K. Keith, recommended that the government stockpile supplies of manganese, chrome, tungsten, nickel, tin, mercury, and mica to guard against an emergency. The U.S. military services supported this proposed policy. However, President Roosevelt did not support stockpiling of minerals until it was almost too late. As a result, after the United States entered World War II, an expensive crash program had to be launched to acquire strategic minerals domestically and from abroad.[3]

Technological developments during World War II increased the list of vital war materials from one dozen to five dozen. The changing mix of minerals meant that the United States became far more dependent on developing, mineral-exporting nations for a variety of minerals; the United States became a net importer of minerals. This fact led several U.S. mineral experts to proclaim that the United States had become a "have-not" nation with regard to minerals. Harold Ickes[9] pointed out that America's immediate problems were with alloys and key nonferrous metals. Included in his danger list were manganese, vanadium, copper, lead, zinc, tin,

nickel, bauxite, chrome, and cadmium. He also predicted that the United States would soon be dependent on foreign sources of oil. He recommended vigorous stockpiling programs, exploration for new reserves, improvement of mining technology, and the promotion of greater public awareness of the country's growing mineral import dependence. Although his views were shared by a number of other mineral experts in the public sector, the domestic mining industry strongly disagreed, claiming that the mineral-shortage "myth" was a brainchild of international economists more familiar with trade theory than with the nature of the mining industry.[3] C. K. Leith[4] and others called for a strategic minerals stockpiling program to avoid the mistakes made during the two prior wars. The stockpiling act passed in 1946 was a compromise between those favoring the purchase of domestic minerals and those promoting the purchase of foreign materials.

The Department of the Interior completed a confidential study of America's minerals import dependence in 1948. They concluded that a dangerous situation existed. The problem required immediate action, particularly in the case of chrome and manganese, which the United States imported almost solely from the USSR. The report further recommended diplomatic action to obtain assurances regarding continued political availability of foreign sources of supply, provision of adequate transportation facilities, and firm military plans to protect sea-lanes.[3]

In the late 1940s and early 1950s, it became increasingly evident that the USSR was preparing to challenge the United States in a quest for global domination. The danger that the USSR would attempt to cut off Third World mineral supplies to the United States and its European allies was perceived.[10] One such indication occurred late in 1948 when the USSR reduced shipments of chrome and manganese to the United States. The Truman administration attempted to improve the climate for U.S. mining investment overseas. However, this program was hampered by increasing nationalism in the Third World. With the granting of independence to former colonies, this problem was further exacerbated.

The Korean War further aroused concerns about the adequacy of long-term mineral supplies. President Truman appointed a five-member presidential commission to consider the implications of materials policy in 1951. The commission rejected a policy of national minerals self-sufficiency, which they regarded as a self-imposed blockade, and favored instead a policy of interdependence. The commission also rejected earlier fears about physical exhaustion of mineral resources, believing rather that minerals would become increasingly difficult to win, and therefore, more expensive. To counteract these perceived higher costs, they recommended a policy of least-cost acquisition of minerals. However, a strong domestic mineral program, with particular emphasis on improved exploration, mining, and beneficiation, was also proposed and a substantial strategic stockpile policy supported.

The report received generally favorable comment, although it was criticized in mining circles for overstating America's "have-not" status.[11] The mining industry feared that this would lead to an excessively pessimistic view of U.S. mineral potential. Other critics pointed out that foreign mineral ventures were increasingly subject to disruption, confiscation, and nationalization and warned against excessive dependence on remote and insecure sources of supply.

Eisenhower's approach to minerals issues was influenced by the Cold War. He was more familiar with the importance of strategic minerals in wartime than were any of his predecessors, and favored a substantial strategic stockpile supplied from domestic and foreign sources. However, some of the younger people in his administration regarded this as a wasteful policy, believing that the next war was likely to be a thermonuclear holocaust lasting no more than 60 days. The launching of *Sputnik I* in 1957 had a substantial effect on Eisenhower's mineral policy. According to Eckes,[3] it enhanced the strategic school of thought which held that future U.S.-Soviet conflicts would involve nuclear weapons, and resulted in a higher funding priority for education and technology than for stockpile purchases.

In the 1960s, three U.S. presidents sold huge quantities of stockpiled minerals. The practice started with President Kennedy and reached its zenith under President Johnson, who used sales of stockpiled copper and aluminum to reduce inflationary pressure during the Vietnam War.[3] The grave concern about adequate supplies of strategic minerals that prevailed during the Truman and Eisenhower administrations subsided, while U.S. import dependence continued to increase.

More recently, there has again been growing concern about America's vulnerability with regard to a number of strategic nonfuel minerals. Several members of Congress, including Representative James D. Santini, urged President Carter in 1977 to initiate a high-level task force to investigate this problem. As a result, he appointed a cabinet-level coordinating committee to conduct a Non-Fuel Minerals Policy Review to analyze the problems and prepare policy option recommendations.

The draft report of the panel, released in August 1979, did conclude that future supplies of several imported minerals critical to the United States and its allies are becoming less secure. Chromium, manganese, cobalt, and the platinum-group metals were identified as the commodities of greatest concern. The major sources of these metals are in central and southern Africa. The report took the complacent view that the USSR represents an alternative source of supply. However, some mineral economists now believe that the USSR has not developed enough production capacity for these minerals to support its own industrial and defense needs.[12]

The Non-Fuel Minerals Policy Review report was also criticized because it failed properly to analyze and evaluate the key problem areas and the long-term impact of import dependency. Congressman Santani[13] commented that the report failed to look at the national security aspects of mineral import dependence, the related problems of disruptions, and the increasing loss of flexibility of the United States to cope with an emergency. The report also failed to identify any practical solutions to the problem.

Under the chairmanship of Congressman Santini, the Subcommittee on Mines and Mining of the House Committee on Interior and Insular Affairs held a number of hearings on America's mineral import dependence. Prominent members of the new Reagan administration (including Alexander Haig) testified before the committee, and expressed their concern about this matter. In a report entitled "Sub-Sahara Africa: Its Role in Critical Mineral Needs of the Western World,"[13] the subcommittee confirmed that the United States and its allies are dependent on South African mineral supplies, and that the interruption of supply from South Africa would directly disrupt strategic and nonstrategic sectors of the U.S. economy. This disruption could be so severe in the case of certain minerals that the President of the United States would have almost no other choice but to assume economic mobilization powers, impose resource use priorities, and provide for domestic production capacity if possible. However, the report concluded that neither the stockpile nor substitution would compensate even in the near term for the loss of South African mineral exports to the West in the current post-Afghanistan invasion period in which international resource politics have emerged as a threat to peace.

The committee recommended that diplomatic communication should be opened with South Africa.

At a meeting in Pittsburgh held in June 1980, and at subsequent meetings elsewhere in the United States and in France, the Council on Economics and National Security and the National Strategy Information Center have expressed the view that the United States and its allies are already locked in a "resource war" with the Soviet bloc. They believe that Soviet interest in waging such a war against the West has its roots in the thoughts of Lenin, who had perceived the connection between the West and its colonial empires as the "weak link" of the West. They are also convinced that attempting to debilitate the Western industrial economies by depriving them of their raw material imports is now firmly implanted in Soviet doctrine.[14] Similar groups in Britain, West Germany, and France have cooperated with the National Strategy Information Center in sponsoring conferences on the resource war.

Former President Nixon in a book entitled *The Real War*[15] stated that Leonid Brezhnev had told the President of Somalia in 1968 that it was the objective of the USSR to deprive the West of its two treasure troves: the oil of the Middle East, and the strategic minerals of central and southern Africa. Moss[16] stated that Brezhnev had also told a secret meeting of Warsaw Pact leaders in Prague in 1973 that the Soviet objective was world dominance by 1985, and that the control of Europe's sources of energy and raw materials would reduce it to the condition of a hostage to Moscow.

On the other hand, the subcommittee on African Affairs of the Committee on Foreign Relations of the U.S. Senate published a report entitled "Imports of Minerals from South Africa by the United States and the OECD Countries" in 1980, in which they concluded that, while South Africa was indeed an important source of a number of key strategic minerals, "It is fortunate that in the case of each of the critical minerals imported from South Africa, means are available for dealing with an interruption without depending on the Soviet Union as an alternative supplier. These means may be costly, and they cannot in all cases be implemented without disruption. But in general, the disruptions can be minimized if preparations for a possible cut-off in South African supplies are made in advance."[17]

Advance preparations suggested include stockpiling, conservation, process changes, alternative technologies, use of functionally acceptable substitutes, providing incentives to encourage design changes, recycling, and exploiting untapped reserves. There was a lack of appreciation of the lead times involved in all such actions.

The subcommittee's report was essentially a political document rather than an analysis of economic impacts. The report also failed to allow for the contingency of the United States sharing its strategic and critical stockpiles with its allies in the event of serious, prolonged disruptions of supply. The report referred to such phrases as "alternative technologies," "functionally acceptable substitute," and "process improvements" which detracted from its technical credibility.

Although there was increased interest in strategic minerals in the latter part of the Carter administration, it still appeared that a nonfuel minerals policy was not an issue in the administration. By contrast, several senior members of the Reagan administration have, over the years, expressed concern about America's growing mineral import dependence. This issue is reflected in the administration's defense and foreign policies, and that there is a greater appreciation of the fact that *availability* rather than cost of strategic minerals such as manganese, chrome, cobalt, vanadium, and platinum-group metals is the major problem facing the United States and its allies. It should also be appreciated that this is not solely a U.S. problem, but a problem of the Western alliance.

TECHNOLOGICAL INNOVATION IN THE U.S. MINERAL INDUSTRY

Most of the technology used by the mineral industry is generated outside the industry, primarily by equipment manufacturers and other suppliers, although government organizations like the Bureau of Mines have made valuable contributions. Research and development done within the U.S. mineral industry is largely oriented toward short-term objectives involving improvements to existing production operations. Most of a mining company's new development funds are spent in exploring for new mineral deposits that will assure a continuing company operation. In a mature mining province like the United States, locating new economic mineral deposits has become increasingly difficult and expensive, a situation further exacerbated by increasing government regulation and restrictions to access on federal lands. During periods of depressed or highly volatile prices, there is little incentive to explore for or develop marginal deposits.

Although it is now generally agreed that the

United States will have to rely on foreign sources to supply many of its major mineral raw material needs, there is clearly also a need to minimize this dependence on foreign mineral supplies. One way is by applying superior technology to exploit low-grade or unconventional domestic resources. The means to implement new technology in a timely manner must be established if significant improvement in the U.S. domestic mineral position is to be achieved. The National Commission on Materials Policy recommended in 1973 that agencies having to do with materials and resources undertake appropriate research and development to generate new knowledge and technology. The commission also recommended that these agencies intensify their efforts to capitalize on available knowledge in the development of domestic raw material supply. However, a National Research Council report in 1969 noted that the state of mineral technology in the United States was "wretched."[18] Since then, the situation has not improved to any significant extent.

In the university sector, research in the areas of mining, mineral processing, extractive metallurgy, and mineral exploration is practically nonexistent due to lack of funds, manpower, and industry and government programs to provide guidance and direction.[18] This lack of research opportunity has led to periodic shortages of adequately trained faculty members in these fields. Most of those now in faculty positions were educated before and during World War II and will be retiring within the next 10 to 15 years. Many positions have remained vacant for several years. Several mining schools in the United States are searching for departmental chairmen. Thirty-one Mining and Mineral Resources Research Institutes have been established in terms of the Surface Mining Control and Reclamation Act of 1977. However, the level of funding of these institutes is totally inadequate.

In the industrial sector, research and development budgets as a percentage of sales have been lower in the mineral industry than in most other segments of American industry. There are also certain institutional disincentives to technological advance in the U.S. mineral industry. When exploration was open in most areas of the world, and when the costs of foreign exploration were below the technology cost required to increase domestic exploration, capital was devoted to exploration rather than research. Given the demand inelasticity for minerals, pressures for technological developments to decrease mineral costs have been insufficient. Technological developments in the mineral industry, although patentable, generally have not been protected due to difficulties in enforcement of international law.[18] The mineral industry is also highly fragmented and the return on innovation has been regarded as poor. Above all, as long as foreign mineral producers enjoyed higher-grade deposits, lower labor costs, higher governmental incentives for mineral development, and much less stringent environmental regulations than U.S. producers, and as long as there was at least a perception of security of supplies from abroad, there was little incentive for technological innovation in the domestic mineral industry. Clearly now there is a need for technology, particularly for the exploitation of low-grade and unconventional ore deposits, and for the discovery of rich deep-seated deposits without surface signatures.

In a recent study by a committee of the National Research Council,[18] it was concluded that currently available technology does not provide the capability for meeting the problems anticipated for the mid- and long-term future, when ore grades are likely to be lower and environmental constraints have been made perhaps even more severe. With regard to metallurgical technology, it was concluded that in the face of anticipated future trends in ores available for treatment, today's technology could be totally inadequate. This is very serious, considering that present technology is already too expensive in capital and in energy requirements to make new operations feasible in many of the primary metal commodities. In addition, there are long lead times in the application of new techniques.

The committee identified a number of additional constraints on technological innovation in the U.S. mineral industry. Among these are uncertainties regarding U.S. government policies, law, and regulations; the shortage of well-trained personnel; the heavy financial requirements of the industry; and the industry's conservative management philosophy. Despite all these problems, it is now becoming clear that political uncertainties in many foreign areas will force the mineral industry to give more attention to the development of U.S. deposits. The need for a concerted federal program to develop new exploration, mining, and metallurgical technology is abundantly clear.

As the United States becomes more import dependent, it is faced with the prospect of greatly reduced security of foreign mineral supply and the prospect of a new cold war with the USSR. At the same time, even with a heroic effort, it will take years to repair American mining and metallurgical training and research facilities to a reasonable level of operation. Meanwhile, there is growing evidence that the United States has lost its lead in the area of minerals technology and that available facilities have little immediate potential in assisting in a material improvement of the domestic mineral industry.

A technological fix is very unlikely to solve America's growing problem of minerals import dependence. Instead, more companies are being forced to increase their legal staffs to comply with a bewildering array of conflicting regulations, at the expense of any meaningful research efforts. There is a need for incentives to encourage more meaningful mining and metallurgical research in the United States. However, the decline in mining and metallurgical research in the United States has already led to a dramatic change in the nature of its mineral import dependence. Where once America imported ores and concentrates and jealously defended domestic processing facilities, she now increasingly imports minerals in processed forms.

INCREASING U.S. RESOURCE IMPORTS

The trend toward increasing dependence on foreign sources of minerals has not significantly changed since it first became apparent after World War II. It has only been in the last decade or so that alarm has been expressed over the extent to which the U.S. now relies on mineral imports. Petroleum is the most obvious example, for in the late 1970s, the U.S. relied on imported oil for roughly 40% of its needs. In fact, the decline in U.S. petroleum production relative to consumption since 1970 is the major factor behind the increase in U.S. fuel imports. This is graphically illustrated by Figure I-3-1, which shows the levels of U.S. energy production and consumption from 1952 to 1981.

However, this situation is not restricted to fuels. In fact, in the case of a number of nonfuel minerals, the United States is totally (or almost totally) dependent on foreign supplies, for these minerals simply do not occur in commercial quantities in the United States. The extent of current U.S. import reliance as a percentage of apparent consumption is shown in Table I-3-1. It is obvious that there are many minerals for which the United States is extremely dependent on imports. These minerals include bauxite, chromium, columbium, cobalt, fluorspar, manganese, nickel, platinum group metals, potash, tantalum, tin, and the titanium ores of ilmenite and rutile.

Table I-3-1 also shows that U.S. dependence levels for certain minerals have been increasing. For some, such as chromium, manganese, and tantalum, the United States has always relied on imports. For

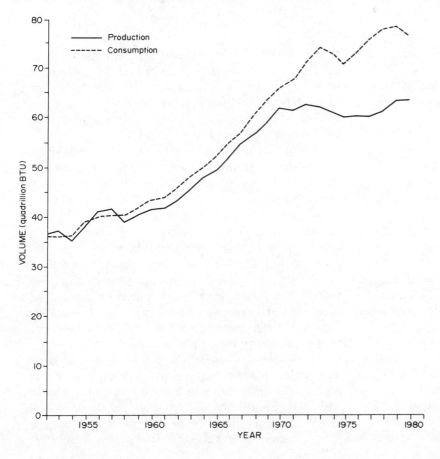

FIGURE I-3-1 U.S. energy production and consumption, 1952–1981. (From Ref. 5.)

TABLE I-3-1   U.S. Import Dependence for Selected Minerals, 1976–1983[a]

| | Year | | | | | | | |
Item	1976	1977	1978	1979	1980	1981	1982	1983
Aluminum metal	9	7	11	4	E	E	7	18
Asbestos	84	85	85	83	78	78	74	65
Bauxite, alumina	91	91	93	93	94	94	96	96
Chromium	89	91	91	90	91	90	85	77
Cobalt	98	97	95	94	93	92	92	96
Columbium	100	100	100	100	100	100	100	100
Copper	12	13	20	13	16	6	1	17
Fluorspar	77	80	82	86	87	83	83	W
Gold	60	61	53	50	18	15	32	21
Ilmenite	28	39	41	39	32	52	75	W
Iron ore	31	48	29	25	25	22	34	37
Lead metal	15	13	9	4	E	1	3	11
Manganese	98	98	97	98	98	98	99	99
Nickel	70	70	77	69	73	75	76	77
Platinum-group metals	90	91	90	89	88	84	80	84
Potash	61	63	64	66	62	65	65	75
Silver	45	31	48	42	7	53	55	61
Tantalum	96	97	97	96	90	92	92	91
Tin	85	82	79	80	79	77	68	72
Tungsten	53	52	56	58	53	50	46	39
Vanadium	37	35	36	28	17	34	24	52
Zinc	58	57	66	63	60	64	58	66

[a]E, exporter; W, withheld.

Source: Ref. 19.

others, most notably copper, lead, and zinc, the currently high levels of U.S. imports are new. That is, the United States was once practically self-sufficient in these minerals. That this is no longer true is a matter of some concern.

UNDERLYING FACTORS

It is apparent that the United States has grown increasingly dependent on imports of minerals. There are a variety of reasons for this trend.

The first and perhaps most obvious reason is that domestic consumption is increasing. This is indeed the case with certain minerals, especially those which are not usually produced in the United States. Good examples of this are the platinum-group metals. Although their consumption fell in the early 1980s due to the overall economic recession, their importance to a variety of chemical and industrial applications means that their use will continue at high levels. Thus U.S. reliance on imports will be maintained or even increased.

A second reason is that domestic production of minerals has fallen over the years. This is not always bad, for sometimes factors such as low ore

grade dictate reliance on foreign imports. However, in many cases, U.S. production has declined or even ceased for reasons that have little to do with market forces. These factors have attracted a great deal of attention, partly because so many of them are based on governmental actions.

Perhaps the most widely known action is the variety of governmental regulations that have been adopted on the basis of safety and environmental considerations. No attempt is made to discuss in detail the usefulness of these regulations. It suffices to note simply that these regulations have in many cases resulted in significant cost increases to mineral producers. Unfortunately, these cost increases are not easily passed on to the consumer.[7] Thus domestic producers must find a way to make ends meet. One frequently chosen response is to reduce exploration expenditures, which has an obvious and serious long-term effect.

Sometimes, though, the inherent cost in meeting environmental regulations is such that a firm will simply close down rather than meet these costs. This has frequently occurred in the U.S. mineral industry, most notably in the mineral-processing sector. Processing capacity for many ferroalloys, especially chrome, manganese, and silicon, has declined dramatically since the 1960s. Domestic production capacity of zinc dropped by more than 50% from 1969–1979 alone.[8] Production capacity in lead, copper, steel, and other products also fell. In all these cases, much of the decline in capacity can be attributed to the costs of meeting environmental control standards. Some would argue that the decline in domestic processing capacity is not a good trade-off for some improvement in air and water quality. Certainly, it has had an adverse effect on the U.S. balance of payments.

Another type of governmental action focused on the taxation of domestic mineral producers. U.S. tax laws are not as favorable as those in many other mineral-producing nations. As a result, otherwise economic projects are delayed or not started at all.[20] This has happened with minerals such as coal, copper, petroleum, and zinc. This issue is particularly important because mineral taxes have not been lowered to balance the increasing costs of environmental controls.

A final government action involves the withdrawal of certain public lands from mineral exploration and/or development. As it turns out, most of these withdrawn lands are in the western United States, which is also the land most geologically favorable for mineral discoveries.[21] This issue is discussed more fully in Chapter I-4, but a policy of locking up land likely to contain needed minerals is a questionable one for a nation as import dependent as the United States.

The discussion above might present an overly pessimistic view of government actions as they relate to the U.S. mineral industries. Indeed, some of the problems currently facing the industries were brought on by their own actions. For example, one major reason operating costs are as high as they are for U.S. mineral producers is that labor costs are very high. By 1980, hourly wages in the mineral industries exceeded the average of all U.S. manufacturing by some 25%.[21] The wages do reflect a somewhat more hazardous occupation, but some of this high labor cost could possibly have been avoided had companies taken less adversarial stands against unions in contract negotiations.

A second industry-based problem is a lack of emphasis on research and development (R&D) expenditures. It was recently estimated that average R&D expenditures for the domestic mineral industries amount to only one-fourth that of the average for all U.S. industries.[21] A great deal of this can be attributed to the higher costs and lower profitability that items such as pollution control lead to for the mineral industries. However, the companies have exhibited such a conservative approach to research and development that it seems that low expenditures would continue even if pollution control expenses and other costs eased. Unfortunately, the attempts of the Reagan administration to spur on increased mineral industry research and development seem doomed to fall far short of their goal.[10]

A third industry action has been the attempt to exploit low-cost deposits in other countries. Copper in Chile, nickel in Canada, bauxite in Guinea, Jamaica, and Surinam; these are just a few examples of the minerals U.S. firms have helped exploit in other countries. Often, these actions have been undertaken with financial support from the U.S. government. Development loans through the United Nations, the World Bank, and so on have fostered rapid development of mineral projects in many countries. In some cases, these new producers have come back to compete with domestic industries in the United States. This is especially true of minerals produced in countries that have few if any other sources of foreign exchange.

A final problem of U.S.-based mineral industries is that they have often failed to keep pace with changes in their own industry. A particularly good example is the way Japan in the early 1960s set out to develop a domestic mineral industry that imported raw materials rather than finished or intermediate products. This took place with copper, iron ore, and many other minerals. In all cases, U.S. producers failed to see this as a challenge to their previous practice of exporting finished or intermediate goods. By not recognizing this, U.S. producers were unable to hold off similar actions by Taiwan and other nations. The United States, on the other hand, neglected its mineral-processing industries and allowed them to become obsolete. As a result, the United States now increasingly has to import minerals in processed forms.

CURRENT CONCERNS

Since the early 1970s, a variety of industrial, academic, and governmental bodies have issued warnings about the gradual increase in U.S. dependence on foreign sources of minerals. These warnings are not entirely new. Most have been heard in one form or another since early in this century. However, several of the issues brought up by these institutions deserve closer attention.

Certain groups have concluded that any increase in the level of U.S. mineral imports is bad. This is an overstatement and an overreaction. Some increases in U.S. imports cannot be avoided if the particular mineral is not found in the United States. Moreover, it is wrong to condemn all mineral imports as dangerous. This is because the bulk of U.S. mineral imports come from developed, and hence in most cases friendly, nations.[21] Canada, for example, is the primary supplier of a large number of minerals.

The problem is also not simply one of an import dependence per se. Rather, it is one of ensuring that import dependence does not lead to supply disruptions.[22] This is the key issue. If the United States is adequately prepared, a heavy import dependence need not mean that the United States will be vulnerable to supply cutoffs, and to the economic and social disruption sure to follow.

Most of the groups focusing on the issue of vulnerability have agreed that the United States is currently engaged in a resource war with the USSR. They point to a variety of Soviet activities in many of the developing nations from which the United States imports minerals, arguing that the Soviets are engaged in an attempt to deny the United States access to these minerals.

In a nation as secretive as the USSR, it is difficult to be completely certain of the reasons behind actions undertaken. However, it is certain that the Soviets are heavily involved in a number of mineral-rich nations. Cuba and southern Africa are two examples. A more recent one is Afghanistan, which, as noted by Shroder, is very rich in minerals.[23] The Soviets first involved themselves in Afghanistan's hydrocarbon industry in the early 1960s. Shortly afterward, they replaced all Western technical advisers and then made all geological maps and reports state secrets.

Recent Soviet attempts to subjugate Afghanistan may make people feel that the actions were undertaken simply to control another country on the Soviet border. This fits the pattern of Soviet actions in Eastern Europe at the conclusion of World War II. However, there is some credibility to the notion that the Soviets include access to minerals, or, conversely, denial of minerals to the United States, in their economic and military ventures around the globe. Moreover, as was shown in both world wars, situations analogous to the current presumed resource war have occurred before. Thus this is not a concept that should be ignored, but neither should it be exaggerated.

Soviet involvement in developing nations, which also happen to produce minerals, may simply be to ensure that those minerals are available to the Soviet economy. This is because the Soviet economy is apparently becoming more import dependent. If Soviet involvement is of a more strategic nature

(i.e., denial of resources to the U.S. and its allies), the results could be serious. Production could slow down or stop. Exports could be delayed by infrastructure problems, such as rail lines in southern Africa. In an extreme case, the Soviet navy could sink mineral-laden ships in the world's sea-lanes, which are no longer under British or American dominance.

Even if the specter of a resource war is false, the United States must pay attention to the prospect of supply vulnerability. This would be more likely to occur in the form of price increases in a tight supply situation than as an absolute supply cutoff. This type of vulnerability is far more likely than the resource war cutoff for one major reason: global consumption of many minerals is increasing.

In both the resource war case and the increased competition case, the United States is faced with the prospect of not being able to ensure the availability of needed minerals. This is perhaps the key factor behind designating many minerals as "critical" or "strategic." Minerals such as cobalt, bauxite, chromium, manganese, and platinum-group metals all have the following four characteristics in common:

1. They are critical to defense and our industrial economy.
2. U.S. resources are inadequate or even nonexistent, with little prospect of this changing.
3. Substitution possibilities are limited at best.
4. Alternative, more secure import sources do not currently exist.

One reason why the U.S. imports more minerals now than it once did is because domestic production of certain minerals has fallen. This is a key issue. If the United States were to produce all the minerals now found in this country, current levels of import dependence could fall significantly for many minerals. Of course, geological constraints and economic considerations would limit the extent of a U.S. turn to greater mineral self-sufficiency. Yet the point holds true: The United States could lower its import dependence, and perhaps vulnerability, if it were to produce more minerals domestically.

However, the United States has missed many opportunities to do that. Environmental regulations have drawn out the predevelopment phase of proposed mines to 10 years or more in many instances. As a result of high taxes, high operating costs, uncertain product prices, and returns on investment only decades in the future, many firms have turned away from otherwise worthy mineral development schemes. It is likely that this will continue. Thus, if the United States wishes to reduce its level of import dependence, it must pursue alternative strategies. Chapter I-4 addresses the most important of these strategies.

REFERENCES

1. Cameron, Eugene N. "The Contribution of the United States to National and World Mineral Supplies." In *The Mineral Position of the United States*. Edited by Eugene Cameron. Madison, Wis.: University of Wisconsin Press, 1973.
2. Tilton, John E. "U.S. Policy for Securing the Supply of Mineral Commodities." *Primary Commodities: Security of Supply Conference*, Second Round Table, Washington, D.C., 11–12 December 1980.
3. Eckes, Alfred E., Jr. *The United States and the Global Struggle for Minerals*. Austin, Tex.: University of Texas Press, 1979.
4. Leith, C. K., "Principles of Foreign Mineral Policy of the United States." *Mining and Metallurgy* 27 (1946), p. 14.
5. May, E. R. *Lessons of the Past: The Use and Misuse of History in American Foreign Policy*. New York: Oxford University Press, Inc., 1973.
6. Redfield, W. C. *Dependent America: A Study of the Economic Bases of Our International Relations*. Boston: Houghton Mifflin Company, 1926.
7. Carroll, B. A. *Design for Total War: Arms and Economics in the Third Reich*." The Hague: Mouton Publishers, 1968.
8. Weinberg, G. L. *The Foreign Policy of Hitler's Germany: Diplomatic Revolution in Europe, 1933–1936*. Chicago: The University of Chicago Press, 1970.
9. Ickes, H. "The War and Our Vanishing Resources." *American Magazine* 140 (1945), pp. 20–22.
10. Abbott, C. C. "Economic Penetration and Power Politics." *Harvard Business Review* 26 (1948), pp. 410-424.
11. U.S. President's Materials Policy Committee. *Paley Report*, June 1952, Vols. 1–5.
12. Kroft, D. J. "The Geopolitics of Non-energy Minerals." *Air Force Magazine*, June 1979, p. 76.
13. Santini, J. D. Response to McGovern's Report on "Imports of Minerals from South Africa by the United States and the OECD Countries," October 22, 1980.
14. Fine, D. I. "Mineral Resource Dependency Crisis: Soviet Union and United States." In *The Resource War in 3-D: Dependency, Diplomacy, Defense*. 18th World Affairs Forum. Edited by James A. Miller, Daniel I. Fine, and R. Daniel McMichael. Pittsburgh, Pa.: World Affairs Council of Pittsburgh, 1980.
15. Nixon, R. M. *The Real War*. New York: Warner Books, 1980.
16. Moss, R. A White Paper. Council on Economic and National Security, August 1980, p. 42.
17. McGovern, G. Imports of Minerals from South Africa by the United States and the OECD Countries. Subcommittee on African Affairs, September 1980.
18. Dresher, W. H. *Technological Innovation and Forces for Change in American Industry*. Washington, D.C.: National Research Council, 1978.
19. U.S. Department of the Interior, Bureau of Mines. *Mineral Commodity Summaries*. Washington, D.C.: U.S. Government Printing Office, various years.
20. U.S. General Accounting Office. *The U.S. Mining and Mineral Processing Industry: An Analysis of Trends and Implications*. Washington, D.C.: U.S. Government Printing Office, 1979.
21. U.S. Department of the Interior, Bureau of Mines. *The Domestic Supply of Critical Minerals*. Washington, D.C.: U.S. Government Printing Office, 1983.
22. Landsberg, Hans. "Is a U.S. Materials Policy Really the Answer?" *Professional Engineer*, September 1982.
23. Shroder, John F., Jr., "The USSR and Afghanistan Mineral Resources." In *International Minerals: A National Perspective*. Edited by Allen F. Agnew. Boulder, Colo.: Westview Press, Inc., 1983.

Strategies to Reduce U.S. Import Dependence and Vulnerability to Supply Disruptions

U.S. dependence on imported sources of key strategic minerals is causing growing concern. Factors contributing to this growing import dependence include past neglect of mining and metallurgical research, growing concern with environmental affairs, restrictions on mining on public lands, increasing nationalism and protectionism in the Third World, and growing competition for available supplies from other countries.

The United States has four basic options that can affect the supply of these commodities. The first option is stockpiling—the cheapest and quickest short-term insurance policy against supply disruption. The second option is increased domestic production. That is affected by land withdrawals, the financial requirements of the mineral industry, and the role of education in ensuring future manpower requirements and technological innovation. The third option is conservation, substitution, and recycling; and the fourth is the diversification of, and continued access to, foreign sources of supply.

None of these options alone will solve the problem. The extent to which they are feasible will vary from one commodity to another. For instance, increased domestic production of zinc is technically quite feasible, whereas this is not a realistic option for manganese. However, increased domestic production of any commodity will probably require the exploitation of deposits that are low in grade, small, complex, unconventional, occur in remote areas, or which suffer from a combination of these features. Only through the application of superior exploration, mining, and metallurgical technology will such deposits be exploitable economically.

Substitution is possible for some commodities, for instance, aluminum or fiber optics for copper. For other minerals, such as chromium and manganese, substitution possibilities are much less likely. Similarly, increased recycling of chromium, aluminum, and platinum-group metals could be achieved. However, this does not appear to be a feasible option for manganese.

Diversification of foreign sources of supply is possible for those commodities that have a fairly even distribution of resources on a global basis, but much less feasible for commodities having an extreme concentration of resources in one or two countries, such as chromium or the platinum-group metals.

The only option applicable to all commodities is stockpiling. This option is examined in much greater detail than any of the others.

STOCKPILING

Important Concepts

There are a number of important concepts that have to be considered in the formulation of strategic stockpile policy. The type of contingency that the stockpile is held for will affect the content and amount of the stockpile. A purely defense stockpile will put a higher priority on the defense uses of materials and assume a high degree of austerity regarding the civilian uses of materials. A stockpile formed to counter the economic effects of a supply disruption, or to discourage formation of a producer cartel, will consider the civilian uses of materials and have the goal of maintaining commercial consumption of them throughout the supply disruption.

Stockpile Types. In general, stockpiles can be likened to an insurance policy, where a certain amount of resources is diverted from present use in order that it will be available during a future emergency. Stockpiles can be divided into two basic categories, depending on the type of emergency to which they are designed to respond: defense stockpiles and economic stockpiles. A purely defense stockpile will be controlled by government, and the use of it will be subject to a rigid interpretation of what constitutes a defense use (e.g., it might require a direct attack on the country holding the stockpiles before materials are released). A purely economic stockpile, on the other hand, will be controlled by private industry, and its materials will be released upon trigger mechanisms related to price fixing, cartel formation, and nonmilitary conflict. Most stockpiles, whether defense or economic, require some form of government subsidy to be economically feasible.

Defense Stockpiles. A defense stockpile is a stock of resources held by government only for release during a protracted military conflict which directly involves the country in possession of the stockpiled resources; the Strategic Stockpile of the United States is such a stockpile. The Strategic and Critical Materials Stock Piling Act stipulates that "the purpose of the stockpile is to serve the interest of national defense only and is not to be used for economic or budgetary purposes."[1] The interest of national defense is assumed by stockpile planners to involve the contingency of a "conventional" military conflict that is of protracted duration (up to three years). In such a scenario, the level of civilian austerity is high. The policy parameters are an extension of our experience during World Wars I and II.

A cogent argument can be made for a modified defense stockpile that allows for the release of stockpiled materials during a military conflict which disrupts the supply of a strategic or critical material, but which does not directly involve the country holding the stockpiled resources. This contingency would assume less sacrifice by the civilian population, and less direct need for military hardware, since the country would not be involved in the conflict, and therefore would not experience the attrition associated with such involvement. So-called "brushfire" wars have become the norm in recent years, and seem very likely to persist into the future. Many regions rich in natural resources, such as Latin America, the Middle East, and southern Africa, are prime locations for such conflicts.

Economic Stockpiles. An economic stockpile is one that is intended to respond to an economic emergency. An economic emergency is a supply shock that threatens the normal functioning of the marketplace for a commodity or group of materials. Implicit in the notion of an economic stockpile is the maintenance of as much civilian economic well-being as possible. Under these circumstances the contents of the stockpile could differ markedly from the defense stockpile, with more of an emphasis on the civilian uses of materials reflected in the goals of the economic stockpile. For example, three years of war would require considerably more titanium than would three years of relatively normal civilian uses of titanium.

Often, economic emergencies are closely linked to military conflict, and there is a thin line between what is strategic and what is economic. In recent years the most far-reaching economic emergency was the Arab Oil Embargo of 1973–1974, and the second oil supply shock of the late 1970s. In the nonfuel sector, the cobalt supply shock of 1978–1979 was an economic emergency which, whether grounded in reality or not, touched off a rampage of panic buying. A rapid price jump resulted, increasing the average price per pound in 1977 of under $6 (constant 1978 dollars) to almost $23 in 1979.[2]

Both the oil supply and cobalt supply disruptions were either directly or indirectly the result of military conflict. The Arab Oil Embargo resulted from U.S. support of Israel during the Yom Kippur War. The oil disruption of the late 1970s was caused by the Iranian Revolution. The cobalt scare was the result of a rebel invasion of the Shaba Mining Province of Zaire.

Producer Stockpiles. There is another generic category of stockpile which is a producer stockpile. Private producers rarely desire to accumulate a large stock or surplus inventory. This is the case not only because of the cost of maintaining a large inventory in physical terms, but also because of the opportunity cost associated with holding inventory instead of turning that inventory into cash. For each time period that the inventory is held, the value of the cash received at a time in the future is discounted. Discounting can be expressed by the equation

$$PV = \frac{CF}{(1 + i)^N}$$

where PV equals the present value, CF the cash flow at year n, i the discount rate per year, and N the number of years.

There are different considerations when the producer of a given commodity is a national entity. Public enterprises will normally have a lower discount rate (opportunity cost) than will a private organization. This is because society has a longer time horizon; that is, it puts a higher value on maintaining resources for future use. Also, public enterprises do not usually have the problem of running out of cash that private enterprises have. Public enterprises can operate on a deficit basis for much longer than can private organizations. For example, a country might maintain production of a mineral while demand is down, in order to keep mines open and people employed. By maintaining production, they will inevitably accumulate inventory. If demand increases, they will be in a better position to sell the commodities on the market than if they had closed down the capacity.

Buffer Stocks. Buffer stocks are used to regulate international commodity prices. An organization such as the International Tin Council will hold a buffer stock of tin. The buffer stock will be increased when commodity prices fall below a trigger price, and sold off when the price goes above a certain level. Between October 1981 and March 1982, the United States contributed 1500 metric tons of tin to the ITC buffer stock, or about 0.7% of the U.S. tin stockpile inventory.[1]

Resources and Reserves. There are some basic concepts without which a discussion in mineral economics cannot progress beyond the most rudimentary level. Central to these concepts is an understanding of the relationship between resources and reserves.

Resources. Natural resources can be divided into two categories, renewable and nonrenewable. Renewable resources include such things as agricultural products and fisheries. Nonrenewable resources are materials that are mined from the earth and are depleted over time. Metals such as iron, copper, aluminum, chromium, and platinum, and nonmetals such as fluorspar, diamonds, and rare earths, are all nonrenewable resources. Although the U.S. Defense Stockpile does contain some renewable resources, most notably rubber, this study is concerned primarily with the nonrenewable, mineral resources.

The U.S. Bureau of Mines defines a mineral resource as[3]

> a concentration of naturally occurring solid, liquid, or gaseous material in or on the Earth's crust in such form and amount that economic extraction of a commodity from the concentration is currently or potentially feasible.

Since it is hard to know what is potentially feasible until it is actually accomplished, it is probably easier to think of mineral resources as all of an element that exists in and around the earth's crust, in sufficient concentration to warrant further investigation.

Reserve Base. The term "reserve base" has been developed to describe that portion of a resource which is known or estimated from specific geologic evidence, and can be produced with current mining methods. Although this seems simple enough conceptually, establishment of criteria can prove to be much more complicated, so much so that the Bureau of Mines carries a footnote attesting to the lack of criteria with every reserve base estimate.[3]

It is important to note that the entire reserve base is not economically producible. Only a portion of the reserve base is economically producible, and, in fact, there may be no portion of a reserve base that is currently economical to produce.

Reserves. The portion of a reserve base that is economically, technically, and legally producible at the time of a given study, is a reserve. It is evident that reserves can change with price, with changes in legislation, and with changes in technology that make the cost of production lower. Also, changes in the price of major inputs, such as energy, can change the magnitude of reserves. Resources are not mined; only reserves are mined. Much time, money, and geologic research must be expended to turn a resource into a reserve.

Strategic and Critical Materials. The terms "strategic" and "critical" conjure up images of high-technology, esoteric metals, known only to a small handful of specialists and defense analysts. Although some strategic and critical materials might fit this description, others, such as manganese and chromium, are basic to our everyday lives, for there would be no stainless steel without chromium, and no steel, at all, without manganese. In reality, the common use of a material makes it more strategic and critical than a metal with limited application.

Strategic and critical materials are those which are important to the industrial and military strength of a country, and which, at the same time, are obtained from foreign sources, particularly those in remote locations. Supplies that may face potential threats, due to political unrest, are considered especially strategic and critical. Joel Carl, professor of materials systems at the Massachusetts Institute of Technology, has stated that "critical" and "strategic" have different connotations from each other.[4] "Critical" implies that a material is very important in terms of the defense and economy of a country, while "strategic" indicates that the supply of a critical material is from foreign and insecure sources.[4] Clark's interpretation points out the importance of both supply and demand characteristics. Since most of the literature, and the language of the law, treat strategic and critical materials more or less synonymously, for the sake of clarity they are not separated in this book.

What is strategic to one country may not be to another. By the same token, a private corporation may have a different interpretation of vulnerability than that held by a government. However, there are some general characteristics which are central to why a material is strategic and critical. These characteristics in terms of demand are: gross consumption, military importance, industrial importance, complexity of demand, and demand substitution. On the supply side, the factors consist of import dependence, domestic reserves, substitution of supply, concentration of supply, world reserves, distance of supply routes, infrastructure, political stability, import dependence of U.S. allies, marginal resources, and recycling potential. To one degree or another, these factors are present in most analytical approaches to forming a working definition of strategic and critical.

As pointed out by Szuprowicz,[5] additional factors, such as formation of cartels and price instability, are also important. These additional factors relate to an economic stockpile more than to a defense stockpile. Since the U.S. Strategic Stockpile is a defense stockpile, the military factors are emphasized. The factors presented, although of a general conceptual nature, draw heavily on works such as that by Szuprowicz.

Demand Characteristics

1. *Gross consumption:* Certain metals, such as iron, aluminum, and copper, dominate the economy in terms of the gross amounts of production and consumption. In 1981, 134.4 million metric tons of raw steel, 74 million tons of iron ore, 4.4 millon tons of primary aluminum, and 1.5 million tons of primary copper were produced in the United States. Apparent consumption for these metals was 135.5 million tons of raw steel, 107 million of iron ore, 5.1 million of aluminum (including scrap), and 2.4 million metric tons of copper.[3] Metals such as these are obviously very critical in terms of their importance to society. Many jobs are related to these commodities, and much capital is invested in the capacity to produce and utilize them. They are essential raw materials for a wide range of civilian and defense manufacturing industries.

2. *Military importance:* A material that has important military applications has high status in terms of the U.S. Defense Stockpile, and is considered a strategic and critical material. Figure I-4-1 shows the percentage of consumption constituted by defense in 1979 for some selected nonfuel minerals. As can be seen, titanium is particularly important in terms of military consumption. This is reflected in the stockpile goals. Consumption of titanium sponge, in 1981, was approximately 32,000 tons. Although the stockpile is based on a three-year war scenario, the stockpile goal is about six times the rate of 1981 consumption.[3]

Much strategic planning goes into determining what materials will be most important in terms of both current and future military conflict. Materials can have a direct impact on the balance of strategic

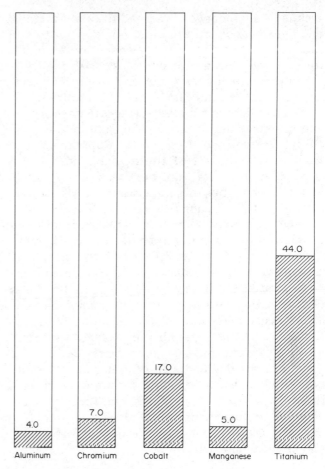

FIGURE I-4-1 Percentage of 1979 consumption used in defense for selected minerals. (From U.S., Congress. House. Committee on Banking, Finance and Urban Affairs. *U.S. Economic Dependence on Six Imported Strategic Non-fuel Minerals.* 97th Congress. 2nd Session.)

power. An example of this would be that the United States uses cobalt in its jet fighters, whereas the USSR does not. This, according to analysts, is one of the contributing factors to the superiority of U.S. aircraft.[4]

3. *Industrial importance:* It is wrong to consider strategic and critical materials only in terms of their direct military applications. Any war effort must be supported by industrial activity which is not directly involved in the machinery of war. Much austerity is assumed in war-planning scenarios. Materials that remain essential to the civilian sector of the war effort should be considered strategic and critical, almost to the same extent as the purely military materials. Many materials have applications to both sectors, and are therefore very important.

4. *Complexity of demand:* Complexity of demand can be likened to a multiplier effect in an economy. Those materials that are related to the

greatest subdemands for additional materials, energy, or labor are the most complex in terms of their interactive effects on the economy. Aluminum, for example, has a high complexity of demand, as it requires much energy per unit of production (an average of 8 kilowatt-hours of electricity per pound of metal[6]), has several stages in its processing, and has many varied uses in the economy.

5. *Demand substitutability:* Substitutability is an important aspect of the demand side of materials. If a material can be replaced by another material which performs the same general function, it is subject to demand substitution. The distinction between military and civilian demands is important in relation to substitutability. In the civilian sector, there is far greater flexibility in terms of material substitution, since a decrease in performance may be frustrating but tolerable. However, in the military sector, a decrease in performance may mean the difference between a battle won and a battle lost. In this way, only materials that can substitute without loss of performance can be used.

From the demand side of materials, the important thing to consider is that the commodities with the most military applications, and those which are essential to the civilian contribution to a war effort, have the greatest potential for being most strategic and critical materials.

Supply Characteristics

1. *Import dependence:* The level of import dependence is at the heart of strategic minerals stockpiling policy. The United States has an import dependence of 50% or more for 30 different metals and minerals.[5] The import dependence level in 1981 was 100% for columbium, corundum, industrial diamonds, natural graphite, natural mica sheet, and strontium; and 90% and more for alumina and bauxite, cobalt, gemstones, manganese, and tantalum. Other important minerals or materials with high import dependence include asbestos, 80%, fluorspar, 85%, nickel, 72%, platinum-group metals, 85%, and tin, 80%.[3]

Import dependence has been a concern in regard to oil since the early 1970s. Yet the historic U.S. import dependence for oil is between 30 and 45%. Fortunately, none of the highly import dependent minerals taken alone is as critical as oil, in terms of total use in the economy. However, taken as a group, the supply cutoff of the most critical minerals would have consequences for the economy as a whole, and certain key industries in particular, especially when so many of the most critical materials are geographically limited to certain regions of the world.

2. *Domestic reserves:* There is a high correla-

tion between import dependence and lack of domestic reserves. Yet there are some metals, such as nickel, where the U.S. reserves are fairly substantial, but where import dependence ran between 68 and 80% from 1978 to 1982.[3] Since the United States is a large consumer of metals such as nickel, its reserves and production would have to be enormous to satisfy demand requirements. Consequently, while domestic production may be higher than in many other smaller countries, import dependence remains high. At least, where there are some domestic reserves, the possibility exists that supply disruptions could be somewhat mitigated. Table I-2-1 gives reserve figures for some key minerals in the United States.

3. *Substitution of supply:* When the demand for a metal or a mineral can be satisfied by a source other than the conventional ore for that metal, supply substitution is possible. Obtaining aluminum from aluminous clays, such as kaolinite, instead of the usual ore bauxite, would be an example of supply substitution. The technology of such substitutions can have very profound effects on the markets for these commodities. For example, a breakthrough in the technology of the supply substitution mentioned above would have the effect of making aluminum a much less strategic and critical mineral, since there is a ready supply of aluminous clays in the United States and throughout the world. In the case of aluminum, the technology to obtain it from sources other than bauxite is already available, but the amount of energy required makes the cost prohibitive,[7] thus pointing out the important link between fuel and nonfuel minerals. Along the same lines, if a breakthrough in energy technology, such as fusion-powered electric generation was to occur, the feasibility of highly energy intensive substitutes would become more practical and economical.

4. *Concentration of supply:* Supply concentration can be looked at in two ways. Concentration of production will indicate the current level of supply concentration, and concentration of reserves will indicate the future level of supply concentration. Concentration can be measured by the percentage of production by, or reserves in, a given country or region. Minerals that are highly concentrated will be more vulnerable to a supply shock. A German study pointed out that there are 11 raw materials for which greater than three-fourths of the world's production are concentrated within those materials' three largest producing countries[8] (see Table I-4-1). Of the minerals on this list, the United States is included in the top three only for lithium, magnesium, and molybdenum.[3]

Tables which show countries that supply mate-

TABLE I-4-1   Concentration of 1980 Mineral Production,
Showing Mineral's Three Largest Producing Countries (Percent)

Raw mineral	1st	2nd	3rd	Total
Platinum-group metals	USSR (47.6)	South Africa (45.4)	Canada (5.9)	98.9
Columbium (niobium)	Brazil (82.2)	Canada (10.6)	USSR (4.6)	97.4
Lithium	US (72.5)	USSR (17.8)	China (5.6)	95.9
Rutile	Australia (65.7)	Sierra Leone (11.8)	South Africa (10.7)	88.3
Molybdenum	US (63.3)	Chile (12.4)	Canada (11.3)	86.9
Zirconium	Australia (64.0)	South Africa (11.1)	USSR (10.5)	85.6
Magnesium	US (48.1)	USSR (23.4)	Norway (13.8)	85.3
Gold	South Africa (55.7)	USSR (21.4)	Canada (4.0)	81.0
Asbestos	USSR (43.9)	Canada (27.2)	South Africa (6.1)	77.2
Vanadium	South Africa (33.4)	USSR (29.4)	US (13.5)	76.3

Source: Ref. 8.

rials to the United States can give a false impression of the dispersion of supply. For example, the Bureau of Mines lists the following countries as suppliers of U.S. cobalt: Zaire, Belgium-Luxembourg, Zambia, and Finland.[3] However, Belgium-Luxembourg, is not a region with its own reserves of cobalt, but is a producer of processed metal from ore mined in the former Belgian colony of Zaire. Other examples include ferromanganese from France (listed as 25% of the supply of U.S. imports), and platinum-group metals from the United Kingdom (listed as 11% of the U.S. import supply for 1982).[3] Neither country has mine production of these metals. France must import manganese ore from primary producers such as Gabon, South Africa, and the USSR. The United Kingdom obtains its platinum from South Africa, refines the metal, and sends it on to countries such as the United States. When this reality is recognized, it becomes more obvious that the current supply of many important minerals is dominated by one region: southern Africa.

5. *World reserves:* Geologic abundance is a fundamental component of the supply of any mineral. Since minerals are nonrenewable resources, the supply is constantly being depleted over time. As supplies are depleted, there is more competition for the lower-quality ores that remain. Thus, over time, there is a built-in mechanism which decreases the supply of a given commodity. However, there are two circumstances that can mitigate this effect:

changes in technology can make the exploitation of lower-grade ores more economic and therefore increase supply, or substitute minerals can replace the depleted mineral's use in society. The effect of depletion of resources over time is either a major constraint to economic growth, or simply a minor challenge to be overcome by research and development, depending on the effects of the mitigating factors just mentioned.

The pessimistic viewpoint is represented by the early work of Thomas Malthus,[9] and more recently by works such as *The Limits to Growth*.[10] The optimistic viewpoint is incorporated in the works of such writers as Herman Kahn[11] and Robert Solow.[12] It should be noted that the Strategic Stockpile is intended to respond to short- or intermediate-term supply shocks (three years or less), not long-term developments. However, long-run changes in supply affect the total environment in which stockpile policy must be made.

6. *Distance of supply routes:* During a period of military conflict, the distance of supply routes becomes a vitally important aspect of material supply. For this reason, minerals that are supplied principally by Canada would be considered less strategic than those supplied by Australia or South Africa. In addition to the distance of the routes themselves, there are strategic considerations concerning the perilous nature of the routes through which the materials must pass. The Straits of

Hormuz is known as the "jugular vein" of the Western world because of the amount of oil that must pass through it daily. The Cape of Good Hope of Africa is a de facto "strait," since the tankers and supply ships must pass very close to the southern tip of Africa to avoid the "Roaring Forties" to the south.

Many ships that carry U.S. materials are not registered in the United States but are "flags of convenience" vessels, registered in countries such as Liberia, Panama, and Honduras. A study by the U.S. government[13] points out that in the past, Liberia has refused to make some shipments, such as arms to Israel. It points out that such events as possible nationalization, and reluctance of foreign sailors during hostilities, could impair U.S. strategic minerals supplies. Obviously, this would make those minerals which are supplied from a long distance even more susceptible to supply interruption.

7. *Infrastructure:* Systems of transportation, the generation of electric power, the availability of water, and facilities for the families of miners and other personnel associated with mineral ventures are all factors that can influence the supply of minerals. Many of the world's important mineral deposits, and probably many more future finds, are located in relatively remote regions of the world. As some analysts have pointed out, in places such as Australia, the iron ore, coal, and nickel developments of the 1960s required $1.8 of infrastructure for every $1 spent on mine development.[14] The disequilibrium between the location of capital and the location of raw materials in the world, and the restraining impact on the abilities of several producing nations, may prove to be one of the most important inhibitors to supply in the future.

8. *Political stability:* The infrastructure mentioned above is subject to more than natural disasters and mechanical breakdowns. Political insurrection and terrorism have become important factors in the disruption of mineral supplies. The political stability of a region is perhaps the hardest variable to quantify and the most important one to predict. As alluded to previously, southern Africa is a region rich in mineral resources. At the same time, it is a region characterized by terrorist activity, tribal disputes, and intervention from foreign powers in the name of national liberation. A report by the Subcommittee on Security and Terrorism of the Committee on the Judiciary, U.S. Senate,[15] stated that "the original purposes of the African National Congress and the SouthWest Africa People's Organization, have been subverted, and that the Soviets and their allies have achieved alarmingly effective

control over them," and later, "the evidence has thus served to illustrate once again the Soviet Union's support for terrorism under the guise of aiding struggles for national liberation."[15] Such activity, whether intended primarily for the disruption of mineral activities or not, could have devastating effects on the supply of materials to the United States and other nations of the Western world.

Aside from the obvious problems created by politics, there are other, more subtle problems created by political entities. Warnuck Davies points out the problems in dealing with some new political leaders when he states:[16]

> It was their apparent disregard for sanctity of contract that caused more concern to potential investors than any other factor, more than coups and more than other colorful characteristics of developing countries that are dramatized in the press.

In a more general sense, political stability can apply to all factors of a political or legal nature. Under this interpretation, high taxes can impede the supply of minerals as much as rebel insurgency.

In the United States, political instability may be understood in terms of uncertainty concerning policy changes from one administration to the next. A pertinent policy arena is environmental protection. Mining ventures are long-term projects that have long lifetimes and payback periods. If a company believes that environmental policy will be changed with each new administration, it will be reluctant to embark on a project of this nature. The world supply of molybdenum, produced dominantly by the United States, is feared by many European analysts to be subject to this kind of political instability.[17]

9. *Dependence of U.S. allies:* U.S. allies are much more dependent, and less prepared for confrontation, in terms of material readiness than is the United States. France has probably the most ambitious stockpile of the Europeans, and even so, has allocated only $300 million between 1975 and 1980, or enough for a two-month supply of 13 materials, including titanium, cobalt, copper, and tungsten.[18]

West Germany has stockpiles of chromium, vanadium, cobalt, manganese, and blue asbestos. The West Germans stockpile their materials in processed form, which offers easy access and availability during an emergency. But it must be remembered that Europe, Germany in particular, is within easy striking range of the communist bloc; this has been a factor in limiting Germany's domestic stockpile.[18]

Japan has not historically held government stockpiles, but has encouraged industry to stockpile through commodity associations. In the early 1980s, Japan made its first moves toward establishment of a government stockpile. The initial stages called for a stockpile of 13 metals held in a 60-day supply.[19] Reports indicate that possibly due to economic considerations, the list has been trimmed to seven metals: nickel, tungsten, chromium, manganese, vanadium, molybdenum, and cobalt.[19]

After the Falklands War, the United Kingdom decided that stockpiles would be a necessary provision for times of military conflict. In late 1982 and early 1983, they began making provisions for the stockpiling of chrome and manganese. Initial outlays were thought by analysts to be between 5 and 10 million pounds sterling.[21]

10. *Marginal resources:* Because reserve is a dynamic concept, it is important to know what resources are available that could become reserves during an interruption of normal supply and subsequent price increases. It could very well be that the United States has no present reserves or production of a material that exists in fairly large quantities in marginally economic amounts. Such a metal is cobalt, which exists in geologic concentration in several locations throughout the United States, including Idaho, Missouri, and California.[2] Minerals in marginal economic supply in the United States offer alternatives to stockpiling, through the development of domestic capacity.

An understanding of the lead times required to exploit domestic resources is crucial to the parameters of stockpile formulation. If the lead times are relatively short, the amount of material in the stockpile can be reduced. However, if the capacity is not already in place, the stockpile must contain at least enough material to cover the lead time to bring the capacity into production.

11. *Recycling potential:* Recyclable metals represent floating stockpiles since they can be used again when needed. The price increase that would take place during a supply interruption would induce people to save and recycle such items as aluminum cans, and cause industry to recycle industrial scrap. Yet the same caution must be applied to recycling as to substitution. The use of recycled materials is often restricted to non-defense-related areas. There is often an effect on the physical properties of a metal that has gone through the recycling process, that makes it unsuitable for certain high-stress situations.[22]

The terms "strategic" and "critical" are dynamic; minerals and materials change not only in terms of their strategic importance, but different nations will have differing views on what is strategic and critical. Changes in technology will create new demands, just as new mineral finds, or events of a political nature, will bring about changes in materials availability. The size of the stockpile itself could become a factor in determining which materials are strategic and critical.

Historical Overview of Stockpile Policy

Current stockpile policy was not made in a vacuum, but evolved over many years. If an understanding of current policy is to be gained, one must first review the events and people who have shaped its development. There are three basic periods in the development of the stockpile: the period preceding the 1946 Stockpile Act, the time between the 1946 Act and the 1979 Act, and the period since the 1979 Act. The primary actors were the Presidents, members of advisory panels on mineral policy, Congress, and specialists in the field such as Charles Leith and Alan Bateman. This chapter borrows heavily from the important work by Alfred Eckes, *The United States and the Global Struggle for Minerals.*[23] The reader is advised that a more in-depth analysis can be gained from Eckes's excellent book.

Important Legislation

Strategic and Critical Materials Stockpiling Act of 1946. The single most important piece of legislation concerning stockpiling of strategic and critical materials is the Strategic and Critical Stockpiling Act of 1946. This act, with some important revisions, is the legislative instrument that directs current stockpile policy. As could well be expected, World Wars I and II played the most important role in the call for a national defense stockpile. In 1919, the War Department was reorganized to include an industrial mobilization planning unit.[22] This was a recognition of the relationship between industrial and military readiness. In 1921, a Colonel Harbord developed a list of 28 materials that had presented supply difficulties during the war.[22] The list was probably the first attempt to define the materials that were critical and strategic to the U.S. economy and to the national defense.

Prior to World War I there was the underlying feeling that America had abundant resources. Under President Theodore Roosevelt, conservation of natural resources became a national priority. However, even studies conducted during his administration, such as one by Gifford Pinchot, did not see any imminent danger of metals shortages. On the contrary, it recommended that the United States "extract the ore at as rapid a rate as possible when profit is assured."[23]

After the end of World War I, analysts were able to make the observations that hindsight affords, and better illustrate the important role that minerals played in the war. C. K. Leith, writing in the late 1920s made note of the German buildup of certain materials just prior to the outbreak of hostilities[24] (see Figure I-4-2). Had the world been more cognizant of the importance of this substantial minerals buildup, it would have been better prepared to counter German aggression.

After the war, Germany's access to minerals was greatly diminished, as it lost colonies that were important mineral suppliers. In addition, the iron-rich region of Lorraine was lost to the French, and mineral-abundant Silesia was lost to Poland.[24] Particularly ominous was Leith's warning that the situation was "one of unstable equilibrium" that would "require wise and delicate handling if future trouble is to be avoided."[24] If history is the judge, the situation was handled neither wisely nor delicately, for Germany quickly moved on the coal fields of Poland and the iron ore of the Minette in France. By 1942, the Axis powers had control of 46% of the world's iron ore industry.[23]

Also in 1942, about 25% of all inbound vessels carrying bauxite from British Guiana and Surinam to the United States were sunk by German submarines, many just offshore.[25] If politicians and strategists had not realized it beforehand, they were made painfully aware of how tenuous are the supplies of critical and strategic materials during wartime.

The list of materials that had presented supply problems during World War II grew from the 28 at the end of World War I to 298.[22] This was evidence of the massive change that had taken place in both the nature of warfare and in the variety of materials used by society. Although the list of disrupted materials had grown since World War I, amazingly enough, prices were kept fairly stable compared to the disruptions of World War I. The price of oil, in constant dollars, fell 12% during World War II compared to a 61% increase during World War I. Copper, lead, and tin cost more in constant dollars in 1937 than in 1944.[23] However, the relatively stable prices were probably due more to the Great Depression that preceded the war than to insightful government planning. In fact, the government would have saved millions of dollars had it purchased the materials during the depression years. The government had to spend much money to obtain minerals from other countries and in subsidies to domestic producers of low-grade deposits.

The drain of resources on the national storehouse was immense. Secretary of the Interior Harold Ickes attempted to make the public appreciate this reality when he stated that, "the prodigal harvest of minerals that we have reaped to win this war has bankrupted some of our most vital mineral resources."[23]

In response to the notion that the United States was in short supply of key materials, the mid-1940s saw various versions of stockpile bills. Some proposals, such as those by Senator James Scrugham of Nevada, placed a strong emphasis on buying American products for the stockpile and thus serving as a subsidy for the domestic mining industry. Senator Scrugham's bill also proposed that the stockpile be controlled by a board composed solely of representatives of the mining industry.[23] Other proposals put more emphasis on the purchase of high-quality, low-cost ore from overseas. The final version, which passed into law in July 1946, contained elements of both approaches.

The 1946 Act was technically an amendment to an Act of June 7, 1939 (53 Stat. 811). However, the 1939 Act got little attention, as the war itself quickly became the top government priority. Min-

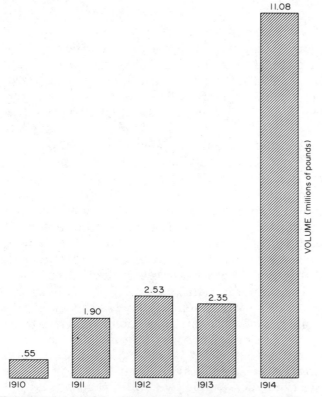

FIGURE I-4-2   U.S. nickel exports to Germany in millions of pounds—prior to World War I. (From Ref. 24.)

erals that the United States purchased had more immediate uses than stockpiling for future conflict. The 1946 Act was intended[26]

to provide for the acquisition and retention of stocks . . . to encourage the conservation and development of sources of these materials within the United States, and thereby decrease and prevent wherever possible a dangerous and costly dependence of the United States upon foreign nations for supplies of these materials in times of national emergency.

The materials referred to are those deemed strategic and critical.

The law gave the power to the Secretary of War, the Secretary of Navy, and the Secretary of Interior, acting in conjunction with the Army and Navy Munitions Boards, to determine "from time to time" which materials fit the description of critical and strategic. In this way the military was given priority in the establishment of these criteria, with the Department of the Interior acting as a technical adviser. Feedback was also provided by the Departments of State, Treasury, Agriculture, and Commerce.

To allow for industry input, the departments involved were directed to "appoint industry advisory committees selected from the industries concerned with the materials to be stockpiled."[26] By including industry representatives, policymakers hoped to soften criticism by mineral producers that the stockpile would interfere with the flow of materials on the free market. The industry panel was used primarily for advice on technical matters such as storage and stockpile specifications. As an industry publication points out: "Industry views on appropriate goals have not been sought."[25] Indeed, until 1962, when the goals were made public, industry had no better idea of the stockpile goals than the average person in the street.

Section 3, paragraph (a), stipulates that purchases should be made "from supplies of materials in excess of the current industrial demand. . . ."[26] This section limited the amount of material purchases immediately after the act became law, since there was a general material shortage in the world.

The act also provides for the refining and processing of materials into a form best suited for the stockpile. This form was to be decided by the Departments of War and Navy. This provision has important implications for today's stockpile, as there is a general need for upgrading forms.

Notice of proposed disposition must be given to Congress under the act. Congress must then give approval for the sale of material from the stockpile. This gives added protection to producing industries, as they have the ability to block such sales through their congressional lobby.

Although the act provided that Congress should appropriate the necessary funds from time to time, there was no money specifically appropriated, nor was a timetable set. Whereas the authority was created for a stockpile, the budget mechanism had not been put into place. Congress eventually appropriated $100 million for fiscal 1947, when it had been estimated that obtaining a minimum goal would require $2.1 billion.[23]

To make budgetary matters even worse, Section 9 of the act stated that any funds obtained from the sale of stockpiled materials judged to be in excess of goal would be returned to the national treasury. Because of the provision that revenue be returned to the general treasury, the stockpile became a source for generating funds to finance other government projects. Such manipulation made the stockpile more of a political pawn than a strategic tool. There remained a significant gap between the language of the law and the political reality of the budgeting process.

The Munitions Boards calculated that it would take five years to reach the $2.1 billion level in stockpiled materials. This included about $300 million in materials that were accumulated through the Reconstruction Finance Corporation.[23] This meant that $360 million a year would have to be accumulated. However, during the first two years of the plan, a total of $275 million was appropriated, far short of the funding level needed to achieve the goal.[23]

Unfortunately, it took the reality of material shortages during the Korean War, exaggerated by the Soviet embargo of chrome and manganese, together with the deep-seated fears of Communist aggression fostered during the Cold War years, to accomplish the bulk of the stockpile purchases. The result was that the government paid a premium for the stockpiled materials.

Another factor that contributed to the high price of stockpile materials was the use of the stockpile to bolster domestic producers. The result was a temporary boom to domestic producers. Tungsten was procured from domestic sources at four times the world price.[22] However, once the boom died down, many of the domestic industries failed and there were few long-term boosts to domestic capacity.

However complicated the motives, and inefficient the economics, the early 1950s did see the procurement of large quantities of stockpiled materials. During this period, the scenario for the potential conflict was increased from three years to five years, thus increasing the stockpile goals for every category. In later years, when the conflict scenarios were lowered again, the materials that had been filled to the five-year goals represented large surpluses, which more often than not, were sold and the revenues used for purposes other than stockpiling. By the late 1950s attention had turned, with the launching of Sputnik, from the Cold War, to the space race, and the United States entered a period of no stockpile purchases. Some agricultural products were bartered for materials as late as 1967, and these were eventually transferred to the stockpile. It was not again until the early 1970s, with the activities of OPEC, that concern was expressed by Congress about U.S. dependence on foreign sources of minerals.

The Strategic and Critical Materials Stockpiling Act of 1979. The Revision Act of 1979 made some important changes in the 1949 Act. The changes included the transfer to the president from the executive departments the power to determine which materials should be deemed strategic and critical. In addition, the conflict scenario was set to a period of three years so that it could not vary from administration to administration. Finally, a Stockpile Fund was set up to provide funds for future purchases from revenues derived from the sale of materials deemed to be in excess of goal.[27]

The Fund was established as recognition of the great discrepancies between the holdings and goals from one material category to another. The 1946 Act stipulated that moneys derived from the sale of excess materials should be returned to the general treasury. This meant that assets would be lost in the attempt to balance the holdings and goals of the various categories. By making the Fund separate, the U.S. Congress wanted to ensure that the stockpile program would maintain its assets and bring about a balance among material categories. Although the Stockpile Fund is separate from the general treasury, the money must still be appropriated by Congress. This ensures that Congress has control over the Fund, while allowing the assets of the stockpile to be protected.

Several government stockpiles were transferred to the National Defense Stockpile under the 1979 Act. The transfers included the supplemental stockpile established in section 104(b) of the Agricultural Trade and Development Act of 1954,

materials acquired under the requirements of section 303 of the Defense Production Act of 1950, and procurements obtained under section 663 of the Foreign Assistance Act of 1961.[27]

Significance of the Stockpile Acts. Procedures and problems of the Stockpiling Acts as they apply today are discussed in the section on current strategies. From the point of view of historical analysis, the Stockpiling Acts were important, as they:[26]

Were a formal and concrete recognition of the United States' potential vulnerability during national emergency due to its increasing dependence on supplies of foreign minerals.

Resulted largely from war experiences, since, both WWI and WWII brought about legislative action intended to build a defense stockpile. In addition, the Korean War was a major impetus for the purchase of materials. The result of this is that the military is the most important actor in setting stockpile policy.

Contain somewhat contradictory language, and have fostered considerable debate over the purchase of stockpiled materials from domestic sources vs. purchase of lower cost supplies from overseas, since, at one point, the acts call for the encouragement of conservation and development of sources within the United States, while elsewhere it is stated that the purpose of the stockpile is to serve in the interest of defense only and is not to be used for economic or budgetary purposes.

Defense Production Act of 1950. The Defense Production Act (DPA) of 1950 is important, as it provides a potential source of materials for the stockpile program. Those materials are obtained from domestic producers whose production of strategic and critical materials are only marginally economic. Every pound of domestic capacity means 3 pounds are not needed in the stockpile, since it is planned for a three-year conflict. In this way, the Defense Production Act is an alternative to the Defense Stockpile. The support for mineral production capacity is in the form of such mechanisms as floor prices, where the government agrees to buy a certain amount of materials for so many years at an agreed-upon price. Assistance to industry can also come in the form of increased depreciation, loans, and other devices to make industry more amenable to investing in high-risk mining projects which involve strategic and critical materials.

The Defense Production Act was used during the Korean War to give $37 billion in tax write-offs.[7] The last time it was used for nonfuel minerals was a loan of $83 million to the Duval Sierrita Copper Mine in 1967.[7] In recent years the DPA had been used to bolster the domestic synfuels industry. In 1980, $3 billion was provided for the synfuels industry upon the passage of the Energy Security Act of June 30, 1980.[28] In the stockpile report, dated March 1981, the Federal Emergency Management Agency was considering projects under the Defense Production Act which supported domestic production of cobalt, titanium, refractory bauxite, and guayule (natural rubber).[29] In the 1982 report, cobalt was mentioned as the most likely material to receive support.[1]

Key Personalities

Presidential Administrations

1. *President Franklin Delano Roosevelt:* The most significant aspect of President Roosevelt's stockpile policy was the lack thereof. Even with the specter of war looming ominously over Europe, President Roosevelt consistently stopped plans to develop a stockpile. In 1935, State Department economic adviser Herbert Feis proposed the purchase or barter of $25 million annually in strategic materials. In 1936, Congressman Charles Faddis introduced a similar plan that also involved cancellation of European war debt in exchange for materials.[23] However, neither of these plans was adopted. In March 1939, President Roosevelt agreed to allow congressional initiative for the stockpiling of strategic and critical materials. He supported an allotment of $10 million per year, while executive agencies pushed for $25 million. Eventually, an act was passed authorizing $100 million over a four-year period. However, by the time the appropriations committees finished with the act, it was restored to the $10 million annual figure that President Roosevelt requested.[23]

Before the four-year period had elapsed, the United States was deeply immersed in World War II. In the struggle for strategic and critical resources that ensued, $10 million was an insignificant amount of money. The excess inventories and lowered prices which were remnants of the Great Depression proved to be the only factors that kept prices from getting out of hand.

2. *President Harry S. Truman:* If President Roosevelt's administration can be characterized by its lack of stockpile initiative, the administration of his successor, Harry Truman, can be noted for its flurry of stockpile activities. The Stockpile Act of 1946 was enacted during President Truman's administration. Equally important, in terms of affecting the thinking of those in positions of

power, was not an act of legislation but a panel of important people known as the Paley Commission. The Paley Commission (1951–1952), named after its chairman, William S. Paley, president of the Columbia Broadcast Systems, was probably the most systematic assessment of U.S. mineral vulnerability undertaken until that time. Other members of the panel included George Brown, a powerful Texas industrialist; Arthur H. Bunker, president of Climax Molybdenum, Eric Hodgins, writer and editor of *Fortune* magazine, and Edward S. Mason, a prominent economist.[23] The gist of the Paley Report was that stockpiling should be continued, and that in some circumstances, government should use the newly enacted Defense Production Act to keep domestic capacity in place in the event that it would be needed in time of war.

There emerged under President Truman a clear unwillingness to use stockpile policy as a way to support domestic industries. While signing the 1946 Act, he made it known that he had no intention of "subsidizing those domestic producers who otherwise could not compete successfully with other domestic or foreign producers."[23] Although the Defense Production Act was an alternative explored by the Truman administration, his natural inclination was to support the stockpiling of least-cost, foreign-supplied materials as an insurance against wartime disruption. At the same time, he encouraged free trade with the nations of the world, emphasizing interdependence over self-sufficiency.

3. *President Dwight D. Eisenhower:* President Eisenhower, in a turnaround from President Truman, favored a more flexible interpretation of the use of the national defense rationale for the support of domestic minerals industry. In 1953, President Eisenhower's cabinet committee urged that stockpiles be enlarged to help the domestic industry. A more formal proposal was made in 1954 that recommended increasing stockpiles to five years of domestic consumption to achieve these results.[22]

The Paley Commission bridged the two administrations by delivering its report to then President-elect Eisenhower in December 1952. Paley wrote President Eisenhower personally, stating: "Our country has moved into a new era in its economic history, an era in which we can no longer produce enough of many materials at a cost to satisfy our expanding economy or security needs."[23]

President Eisenhower had to contend with the conservative ranks of his own party, especially politicians from the western mining states, who favored protectionism of American firms and isolationism from the world at large. As is often the way in Washington, Eisenhower appointed a new commission, chaired by Clarence B. Randall, a steel executive, and vice-chaired by Lamar Fleming, Jr., a businessman from Houston, Texas. The Randall Report was submitted in January 1954. Although it was officially a "Commission on Foreign Economic Policy," in general, the report gave specific attention to the international flow of mineral resources.

The Randall Report agreed with Paley's findings that the United States would become more mineral dependent, rejecting the idea that the United States could ever achieve self-sufficiency. The Randall Report differed from the Paley Commission over the idea of international commodity agreements. Whereas the Paley Commission had favored such agreements in order to stabilize prices, the Randall Report found them to "introduce rigidities and restraints that impair economic adjustment."[23] In the final assessment, the Randall Report was a victory for expansionist Republicans who wanted to see free trade in the world and U.S. investment in overseas operations.

4. *President John F. Kennedy:* The Kennedy years were to represent a shift in stockpiling policy that became the norm for about 20 years. Certain events in the world shifted the emphasis regarding strategic thinking. The growth of nuclear weaponry in general, and the Cuban Missile Crisis specifically, made everyone more skeptical about the possibility of a conventional war of long duration. The other event of significance was the Soviet launching of *Sputnik* in 1957, which focused America's attention on the immediate goal of space exploration rather than providing funds for a future conflict.

There were some very basic and important developments in stockpile policy during the Kennedy years. The contents of the stockpile were declassified in 1962. This was the result of the feeling that too much secrecy was involved with stockpile policy. President Kennedy was very critical of stockpile purchases that took place during the Korean War when prices were high and materials were obtained from domestic producers at higher-than-market prices. As a result, he halted the purchase of stockpiled materials as a way to support domestic industry. In addition, the war scenarios were reduced from five years to three years. The possibility of a number of smaller wars, as well as a mobilization short of war which interrupted material supplies, was also considered.[22]

5. *President Lyndon B. Johnson:* Following President Kenney's lead, President Johnson began selling off materials from the stockpile. He had the Great Society programs to finance, as well as a

growing involvement in the Vietnam War. Between 1958 and 1975, $7 billion was generated from the sale of stockpiled materials.[23] During the same period, the only additions to the stockpile came indirectly through other programs, such as the supplemental stockpile, which functioned until 1967 and which was later absorbed by the National Defense Stockpile.

6. *President Richard M. Nixon:* Policy analysts and advisors recommended to President Nixon that the planning scenario be reduced to one year of war.[23] At this time, there was not a legally binding war scenario (a three-year scenario was added in 1979), and President Nixon was able to make the policy change that made additional materials available for sale as "excess" inventory. Toward the end of the Nixon administration an event occurred that changed drastically America's perception of its mineral self-sufficiency. The Arab Oil Embargo of 1973–1974 made the general public aware, in a very dramatic fashion, of how much America's dependence on foreign supplies of raw materials could make the United States vulnerable to events beyond its control. However, while the above was occurring, President Nixon became very concerned with events that were taking place within his administration. When he left office in August 1974, the public was both confused by the "oil crisis" and skeptical of government in general.

7. *President Gerald R. Ford:* The Ford administration inherited many problems from President Nixon's administration. Most notable of the problems was the energy crisis and a lack of confidence in government. The National Commission on Supplies and Shortages was created in response to the oil embargo, and represented a renewed interest by government in the issue of U.S. mineral vulnerability.

In 1975, the Subcommittee on Seapower and Strategic and Critical materials of the House Armed Services Committee voted to stop further stockpile disposals until a policy analysis could be made.[30] As a result of these investigations, and others inside and outside the Ford administration, the goals for the stockpile were moved in 1976 to three years from the one-year scenario instigated under President Nixon.[23] This, of course, reduced the amount of material available for sale and expanded the stockpile targets for many commodities.

8. *President James E. Carter:* The Carter administration inherited the stockpile when there was considerable debate over how much was needed in the stockpile and how inventory should be purchased and disposed of. In February 1977, President Carter received a letter from 43 House mem-

bers, requesting an in-depth analysis of nonfuel energy issues in relation to the security of the United States.[31] As a result, President Carter announced that there would be a moratorium on stockpile acquisitions and disposals. The Carter administration was responsible for the Non-Fuel Mineral Policy Review, which attempted to identify problems and develop policy options that could address important areas of mineral policy. Although many important areas, such as import dependence, environmental regulations, federal lands resource potential and competitiveness of U.S. minerals industries were touched on, stockpile policy was notably absent. The justification was that administration policy had been set on that matter.

The most important event regarding the stockpile during the Carter administration was the 1979 Stockpile Revision Act, which was discussed above. The act was an important step toward renewal of stockpile efforts, but came too late to renew stockpile purchases during President Carter's administration. This was the case despite President Carter's request for $245 million for stockpile acquisitions for fiscal year 1979.[32] So, as 1980 approached, the United States had still not made a direct stockpile purchase in over 20 years.

9. *President Ronald Reagan:* On April 5, 1982, President Reagan submitted to Congress the National Materials and Minerals Program Plan and Report, which was a requirement of the National Materials and Minerals Policy, Research and Development Act of 1980. The report addresses five areas deemed important to minerals policy: land availability, minerals data research and development, regulatory reform, and stockpile policy.[33]

The stockpile policy statement gave firm support for increased stockpile purchases. President Reagan initiated the purchase of $78 million in cobalt for the stockpile, the first major stockpile purchase in over 20 years. Additional plans were made for the attainment, partially through barter, of 1.6 million tons of Jamaican bauxite.[33] In another report to Congress,[1] the materials that are given the highest priority by planners (presumably in consultation with the administration) are: agricultural-based chemical intermediaries, aluminum oxide, bauxite (refractory grade), cobalt, columbium, fluorspar, medicinals (including opium salts), nickel, platinum-group metals, rubber (including guayule), tantalum, titanium (including rutile), and vanadium. Other areas stressed by the administration included examination of materials in the stockpile for proper form and quality, support of domestic capacity as an alternative to

stockpiling, and coordination of international mineral policy. The April 5 report also announced the formation of a cabinet Council of Natural Resources and the Environment. The purpose of the council, which is chaired by the Secretary of the Interior, is the overall coordination of administration mineral policy.

Mineral Specialists. There are several personalities who stand out as having been influential in the development of mineral policy and in voicing strong support for a stockpile program. The efforts of two such persons have had substantial influence on how U.S. mineral stockpile policy has evolved. Charles Leith and Alan Bateman were both geologists, pioneers in the field of mineral economics, and able to see the important relationship between geologic reality and geopolitical uncertainty.

1. *Charles K. Leith:* In 1931, Charles K. Leith published *World Minerals and World Politics,*[24] which was probably the first in-depth analysis of the mineral dependency among nations. Leith wrote:[24]

> Not even the most favored nation is entirely self-sustaining in minerals, nor can it be made so. The interdependence of nations and specialization in mineral production have been determined once and for all by nature's distribution of minerals.

The book was written between World Wars I and II, during which time Leith served as minerals advisor to President Wilson for the Paris Conference, which took place after World War I. Leith was among many economists and planners sent to Paris to discuss strategies for the maintenance of world peace. He was impressed with the understanding that the British and the French demonstrated regarding minerals and their importance in maintaining economic stability and international security.[23] Leith argued for an internationalist approach to the problem of mineral resources. At the same time he realized that given a high degree of international interdependence, each nation must develop contingencies to protect against interruptions of supply. Unfortunately, Leith came away from Paris disillusioned by President Wilson's seemingly grandiose scheme for world harmony at the expense of more pragmatic economic agreements.

As mentioned before, Leith saw the danger in a Germany stripped of its mineral wealth and influence in the world. As soon as the Axis powers began exhibiting aggressive behavior, Leith was one of the first to advocate strict economic sanctions to keep them from obtaining important minerals for their war effort. In addition, he advocated that the

United States should immediately begin a large-scale stockpiling effort. Instead, the Axis powers were able to obtain important minerals, such as rutile (titanium ore) from Brazil, which was transported for part of its journey through the "neutral" United States. The effort to establish stockpiles came too little and too late.[23]

2. *Alan M. Bateman:* In the spring of 1942, Alan Bateman was sent to Mexico by the U.S. government to obtain minerals desperately needed for the war effort, such as antimony, lead, zinc, and mercury. The expedition was successful in that despite problems in rail transportation, imports from Mexico increased by 236%.[23]

As a result of his experiences during World War II, Bateman was asked to address the meeting of the Geological Society of America and Society of Economic Geologists, which was held in Pittsburgh in 1945. The title of that presentation was "Wartime Dependence on Foreign Minerals." This presentation proved to be influential in pointing out the important role that minerals, and the lack thereof, played during the war. It is probably no coincidence that the Stockpile Act was passed the following year.

As late as 1940 the Army-Navy Munitions Board stated that "with only the United States to consider, an ability to produce copper in war is sufficient for all foreseen requirements."[23] Yet as Bateman pointed out, during the war years, the United States needed to acquire between 35.4 and 40.3% of its primary copper supply from foreign sources. Some of these supplies had to be obtained from as far away as the then Belgian Congo.[34]

Bateman's major findings as the result of his war experiences were the following:[34]

Strategic minerals numbered many more than the dozen that were envisioned at the beginning of the war.

The quantities required for many minerals, during the war, exceeded the entire available world supply.

With the exception of coal, iron, and salt, some quantity of every mineral had to be imported to the United States during World War II.

Demands for minerals, which previously had little consumption, grew tremendously (an example being optical calcite for gunsights).

Trading during the war paved the way for better mineral trade between nations after the war was over.

Geologists had to extend themselves beyond geology and adopt a more political world view.

The costs of obtaining minerals during the war were vastly greater than if they had been obtained during peacetime and saved for an emergency (these costs were not only in terms of money, but in terms of manpower, time, equipment, and lives).

Most important, the foregoing findings led Bateman to the conclusion that "mineral stockpiles are more basic than battleships and bombers," and that "we should plan immediately toward the accumulation of emergency mineral stockpiles."

Changing Stockpile Objectives. The stockpile goals for individual commodities have changed over the years as the result of two basic policy variables: changes in the length of war scenarios, which affected all commodities, and changes in circumstances that were unique to specific commodities. Many of the changes regarding war scenarios and in specific commodity goals were political, not primarily strategic considerations.

In 1954, domestic zinc producers were looking for protection from foreign suppliers. The end of the Korean War had left U.S. producers with a surplus of zinc; the price fell from a high of $0.24 per pound (1972 constant dollars) in 1951 to $0.15 per pound in 1954.[23] President Eisenhower had two options to help domestic producers. He could institute tariff protection, which would interfere with the international market for zinc, or he could increase the stockpile goal for zinc and purchase the metal from domestic producers. The goal was increased from 740,000 tons on April 30, 1953, to 1,100,000 tons on August 3, 1954.[35]

It was not only domestic politics that played an important role in the formulation of stockpile goals. Goals for materials that were produced by countries outside the United States were also changed for political reasons. Nineteen fifty-four was an important year for changes in foreign mineral stockpile goals, which corresponded very conveniently with the Agricultural Trade and Development Act of 1954. The act provided the authority for trading surplus agricultural goods for minerals to be stockpiled. This had a particular influence on mineral producers of the Third World. For example, the goal for bauxite from Surinam increased from 5,000,000 dry metric tons on August 3, 1954, to 12,000,000 by September 29, 1954.[35] The Agricultural Act is dated August 28, 1954.

In the final analysis, changes in stockpile objectives since 1962 have been detrimental to the stockpile inventory. A report by the U.S. Senate Republican Policy Committee compares the inventory held in the stockpile in 1962, when the contents were first declassified, with the 1980 holdings. For example, the report shows that cobalt was held in the amount of 103.1 million pounds in 1962. By 1980 the amount had dropped to 40.8 million pounds, less than half the 1980 goal of 85.4 million pounds.[36]

In retrospect, the most important flaw in the 1946 stockpile law was the lack of a transaction fund to protect the assets of the stockpile, combined with vagueness as to the number of years to be included for stockpile planning purposes. This led to a pattern of political manipulation where one administration would increase the war scenario and use the increased goals to bolster the mineral industry, while another would decrease the goals and sell off the "excess" materials to finance other government programs. The creation of the Fund in 1979, as well as the legally binding three-year war scenario, have aided substantially in eliminating much of the political manipulation that the stockpile suffered from in previous years. Future development of the stockpile will be enhanced by these changes.

Stockpile Strategy under the 1979 Act

Following the passage of the 1979 Act, there was a considerable change in the direction of stockpile policy. This was due to the priorities set by the Reagan administration coupled with the key components of the act: the Transaction Fund and the statutory three-year-goal scenario. These factors have greatly aided the mechanics of stockpile planning. The Transaction Fund has meant protected assets which give materials buyers at the General Services Administration and policy planners at the Federal Emergency Management Agency a much better idea of the monetary resources they will have to work with. The three-year scenario means that a consistent framework for planning has been given to the policymakers.

Mechanisms of Stockpile Planning

Annual Materials Plan. The Annual Materials Plan (AMP) is the basic interagency mechanism for the coordination of current stockpile strategy. The AMP is developed through an Annual Materials Plan Steering Committee, which is chaired by the Federal Emergency Management Agency (FEMA) and includes the Departments of Defense, Commerce, Interior, Energy, Agriculture, State, and Treasury, as well as the Central Intelligence Agency (CIA), General Services Administration (GSA), and the Office of Management and Budget (OMB). This committee is then further broken down into subcommittees which are responsible for areas perti-

nent to the concerns and expertise of their member organizations.

FEMA initiates the process by submitting a list of materials for purchase or sale that it has developed through extensive modeling techniques, that are discussed in more detail later. This list is reviewed by the General Services Administration for its potential impact on the market. After the market constraints are added to the list of materials, it is submitted to the various subcommittees.[33]

Strategic Implications Subcommittee. The Strategic Implications Subcommittee, which is chaired by the Department of Defense, looks at how the latest changes in the technology and nature of warfare may affect the materials needs for defense purposes. The Department of Defense is assisted in this analysis by FEMA and the CIA. Abrupt changes in the demand for materials during wartime caused problems during World War II. As Alan Bateman pointed out:[34]

> New uses of radios, radar, and others created demands for steatite talc, quartz crystals, optical calcite, and other minerals that formerly we had thought of mining as adornments of museum collections rather than as pressing industrial materials.

The rapid changes in computer electronics, laser technology, and sophisticated weapons systems makes the danger of stockpile obsolescence greater today than at any time in the past. Thus the Strategic Implications Subcommittee is vitally important to security considerations.

International Economic and Political Impacts Subcommittee. The International Economic and Political Impacts Subcommittee, which is chaired by the Department of State, and includes the Departments of Treasury, Commerce, Interior, Defense, as well as FEMA, has responsibility for determining the impact of the Annual Materials Plan on producing nations, multinational companies, the international flow of materials, and on international agreements. In an interview with State Department officials,[37] it was brought to light that there is sometimes conflict between U.S. policy toward Third World mineral-producing nations in regard to economic development and the barter mechanism for stockpile transactions. The United States has been encouraging many countries to move away from bartering systems within their own economy. The offer to barter on an international scale sends confusing signals to these nations. Such are the factors that the State Department examines.

Market Impact Subcommittee. The Market Impact Subcommittee, which is chaired by the Department of Commerce, evaluates, on a case-by-case basis, the effects that stockpile acquisitions and disposals will have on the international commodity market and on domestic suppliers and consumers. Commerce works with Interior, State, Treasury, GSA, and FEMA. Their main goal is to find specific circumstances where serious market disruptions could occur. The 1979 Act requires that stockpile transactions "avoid undue disruptions of markets."[27] Given this rather general charge of responsibility, it is the work of policy and economic analysts within the Department of Commerce to establish this criterion. Market impact statements are prepared by the Bureau of Industrial Economics; these findings are processed through the Market Impact Subcommittee to the General Services Administration in the form of advice on the timing and methods to achieve optimum market stability.[38]

Economic and Budgetary Impact Subcommittee. The Economic and Budgetary Impact Subcommittee, chaired by FEMA, and including the Departments of Commerce, State, and Treasury, as well as the Office of Management and Budget and the General Services Administration, examines the projections of revenues generated by sales of materials in excess of goal and the costs of purchase of materials to be stockpiled. They also look at how the budgetary process will affect the rate of procurement of material in the near future. The Transaction Fund has made the procurement process more consistent than in the past, thus aiding in the budget planning process.

Finalization of the Plan. After each subcommittee has reviewed the possible materials for sale and purchase and made recommendations, the list is reviewed by the full committee and its member agencies. The director of FEMA then submits the AMP plans with revisions to the National Security Council. An additional copy is provided to the Office of Management and Budget (OMB) for review. If any changes need to be made beyond this point, the National Security Council (NSC), OMB, and FEMA work together to make the revisions.

Longer-term planning periods of five years are also conducted as a function of the AMP process. Such a period runs from 1983 to 1987. The plan provides for disposals of materials authorized by the Budget Reconciliation Act of 1981, and other materials previously authorized. Over the five-year period, about 20 materials will be acquired in varying amounts; their exact identity remains classified until the invitations to bid are announced.[33]

Stockpile Planning Models. As has been discussed, the contents of the stockpile have evolved over many years of acquisitions, changing goals, and selling of excess materials. The methods of analysis have involved everything from simply making lists of materials in short supply during war, to the Annual Materials Plan. The common thread that runs through all the analyses is that first a method is developed to determine which materials are the most strategic and critical, then the requisite quantities are determined. To one degree or another, the components of supply and demand that were mentioned above are incorporated into the planning process. To this end, elaborate models have been established which consider the interaction of these many factors.

The National Security Council submits to FEMA planning assumptions which are the major policy drivers for an elaborate modeling scheme which is used to estimate the dynamic changes that would take place in the U.S. economy during a wartime emergency. The purpose of the model is to establish where the deficiencies would exist between supply and demand for strategic materials during a war. To achieve this, both the demand and supply characteristics must be examined. On the demand side, the product market is forecast for all sectors of the economy, including consumption, investment, government spending, exports, and imports. These components of aggregate demand are classified as Defense, Essential Civilian, or Basic Industry.[38] The demands are disaggregated by National Income Account categories (NIA) by applying historical data in Stipulation Table Ratios. These (NIA) final demand estimates are converted from a product to an industry basis using Defense Interindustry Transformation Tables.[38]

The government has stated that stockpile goals should reflect detailed assumptions regarding changes in a wartime civil economy, wartime political and economic stability of foreign nations, and alternative foreign and domestic production levels for stockpile materials.[33]

Important Policy Parameters. Three important variables to be considered in relationship to the changes noted above are the lead times for bringing on-line additional production from domestic capacity, the degree of substitution from strategic materials to materials that would be in better supply during an emergency, and the level of civilian austerity prevalent during a wartime economy.[39]

1. *Lead times:* Lead time can be defined as the time it takes from the decision to expand capacity until the added capacity is in operation. Increases in

production can be achieved through three approaches:[39]

1. Higher utilization rates for existing facilities. This includes using more employees, working more shifts, bringing old equipment back onstream, and generally stretching the boundaries of what would be normally acceptable production procedures. It is important to note that this is the only method for increasing production that is considered in the formation of the stockpile goals.

2. Capacity expansion at existing sites, which entails the building of new facilities, but at locations which currently have facilities. There are, in many cases, already plans for expansion associated with the production facilities.

3. Building of new facilities where none presently exist. This is the most time consuming of the three methods and probably offers little relief for the stockpile goals, since the time to build totally new facilities usually exceeds the time frame for the stockpile.

2. *Substitution:* The replacement of a material, which is subject to a supply disruption during an emergency with one that is secure is the operational definition of substitution in terms of the stockpile. The following assumptions are made by the government in relation to substitution in the stockpile:[39]

Only substitution that has proven to be technologically and economically feasible with no reliance on new substitution technology

No substitution until the end of the first year

No substitution for high-temperature materials (these materials have a high correlation with military importance)

No substitution from one stockpile material to another

Availability of substitutes based on relative geographic location

3. *Civilian austerity:* Austerity measures can be divided into two basic categories, nonselective and selective. Nonselective austerity concerns personal consumption expenditures (PCE) in major aggregate categories such as consumer durable goods. Selective austerity utilizes detailed components of gross national product (GNP) and the specific relationship between a demanded good or service and strategic and critical materials. An example of selective austerity would be the effect of a reduction in the purchase of new automobiles on aluminum demand.

Stockpile planners have concluded that selective austerity is more effective in reducing final demands for strategic materials since reductions in (PCE) will not always decrease demand for specific materials to the level desired by planners. Selective austerity can be expressed in one of two ways, either as a dollar reduction in expenditures, or as an expected percent reduction in a final demand category. Multipliers are estimated from final demand and material consumption data. Using the multipliers, planners can estimate the demand for each material per dollar change in expenditures for each final demand category.

Based on 1972 data, the largest multiplier exists between investment in the ferrous metals industry and demand for aluminum. A $1 million reduction in ferrous metals production yields a reduction in aluminum requirement of 77.9 tons. In 1972, 116,870 tons of aluminum were consumed as inputs in the ferrous metals industry.[39]

When the policy parameters just discussed are applied to the demand categories of Defense, Essential Civilian, and Basic Industry, together with the additional assumptions regarding private domestic inventory and the amount of imports that will survive hostilities at sea, there can be developed a deficiency or shortfall for each category corresponding to the three years of war. These numbers become the stockpile planning matrix. The total of all nine numbers in the matrix is the stockpile goal.

Stockpile Matrix. The matrix shown in Figure I-4-3 is a hypothetical matrix developed for cobalt. The actual matrices are classified by the U.S. gov-

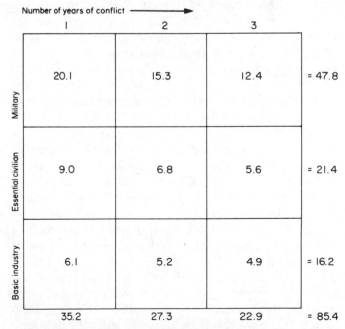

FIGURE I-4-3 Hypothetical cobalt planning matrix showing decreasing priority over time and within demand categories. (From Refs. 21 and 40.)

ernment. The only published information included in the matrix is the overall goal of 85.4 million pounds[1] and the estimate that about 46% of the goal is allocated to the military tier.[40] Given this information, the other amounts were estimated from policy parameters. The discussion that follows can be applied generally to all commodities. Cobalt is used as an illustrative example.

The Federal Emergency Management Agency (FEMA) has developed a three-by-three matrix which prioritizes the contents of the stockpile for each commodity. The rows, or tiers, are defined by demand categories, and are arranged in decreasing priority. The columns are defined by the number of years of conflict for which the stockpile provides. The top row is the military demand category, followed by essential civilian, then basic industrial demand.

It should be pointed out that while basic industrial is the lowest of the three tiers, it does have importance as far as the stockpile is concerned. There is, in essence, a fourth, or phantom tier, which contains those industrial and commercial activities that could be curtailed during a time of national emergency.[38]

1. *Column vectors:* Since the number of years of assumed conflict determines the number of column vectors, it is an important assumption in the goal setting of the stockpile. The current assumption of a three-year war, of course, indicates that three column vectors should appear in the matrix. The figure has been moved in the past to a one-year scenario. The nature of the warfare assumed can, of course, affect not only the duration but also the types of demands put on an economy and its defense-industrial base. Smaller, brushfire wars can last longer than three years, while disrupting the flow of materials to the United States, but having no direct effect on the infrastructure of the United States, nor involving the U.S. militarily. The current policy does not allow for the release of stockpiled commodities during an isolated conflict in which the United States is not directly involved.[26] The wisdom of such a policy will be discussed later.

2. *Military tier:* The top tier in the matrix, the military demand vector, is the highest priority of the stockpile, and the assumptions that go into it have a profound effect on all other aspects of the stockpile as well. As alluded to previously, the changes that occur in an economy during an emergency will have the least effect on military demands with regard to reductions in its needs. In reality, the demands will grow. Military demands are based on performance, making them inelastic to changes in price, and not amenable to substitution.

The experience with cobalt in the late 1970s and early 1980s is a good example of the inelasticity of military demand. During 1976 the total apparent U.S. consumption of cobalt was 9920 tons of metal, of which 12% or 1190.4 tons were used by the aircraft industry.[3] By 1980, apparent consumption had dropped to 8527 tons, of which about 40%, or 3410.8 tons, was accounted for by superalloys for the aircraft industry.[3] Although cobalt proved to be very elastic in response to price in the civilian sector, going from 3273.6 tons used in electrical magnets to 1279.05 tons, during the period mentioned, military demand was obviously less affected. This was the case even though the prices increased about fivefold during the five-year period.[3]

The important points to consider when formulating the military demands for a strategic material are:[37]

The length and magnitude of possible wars

The nature of a potential conflict (e.g., nuclear or conventional and the chances of each type of conflict)

Number of fronts on which a potential conflict will be fought

The size of the military force

The availability of imports

The possibility of internal or foreign supply disruptions

The amount of domestic producer and consumer inventory on hand at the start of the conflict

The amount of military requirements that can be met from domestic production

3. *Essential civilian tier:* Whereas the policy parameters of substitution and austerity have little effect on the military tier, they are very important in determining the goals for the essential civilian tier. A goal of 90% is set for the third year of conflict in terms of the personal consumption index, with the mobilization year serving as the 100% bench mark.[38]

4. *Basic industrial tier:* It has been observed by a government official that the order of the tiers of essential civilian and basic industrial is actually the reverse of what seems to be the case.[41] The essential civilian tier is in reality the basic industry that is needed to support a war effort. The basic industrial tier consists of those industries that are essential to the civilian population even during times of severe austerity.

Quality and Form of the Stockpiled Materials. The stockpile must contain materials appropriate to the current needs of the defense-industrial complex of the United States. These needs are not static, but change over time with new technologies, differing levels of domestic processing capacity, and shifts in final demands.

Assessing these needs is a complex and ongoing issue which has implications for the planning cycle and maintenance of the stockpile. There are two parameters that are important in terms of meeting the needs of the current conditions of the U.S. economy. First is the quality of the commodity held in stockpile. Changes in technology can make the level of impurities in the stockpiled minerals very important, and can even make the stockpile obsolete. A case in point would be cobalt. The level of strontium in the cobalt purchased over 20 years ago is higher than what is now desirable for current cobalt applications.[42] This is an obvious hindrance to the rapid employment of this metal during an emergency since it may well have to go through reprocessing, and, as is shown below, the only processing facility in the United States is in Louisiana, a long distance from the cobalt stockpiles in the Northeast.

The second important aspect is related to the level of processing or form that the stockpiled material takes. The capacity in the United States to process many metals has been deteriorating in recent years.[21] The level of processing in the stockpile should match the capacity to process those materials. For example, the amount of ore in the stockpile for a given metal should not exceed three times the amount that could reasonably be processed in a given year. Otherwise, there will be materials that simply cannot be used within the time frame of the stockpile planning, or that will have to be processed overseas, thus defeating the entire purpose of the stockpile.

Four considerations relate to the choice of form of the stockpile:[21]

1. *Availability:* Such characteristics as end-use flexibility, quickness of access, and technological obsolescence can affect the availability of the material. End-use flexibility favors the storage of ores and less processed materials, while quickness of access favors the more highly processed forms. Technological obsolescence can have an equal effect on either processed or nonprocessed materials and as indicated earlier is a function of quality more than form.
2. *Economic feasibility:* This is not as important as the availability since it is essentially a national security stockpile and not an economically dominated project. However, there are economic considerations, such as the cost at time of purchase, the cost of storage and resultant deterioration, and the administrative costs involved.

3. *Energy requirements:* The more highly processed forms represent additional energy stockpiles since the energy it takes to process them is imbedded in the material itself. Energy costs would increase during an emergency and would therefore represent an added burden.
4. *Environmental effects:* Of low priority during an emergency would be environmental impacts. However, during the storage process, which hopefully, will be the entire existence of the stockpile, certain forms can leach into the surrounding soil. This can be avoided by stockpiling processing metal instead of the less stable ore.

These four characteristics can be combined into an evaluation index such as the one shown for titanium in Table I-4-2.

Each form is evaluated and given a score based on a scale of 0 to 100 for each major heading. This figure is then multiplied by a weighting from 4 to 1, with availability given a 4, economic feasibility given a 3, energy conservation a 2, and environmental concerns a 1. The weighting is based on military and strategic considerations, availability being the most important factor during a period of mobilization. The higher the total score after adding each category, the more desirable the form.[21]

Stockpile Purchases. After the content, quantity, and quality have been determined, it is the task of the General Services Administration to do the actual purchases. It is given a "shopping list" for the fiscal year by the steering committee and the participants in the Annual Materials Plan. There are commodity specialists at the GSA who are responsible for analyzing the market and finding the best buy. Because the quantities are relatively large, the government can make specific purchase contracts rather than buying on the open market.

In the case of cobalt purchases in the summer of 1981, the GSA first asked for sealed bids; it received 17.[43] The contract price, offered by Sozacom of Zaire, was $15 per pound, about $5 below the going price at the time. The total cost of the acquisition was $78 million for 5.2 million pounds of contained cobalt.[43]

Barter Transactions. In the 1981–1982 stockpile report, the first agricultural barter since 1967 is mentioned. Surplus dairy products from the Commodity Credit Corporation were exchanged for metal-grade bauxite from Jamaica. Figure I-4-4 outlines the procedure for such a transaction, which also included direct cash transfer as well as sale of excess materials on behalf of the Jamaican govern-

TABLE I-4-2   Quality Index: Titanium

	Rutile	Sponge	Ingot	Mill products	Scrap
National security/availability					
End-use flexibility	20	20	5	2	5
Technological obsolescence	20	15	5	2	5
Processing facilities	0	5	10	20	0
Timing	5	5	18	20	12
Number of forms	20	15	5	1	5
Total × 4	260	240	172	180	108
Energy conservation: energy content	5	80	90	100	80
Total × 2	10	160	180	200	160
Environmental effects	—	—	—	—	—
Economic feasibility					
Physical deterioration	18	20	20	20	20
Disposal programs	20	20	10	5	10
Initial acquisition cost	20	10	8	1	7
Storage cost	10	5	20	10	15
Number of forms	20	20	10	1	5
Total × 3	264	225	204	111	171
Total index	534	625	556	491	439

Source: Ref. 21.

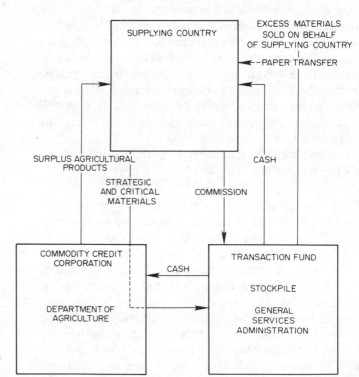

FIGURE I-4-4 Schema for barter transactions. (From Ref. 1.)

ment.[1] Because the Commodity Credit Corporation must, by law, protect its assets, it must be reimbursed in full for the surplus products. Because of the need to reimburse the CCC, the barter system does not represent any increase in the assets available to the stockpile.

Developing stockpile strategy has become a sophisticated process involving computer modeling and complicated assumptions regarding changes in the supply and demand of materials during a military conflict. Given this level of sophistication, it is ironic that the principal scenario is based on a conventional war of duration, along the lines of World Wars I and II. If future strategy is to conform to the realities of modern warfare and the dynamics of geopolitical developments in the world, the central assumptions on which the policy is based should be examined in more detail, and revised from time to time.

Condition of the Current Stockpile

The changes that resulted from the 1979 Act were important in shaping the legal environment of stockpile policy. The protection of assets, and the stabilization of goals, have set important parameters for the conduct of stockpile planning. However, the de facto characteristics must also be considered. Stockpile planners must consider factors pertaining to market conditions and relations with foreign governments. These factors can have a limiting effect on the range of policy options that are available, as much as any legal limitation. Additionally, the condition of the stockpile, as it stands in recent years, becomes a vital part of the strategy of future policy. To understand where the weaknesses and strengths of the U.S. Defense Stockpile lie, this section analyzes the goals and inventories of specific commodities, the location of the important commodities throughout the United States, and the stockpile depots that contain the most strategically important commodities.

Quantitative Characteristics. There are 61 commodity families represented in the National Defense Stockpile, composed of 93 specific commodities. For example, manganese is divided into two families: manganese dioxide, battery-grade group; and manganese, chemical and metallurgical group. The battery group is further broken into two specific commodities, natural ore and synthetic dioxide. The chemical and metallurgical group is broken into seven specific commodities, including both chemical- and metallurgical-grade manganese ore; high, low, and medium carbon ferromanganese as well as silicon ferromanganese; and manganese metal, electrolytic.[1]

Stockpile Goals versus Current Inventory. Information on stockpile goals and inventory can be misleading when the data are aggregated to the family level. A case in point would be the chemical

and metallurgical chromium group, which is about 98% of goal when taken as a family group. High-carbon ferrochromium is 218% of goal, and low carbon is 425%, while ferrochromium silicon is 65% of goal and chromium metal is slightly less than 19% of goal.

Twenty-four groups and individual materials are equal to or greater than the goals for their respective categories. Thirty-seven groups and individual materials are less than the goal. The value of the materials in excess of the goal is $3.8 billion in 1982 dollars at the 1982 market value for the commodities. At the same time, the value of the materials that are needed to reach goal for all categories is $10.3 billion. Theoretically, the dollar value of the excess can be subtracted from the amount needed to reach the goal, leaving $6.5 billion as the amount needed to reach the goal. However, there is often a difference between the figure on paper and the amount that can be practically disposed of at any given time.

When discussing goals and inventory, it must be understood that the concept of crediting offset is used to equalize the goals within the commodity families. That is, the materials in excess in one category may be held as a credit to offset the deficit of materials in another category. For example, 217,695 tons of high-carbon ferrochromium is held against a shortfall of 544,238 dry tons of specification-grade ore.

Care must be taken when the crediting offset runs in the other direction, that is, when an ore is held in credit for a more highly processed material. For example, where 56,830 short tons of non-specification-grade chromite ore is held against a shortfall of 16,237 tons of chromium metal, there must be both the domestic capacity to process the ore that is held for the goal, and capacity to process the ore that is held as a credit offset.

Stockpile Excesses. Although there is $3.8 billion worth of materials in excess of goal, this excess is concentrated heavily in two commodities: tin and silver. As can be seen graphically in Figure I-4-5, these commodities dominate the excess value, with tin accounting for $2.120 billion and silver comprising $1.002 billion, for a total of $3.122 billion. Of the remaining amount, over half is held in tungsten and industrial diamonds.

The concentration of the stockpile excess is very important in terms of the potential market impact caused by the sale of the stockpiled materials. It is intuitively obvious that if one is to minimize market impacts from the sale of a certain dollar value of commodities, then the more varied

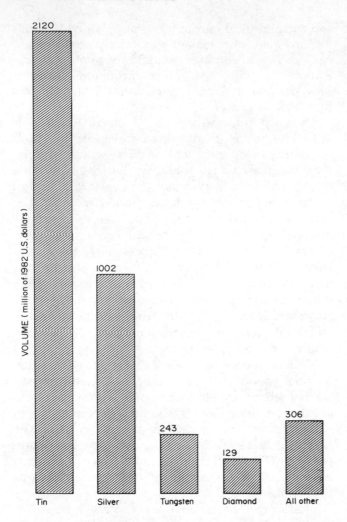

FIGURE I-4-5 Stockpile excesses in millions of 1982 U.S. dollars. (From Ref. 1.)

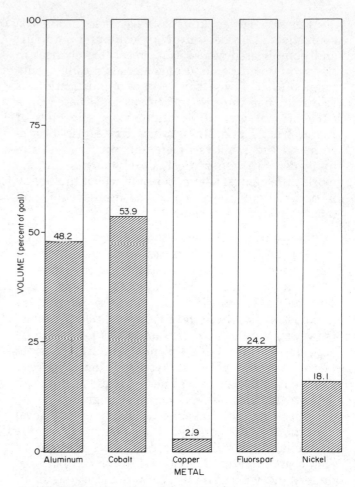

FIGURE I-4-6 Stockpile deficiencies for aluminum, cobalt, copper, fluorspar, and nickel—as a percent of goal, 1983. (From Ref. 1.)

the commodities sold, the less the market impact on one particular commodity market. But since the number of commodities that account for the majority of the stockpile excess is limited, the amount that can be sold off without causing a large impact on the market is relatively small.

Stockpile Deficiencies. Thirty-seven groups and individual materials are held in quantity less than the specified goal. Of these, 6 are between 80 and 90% of goal, 3 are from 60 to 79% of goal, 10 are from 40 to 59% of goal, and 12 are less than 20% of goal.[1]

Unfortunately, while the excesses occur in the less important minerals, the deficiencies are predominately in the more important ones. For the entire aluminum metal group there is an inventory of 48% of goal, or 3,705,936 tons of contained metal. Cobalt stands at 46,002,305 pounds or 53% of goal (see Figures I-4-6 and I-4-7).

A very important, and possibly dangerous

deficiency is in metallurgical-grade fluorspar, which is only 24% of goal, at 411,738 dry tons. Acid-grade fluorspar is, fortunately, in a better position, at 64% of goal. One must bear in mind that acid grade is used in the aluminum industry, while metallurgical is used in the steel industry. Acid grade can be substituted for metallurgical, but not vice versa.

The assumption relating to a small fluorspar stockpile is probably that Mexico has historically been the leading supplier of fluorspar to the United States. However, as Figure I-4-8 shows, South Africa had become an increasing supplier of fluorspar to the United States. This trend is likely to continue, as South Africa has larger reserves of fluorspar and has been actively seeking to develop this trade pattern.

Locations of Stockpiles by Commodity. The stockpile locations for 10 key commodities provide a good illustration of the concentration of the stockpile value, within certain commodities. Addi-

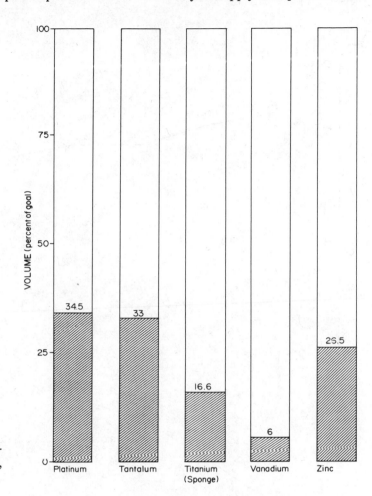

FIGURE I-4-7 Stockpile deficiencies for platinum, tantalum, titanium, vanadium, and zinc—as a percent of goal, 1983. (From Ref. 1.)

tionally, the stockpile is concentrated by depot locations. Nine of these are the commodities held in the highest dollar value. The tenth, platinum, is included because of its strategic importance; lead is not included, as it is low in importance. As Table I-4-3 indicates, tin constitutes 25% of the total value of the stockpile. The next three commodities account for the next 25%. They are chromium, the chemical and metal group; silver; and tungsten. Six commodities and commodity groups make up about the third 25% interval: the aluminum metal group; cobalt; industrial diamonds; manganese, chemical and metal group; titanium sponge; and lead. This means that over 75% of the worth of the stockpile is concentrated in the top 10 commodities and groups.

Tin. Tin is distributed throughout 15 stockpile depots (see Figure I-4-9). However, the two largest depots contain almost 20% of the tin each, and the top three combined contain over 50% of the total amount. The top eight depots contain over 82%, with the remainder in the last seven locations.[44]

The importance of tin as far as the stockpile is

currently concerned is in its value as a generator of revenues. Since tin has a relatively high unit value, the locations are not extremely important with regard to market access, which for most tin will mean direct sales. The largest tin stockpiles are also the locations for many other important metals.

Chromium. Six different commodity types are included in the chemical and metallurgical chromium group. The largest dollar value is contained in the low-carbon ferrochromium category, with $418.1 million of the $1063 million held in total. The next largest is high-carbon ferrochromium at $294.5 million, followed by $260.8 million worth of metallurgical-grade chromite ore. The other categories combined are only valued at $89.6 million.

The largest chromite depot is located in Nye, Montana, far away from the closest processing facility, as shown in Figure I-4-10. The only explanation for this is the depot's proximity to the Stillwater Complex in Montana. The government could have anticipated that the complex would become a chromite producer in quantity large enough to merit processing facilities in the region.

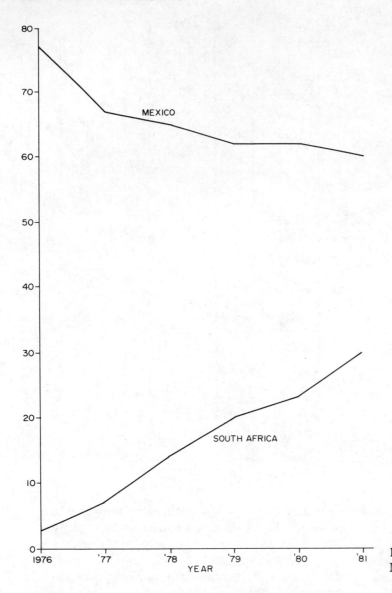

FIGURE I-4-8 Fluorspar imports as a percent of total—Mexico and South Africa. (From Refs. 2 and 3.)

TABLE I-4-3 Largest Commodity Holdings by Value

Commodity	Million Dollars	Percent	Cumulative
1. Tin	2715.9	25.0	25.0
2. Chromium	1063.0	9.7	34.7
3. Silver	1001.7	9.1	43.8
4. Tungsten	665.1	6.0	49.8
5. Aluminum (metal group)	629.8	5.7	55.5
6. Cobalt	588.0	5.3	60.8
7. Diamonds	469.6	4.3	65.1
8. Manganese (chemical and metallurgical)	466.7	4.2	69.3
9. Titanium sponge	458.8	4.1	73.4
10. Lead	342.6	3.1	76.5
11. Zinc	293.2	2.7	79.2
12. Platinum	215.0	2.0	81.2
13. Nickel	211.9	1.9	83.1
14. Tantalum	165.7	1.5	84.6
15. Antimony	162.9	1.5	86.1

Source: Ref. 1.

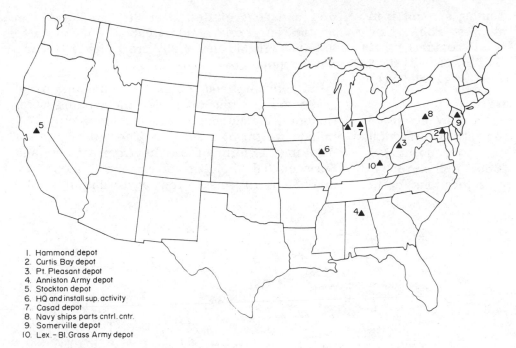

1. Hammond depot
2. Curtis Bay depot
3. Pt. Pleasant depot
4. Anniston Army depot
5. Stockton depot
6. HQ and install sup. activity
7. Casad depot
8. Navy ships parts cntrl. cntr.
9. Somerville depot
10. Lex. – Bl. Grass Army depot

FIGURE I-4-9 Ten largest tin stockpile locations ranked in order. (From Ref. 44.)

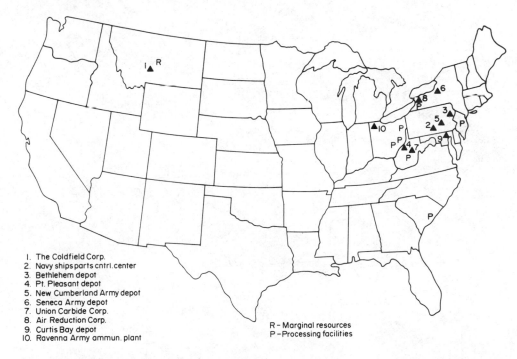

1. The Coldfield Corp.
2. Navy ships parts cntrl. center
3. Bethlehem depot
4. Pt. Pleasant depot
5. New Cumberland Army depot
6. Seneca Army depot
7. Union Carbide Corp.
8. Air Reduction Corp.
9. Curtis Bay depot
10. Ravenna Army ammun. plant

R – Marginal resources
P – Processing facilities

FIGURE I-4-10 Ten largest chromite ore locations by ranking. (From Ref. 44.)

This stockpile would provide high-grade ore material that could be mixed, in times of emergency, with the lower-grade chromite from the Stillwater Complex. Unfortunately, the Stillwater Complex has not proven economical at prices that have been prevalent in recent years, leaving the stockpile far from processing facilities. This would mean an additional transportation cost if the ore were to be processed.

Another example of how things have changed since the original stockpile planning took place is the amount of low-carbon ferrochromium in the stockpile. When the original planning was done,

low-carbon ferrochromium was needed to make high-quality metal products. Since the advent of the argon-oxygen decarburization (AOD) process, this is no longer a requirement. Since the low-carbon ferrochromium costs more, it has not been the preferred type since the AOD process was discovered.

Silver. Because silver is a high-unit-value commodity, it must be stored in secure locations. It is stored in only two locations. The largest silver depot (about 65%) is located in the San Francisco Assay Office. Silver is the only stockpiled material at the Assay Office. The remaining amount is in the U.S. silver depository at West Point, New York.[44] The West Point location contains other high-unit-value materials and is the second largest depot in terms of market value (see Figure I-4-11).

Tungsten. Tungsten is stored in four forms: ores and concentrates, carbide powder, metal powder, and ferrotungsten. The vast majority of it is stored in the form of ores and concentrates, some $538.7 million of the $665.1 million total in the tungsten group. There are 18 locations for the

ores and concentrates. Over 30% are located in Scotia, New York. A total of about 55% is located in New York and almost 90% are located in the top seven depots[44] (see Figure I-4-12).

Aluminum Metal Group. The aluminum metal group contains four categories: the alumina sector is empty, the aluminum section contains only $2.7 million worth, or 1733 tons. The vast majority is held in the form of bauxite from Jamaica and Surinam. The holdings of Jamaican bauxite are 10,058,881 dry metric tons at a value of $399.2

1. U.S. Silver depository
2. Assay office
3. U.S. Bullion depository
4. Federal Center
5. Citibank
6. Dept. of Army
7. Geologic Survey

FIGURE I-4-11 Precious metal stockpile depot locations. (From Refs. 1 and 44.)

1. Scotia depot
2. Griffiss AF Base
3. Lex.-Bl. Grass Army depot
4. Binghampton depot
5. Defense constr. supply cntr.
6. Navy ships parts cntrl. cntr.
7. Somerville depot
8. Federal depot
9. Casad depot
10. Mang. and Chrome depot

FIGURE I-4-12 Ten largest tunsten ore and concentrate locations by ranking. (From Ref. 44.)

million. The Surinam bauxite holdings are 5,299,597 dry metric tons, valued at $263.9 million.[44] The Jamaican bauxite is kept at four locations. They are all company sites. The largest is the Reynolds location in Gregory, Texas, with 5,630,054 dry metric tons. Kaiser has two locations in Louisiana, totaling 5,572,792 dry metric tons, and ALCOA has one location at Point Comfort, Texas, with

313,806 dry metric tons. Surinam bauxite is more dispersed in its locations, with 11 total depots. Most of these are government sites, as shown in Figure I-4-13.

Cobalt. As can be seen from Figure I-4-14, there is a fairly high correlation between the cobalt depots and the states where most of the cobalt is

JAMAICAN

J1. Reynolds Metals Co
J2. Kaiser Aluminum Co
J3. Kaiser Aluminum Co
J4. Alcoa

SURINAM

S1. GSA Storage site
S2. Naval Constr. Battal. Center
S3. GSA Depot
S4. Reynolds Metals Co
S5. Red River Army Depot
S6. Anniston Army Depot

FIGURE I-4-13 Bauxite stockpile depot locations. (From Ref. 44.)

1. Casad depot
2. Warren depot
3. Somerville depot
4. Ravenna Army amun. plant
5. Scotia depot
6. Binghampton depot
7. Pt. Pleasant depot
8. Navy ship parts cntrl. center
9. Naval constr. battal. center
10. Curtis Bay depot
11. Hammond depot
12. Seneca Army depot

A,B,C,D,E – State rank

R – Marginal resources
P – Processing facilities

FIGURE I-4-14 Cobalt stockpile depot locations. (From Ref. 44.)

industrially consumed. The largest depot, in New Haven, Indiana, contains 7906 tons of cobalt or 34% of the total cobalt stock. The second largest, in Warren, Ohio, contains 3728 tons or about 16% of the total.[44] Although neither state is the largest industrial consumer of cobalt, both have substantial consumption. The third largest depot, Somerville, New Jersey, containing 3356 tons, is in the largest consuming state.[3]

One must also remember that during a supply interruption, the price of cobalt would rise to a level that would make transportation by air economical. A problem might arise if the cobalt needs further processing, since the only processing facility in the United States is located in Louisiana.[3] By the same token, if support is given to the domestic cobalt mining industry in such places as the Blackbird Mining District of Lemhi County, Idaho,[2] there would be a need to increase the domestic processing capacity to facilitate the cobalt production. This has important implications for support of the cobalt industry in the United States, since 66 to 75% of the cost of cobalt is accounted for by the refining process.[40]

Industrial Diamonds. The industrial diamond group, like other high-unit-value items, is stored in secure locations. There are three commodity categories within the group. Most of the group is in the form of industrial stones, $421.1 million, with crushing bort comprising $47.4 million and small dies only $1.1 million.[44]

The diamond dies are only at the U.S. Silver Depository in West Point, whereas the stones and bort are both stored at the U.S. Silver Depository,

Citibank in New York, and the U.S. Bullion Depository at Fort Knox, Kentucky.[44] Diamond locations are shown in Figure I-4-11.

Manganese. Manganese is divided into two groups: the manganese dioxide battery group and the chemical and metallurgical group. The group that is of strategic concern is the chemical and metallurgical group. The chemical and metallurgical group is broken into seven specific commodities. These are dominated by two commodities, high-carbon ferromanganese ($262.5 million) and metallurgical-grade ore ($186.1 million). The remainder has a combined value of $73 million.[44]

The high-carbon ferromanganese is limited to only seven locations (Fig. I-4-15), with 47.4% located in one depot at Curtis Bay, Maryland (284,657 tons). The metallurgical ore is distributed throughout 34 stockpile locations. However, the 10 largest account for almost 78% of the total holdings. As Figure I-4-16 illustrates, the two largest are in Pennsylvania.[44]

Titanium Sponge. The stockpile contains $458.8 million worth of titanium sponge. As Figure I-4-17 indicates, there are three locations that dominate: Warren, Ohio, with 7233 tons, or 22%, Clearfield, Utah, with 6953 tons, 21.5%; and Curtis Bay, Maryland, 6617 tons, and 20.5%. The remaining 36% is in eight other depots.[44]

Platinum-Group Metals. Three of the six platinum-group metals are stockpiled by the U.S. government. The three include iridium, held in a quantity of 16,990 troy ounces; palladium, which

FIGURE I-4-15 High-carbon ferromanganese stockpile locations by rank. (From Ref. 44.)

1. Curtis Bay depot
2. Seneca Army depot
3. Casad depot
4. Belle Mead depot
5. Warren depot
6. Federal depot
7. L. and N. Railroad depot

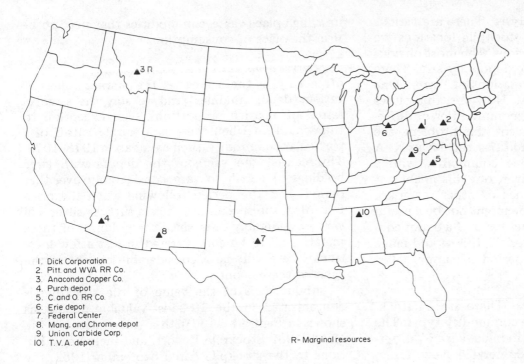

1. Dick Corporation
2. Pitt. and WVA RR Co.
3. Anaconda Copper Co.
4. Purch depot
5. C. and O. RR Co.
6. Erie depot
7. Federal Center
8. Mang. and Chrome depot
9. Union Carbide Corp.
10. T.V.A. depot

R- Marginal resources

FIGURE I-4-16 Ten largest metallurgical manganese ore locations by rank. (From Ref. 44.)

1. Warren depot
2. Federal depot
3. Curtis Bay depot
4. Somerville depot
5. Ravenna Army ammun. plant
6. Sharonville depot
7. Pt. Pleasant depot
8. Gadsden depot
9. Stockton depot
10. Casad depot

FIGURE I-4-17 Ten largest titanium sponge locations by rank. (From Ref. 44.)

amounts to 1,255,003 troy ounces in stockpiled weight; and platinum, which is held in a quantity of 452,642 troy ounces.[44] Like other high-unit-value materials, the importance of storage is in the security of the location rather than the proximity of the depot to market demand. For this reason, the platinum-group metals are stored in relatively few locations throughout the United States, but these locations are well guarded (refer to Figure I-4-11).

The largest stockpile for the platinum-group metals, for all three combined, is the U.S. Silver Depository at West Point, New York. Interestingly enough, as mentioned earlier, it is not the largest silver depot, containing 35%. West Point contains 302,054 troy ounces of palladium, again about 67%, and 3154 troy ounces of iridium, almost 19%.[44]

Fort Knox and the Federal Center in Denver contain about equal portions of each metal. The two locations in Virginia contain rather insignificant amounts of platinum-group metals.

Methods of Depot Analysis. There are various ways to determine which stockpile locations are the most strategically important. The most obvious is to compare the stockpile depots by size. There are several different interpretations of "size." One can look at the total weight. However, this causes a problem when comparing commodities of different unit value. To circumvent this problem, the market value of the commodities can be used. A stockpile might contain one commodity with a high unit value, such as silver, but this might not have the strategic importance of other materials, and as such the market value might not be a useful overall measure. On the other hand, if a depot contains a wide variety of materials, this would make it more strategic during a period of supply interruption.

Variety of Commodities. There are 14 stockpile depots with 15 or more commodity types. The most varied is New Haven, Indiana, with 57 different types of commodities. Somerville, New Jersey, contains 50 different commodities; Warren, Ohio, contains 48; and Curtis Bay, Maryland, has 32 different types.[44]

What is obvious is that relatively few stockpiles contain a varied amount of materials. Most contain from three to seven different commodities, and many only one or two. The commodity-specific stockpiles tend to be company sites where only one or two commodities that are important to the company are stored. This is especially true of low-unit/high-place-value commodities that need to be near the place of consumption.

Current Market Value. Since commodity prices often vary in the aggregate, the market values of various depot holdings tend to stay the same in terms of relative proportions. The exception to this would be when there is a supply cutoff of a particular commodity, such as cobalt in 1978–1979. This caused the value of the depots with large holdings of cobalt to vary considerably over that period, and the period following when the price fell. Also, the speculative metals such as silver will vary considerably, and since the holdings of these metals tend to be concentrated in just a few locations, there will be a considerable change in the value of these depots.

About 62% of the value of the stockpile is concentrated in the 10 most valuable depots, as shown in Figure I-4-18. With a value based on the March 1982 Stockpile Report, and including additions to the stockpile dated September 1982, the most valuable stockpile depot is Curtis Bay, Maryland, at about $1020 million. Close behind is the U.S. Silver Depository at West Point, New York, at about $1000 million. These depots contain 9.3 and 9.1% of the stockpile value, respectively. The top two, together with the third in value ranking, New Haven, Indiana ($700 million, 6.4%), contain a total of 25% of the worth of the stockpile. The other seven depots of highest value contain 37% of the total value. This brings out the reality that

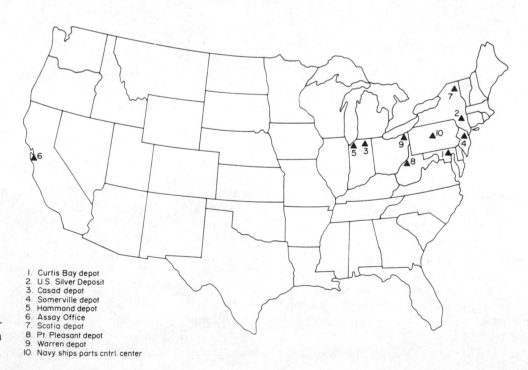

1. Curtis Bay depot
2. U.S. Silver Deposit
3. Casad depot
4. Somerville depot
5. Hammond depot
6. Assay Office
7. Scotia depot
8. Pt. Pleasant depot
9. Warren depot
10. Navy ships parts cntrl. center

FIGURE I-4-18 Stockpile locations by market value. (From Refs. 1 and 44.)

although there are technically 122 stockpile locations, many are relatively unimportant.

The Most Important Depots. There is no simple way to look at a stockpile location in terms of its importance, but there are many factors to consider. When one combines the factors, Curtis Bay, Maryland, emerges as the most important single stockpile location. Not only does it have the highest relative dollar value, it also has a diverse selection of commodities, and contains some of the larger amounts of the most important materials.

Other important depots are New Haven, Indiana; West Point, New York; Somerville, New Jersey; Hammond, Indiana; and Scotia, New York. The most strategic stockpiles merit a closer examination of the specific facilities, the proximity of the depots to transportation systems, and other important considerations.

Curtis Bay. The Curtis Bay depot is located in Baltimore Harbor, just outside the Baltimore city line (Anne Arundel County, Maryland). It is bordered on the northeast by Curtis Creek and on the southeast by Furnace Creek. Governor Ritchie Highway runs to the west of the depot, on a north-south axis.[45]

As shown in Figure I-4-19 the individual depots are serviced by a web of rails and access roads. Since it is located in Baltimore Harbor, the depot has excellent access to shipping channels, and the area is serviced by numerous railroads.[45]

U.S. Silver Depository. The U.S. Silver Depository is located on the U.S. Military Academy, adjacent to the Hudson River in Orange County, West Point, New York. The depository is on the western border of the campus, about 700 feet from Storm King Highway (New York Route 218). Since the contents of the depository are high in unit value, access to ready transportation is less important than for the bulk items. However, if needed, there is a Conrail track which runs onto the campus, there are docking facilities on the Hudson River, and U.S. Route 9W runs close to the campus. A comparison of Figure I-4-20 with the Curtis Bay depot shows the marked difference between the facilities needed for a bulk depot and a high-unit-value depot.[46]

Casad Depot. The Casad depot is located on a U.S. military reservation, the Casad Ordnance Depot. The depot is on Indiana Route 14, 3 miles due east of New Haven, Indiana, a suburb of Fort Wayne, Indiana. The depot is serviced by the Norfolk and Western Railroad. Maps indicate that the station is in New Haven, and that there is not a

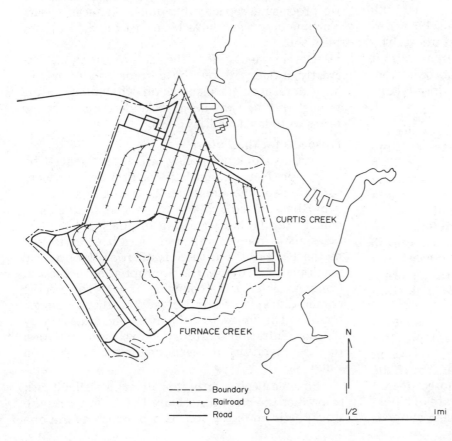

CURTIS CREEK

FURNACE CREEK

N

- - - - - Boundary
+—+—+ Railroad
——— Road

0   1/2   1 mi

FIGURE I-4-19 General Services Administration Curtis Bay, Maryland, depot. (From Ref. 45.)

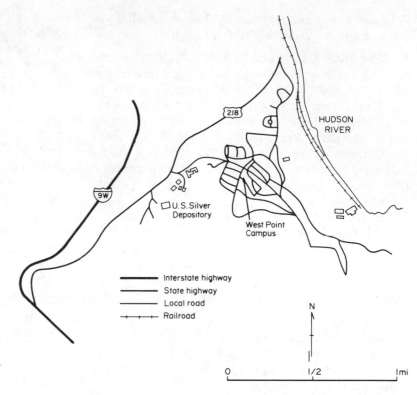

Interstate highway
State highway
Local road
Railroad

N

0   1/2   1mi

FIGURE I-4-20 United States silver depository, West Point, New York. (From Ref. 46.)

stop at the depot. The depot is bounded on four sides by small access roads.[47]

Since the depot is on a military installation, the exact configuration is unavailable for publication. The configuration is probably similar to other bulk locations, with the individual depot sites being serviced by a network of access roads and tracks. The entire facility is 2 miles from east to west and about ½ mile north to south. The depot is approximately 120 miles from Lake Michigan, the closest large body of water.

Somerville Depot. The Somerville, New Jersey, depot is listed as the Veterans' Administration Supply Depot on U.S. Geological Survey maps. The facility is serviced by Conrail. The depot is located about 2 miles south of Somerville, and less than a mile west of Manville, New Jersey. U.S. Route 206 runs through the eastern corner, north to Somerville and south to Belle Mead. The individual storage sites are serviced by the network of railway and access roads that is indicative of the bulk storage facilities. The facility is about 1½ miles east to west, and ½ to ¾ mile north to south.[48]

Some problems emerge when the stockpile is examined in detail. The most important problem is the high concentration of stockpile excesses in two commodities, tin and silver. This limits the amount of materials that can be disposed of in a given year, and as a consequence, the rate of stockpile transactions. Additionally, important minerals

are in deficit, such as aluminum, cobalt, fluorspar, nickel, platinum, tantalum, titanium, and vanadium. Other materials, such as manganese and chromium, need to be upgraded in form, as domestic processing capacity has fallen in recent years. Fortunately, plans have been instituted to accomplish this.

Developments for the future stockpile are greatly influenced by the current conditions. To have a realistic perspective on what those future trends will be, we have to understand them in terms of today's realities.

Prospects for the Future

Future stockpile policy will be influenced by two basic factors: legislative action and de facto conditions of the stockpile. To a degree, the problems that the stockpile experienced in the past, namely the periodic shifting of goals and the loss of assets to the general fund, have been addressed by the 1979 Act. But there is still room for political manipulation through the appropriation process in Congress, and by the changing of goals for specific commodities, for political rather than strategic reasons. The de facto influences are indicated by the concentration of stockpile excesses in relation to the generation of revenues for the Transaction Fund.

Possible Legislation. It is always a difficult task to predict the direction of any U.S. policy, and the stockpile is no exception. If a bill such as the one proposed by Representative Bennett on January 3, 1983, becomes law, there could be a shift in the administration of the stockpile.[49] Bennett's bill would move all the stockpile functions, which are now dispersed through various agencies and depart-ments, to the Department of Defense. In addition, personnel would be transferred from various agen-cies to the Department of Defense. Such a move would be intended to make policy less cumber-some, and hopefully, to make strategic considera-tions the top priority. However, this proposal has been opposed by representatives from the Depart-ment of Defense, who have indicated that a poten-tial internal conflict could exist. During an emergency, the Department of Defense would be both a consumer and an allocator of materials.[50] The American Mining Congress has expressed sup-port for an independent government organization.[51] Such an organization would have its own balance sheet, apart from the federal budget. In this case, the assets of the Transaction Fund would not be considered in congressional appropriations. This would free the Fund from the political manipula-tion that comes with being included in the budget. However, Congress would then lose control over the stockpile. Both liberals and conservatives have reasons for wanting to maintain that control. There is still the problem of obtaining funding for the stockpile under any legislation.

De Facto Stockpile Influences. The long-run results of stockpile policy must always be con-sidered. If, for example, the United States were to sell large amounts of tin to finance the purchase of other, more strategically important materials, this would have a depressing influence on the price of tin. Tin sales are invariably opposed by the produc-ing countries. Such a drop in the price of tin could have a destabilizing effect on countries such as Bolivia, Indonesia, Malaysia, and Thailand, which might provoke aggression from both internal and external forces. Such considerations will limit the sale of tin, and thereby limit the amount of revenue that can be generated for the Stockpile Transaction Fund. Additionally, the government would not want to sell such an amount of tin as would depress the price, for the strategy is to maximize the reve-nues from tin sales.

The other commodity that is in large excess value is silver. There has been an ongoing contro-versy regarding sales of silver from the stockpile. As would be expected, the issue pits producers against consumers. In a 1982 report by the Comp-troller General of the United States,[52] it was pointed out that since the authorization of disposals, based

on 1978 data, the defense-related needs for silver have increased, while the domestic production of silver has decreased. The General Accounting Office also investigated the possibility of minting coins instead of direct silver sales. This would create a new demand (in collectors' coins) for the silver and disrupt the market less than direct sales.[52] Since silver and tin make up such a large proportion of the commodity excess, the limitations on their disposal have important implications for future stockpile funding. If the stockpile purchases are to increase to a level that would make it possible to reach the three-year goal within a reasonable period of time, there must be funding from outside the Transaction Fund.

Relations with Allies. As mentioned above, U.S. allies possess much less in the way of stock-piled materials than does the United States. Al-though there are no formal treaties for the sharing of stockpiled materials, there is certainly a strategic importance created by these conditions. One of the inhibitions to the European stockpile is proximity to the Communist bloc. If Western Europe were invaded, the materials could quickly fall into the hands of the Warsaw Pact nations. A possible al-ternative would be the formation of a NATO stock-pile in the United States.

Alternatives to Stockpiling

Defense Production Act of 1950. Use of the Defense Production Act to encourage domestic production of minerals that the United States must import is one of the feasible alternatives to stockpil-ing in the future. The most probable mineral indus-try to receive support from the Defense Production Act is cobalt. Although the United States has no reserves of cobalt at 1983 prices, there is a reserve base of 350,000 tons of contained cobalt from ore grades of 0.01 to 0.08%.[3]

Since 1971, all the production in the United States has come from secondary sources; in 1982 there were 500 tons of nonprimary cobalt produc-tion. A report by the U.S. government proposes production under Defense Production Act support from four onshore locations and one offshore site of deep-sea nodules. (See the section below on sea mining for details.[40])

The onshore locations would provide cobalt production of 2 million pounds annually from Madison, Missouri, where the ore grade is 0.27%. Four million pounds of cobalt would be produced from 0.6% ore in the Blackbird, Idaho, mining dis-trict. (The Blackbird region contains the highest ore grade of U.S. deposits, and would become re-serves at a lower price than other domestic deposits.)

An additional 2 to 4 million pounds would be mined yearly from Gasquet Mountain, California; the ore grade at Gasquet Mountain was not determined as of the 1980 study. Finally, 1 to 2 million pounds of cobalt would be produced under the plan from the 0.01% ore at the Duluth Gabbro in Minnesota. (It should be noted that mineralization is not technically an ore body until it can be economically produced.)

The onshore sites would produce 9 to 12 million pounds of cobalt. The average apparent consumption of cobalt between 1978 and 1982 was about 16 million pounds of contained cobalt, annually. The project, which is dated November 26, 1980, presents a floor price of $15 per pound to support domestic production. The price of cobalt by March 1983 had dropped to under $5 per pound. Such price fluctuations illustrate the problem in price support schemes such as that mentioned above. If the price remains low after the initial contract period, the government must continue the subsidy, or the mine will close and the domestic capacity will be lost. In addition, domestic processing capacity must be added so that the cobalt does not have to be sent overseas for processing. Sixty-six to seventy-five percent of the cost of cobalt is accounted for by refining.[40] Congressional reports have speculated that the low-cost cobalt producers, such as Zaire, will attempt to keep the world price below the level that would make U.S. production economic.[53]

Sea Mining. The future prospect of sea mining could have an important impact on U.S. mineral dependency. The Cobalt Project proposes production of 38 million pounds of cobalt from manganese nodules with a grade of 25% manganese, 1.25% nickel, 1% copper, and 0.25% cobalt. The project would begin sometime in the 1990s and require an investment of $1500 million ($1980).[40]

The United States declined to sign the Law of the Sea Treaty on December 10, 1982. On March 10, 1983, President Reagan proclaimed U.S. sovereignty over an Exclusive Economic Zone extending 200 nautical miles. This zone contains manganese nodules and sulfide deposits, as well as oil and gas.[50] These moves indicate that the United States is moving toward a mining program on its own accord. The basic question will be whether this can be accomplished with private funds, or whether it will require government subsidy under the Defense Production Act.

The problem with sea mining is that although it would increase domestic capacity during peacetime, it would represent an easy target during wartime.

Sea mining would be effective during a supply cut-off caused by a regional conflict that was not in the vicinity of the mining operation. However, if the conflict spread, the mining operation could not be considered secure.

Conclusions

Prospects for the stockpile in the future depend not only on laws that direct stockpile policy, but also on legislation relating to stockpile alternatives. The biggest question concerning the stockpile relates to funding. The funding pattern for fiscal 1983 and 1984 has indicated that the annual amount appropriated for stockpile purchases is $120 million. The amount needed to reach goal is $10,300 million,[1] about 86 times that funding level. Clearly, if the goals are to be taken seriously at all, future appropriations must increase.

In summary, certain points merit emphasis. The concepts "strategic" and "critical" are dynamic and change with technological, geological, and political reality. Among nations, the list of strategic materials varies significantly. Japan stockpiles molybdenum, which is produced in large quantities by the United States. Also, a private corporation will have a different set of criteria with which to determine material vulnerability than will a public organization. Factors relating to cartel formation, the structure of industry, and trends in mineral pricing will be considered by the private organization to a greater degree than by governments. Governments will be concerned primarily with strategic (defense-related) developments. To the degree that strategic and economic considerations overlap, there is a gray area as to where valid strategic policy ends and undue disruption of the marketplace begins.

Specific events have shaped not only stockpile policy, but the actual contents of the stockpile. For example, the large amounts of tin and tungsten in the stockpile resulted from purchases made during the Korean War, when there was the threat of supply interruptions from the Far East. At the same time, current war scenarios are based more on conventional war experiences that took place during World Wars I and II than on the more possible brushfire wars that have exemplified warfare during the past three decades.

The 1946 Act contained significant flaws relating to goal setting and the maintenance of stockpile assets. By allowing the changing of goals, the stockpile became a political football that was used by one administration to support the mineral industry, and by another, by lowering goals and selling material, to obtain revenues for the general fund (since assets were not protected by the 1946 Act).

The direction of future stockpile activities are greatly influenced and even inhibited by the current stockpile conditions. Specifically, the large concentration of stockpile excesses in two commodities, tin and silver, make the options for disposals rather limited. This should be examined closely in the light of future funding.

Recommendations for stockpile policy should incorporate the foregoing observations into a workable stockpile plan. Such a plan should include the following:

A more technical definition of "strategic" and "critical" should be developed which links the term "critical" to demand factors within the economy, and bases "strategic" on the availability of critical material supply.

Careful examination of the conventional war assumption with the understanding that changes have taken place in the nature of warfare and of international politics which make other scenarios more likely, but no less dangerous in terms of their impact on U.S. security and the U.S. economy. A specific example would be the outbreak of hostilities in the southern Africa region. Although the United States would not be directly involved, the impact on the lives of Americans would be large. It is unrealistic not to see such an impact as having strategic implications for the security of the United States and its political and economic allies.

An effort should be made to appropriate some funds apart from those contained in the Transaction Fund. For, compared to the defense budget, such funds would be practically insignificant, yet their return on investment in terms of U.S. security would be large. At 1983 and 1984 levels of appropriation, it will take 85 years to achieve stockpile goals for all commodities.

There should be a considerable effort on behalf of government to develop the minerals and metals of which the United States has a ready domestic supply. Such projects would include substitution of molybdenum for chromium and columbium in alloys and superalloys, or work on reducing the energy requirements for non-bauxitic sources of aluminum.

The stockpile issue has become a "liberal" versus "conservative" debate. It is seen as repugnant by some that there should be consideration of a future conflict. However, everyone, regardless of political affiliation, is a consumer of strategic and critical materials. The quickest way to become involved in a war would be to find that the United States was unprepared for a supply interruption from a specific region, and that it would require military intervention to secure that supply. Everyone looks forward to a time free from geopolitical unrest. Unfortunately, that world does not appear close on the horizon.

DIVERSIFICATION OF FOREIGN SOURCES OF SUPPLY

Supply diversification is a strategy that could yield short- to medium-term reductions in U.S. vulnerability to supply disruptions. This strategy has been pursued with some success by Japan and other nations, so it certainly deserves attention from U.S. authorities. However, this strategy is not without constraints.

A first constraint concerns the length of time over which a mineral-exporting nation will be capable of exporting a mineral required by the United States. This obviously is the case when currently operating mines are depleted and no new deposits are discovered. However, a situation more likely to occur centers on the industrial development plans of a mineral-exporting nation. Intensive industrial development in that nation may consume all the production once destined for export markets. A good example of this case is the decision by both Brazil and India to reduce manganese exports to the United States to meet the needs of domestic steelmakers.[31]

A second constraint is simply that geology limits the possible range of potential supply sources for U.S. import needs. Cobalt is a good example. The fact that cobalt has been discovered in only a few nations severely restricts U.S. flexibility in trying to diversify supply sources.[54] Chromium and the platinum-group minerals are other examples of this phenomenon.

A third constraint involves the inability of the United States to diversify without first investing in mining projects. This has been done successfully by Japan to provide coal, iron ore, and other minerals to Japanese industries. However, it is likely that the United States may not be as fortunate in finding desirable projects in which to invest. Since the United States is more self-sufficient in minerals than is Japan, and is critically dependent largely on exotic minerals, the United States will have a smaller range of potential investments from which to choose. In addition, many of these projects may be located in insecure and/or developing nations where the chances are good that the U.S. investor will be forced out. Past experiences with expropriation soured many Americans on investments that may otherwise look good due to low energy and labor costs, and other factors.

A final constraint on the ability of the United States to diversify its supply sources centers on the level of processing of U.S. imports. There are fewer nations that process a mineral to an intermediate stage than there are those which produce ores or concentrates. For example, many more nations produce bauxite than produce alumina. This is a critical point precisely because, as shown in Chapters I-2 and I-3, the U.S. mineral-processing industries are closing down. Without the domestic capacity to upgrade raw ores and concentrates, the United States will have fewer alternative supply sources for a needed mineral. Chromium is another example of this phenomenon. More than 20 countries produce chromite, but only four or five have the capacity to export ferrochromium. The demise of the U.S. mineral-processing industry will limit our ability to diversify sources of supply.

This raises another concern about the potential usefulness of supply diversification. Use of this strategy might indeed reduce U.S. vulnerability to supply cutoffs, but it does nothing to resolve the domestic problem that forced diversification in the first place. That is, no attention would be paid to such issues as taxes, environmental regulations, industrial reform, mineral exploration, and all the other factors which determine if the United States can *produce* a mineral domestically.

Even with all these questions about supply diversification, there are some attractive features. Above all, the strategy might actually reduce vulnerability. It *may* provide minerals at lower cost than would be the case were the minerals produced domestically. The United States has some leverage in dealing with potential suppliers. The U.S. economy is diversified, whereas many developing nations depend on one industry for the bulk of their foreign exchange. Many nations must export minerals or face economic ruin. Moreover, even today, the United States is one of only a few nations capable of processing minerals into certain desired forms. It seems at least possible that the United States could gain additional supplies of needed minerals if it, in turn, exported finished or intermediate products to producer nations.[55]

Supply diversification is a strategy that should be pursued. It has its risks, and it places the United States in a potentially awkward situation when dealing with producer nations. But if offers distinct, although limited-term advantages, such as spreading the risk of supply disruption and minimizing the effects of supply disruptions from any individual nation.

This option requires that the United States adopt a realistic foreign policy which recognizes U.S. mineral import dependence. The United States should adopt policies that encourage U.S. industries to invest in foreign minerals projects. As a result of such participation, the United States might receive preferential treatment from these countries because negotiations for prices, tax concessions, and supply details are formulated in the early stages of development projects.

Newer arrangements between producer and host countries contain provisions for greater domestic employment, employee training programs, and community facilities. Materials processing capabilities and environmental protection are also being included in the arrangements.

Foreign investment incentives for U.S. companies have a higher profit potential because of richer deposits and lower capital and labor costs. Foreign governments may add other incentives, such as concession exclusivity, infrastructure, subsidized energy supplies, and tax concessions. The latter may include tax moratoriums on revenues, pending recovery of investment outlays, and low future taxes.

Any U.S. foreign policy toward mineral supplies should include consultation with its allies. This would help prevent duplication of efforts and harmful competition for scarce materials.

In recent years, the concept of a Tri-Oceanic Alliance has been advanced by several strategists. Such an alliance should include the NATO countries and Australia, Saudi Arabia, Egypt, Indonesia, Singapore, Brazil, Mexico, South Africa, Nigeria, and Zaire. It would provide greater security for the sea-lanes by which most of the world's resources traverse. The alliance would also contain a virtual monopoly of the world's oil reserves, food supplies, and scientific knowledge. It would formalize the mineral supply diversification option.

International Developments Affecting Mineral Trade

The development of world trade in minerals, which grew steadily between about 1950 and 1970 (except for minor disruptions such as the Korean War, minor recessions, and a growing tendency toward economic nationalism in the countries of the Third World), was rudely disrupted by the 1973 oil crisis. This event was followed by the firm entrenchment of the most successful mineral cartel in the history of world trade (OPEC), by massive balance-of-payments deficits, coupled with lower economic growth rates in the Western industrialized countries, and by almost catastrophic conditions in most Third World countries. A second wave of massive oil price increases in 1979 had similar effects. The long-term effects of increased energy costs on the magnitude and pattern of world mineral trade cannot yet be effectively evaluated. However, it is clear that apart from the political effects, the resulting inflation and inadequate capital formation will force a slowdown in new mineral ventures, particularly in the Third World.

The attitudes on the part of the developing countries, as well as those of countries that are fairly developed, such as Canada, Australia, Venezuela, and Chile, have changed during recent years. They are no longer prepared to accept unrestricted foreign investment in their mineral industries or to allow foreign companies to exploit unlimited quantities of their unprocessed minerals. There has been a growing tendency toward governmental restrictions on foreign investment, control of mineral exports, expropriation and forced changes of agreements, and nationalization of natural resources. These developments tend to make it much more difficult for the United States and its allies to obtain supplies of minerals from abroad.

There is growing opposition to the role of multinational corporations in the Third World, and the corporations are facing confrontation by such bodies as the United Nations and also by many of the developing countries, on the grounds that they do not promote optimum utilization of the mineral resources of the producer countries. Activities of multinational corporations are expected to come under increasing scrutiny during the next few years, when more countries may impose restrictions on the export of unprocessed minerals and on foreign investment in their mineral industries.

The fundamental issues and problems relating to international minerals trade are not new. During World Wars I and II, the United States and its allies were confronted with problems similar to those of today: national security, import dependence, and access to minerals in times of crisis; uneven distribution of mineral resources throughout the world; complexities of foreign trade and investment in a changing international market structure; the lack of reliable information on foreign mineral policies; and the need to address the problem of assured minerals availability. What is new is the fact the international setting has become much more intense and complex. The dismantling of colonial empires led to the establishment of numerous new sovereign states that have become polarized into a loosely unified entity. At the same time, Western colonial dominance as it was known for centuries has ended. To make matters worse, it now appears that the Soviet bloc will compete increasingly for available mineral supplies.

Numerous functionally oriented international conferences dealing with mineral-related matters were held during the 1970s. For instance, the U.N General Assembly in 1969 declared a moratorium on the exploitation of seabed resources pending the establishment of an international seabeds regime, and in April 1974, the Sixth Special Session of the U.N. General Assembly met, on the initiative of Algeria, to discuss problems of raw materials and development. This session approved a declaration and a program of action on the establishment of a New World Economic Order, which in effect constitutes a demand for a changed economic system, biased in favor of the developing countries, and intended to offset what these countries regard as an imbalance of wealth enjoyed by the Western world. The subsequent regular session of the U.N. General Assembly in December 1974 approved far-reaching demands by the developing countries for preferential trade and monetary arrangements and for an increased role in world decision making. Subsequently, there have been similar meetings where the same or even stronger demands were made. The United Nations Conference on Trade and Development (UNCTAD) has been particularly active on behalf of the developing countries.

When the Third World countries began to realize their strength in numbers and the extent of the Western world's dependence on their resources, many adopted a variety of tactics in their relations with developed nations. This led to the disruption of conventional trade patterns and institutions, as well as the disruption of certain traditional aspects of international law. Unfortunately, many of the Western nations put their own interests before the common good, and the Western world was not unified in its opposition to these actions.

Paralleling and contributing to the abrupt shifts and imbalances in foreign economic policy during the 1970s has been the worldwide state of political instability. Many of these developments have threatening and disruptive overtones for the West: the American defeat in Vietnam, open warfare and Communist intervention in Africa, instability in the Middle East culminating in the overthrow of the Shah of Iran, war between Iran and Iraq, and direct Soviet intervention in Afghanistan.

All these developments indicate that at a time when the United States has become much more dependent on imported sources of minerals than it was during World War II, the United States no longer enjoys the dominant control over world mineral resources that it, together with the British Empire, had during World War II.

Two developments have cast a shadow on future mineral supplies from abroad. They are (1) a lack of sufficient capital investment in new mining ventures, resulting from the increased nationalism of many host countries; and (2) the investment activities of national mining enterprises, which are making the markets more competitive for many strategic minerals and replacing the multinational companies in some of these markets. These new entrants are attuned to national political and social objectives, such as employment opportunity, and are likely to maintain high levels of production even in depressed markets. This reduces the incentive for an orderly long-term investment policy on the part of the multinational. The huge debts of many Third World countries has further exacerbated this problem.

It is clear that even without the disruptive attempts by the USSR to deny mineral supplies from abroad to the United States and its allies, the United States is faced with a far more difficult task than ever before in obtaining adequate and secure supplies of nonfuel minerals from abroad, even in peacetime. In the event of a serious conflict, the opposition would obviously become much worse.

It is necessary for the United States and its allies to coordinate their strategies for dealing with the mineral-exporting nations. Their efforts in protecting the supply routes for these minerals should be expanded. In this regard it is obviously necessary that NATO's sphere of influence be expanded to include the South Atlantic and Indian Oceans. Although favored by some, it does not appear that international commodity agreements represent an effective means of ensuring adequate supplies of strategic minerals. In light of the obvious trend toward the import of minerals in more highly processed forms, present U.S. tariffs on imported processed minerals do little to protect the domestic industry and merely fire the embers of inflation. Protection of the domestic processing industry should go hand in hand with economic incentives for their development.

There should be an appreciation of the fact that dependence on imported sources of nonfuel minerals is not a U.S. problem—it is a problem of the Western alliance, and should be treated as such. This means that this dependence should be reflected in foreign policies toward the major mineral-exporting countries.

INCREASED DOMESTIC PRODUCTION

Any serious commitment to improve U.S. domestic mining capabilities must include a redirected and accelerated effort by government and industry to survey federal lands for their mineral potential. This should provide an adequate data base for mineral policy decisions and targets for enhanced exploration. New tax incentives are needed to revitalize the domestic mining industry. A review of environmental policies using cost-benefit analysis is required to ascertain the economic effects of current environmental policies on the domestic minerals industry. In addition, increased funding for research and development is required to improve both the efficiency and the innovative capabilities of the industry.

Without a sufficient and reliable inventory of the nation's mineral resources and reserves it is difficult to formulate appropriate policy options for nonfuel minerals. The closure of such a large proportion of the federal lands makes this difficult if not impossible. In 1968, only about 17% of public-domain land was denied to mineral use. Today it is estimated that nearly 550 million acres, or 70%, of federal land has been either totally withdrawn or restricted from mining considerations.[57] This represents a 300% increase in land withdrawals in a decade. Continued land withdrawals will further frustrate efforts to determine the extent of U.S. mineral reserves and will obviously delay any attempts at increasing domestic production. These facts are especially significant because most public lands are in the mineralized areas of the western United States and Alaska.

From 1930 to 1971, the amount of land used for mining in the United States was estimated at a little over 3 million acres, or 0.16% of total U.S. land area.[54] Fears that opening federal lands to mining would destroy this land therefore seem unrealistic. Yet, when former Secretary of the Interior James Watt proposed increased access to federal lands for exploration and development, more than a million people signed petitions calling for his dismissal. Comprehensive exploration of these lands should be undertaken to establish a reliable data base. Geophysical exploration techniques should include electrical, gravimetric, magnetic, and seismic surveys. In addition, geochemical methods are especially important for detecting precious metals. Classical geological mapping and subsurface data should also be used.

In 1978, the Comptroller General recommended an accelerated rate of federal mining surveys, warning that the current rate of federal mineral surveys would require 50 years for completion.

New tax incentives to increase the rate of return on mineral investments are needed to induce new capital flow into the minerals industry. These could include accelerated depreciation allowances, higher depletion allowances, sliding-scale tax rates, and perhaps subsidies. Government support for research and development, and guaranteed purchase of products, would also help in some industries. Antitrust legislation should be revised to allow companies to cooperate in massive, capital-intensive ventures, and to allow more horizontal and vertical integration in the industry. Tax credits and more flexibility in write-off times for exploration costs and pollution control equipment should also be considered.

An area of growing cost in terms of money and time to the mineral industry is the increasing burden of regulation. Federal laws affecting the mining industry include the National Environmental Policy Act (NEPA), Clean Air Act (CAA), Clean Water Act (CWA), Resource Conservation and Recovery Act (RCRA), Toxic Substance Control Act (TSCA), Safe Drinking Water Act (SDWA), Occupational Safety and Health Act (OSHA), Mine Safety and Health Act (MSHA), Wild and Scenic Rivers Act (W&SR), Endangered Species Act (ESA), and the Coastal Zone Management Act.

The increased cost of mineral production caused by excessive regulations has become obvious, and these and other acts have had a highly adverse effect on the U.S. domestic mineral and processing industries. Consistency of regulation should be a primary goal of nonfuel minerals policy.

The federal domain represents about one-third of total U.S. land area. Most of the federal lands occur in the western United States and Alaska, in areas of high potential for mineralization. Although quantitative evidence is lacking, it appears that these lands probably have the potential to supply at least 50% of all additional domestic mineral production in the United States. For this reason, the federal lands issue, as it relates to potential increased domestic production of minerals, is examined in some detail.

Federal Lands

History of the Federal Lands. The total land area of the United States is about 2.3 billion acres. Approximately 80% of this was under the control of the federal government at one time. The federal government obtained land in several ways: state cessions, purchases, treaties, and other special agreements.

When the United States was recognized as a country after the Revolutionary War, several states ceded some of their lands to the new government. The lands ceded were mainly frontier. Congress accepted these lands and pledged that "all lands ceded by the states would be disposed of for the common benefit of the United States."[58]

More land was acquired when the federal government made several large land purchases from 1800 to 1870 with the intent of expanding U.S. boundaries. These purchases included the Louisiana Purchase from France in 1803 of 529.9 million acres for $23.2 million, a purchase from Texas of 78.9 million acres in 1850 for $15.5 million, and the Gadsden purchase from Mexico in 1853 of 19 million acres for $10 million.[59] In addition, perhaps one of the best deals made by the United States was the purchase of 378.2 million acres of land in 1867 from the USSR for $7.2 million. This area, now known as Alaska, is suspected to contain some of the richest mineral deposits in the United States.

Other means by which land was acquired include treaties with Great Britain and Spain, a cession from Mexico in 1848, and the state of Texas joining the Union. Table I-4-4 shows the acquisition of the public domain from 1781 to 1867. There is a distinction between acquired lands and public-domain lands that is important in the application of mining and mineral leasing laws. By U.S. government definition, acquired lands are[59]

lands in federal ownership which were obtained by the government through purchase, condemnation, or gift, or by exchange for such purchased, condemned, or donated lands, or for timber on such lands.

TABLE I-4-4   Acquisition of the Public Domain, 1781–1867

Acquisition	Land (acres)	Cost
State cessions (1781–1802)	233,415,680	$ 6,200,000
Louisiana Purchase (1803)	523,446,400	23,213,568
Red River Basin (1782–1817)	29,066,880	—
Cession from Spain (1819)	43,342,720	6,674,057
Oregon Compromise (1846)	180,644,480	—
Mexican cession (1848)	334,479,360	16,295,149
Purchase from Texas (1850)	78,842,880	15,496,448
Gadsden Purchase (1853)	18,961,920	10,000,000
Alaska Purchase (1867)	365,333,120	7,200,000
Total public domain	1,807,533,440	$85,079,222

Source: U.S. Department of the Interior, Bureau of Land Management, *Public Land Statistics 1982.* April 1983, p. 3.

Acquired lands may also be described as lands that were obtained for a specific purpose. Public-domain lands are those lands which have never left federal ownership.[59]

The United States began to distribute some of the federal lands for various reasons. The primary reason was to aid in settling the frontier; however, generation of revenue was another reason. Certain lands were granted to individuals and private corporations such as railroads. Land was also granted to veterans as military bounty. Some states received land for construction of roads, canals, railroads, schools, and other minor public uses. Finally, land was sold to those who fulfilled the requirements of certain laws, such as the Homestead Act. The Homestead Act of 1862 was designed to aid settlement of the frontier. Individuals were able to purchase acreage for a nominal fee provided that they built a house and lived on the land for five years. The purpose of homestead laws was to provide land to those who could improve vacant agricultural lands by settling on them and farming the land.

There are several other laws that provide for the disposition of public land. The Timber and Stone Law aided the sale of land that was unfit for agricultural purposes but valuable for timber and stone. Timber culture laws granted land to settlers on the condition that they plant and cultivate trees. Arid agricultural lands were sold to settlers under desert land laws on the condition that they be irrigated and farmed. Table I-4-5 shows this disposition of public lands.

The federal government worked actively to acquire and dispose of land. This era of free access and disposal of public lands ended in 1934 with the Taylor Grazing Act. This act ended free access of public lands to ranchers for grazing purposes and marked the beginning of permanent federal land management.

Currently, federal lands comprise 729 million acres, which is equivalent to nearly one-third of the total land area of the United States. Approximately 93% of these lands lie in the 11 western-most continental states and Alaska. Table I-4-6 lists federal land acreage in the United States. One should note that the federal government is a major landowner in many of the western states, owning over 50% of the land in Utah, Nevada, Oregon, Idaho, and Alaska.

The location of these federal lands is significant for several reasons. First, the federal government's land holdings are both extensive and potentially rich in minerals. Second, large quantities of land are being withdrawn to mineral exploration and development. In 1965, only 17% of the federal

TABLE I-4-5   Disposition of Public Lands, 1781–1982[a]

Type of disposition	Acres
Disposition by methods not elsewhere classified[b]	303,500,000
Granted or sold to homesteaders[c]	287,500,000
Granted to states for:	
Support of common schools	77,630,000
Reclamation of swampland	64,920,000
Construction of railroads	37,130,000
Support of miscellaneous institutions[d]	21,700,000
Purposes not elsewhere classified[e]	117,600,000
Canals and rivers	6,100,000
Construction of wagon roads	3,400,000
Total granted to states	328,480,000
Granted to railroad corporations	94,400,000
Granted to veterans as military bounties	61,000,000
Confirmed as private land claims[f]	34,000,000
Sold under timber and stone law[g]	13,900,000
Granted or sold under timber culture law[h]	10,900,000
Sold under desert land law[h]	10,700,000
Grant total	1,144,380,000

[a]Data are estimated from available records.
[b]Chiefly public, private, and preemption sales, but includes mineral entries, scrip locations, and sales of townsites and townlots.
[c]The homestead laws generally provide for the granting of lands to homesteaders who settle upon and improve vacant agricultural public lands. Payment for the land is sometimes permitted, or required, under certain conditions.
[d]Universities, hospitals, asylums, etc.
[e]For construction of various public improvement (individual items not specified in the grant acts), reclamation of desert lands, construction of water reservoirs, etc.
[f]The government has confirmed title to lands claimed under valid grants made by foreign governments prior to the acquisition of the public domain by the United States.
[g]The timber and stone laws provided for the sale of lands to settlers on conditions that they plant and cultivate trees on the lands granted. Payments for the lands was permitted under certain conditions.
[h]The desert land laws provide for sale of arid agricultural public lands to settlers who irrigate them and bring them under cultivation.

Source: U.S. Department of the Interior, Bureau of Land Management, *Public Land Statistics 1982.* April 1982, p. 5.

lands were withdrawn from mineral entry.[60] Today, mineral exploration and development is formally prohibited or severely restricted on over 50% of the land.[61] It is therefore most prudent to know the relationship between these lands and their mineral potentials before making any decisions concerning their uses.

Federal Land Restrictions. Federal lands may be classified according to the degree of restriction placed on their mineral exploration and develop-

TABLE I-4-6 State Lands Owned by the Federal Government

State	Acreage owned by the federal government		Percent owned by the federal government	
	1981	1982	1981	1982
Alabama	1,126,036.7	1,140,954.2	3.446	3.492
Alaska*	307,381,586.5	327,028,961.7	84.103	39.479
Arizona*	32,394,956.7	29,194,570.3	44.567	40.164
Arkansas	4,010,705.3	3,404,254.1	11.937	10.132
California*	45,218,649.8	47,525,524.8	45.125	47.428
Colorado*	23,754,666.7	23,949,523.7	35.729	36.022
Connecticut	9,933.9	9,940.3	0.317	0.317
Delaware	40,852.2	40,683.6	3.227	3.214
District of Columbia	12,584.1	12,278.5	32.234	31.451
Florida	4,183,359.6	3,651,653.0	12.048	10.517
Georgia	2,269,090.2	2,281,041.8	6.084	6.116
Hawaii	693,381.8	786,769.6	16.889	19.163
Idaho*	34,546,150.6	34,281,911.6	65.264	64.765
Illinois	615,337.2	626,480.1	1.719	1.750
Indiana	499,323.2	528,699.8	2.156	2.283
Iowa	227,666.5	228,049.0	0.635	0.636
Kansas	733,067.8	733,155.2	1.396	1.396
Kentucky	1,408,086.9	1,417,715.8	5.519	5.557
Louisiana	1,073,332.2	1,157,436.1	3.715	4.009
Maine	135,913.6	135,882.3	0.685	0.685
Maryland	205,475.1	210,380.8	3.252	3.329
Massachusetts	79,740.1	84,801.1	1.584	1.684
Michigan	3,482,527.1	3,533,418.3	9.543	9.683
Minnesota	3,422,658.1	3,448,659.3	6.684	6.735
Mississippi	1,729,223.4	1,748,821.0	5.722	5.786
Missouri	2,198,473.9	2,253,755.4	4.969	5.093
Montana*	27,611,094.7	27,468,335.1	29.603	29.450
Nebraska	704,614.5	696,120.9	1.437	1.420
Nevada*	60,798,822.1	57,383,816.7	86.516	81.669
New Hampshire	730,929.2	738,417.0	12.670	12.800
New Jersey	144,321.4	146,438.3	2.998	3.042
New Mexico*	25,856,900.6	25,862,337.1	33.250	33.256
New York	232,857.2	248,936.3	0.759	0.811
North Carolina	2,102,569.7	2,169,354.1	6.696	6.908
North Dakota	2,295,582.5	2,246,256.9	5.164	5.053
Ohio	343,984.6	351,104.9	1.312	1.339
Oklahoma	1,737,261.8	1,749,004.0	3.941	3.967
Oregon*	33,314,870.4	30,102,588.0	54.084	48.869
Pennsylvania	675,103.8	695,456.9	2.344	2.414
Rhode Island	6,906.7	6,157.5	1.020	0.909
South Carolina	569,114.2	1,193,610.5	2.938	6.161
South Dakota	3,334,416.5	3,151,989.7	6.821	6.448
Tennessee	1,823,352.7	2,095,787.1	6.822	7.841
Texas	3,407,792.9	3,528,274.1	2.026	2.097
Utah*	33,299,404.2	32,166,670.8	63.190	61.041
Vermont	304,182.3	319,530.1	5.124	5.382
Virginia	3,200,036.9	2,360,367.3	12.551	9.258
Washington*	12,421,200.9	12,104,326.1	29.094	28.352
West Virginia	1,102,586.4	1,113,733.7	7.155	7.227
Wisconsin	1,866,367.1	1,896,917.3	5.331	5.418
Wyoming*	30,194,454.6	30,610,009.1	48.433	49.099
Total federal land	719,521,617.1	729,820,861.4	31.678	32.132
Total federal land in 12 western states (*)	666,783,714.8		92.67	

Source: U.S. Department of the Interior, Bureau of Land Management, *Public Land Statistics 1982*. April 1983, p. 9.

ment. These levels are not clearly delineated. The United States Geological Survey (USGS) has classified land into the following categories: formally prohibited, severely restricted, moderately restricted, and slightly restricted.[41] These classifications are defined as follows:

The "formally prohibited" classification includes lands for which there is a specific congressional act, or executive order, or some other type of public land notice that prohibits mineral exploration and development in any form. This classification applies only to lands that are currently restricted, not to those that may be excluded in the future.

The "severely restricted" category is used to designate land where mineral exploration and development is highly unlikely because of policies that either restrict mineral exploration, leasing, and mining, or create increased risk for mineral mining investments. An example of this type of policy is found in Forest Service lands, which prohibit the building of roads. Exploration may be allowed, but it must be done by helicopter or by means not requiring the use of roads. Over 50% of the federal lands fall into the formally prohibited or severely restricted categories.

"Moderately restricted" lands are limited by policies in very much the same manner as severely restricted lands but to a lesser extent. Mineral development must be quite profitable to overcome the economic burden placed by these restrictions before a company will invest in development.

Lands in the "slightly restricted" category may be restricted or have only slight restrictions on mineral exploration and development. Leases may be considered on a case-by-case basis or may be subject to managing agency discretionary authority.

The only category for which there are formal prohibitions is the first, formally prohibited. The restrictions applied in the other three categories—severely restricted, moderately restricted, and slightly restricted—have to do with the extent to which policies and regulations are applied.

Two other terms that are used in the discussion of withdrawal of federal land to mineral exploration and development are "de jure" withdrawals and "de facto" withdrawals. "De jure" withdrawals, having a literal meaning "by law," are those for which some law, mandate, or congressional or executive order exists. "De facto" withdrawals refer to lands that are closed to mineral exploration and development because of precedence or administrative policies. For example, lands withdrawn by the Wilderness Act of 1964 were legally open to mineral exploration and development

until December 31, 1983. However, in the past 20 years very few leases were issued for mineral activities on these lands; thus this was a "de facto" withdrawal.

Table I-4-7 shows the amount of land that is open and closed to both leasable and locatable minerals. A note should be made at this point because these figures do not exactly match those in other tables which list the amount of federal land managed by the various agencies. This reflects recent changes due to legislative acts. This table is included because it shows a different categorical division of the land. In the literature, the total amount of federal land is given in the range 690 million to 800 million acres. When the various agencies were asked for a total figure of acreage managed, they had difficulty in arriving at one. This exemplifies a problem in the current system of land management. If the agencies themselves do not know how much land they manage, how can they manage it efficiently and effectively?

Public Land Statistics is published annually by the Department of the Interior and is the official document for information on the federal lands. The information in most of the tables came from the latest volume, *Public Land Statistics 1983*.

Managing Agencies. Federal lands are managed by several government departments. The majority of the land, over 510 million acres, is managed by the Department of the Interior through several bureaus and services. The U.S. Forest Service of the Department of Agriculture manages over 192 million acres of land. The Department of Defense controls over 22 million acres divided among the Air Force, Army, Navy, and Corps of Engineers. The Department of Energy controls 2.3 million acres of land. The remaining land, about 2 million acres, is controlled by several smaller agencies. This study focuses on 11 western states and Alaska.

Department of the Interior. The Department of the Interior is the nation's principal conservation agency and has responsibility for managing, conserving, and developing most of the nation's federal lands and resources. It controls over 510 million acres of land and has trust responsibilities for approximately 50 million acres of land, which consist mainly of Indian reservations. The majority of this land is managed by the Bureau of Land Management (BLM), the National Park Service, the Bureau of Indian Affairs, the United States Fish and Wildlife Service, and the Bureau of Reclamation.

1. *The Bureau of Land Management:* Over 341 million acres are managed by the Bureau of Land Management (BLM). Most of these lands are located

TABLE I-4-7   U.S. Federal Land Restrictions

	Leasable minerals	Locatable minerals
Total federal mineral base	747.5	723.6
Acreage closed to mining or mineral leasing		
National parks	76.1	79.0
Fish and wildlife refuges	86.8	86.8
BLM wilderness study areas	23.6	0
Military lands	23.7	30.3
Forest service	43.0	9.1
State of Alaska and native selections	72.1	72.1
Miscellaneous actual and "de facto" withdrawals (e.g., segregations resulting from the filing of withdrawal applications, classifications, in-lieu state selections, etc., nonreservation Indian lands, wild and scenic rivers, pipeline corridors, irrigation projects, water supply and control, DOE, TVA, incorporated towns and cities, incompatible surface occupancy)	21.7	27.5
Total acreage closed	351.1	308.9
Acreage open to mining or mineral leasing		
Surface is privately owned, but minerals are owned by federal government	65.8	41.9
Both the surface and the minerals are federally owned	330.6	372.8
Total acreage open	396.4	414.7
Total federal mineral base	747.5	723.6

Source: U.S. Department of the Interior, Bureau of Land Management.

in 11 western states and Alaska. Lands managed by the BLM are those which were least desirable during the period of disposition. BLM lands are commonly referred to as public lands. In addition, the BLM is the leasing agent for all federal lands.[62]

According to the Federal Register, "bureau programs provide for the protection, orderly development, and use of the public lands and resources under the principles of multiple use and sustained yield."[63] Resources managed by the BLM therefore include timber, minerals (including oil and gas), geothermal energy, wildlife habitats, endangered plant and animal species, rangeland vegetation, recreation areas, wild and scenic rivers, and designated conservation and wilderness areas.[63] The resources delegated to the BLM are quite diverse. This presents several problems. It is difficult to manage such a diversified quantity of acreage. The rules that apply to various areas are different. Finally, it is difficult to balance the value of all resources on the land. For example, mineral activities as well as several other activities are prohibited on wilderness land, but the mandate of

multiple use specifies a variety of activities for land use; therefore, this is in conflict.

In addition to the lands mentioned above, the BLM administers the mineral rights to approximately 66 million acres of land where the surface is privately owned.[63] This has potential for conflict if the BLM wishes to develop some of the minerals while the surface owner feels that mineral development will inhibit use of the land.

2. The United States Fish and Wildlife Service: The U.S. Fish and Wildlife Service is responsible for conserving, protecting, and enhancing fish, wildlife, and their habitats to benefit the nation.[63] This department manages over 81 million acres of land. Wildlife refuges are established to aid in fulfilling this department's responsibilities as stated above. Wildlife refuges may also be used for other things, such as grazing, recreation, and mineral development as long as these uses do not preclude the original purposes of the refuge.[63] Recently, however, conflicts have arisen concerning the compatibility of mineral development with wildlife refuges. Case studies done for a study of the policies affect-

ing fossil-fuel development on federal lands at the University of Texas address issues faced by the Arkansas National Wildlife Refuge and the Big Thicket Reserve in Texas as well as those faced by the Kenai National Wildlife Preserve in Alaska.

3. *The National Park Service:* The National Park Service manages over 68 million acres of land. Its 330 units include national parks, monuments, recreation areas, and historic sites. The goal of the National Park Service is "to administer the properties under its jurisdiction for the enjoyment and education of our citizens, to protect the natural environment of the areas, and to assist states, local governments, and citizen groups in the development of park areas, the protection of natural environment, and the preservation of historic properties."[63] National Park Service lands are not open to mineral exploration and development.

4. *The Bureau of Reclamation:* The Bureau of Reclamation was established to "locate, construct, operate, and maintain works for the storage, diversion, and development of waters for the reclamation of arid and semi-arid lands in the western United States."[63] This department manages over 5.3 million acres. Although the Bureau of Reclamation is not as involved with the mineral aspects as the BLM, it does manage a significant amount of acreage on which minerals may occur.

5. *The Bureau of Indian Affairs:* The Bureau of Indian Affairs manages over 3.3 million acres. The principal objective of this department is to "actively encourage and train Indian and Alaska native people to manage their own affairs under the trust relationship to the federal government."[63] To achieve these objectives, the Bureau provides educational opportunities, promotes improvement of social welfare, develops and implements economic programs, aids in conservation of natural resources, and acts as trustee for lands and money held in its trust. Mineral resources development is included among these objectives. The decisions affecting mineral resource development, however, are tribal decisions which, for the most part, are made on a case-by-case basis.

Department of Agriculture. The Department of Agriculture (USDA) has fairly diverse interests but is mainly concerned with agricultural products and the resources needed to produce and protect them. Nearly all of the USDA's 192.4 million acres of land are controlled by the Forest Service. The Forest Service manages nearly 192.1 million acres, consisting of 154 national forests and 19 national grasslands. The Forest Service is concerned with aspects of forest management such as economics,

natural resources that lie on forestlands, and public education. This land is managed under the principles of "multiple use and sustained yield." Under these principles, mineral development on forestlands is possible. However, significant acreage has been closed to mineral activities because of its designation as wilderness.

Department of Defense. The main responsibility of the Department of Defense (DOD) is "to provide the military forces needed to deter war and protect the security of our country."[63] Responsibility for the nearly 30 million acres of land is divided among the Air Force, Army, Navy, and Corps of Engineers.

It is possible for mineral exploration and development to occur on DOD lands. Leases are considered on a case-by-case basis. The base commander has discretionary authority in allowing mineral activities. Although mineral assessments have not been made on military lands, there are good reasons why they may not be permitted. For example, a tract of land may be closed to mineral activities because it was once a test range and there may still be live shells on the land.

Department of Energy. The Department of Energy (DOE) was created by the Carter administration to "provide the framework for a comprehensive and balanced national energy plan through the coordination and administration of the energy functions of the federal government."[63] The DOE involves itself with research and development of energy technology, marketing federal power, energy conservation, nuclear weapons development, energy regulatory programs, and energy data collection and analysis.[63] Approximately 2.4 million acres of land are managed by the DOE. Fuel minerals—oil, natural gas, coal, and uranium—are a direct concern of this department. The DOE regulates and collects information concerning these minerals.

Special Land Designations. Certain federal lands have been dedicated to specific purpose such as wilderness areas, wild and scenic rivers, and national trails. These areas are designated by Congress, and although they are part of lands administered by another agency such as the BLM, the National Park Service, the Fish and Wildlife Service, or the Forest Service, special limitations are placed on their use. These dedicated federal lands are to be used primarily for recreational purposes.

The largest of these categories is wilderness. Currently, there are 25.4 million acres of designated wilderness.[63] Nearly 19 million acres are in the 11 western states and over 5 million acres lie in

Alaska. An additional 36 million acres have been designated as "wilderness study areas" (WSAs).

Wilderness areas and WSAs make up a significant portion of the federal lands. Because of this designation, these lands are closed to mineral exploration and development without consideration of minerals which may occur on them. Assessments of critical and strategic minerals on these lands do not exist. However, a study conducted by the U.S. Geological Survey on the Petroleum Potential of Wilderness Lands in the western United States revealed that approximately 3 million acres of these lands have high potential for petroleum occurrence.[65] Results of this study are shown in Table I-4-8. How can the value of all resources on the land be considered for decision-making purposes when information concerning minerals, especially critical and strategic, is not available?

Summary. The United States has undergone several trends in the acquisition and disposition of federal lands. The federal government worked actively to acquire land as well as to transfer portions of it to private ownership. A period of reservation followed where certain lands were set aside for specific purposes such as national parks. The most recent trend, which began around 1964, is the preservation of the federal lands.

Federal lands are geographically concentrated in the western United States and Alaska. Twelve states contain 93% of all federal lands. The federal government is a major landowner in these states owning over 50% of the land in Alaska, Idaho, Nevada, and Utah. In addition, it owns over 45% of the land in Oregon, California, and Wyoming.

Federal lands may also be classified according to the degree of restrictions placed on mineral exploration and development. The U.S. Geological Survey has classified land into the following categories: formally prohibited, severely restricted, moderately restricted, and slightly restricted. Formal legal restraints exist only for the category of formally prohibited. The restrictions placed on the other three categories have to do with the degree to which regulations and policies are applied.

Although Congress ultimately controls the destiny of the federal lands through legislation, several agencies have been delegated management responsibilities for these lands. The Department of the Interior manages approximately 70% of the lands through the Bureau of Land Management, the National Park Service, the U.S. Fish and Wildlife Service, the Bureau of Indian Affairs, and the Bureau of Reclamation. Significant quantities of land are managed by the Forest Service of the U.S. Department of Agriculture, the Department of Defense, and the Department of Energy. Several smaller agencies manage the remainder.

Recently, several special land designations have been created to preserve certain lands. These special designations include wilderness, wild and scenic rivers, and national trails. These lands are managed by the managing agencies mentioned previously but in a manner so that their special characteristics are preserved. Wilderness is the largest category with nearly 60 million acres designated as wilderness or wilderness study areas. These special categories are significant because, for the most part, these lands are closed to mineral activities.

The increasing restriction of federal lands to exploration and mineral development surely has had an adverse effect on domestic mineral production. This will not change until the political power of the environmental movements is matched by a public much better informed about the importance of strategic minerals than is the case at present.

SUBSTITUTION AND RECYCLING

Scientific research and development are fundamental to technological innovations that will reduce the use of critical and strategic minerals. New "synterials" research being developed by Westinghouse is an example.[66] Other technological innovations usually take one of two forms: the substitution of one mineral for another, or the more efficient recycling of materials.

Substitution

The substitution of one material for another has several constraints. In many cases the only available substitutes are also heavily imported minerals: for example, nickel (73% imported) for platinum and chromium. In other cases, substitution may result in lower or unacceptable performance standards of efficiency. Although the ideal situation is the substitution of an abundant material for a scarce material, this is not always possible. For instance, recent trends in superalloy technology indicate nickel and ceramic substitutes for cobalt alloys in turbine engines, but other uses for cobalt in some steel and carbide tools remain without substitutes. This is an important consideration in any policy measures dealing with these metals, because substitutability tends to increase or decrease policy options. Industries need to be aware of substitution possibilities and the corresponding abundance or scarcity of the materials considered as substitutes.

In most instances there are long lead times in

TABLE I-4-8 Petroleum Potential[a] by Acreage for Wilderness Lands in the Western United States Summarized by Designated and Proposed Wilderness Categories (Thousand Acres)

State	Wilderness category[b,c]	High	Medium	Low	Low-zero	Zero	Unknown	Total
Arizona	D	0	51.50	72.00	340.33	484.73	0	948.56
	P	0	140.42	1,303.17	3,188.43	603.08	0	5,235.11
Total		0	191.92	1,375.17	3,528.76	1,087.81	0	6,183.67
California	D	0	0	253.43	2,048.60	2,654.48	0	4,956.51
	P	0	36.56	1,056.73	8,200.65	4,598.14	0	13,892.07
Total		0	36.56	1,310.16	10,249.25	7,252.62	0	18,848.58
Colorado	D	0	0	462.44	0	1,434.29	754.50	2,651.23
	P	140.19	94.81	356.97	0	1,010.98	246.94	1,849.90
Total		140.19	94.81	819.41	0	2,445.27	1,001.44	4,501.13
Idaho	D	0	0	0	0	3,945.22	0	3,945.22
	P	115.09	63.38	804.74	1,023.73	1,570.35	36.68	3,613.97
Total		115.09	63.38	804.74	1,023.73	5,515.57	36.68	7,559.19
Montana	D	648.58	367.93	666.91	620.33	836.09	0	3,139.84
	P	725.61	786.90	658.96	591.83	989.35	38.19	3,790.97
Total		1,374.19	1,154.83	1,324.87	1,212.16	1,825.44	38.19	6,930.68
Nevada	D	0	0	65.52	0	0	0	65.52
	P	132.38	1,099.08	3,565.70	0	3,117.93	0	7,915.09
Total		132.38	1,099.08	3,631.22	0	3,117.93	0	7,980.61
New Mexico	D	38.20	0	354.52	68.91	1,009.79	0	1,471.42
	P	58.43	115.32	882.79	89.06	59.84	0	1,205.44
Total		96.63	115.32	1,237.31	157.97	1,069.63	0	2,676.86

State								Total
Oregon	D	0	0	290.55	928.68	0	0	1,219.23
	P	0	138.17	2,579.12	929.22	0	0	3,646.51
Total		0	138.17	2,869.67	1,857.90	0	0	4,865.74
Utah	D	0	0	0	0	32.44	0	32.44
	P	208.94	2,638.26	1,010.35	0	592.74	0	4,450.29
Total		208.94	2,638.26	1,010.35	0	625.18	0	4,482.73
Washington	D	0	0	0	1,620.58	0	0	1,620.58
	P	0	0	952.38	1,319.96	0	0	2,272.34
Total		0	0	952.38	2,940.54	0	0	3,892.92
Wyoming	D	56.99	283.53	362.92	0	574.78	849.10	2,127.32
	P	604.85	108.30	360.35	18.95	1,184.49	1,602.65	3,879.59
Total		661.84	391.83	723.27	18.95	1,759.27	2,451.75	6,006.91
Total acreage	D	743.77	702.96	2,528.29	5,627.43	10,971.82	1,603.60	22,177.87
	P	1,985.49	5,221.20	13,531.26	15,361.83	13,726.90	1,924.46	51,751.14
Total		2,729.26	5,924.16	16,059.55	20,989.26	24,698.72	3,528.06	73,929.01
	%D	27	12	16	27	44	45	30
	%P	73	88	84	73	56	55	70

[a]The geologists evaluated the geological characteristics for the favorability of the occurrence of oil and natural gas and assigned a qualitative rating for the potential for petroleum resources, such as high, medium, etc.

[b]D, Designated wilderness lands included in this study.

[c]P, Proposed wilderness lands in this study include the following five categories: Administratively Endorsed as Suitable (prior to July 1981), Further Planning or Study Areas, BLM Lands Under Appeal, BLM Wilderness Inventory Not Completed, and U.S. Forest Service RARE II under litigation in California. These lands are administered under four agencies: Bureau of Land Management (BLM), U.S. Forest Service (USFS), National Park Service (NPS), and Fish and Wildlife Service (FWS).

Sources: Refs. 7 and 10.

the effective substitution of one mineral for another, and substantial modifications to manufacturing plants may also be required. Clearly, this is unlikely to take place during an emergency, as was pointed out in the section on strategic stockpiles. Exploring and developing the full range of materials substitution possibilities requires a sustained metallurgical and manufacturing research and development program by both the government and private industry. It would be unrealistic to expect dramatic results overnight.

Substitution is a worthwhile strategy for the reduction of import dependence of some commodities. However, for many of the minerals of greatest concern—chromium, manganese, and platinum-group metals—its potential appears to be severely limited.

Recycling

There are several constraints to increased efficiency for recycling of materials. One of the main problems with repeated recycling is the loss of specifications over time. Manufacturers and consumers are sometimes prejudiced against recycled materials. Because of the relative scarcity of secondary materials, manufacturers must rely to a large extent on primary materials to assure a continuing supply. This, in turn, tends to limit the demand for secondary minerals.[54]

Balancing these disadvantages of recycling, however, are some obvious advantagees. Recycling will prolong resource life. Environmental gains from reduction in litter and waste, such as mining wastes and smelter effluents will be realized. Another benefit is partial immunity from market fluctuations caused by price changes, shortages of primary materials, or actions by cartels. A direct economic benefit is that energy requirements for secondary materials are usually an order of magnitude lower than for primary material.

The effectiveness of recycling strategic and critical materials will depend on current recycling rates. Minerals such as copper and lead already had significant recycling rates by 1975: 53.5% for copper and 57.4% for lead.[67] Major increases in the recycling of these metals appear unlikely. However, the potential for increased recycling of platinum, chromium, and cobalt (10 to 12% at present) appears quite promising. With present technology, the recycling of manganese is expected to be insignificant even at double or triple current prices. Obviously, a concentrated research effort toward recycling technology would be a source for major new advances in recycling efficiency. Other measures to expedite and promote secondary materials

use would be better product design for easier recycling, an organized collection system for used materials, and public education of recycling potential and benefits.

A word of caution should be sounded about the potential for recycling of strategic minerals, however. The demand patterns, end uses, and life cycles of final products can place severe limitations on the annual percentage of the commodity that can be recycled. For instance, a high percentage of the aluminum used in beer cans and other containers could be recycled. Aluminum used in windowframes and other construction uses obviously will only be recycled over a much longer time frame. Antimony used as a hardener in lead acid batteries has a high potential for recycling. The same commodity used as a flame retardent has almost no potential for recycling.

CONCLUSIONS

The United States has adequate reserves of minerals such as copper, iron ore, and zinc, yet the import percentage of these minerals has increased significantly in the last decade. This is a result of a preoccupation with environmental concerns, causing a decline in mining and metallurgical technology.

There is a lack of capital inflow into the nonfuel minerals industry because of low returns on investment and uncertainties regarding production capabilities caused by environmental regulations and unstable markets. As a result of large pollution control investments, production equipment in many of our domestic mineral industries is antiquated and inefficient. Competitive ability with foreign sources of supply has decreased. Lack of funds has reduced funds for research and development to produce competitive innovations in the nonfuel mineral industry. The overall effect is lost employment, income, and production, leading to increased imports and a growing U.S. trade deficit. Increasing reliance on foreign minerals makes the United States dangerously vulnerable to events beyond its control. Therefore, a strong domestic minerals program must be developed that recognizes this dependency by placing security of supply for critical and strategic minerals as a priority item.

Without a revitalized U.S. domestic mining industry and an awareness of the basic role that the mineral industry plays in the U.S. economy, the United States will face even larger trade deficits. U.S. foreign policy may become less independent, and will be increasingly subjugated to domestic and foreign policies of other mineral-rich nations. Of equal concern is the sharp deterioration in the U.S.

mineral-processing industry. This development not only increases the cost of imported minerals, but in many instances reduces the number of possible foreign suppliers of a particular commodity.

Although increased domestic production is a technically feasible policy option for a number of commodities, it is not feasible for those commodities, such as chromium, manganese, and the platinum-group metals, for which the United States simply does not have an adequate reserve base. Diversification of foreign sources of supply is possible for some commodities, particularly those having a fairly widespread distribution of world reserves. However, this option should not be used indiscriminantly. It makes little sense to diversify our sources of supply to include unstable or unfriendly nations, or to obtain supplies from remote sources, requiring extensive sea-lane protection. For minerals having an extreme concentration of world reserves (such as cobalt and chromium) supply-source diversification is a limited option.

Stockpiling of critical minerals in adequate quantities, and in the appropriate stage of processing, remains the cheapest, safest, and quickest policy option to reduce our vulnerability to supply disruptions.

Substitution, recycling, and conservation have some potential to reduce U.S. vulnerability to supply disruptions, particularly if coupled with a substantial research and development program. However, the education of the American public about the importance of strategic and critical minerals, and about the declining position of the United States in this regard, is perhaps the most important task that awaits us.

REFERENCES

1. U.S. Federal Emergency Management Agency. *Stockpile Report to the Congress October 1981–March 1982.*
2. U.S. Department of the Interior, Bureau of Mines. *Mineral Facts and Problems, 1980 Edition.* Washington, D.C.: U.S. Government Printing Office, 1980.
3. U.S. Department of the Interior, Bureau of Mines. *Mineral Commodity Summaries 1983.* Washington: D.C.: U.S. Government Printing Office, 1983.
4. Miller, James Arnold. "Cobalt Availability: Thought-Provoking Commentary by M.I.T.'s Clark." *Alert Letter on the Availability of Raw Materials*, No. 19 (October 1982), p. 9.
5. Szuprowicz, Bohdan O. *How to Avoid Strategic Materials Shortages.* New York: John Wiley & Sons, Inc., 1981.
6. Cockran, C. N. "Energy Balance of Aluminum from Production to Application." *Journal of Metals*, July 1981, pp. 45–48.
7. U.S. Department of the Interior, Bureau of Mines. *Alumina Availability—Domestic.* Circular 8861. By G. R. Peterson, R. L. Davidoff, et al. Undated.
8. Schmidt, Helmut and Manfred Kruszona. *Regional Distribution of Mining Production and Reserves of Mineral Commodities in the World.* Hanover, West Germany: Federal Institute for Geosciences and Natural Resources, January, 1982.
9. Malthus, Thomas R. *An Essay on Population.* London: J. M. Dent & Sons Ltd., 1803.
10. Meadows, Donella H. *The Limits to Growth: A Report to the Club of Rome on the Predicament of Mankind.* New York: Universe Books, 1974.
11. Kahn, Herman, and Anthony Weimer. *The Year 2000: A Framework for Speculation on the Next Thirty-Three Years.* New York: Macmillan Publishing Co., Inc., 1967.
12. Solow, Robert M. "The Economics of Resources or the Resources of Economics." *Proceedings of the American Economic Association* 64 (May 1974).
13. Manes, Allan S. *U.S. Merchant Marine, Sealift Acquisition Policy and National Security.* Congressional Research Service, Library of Congress. 1981.
14. *The Changing Relationship: The Australian Government and the Mining Industry* As cited by W. C. J. van Rensburg and S. Bambrick in *The Economics of the World's Mineral Industries.* New York: McGraw-Hill Book Company, 1978.
15. U.S. Congress. Senate. Committee on the Judiciary. *Soviet, East German and Cuban Involvement in Fomenting Terrorism in Southern Africa. A Report of the Chairman of the Subcommittee on Security and Terrorism to the Committee on the Judiciary United States Senate.* 97th Congress, 2nd session, 1982.
16. *Engineering Mining Journal.* February 1982. "Managing Political Vulnerability." An interview with Dr. Warnuck Davies.
17. Warneke, Steven J. *Stockpiling of Critical Raw Materials.* London: The Royal Institute of International Affairs, 1980.
18. "Strategic Stockpiles: Who's Hoarding What?" *The Economist*, 24 May 1980.
19. "Government Measures." *Mining Journal*, 7 January 1983.
20. "Strategic Stockpile of Key Raw Materials 'Set Up by UK.'" *Financial Times*, 14 February 1983, p. 1.
21. National Materials Advisory Board, Commission on Sociotechnical Systems. *Considerations in Choice of Form for Materials for the National Stockpile.* Washington, D.C.: National Academy Press, 1982.
22. Huddle, Frank P. "The Evolving National Policy of Minerals." *Science* 191 (20 February 1976), pp. 654–659.
23. Eckes, Alfred E., Jr. *The United States and the Global Struggle for Minerals.* Austin, Tex.: University of Texas Press, 1979.
24. Leith, C. K. *World Minerals and World Politics.* London: Kennikat Press, 1931. Reissue 1970.
25. *The Stockpile Story.* Washington, D.C.: American Mining Congress, 1963.

26. *Strategic and Critical Materials Stock Piling Act of 1946, P.L. 520.*

27. *Strategic and Critical Materials Stock Piling Revision Act of 1979.*

28. Miller, James Arnold. "Defense Production Act: Relief for Domestic Mining Industry." *Alert Letter on the Availability of Raw Materials*, No. 6 (August 1981).

29. U.S. Federal Emergency Management Agency. *Stockpile Report to Congress October 1980–March 1981.*

30. Nichols, Mike. "Stockpiling for Strategic and Economic Purposes." Term paper. University of Texas at Austin, Spring Semester, 1980.

31. U.S. Congress. House. Committee on Interior and Insular Affairs. *U.S. Minerals Vulnerability: National Policy Implications. A Report Prepared by the Subcommittee on Mines and Mining of the Committee on Interior and Insular Affairs of the U.S. House of Representatives.* 96th Congress, 2nd session, 1980.

32. U.S. General Services Administration. "Budget Includes $245 Million for Stockpile Acquisitions." *GSA News Release*, 23 January 1978.

33. *National Materials and Minerals Program Plan and Report to Congress.* 5 April 1982.

34. Bateman, Alan. "Wartime Dependence on Foreign Minerals." *Economic Geology*, June–July 1946, pp. 308-327.

35. U.S. Congress. Senate. Committee on Armed Services. *Strategic and Critical Materials Stock Piling Act Revision. Hearing before the Subcommittee on Military Construction and Stockpiles of the Committee on Armed Services of the U.S. Senate.* 96th Congress, 1st session, 1979.

36. U.S. Senate Republican Policy Committee. "National Defense Stockpile Inventory of Strategic and Critical Materials 1962-80 Comparison." *Republican Report*, 1980.

37. Gagnon, James, U.S. Department of State, Washington, D.C. Interview, 4 June 1982.

38. Scott, Douglass P. "The Macroeconomic Environment for 1979 Stockpile Goals." Internal report to FEMA.

39. U.S. General Services Administration, Federal Preparedness Agency. *A Study of the Effect of Lead-Times, Substitutability, and Civilian Austerity on the Determination of Stockpile Objectives.* 15 July 1975.

40. U.S. Federal Emergency Management Agency, National Defense Stockpile Policy Division. *Cobalt Project.* By Marilyn B. Biviano. 26 November 1980.

41. Government official, U.S. Department of Commerce, Washington, D.C. Interview, 7 January 1983.

42. Drury, Orcott, U.S. Department of Commerce, Washington, D.C. Interview, 27 May 1982.

43. Miller, James Arnold. "Cobalt Bought from Zaire for the U.S. Stockpile." *Alert Letter on the Availability of Raw Materials*, No. 4 (August 1981).

44. U.S. General Services Administration, Office of Stockpile Transaction, Program Report and Accounting Staff. *Depot Totals per Commodity, Fiscal 1981.*

45. U.S. Department of the Interior, Geological Survey. Map of Curtis Bay Quadrangle. Baltimore County, Maryland. 7.5 Minute Series (Topographic), 1957.

46. U.S. Department of the Interior, Geological Survey. Map of West Point Quadrangle. Orange County, New York. 7.5 Minute Series (Topographic), 1971.

47. U.S. Department of the Interior, Geological Survey. Map of Maples Quadrangle. Allen County, Indiana. 7.5 Minute Series (Topographic), 1971.

48. U.S. Department of the Interior, Geological Survey. Map of Somerville Quadrangle. New Jersey. 7.5 Minutes Series (Topographic), 1969.

49. *H.R. 33.* 98th Congress, 1st session, 3 January 1983.

50. Miller, James Arnold. "DOD Management of the Stockpile." *Alert Letter on the Availability of Raw Materials*, No. 34 (March 1983).

51. Miller, James Arnold. "AMC Supports Transfer of Stockpile to Defense Department." *Alert Letter on the Availability of Raw Materials*, No. 37 (April 1983).

52. U.S. General Accounting Office. Report by the Comptroller General. *National Defense-Related Silver Needs Should Be Reevaluated and Alternative Disposal Methods Explored.* 11 January 1982.

53. U.S. Congress. House. Committee on Armed Services. *National Defense Stockpile Hearings on H.R. 2603, H.R. 2784, H.R. 2912, and H.R. 3364 before the Seapower and Strategic and Critical Materials Subcommittee of the Committee on Armed Services House of Representatives.* 97th Congress, 1st session, 2 and 4 June 1981.

54. Calaway, Lee, and W. C. J. van Rensburg. "U.S. Strategic Minerals Policy Options." *Resources Policy*, June 1982.

55. Landsberg, Hans. "Is a U.S. Materials Policy Really the Answer?" *Professional Engineer*, September 1982.

56. Cameron, Eugene N. "Changes in the Political and Social Framework of United States Mineral Resource Development, 1905-1980." *Economic Geology, Seventy-fifth Anniversary Volume, 1905-1980.* El Paso, Tex.: Economic Geology Publishing Co., 1981.

57. Honkala, R. A., and K. R. Knoblock. "A Cartographic Look at Constraints to Mineral Exploration and Development." *Mining Congress Journal*, February 1980.

58. Gates, Paul W. "The Federal Lands: Why We Retained Them." In *Rethinking the Federal Lands.* Edited by Sterling Brubaker. Washington, D.C.: Resources for the Future, Inc., p. 35.

59. U.S. Department of the Interior, Bureau of Land Management. *Public Land Statistics 1982.* Washington, D.C.: U.S. Government Printing Office, April 1983, pp. 3-4.

60. Bennethum, Gary, and L. Courtland Lee. "Is Our Account Overdrawn?" *Mining Congress Journal*, September 1975, pp. 33-48.

61. U.S. Department of the Interior, Geological Survey. *Principal Federal Lands Where Exploration and Development of Mineral Resources Are Restricted*, NAS-R-0401-75M01. Washington, D.C., January 1981.

62. U.S. Congress, Office of Technology Assessment. *Management of Fuel and Nonfuel Minerals in Federal Land: Current Status and Issues.* Washington, D.C., April 1979.

63. U.S. Office of the Federal Register. *The United States*

Government Manual 1982–83. Washington, D.C.: U.S. Government Printing Office, July 1982, p. 299.

64. Hagenstein, Perry R. "The Federal Lands Today: Uses and Limits." In *Rethinking the Federal Lands.* Edited by Sterling Brubaker. Washington, D.C.: Resources for the Future, Inc., p. 78.

65. Miller, Betty. *The Petroleum Potential of Wilderness Lands in the Western United States.* U.S. Department of the Interior, Geological Survey. December 1983.

66. Irving, Robert R. "How Can We Stave Off an OPEC in Metals?" *Iron Age,* 25 May 1981, pp. 79-85.

67. Grace, Richard P. "Metals Recycling: A Comparative National Analysis." *Resources Policy* 4, No. 4 (December 1978), p. 254.

CHAPTER I-5

Conclusions

The United States relies heavily on foreign suppliers for numerous nonfuel raw and processed minerals, and there is a trend toward the import of processed rather than raw minerals. Net imports now provide 50 percent or more of apparent consumption for at least 20 of the major mineral commodities, compared with 30% for oil. Import reliance is high for some commodities because the United States simply does not have the resource base. For other minerals, the resource base appears to be adequate, and heavy import reliance should be blamed on the low grade of domestic deposits, environmental and other regulations, high costs of labor and energy, low labor productivity, or the decline in America's mineral research and development facilities. More dangerously, the United States no longer has the capability to protect its overseas minerals supply routes.

There are many economic signals that indicate future problems for the U.S. domestic minerals industries. For instance, since about 1960, the ratio of current assets to current liabilities of American mineral industries has been declining. The continued high debt/equity ratios for the U.S. mineral industries mean that these industries must continue to rely on long-term debts instead of retained earnings or equity to finance a considerable share of their new ventures. Since 1970, the primary iron and steel industry and the primary nonferrous metal industries have had a lower return on equity than the average of all industry. The result has been little investment in capacity expansion and virtually no equity financing. This spells problems for the future.

International developments in the 1970s have generally had the effect of making it more difficult for the United States to ensure reliable supplies from abroad than ever before. Increasing nationalism in the Third World, and in countries like Australia and Canada, has acted as a deterrent to American investment in foreign mineral ventures. This trend has been exacerbated by nationalization, forced changes in agreements, increased taxation, and producer-government interference in many of these countries. The political situation in southern Africa and in the Middle East has deteriorated. America has allowed its naval power and other conventional military capability to decline at the precise time that the USSR has vastly increased its military strength. The United States and its allies no longer have control over a dominant share of world mineral resources, or control over vital sea-lanes.

America's energy and minerals problems are closely related. As the United States strives to decrease its dependence on imported oil, so its dependence on imported minerals increases. The United States now imports substantial amounts of energy in the form of energy-intensive processed minerals.

America's mining and metallurgical research facilities have been allowed to deteriorate to the point where the technological lead in these areas is moving elsewhere. The state of mining and metallurgical research in the United States has been described as "wretched" by the National Academy of Science. Much too little funding is being allocated to minerals-related research, such as the Mining and Mineral Resources Research Institutes program.

The U.S. strategic stockpile has serious shortcomings: many stockpiling facilities are located in unfavorable areas; some of the material in the stockpile is either substandard or has deteriorated; many inventories are below target level; and the target levels themselves have been changed for political or short-term financial reasons rather than strategic ones. While the U.S. mineral processing industry has declined, many of the materials in the stockpile are held in the form of ores and concentrates. During a period of huge federal deficits this situation will not improve.

Although considerable efforts are now being made to conserve energy, there is substantial waste of metals and minerals, despite higher prices and increased import dependence. Considerably more could be done to increase the proportion of metals obtained by recycling.

The development of more abundant substitutes for scarce minerals is not taking place to the extent required. The high cost of energy, and a lack of sufficient research and development effort, are further slowing down efforts in this regard.

Efforts aimed at decreasing America's nonfuel

mineral import dependence through the development of improved technology, allowing economic extraction of minerals from low-grade and submarginal domestic deposits, have failed to have the desired effect.

There is a need for legislation and regulation to stimulate domestic mineral exploration and exploitation. Unfortunately, most recent legislation appears to have had exactly the opposite effect, and environmental regulations have had a particularly adverse effect on the American mineral industry. Federal lands should be opened for mineral exploration and exploitation.

Similarly, there appears to be a need for tax incentives to promote domestic mineral exploration and development. Unfortunately, the mining industry has had few champions in Congress or in recent administrations.

Nonfuel mineral policy appeared to be a non-issue in the Carter administration. Social, political, and environmental considerations carried much more weight than strategic or resource requirements. This situation has improved somewhat in the Reagan administration.

The United States has become dangerously dependent on imported sources of nonfuel minerals, as well as oil, and there are few indications that these trends are likely to be reversed soon. Of greatest concern is the high level of dependence on chrome, manganese, cobalt, and platinum-group metals from southern Africa.

The USSR clearly regards the heavy import reliance of the United States and its allies for oil and strategic minerals as the Achilles' heel of the West, and can be expected to make every effort to exploit this situation. There are some signs that the USSR's own mineral industry has not kept up with domestic demands for a number of important minerals, and that its historical policy of self-sufficiency at all costs may be replaced by an outward-access policy. This promises to increase international tension and competition for global supplies of strategic minerals.

Nonfuel minerals import dependence is not merely an American problem, but a problem of the Western alliance. Western Europe and Japan are far more dependent on imports of strategic minerals than is the United States. In the short term, this means that the United States would have to share its strategic stockpile with its allies during times of conflict. In the medium and long term, it means that there will be increasing competition for supplies of these materials.

The adversarial nature of U.S. politics makes it unlikely that compromise on any mineral-related issue will come about easily. Instead, we are likely to have a dichotomy between mineral-related jobs and national security on the one hand, and calls for a clean environment and nature preservation on the other, leading to continuing stagnation of mineral policy. Such developments would have particularly adverse effects on access to federal lands by the mineral industry. The ignorance of the American public about the importance of strategic minerals, and about America's deteriorating mineral position, are of particular importance in this regard.

CHAPTER II-1

Introduction

Mineral trade represents a significant part of the Japanese economy. Almost all of the strategic minerals and energy supplies necessary for its industrial production are imported. Therefore, Japan is vulnerable to import-supply cutoffs. It has sought to reduce this vulnerability through a strong minerals policy, a resource-oriented foreign policy, and close government-industry interaction. Its strategy has been to maintain a steady flow of supplies and to maintain the export markets for its processed materials.

The global situation in the early 1980s affected Japan's minerals-related trade to a great extent. The worldwide recession, decline in metal demand, increase in metal stocks, changes in the minerals industries, trade friction between Japan and foreign nations, and increasing dependence on foreign sources for strategic mineral supplies will all influence Japan's future strategy. The continuing good health of Japan's mineral industries will depend on its ability to respond to cutoffs in mineral supplies and to political and economic factors that affect these industries.

Japan's strategy, both of the past and the future, regarding its mineral industries, is the subject of this chapter. An examination is made of the Japanese mineral industry, with emphasis on business strategies, infrastructure, and environmental concerns. Mineral acquisition strategies, used by both public and private organizations, are also covered. Japan's mineral policy is examined for its effects on various policy options available to a resource-poor, mineral-consuming country. This research also considers some external factors that can affect the flow of foreign mineral supplies to Japan.

In addition, a brief comparison is made between the Japanese and U.S. mineral policies. Similar to Japan, the United States is dependent on foreign sources for several strategic minerals. The proposed U.S. mineral policy offers little in the way of a foreign-supply-diversification goal. This is one option that the Japanese have exercised successfully. Of course, Japan is dependent on foreign sources for a larger number of strategic minerals than the United States, and it has to pursue all possible options to maintain its supplies. However, the United States can still benefit from the Japanese example.

U.S. authors have given only scant attention to the problem of Japan's strategic mineral dependence. Only recently have many of them begun to recognize a similar dependence problem for the United States.

INFORMATION REQUIREMENTS

Information used in this study was obtained mostly from secondary sources. U.S. and foreign trade journals provided much information on the Japanese mineral industry and government mineral policy. Some Japanese government publications were used. Statistics were taken mostly from U.S. Bureau of Mines (USBM) publications, such as *Mineral Facts and Problems*, *Minerals Yearbook*, and *Mineral Commodity Summaries*, and from Japanese sources, such as *Industrial Review of Japan*, *Japanese Economic Journal*, and the *Japan Statistical Yearbook*. All of the materials used were English-language sources.

Although this chapter provides statistics about mineral imports, production, consumption, and exports, the results of the study are of a greater depth than that provided in the many annual surveys available in the literature. The study itself represents an effort to blend together all the available information and to examine major trends in the Japanese mineral industry.

Mineral Industry Strategies

Japan's mineral industry has played a crucial role in this country's rapid industrialization in the 1960s and early 1970s. Production capacity expanded dramatically in response to increased domestic and export demand for basic materials and manufactured goods. By the 1970s Japan had become one of the world's top three steel producers, a leading producer of nonferrous and industrial metals and materials, and of numerous high-quality manufactured goods. All of those industries required significant input of energy and raw materials.

Since the early 1970s, Japan's mineral industry has encountered increased volatility in energy and raw material markets, increased trade-protectionist sentiments in foreign markets, and slow global economic growth. This industry has taken several steps to respond to these changes, in order to retain a leading economic position worldwide.

This chapter examines several factors involving the continued operation of the Japanese mineral industry. The many companies involved in mineral supply acquisition and the extent of their activities are briefly described. A more detailed discussion of the strategies that Japan uses to maintain a secure and steady supply of minerals is provided in Chapter II-5. The various strategies employed by the mineral companies to cope with the changing economic conditions of the 1980s are also evaluated.

MINERAL SUPPLIES

Small amounts of minerals are mined by several mining companies in Japan. The eight major companies are: Dowa Mining Co. Ltd., Furukawa Co. Ltd., Mitsubishi Metal Corp., Mitsui Mining and Smelting Co. Ltd., Nippon Mining Co. Ltd., Nittetsu Mining Co. Ltd., Sumitomo Metal Mining Co. Ltd., and Toho Zinc Co. Ltd. (listed in alphabetical order).[1] Many of the other Japanese mining companies are either partly or wholly owned subsidiaries of one or more of the above eight companies.[2]

The major mining companies had their beginnings in domestic mining operations. The mining activities of four of these firms led to the eventual development of the large Furukawa, Mitsubishi, Mitsui, and Sumitomo business groups.[3] Furukawa

Co. Ltd. has since pulled out of the mining business almost completely.

These companies are involved mainly in nonferrous (zinc, lead, copper, and tin), and precious (silver and gold) metal mining (Table II-2-1). Several ferrous metals are also mined in Japan. Only the production of zinc, tungsten, lead, and silver are significant, accounting for as much as 36%, 25%, 22%, and 21% of domestic ore and concentrate requirements in 1981 (Table II-4-1).

Several of these mining companies maintain overseas operations (Table II-2-2). Frequently, a company will form consortia with other mining companies, manufacturers, metal processors, and trading companies, to spread the risk of overseas operations. The list of present-day overseas operations indicates an emphasis on the development of copper mines. This emphasis is the result of Japanese mineral policy,[4] which encouraged such development as a way to secure copper supplies for Japan's expanding electrical and electronics industries in the 1960s and 1970s.

A large amount of Japan's mineral and energy supplies are acquired through trade by the large Japanese general trading companies (GTCs). GTCs were the original trading arms of Japan's large business groups. Their responsibilities have expanded to include several other areas besides trading, including marketing, maintenance of inventory stocks, transportation and distribution, intelligence, financing, risk assumption, and joint-venture organization.[5]

There are nine major trading companies in Japan: Mitsubishi Shoji, Mitsui Bussan, Marubeni, C. Itoh, Sumitomo Shoji, Nissho-Iwai, Toyo Menka, Kanematsu Gosho, and Nichimen.[6] A tenth company, Ataka, was merged with C. Itoh in 1977. In 1979, Japan's total trade volume was $237 billion and the 10 largest GTCs accounted for approximately 54% of this total.[5]

Japan's large business groups consist of several industrial and manufacturing companies, a central bank, and a GTC. For example, Japan's Sumitomo group includes several basic material, chemical, machinery, warehousing, shipping, and electronics companies (Figure II-2-1). Four of Japan's major

TABLE II-2-1   Major Operating Japanese Strategic Mineral Mines, 1981–1982

Mine, prefecture	Ores mined[a]	Operating company
Akatani, Niigata	Iron (182)	Nittetsu Mining
Akenobe, Hyogo	Copper (1200) + gold, tin, tungsten, zinc	Akenobe Mining (100%-owned subsidiary of Mitsubishi Mining)
Hanaoka, Akita	Zinc (2900) + copper, lead	Dowa Mining
Hirase, Gifu	Molybdenum (30)	Sumitomo Metal Mining
Hitachi, Ibaragi[b]	Copper (550) + zinc	Hitachi Mining (100%-owned subsidiary of Nippon Mining)
Hosokura, Miyagi	Lead (2500) + zinc, copper, gold, silver	Hosokura Mining (100%-owned subsidiary of Mitsubishi Metal Mining)
Iwami, Shimane	Zinc (180) + copper, lead	Iwami Mining (100%-owned subsidiary of Mitsui Mining & Smelting)
Kamaishi, Iwate	Copper (2940), Iron (3840)	Nittetsu Mining
Kamioka, Gifu	Zinc (7000) + lead, silver	Mitsui Mining & Smelting
Kosaka, Akita	Zinc (1750) + copper, lead	Dowa Mining
Nakatatsu, Fukui	Zinc (2000) + copper, lead	Nippon Zinc Mining (99.8%-owned subsidiary of Mitsui Mining & Smelting)
Sazare, Ehime	Copper (750)	Sumitomo Metal Mining
Shakanai, Akita	Copper (930), zinc (70) + lead	Shakanai Mines (100%-owned subsidiary of Nippon Mining)
Shimokawa, Hokkaido[c]	Copper (1500) + zinc	Shimokawa Mining (100%-owned subsidiary of Mitsubishi Metal Mining)
Toyoha, Hokkaido	Zinc (1640) + lead	Toyoha Mines (100%-owned subsidiary of Nippon Mining)
Yaguki, Fukushima	Tungsten (88)	Nittetsu Mining

[a]Numbers in parentheses represent daily capacity in metric tons.
[b]Hitachi mine was expected to close in September 1981, because of depleted reserves. See "The Hitachi Copper Mine . . .," *Engineering and Mining Journal* 182 (September 1981).
[c]Shimokawa mine was expected to close in late 1982, because of depleted reserves. See "Shimokawa Mine to Close," *Mining Journal* 299, No. 7682 (12 November 1982).
Source: Ref. 2.

business groups are the Mitsubishi, Mitsui, Sumitomo, and Fuyo groups.[5]

Japanese GTC's consider the acquisition of natural resources to be one of their most important responsibilities.[7] GTCs act both independently and in consortia to make mineral purchases. Because of the large size of a business group, and the combined efforts of consorting GTCs, huge volume raw material purchases can be made at a lower cost than by individual company purchases. At the same time, GTCs can exert considerable influence on international mineral markets. In addition, because of the diversity within the major business groups, those GTCs are in a favorable position to provide desired products for barter arrangements with overseas mineral producers.

In the examples where Japanese mining companies have established their own overseas mining or processing activities, one or more GTCs are often involved as joint-venture partners. GTCs provide considerable expertise in setting up trade channels.[6] Their established worldwide communication and transportation networks, distribution systems, and marketing skills are essential to the mining company during the early stages of overseas project development. In addition, GTCs can provide the capital required for large-scale mining projects, through their association with large Japanese banks.[5]

In the late 1970s and early 1980s, Japan's GTCs faced a business slowdown largely attributed to their dependence on the lower-value raw and processed material sectors.[8] Their diversification efforts emphasize higher-value sectors, and include either the initiation or expansion of in-house high-technology departments.[9] Diversification efforts will probably have some effect on the basic materials sectors. However, Japanese GTCs will probably

TABLE II-2-2 Major Japanese Overseas Mining Projects, 1981[a]

Mineral commodity	Project, country	Local participants	Japanese participants	Annual exports to Japan (thousand metric tons)[b]	Production startup date
Copper	Bougainville, Papua New Guinea	Bougainville Copper Co.	Consortium of 9 companies which includes Nippon and Furukawa mining companies	78	1977
	Ertsberg, Indonesia	Freeport Indonesia Co.	Consortium of 12 companies which includes Dowa, Furukawa, and Nippon mining companies, and Nippon Steel	40	1977
	Katanga, Zaire	NA	Mitsui Group	1	1974
	Mamut, Malaysia	NA	Mamut Mining Development Co.: Consortium of 7 mining companies, including Nippon, Mitsubishi, Mitsui, Sumitomo, and Furukawa	27	1979
	Mushoshi, Zaire[c]	Government of Zaire	Consortium of 8 companies, including Nippon, Mitsui, Sumitomo Dowa, and Furukawa mining companies	34	1972
Lead/zinc	Huanzala, Peru	NA	Mitsui & Co.	9 (lead) 24 (zinc)	NA
Nickel	Rio Tuba, Philippines	NA	Taiheiyo Metal, Nippon Steel, and two other companies	10	1977
	Soroako, Indonesia	Inco of Canada Ltd.	Tokyo Nickel, Shimura Kako, and 4 other companies	NA	NA
Uranium	Akouta, Niger	Nuclear Fuel Corp. of France, Niger Mineral Resources Corp., Uranium Corp. of Spain	Overseas Uranium Resources Development Co.	0.3 (oxide)	1978

[a]NA, not available.
[b]Annual exports in terms of 1000 metric tons metal content.
[c]In 1983, the Japanese consortium sold its share of the Musoshi project to the Zaire government for $30 million. The consortium suffered losses of $220 million on the project. "Eight Japanese Companies . . .", *Mining Engineering* 39 (August 1983).
Sources: U.S. Department of State, Telegram No. 18,018, Tokyo, October 1982; Ref. 3.

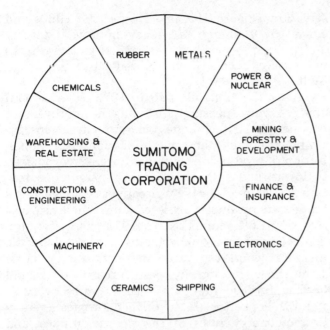

FIGURE II-2-1 Sumitomo business group (Japan). (From Ref. 5.)

continue to be important actors in the acquisition of materials required for Japan's future growth industries.

MINERAL INDUSTRY STRATEGIES

In the face of worldwide recession, higher energy costs, trade-protectionist sentiments, increased substitution for basic materials, and competition from newly industrialized countries (NICs), Japan's industries have had to develop new survival strategies. One strategy is the national switch to higher-technology industries, as discussed in Chapter II-5. Within the minerals industry, several strategies have already been put into operation. Japanese mineral companies have attempted to improve and maintain the security of their energy supplies, using several different tactics, including cutbacks in energy-intensive production capacity. Materials recycling and secondary production have played a role in conserving nonenergy resources. Other strategies include increased emphasis on higher-value production, business diversification, and the export of technology.

Maintaining Energy Supply Security

Although energy and minerals are often considered as mutually exclusive entities, they have an intimate relationship in the minerals-processing industry. Energy is required to fuel the reactions involved in mineral processing. In addition, coke (coal), oil, and natural gas serve as reductants in the smelting processes for some metals. The ex-

ample of a blast furnace well illustrates this relationship, with respect to coal and iron ore (Figure II-2-2). Coal is often used as the energy source for iron ore reduction. In addition, coal is the raw material required to make coke.

Both the availability and the cost of energy supplies are important factors to a mineral-processing company. These factors are extremely important for Japan, whose domestic energy resources are limited. In fiscal year 1979, Japan was 87% dependent on foreign sources for its energy supplies, which include petroleum, coal, liquefied natural gas, and uranium.[10]

Several of Japan's basic industries require very large amounts of energy. The smelting and refining operations for aluminum, ferrosilicon, and zinc, and the production of tungsten metal, are examples of energy-intensive processes. The cost of energy can represent a significant portion of total processing cost, as it does for aluminum metal.

Japan became very dependent on oil-based energy in the 1960s and early 1970s, when its economy was expanding rapidly. Cheap oil was readily available from countries in the Middle East. The OPEC oil price increases during the 1970s were quite significant to the Japanese. In 1975, the average price (based on U.S. dollars) of crude oil in Japan was 250% higher than that in 1973,

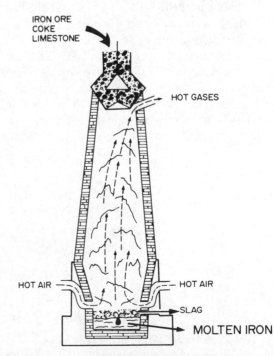

FIGURE II-2-2 Diagrammatic view of the interior of a blast furnace, showing input requirements of energy and raw materials. (From Brian Skinner. *Earth Resources.* Englewood Cliffs, N.J.: Prentice-Hall, Inc., 1976, Fig. 5-2, p. 57.)

after the first set of OPEC price increases.[11] By 1980, after the second price increases, the average price had increased 140% over 1978 levels.

Japan has developed several strategies to deal with its high net import reliance on energy supplies. Four of these, diversification, conservation, cutbacks in energy-intensive production capacity, and stockpiling, are discussed below.

Diversification of Energy Types and Sources. Japan is diversifying its energy mix so that it will be less dependent on oil-based energy. Japan's future energy supplies are expected to have increased input from imported coal, nuclear (uranium) and liquefied natural gas (LNG), and less from imported oil (Figure II-2-3). In addition, domestic input from hydroelectricity, geothermal and solar, is expected to increase.

Japan has already begun to diversify its geographic sources of each of these energy supplies (Table II-2-3). The methods used include involvement in overseas exploration and development projects, and long-term contracts.

Energy Conservation. The strategy of energy conservation has been employed by several of Japan's heavy industries. Energy conservation efforts involve the introduction of new technology, and continued research and development.

One industry that can be used to exemplify these efforts is the domestic iron and steel industry. Iron and steel industry energy consumption represents 15% of national energy consumption.[12] Therefore, this industry's efforts can have significant effects on the nation's total energy supply. Energy conservation has been an important goal of the Japanese iron and steel industry. The decline in this industry's unit energy consumption since the early 1950s exemplifies its success in energy conservation (Figure II-2-4a). During the 1960s and early 1970s energy efficiency improved steadily and production capacity expanded as large, technologically advanced, integrated steelworks were built.

After the first oil crisis in 1973–1974, several energy-savings measures, including the improvement of operational techniques, investments in energy-saving equipment, and modernization and rationalization of production equipment, were incorporated to lower further the unit rate of energy consumption[12] (Figure II-2-4). By fiscal year 1980, Japan's crude steel producers achieved a unit energy savings rate of 11.5%, compared to 1973 energy requirements (Figure II-2-4b). The most successful specific measures employed after the first energy crisis were the reduction in blast furnace and metal-heating furnace fuel rates, waste gas recovery at the LD converter (basic oxygen furnace), reduction in energy rates of the electric furnace, and probably improved steel yield.[12] The latter has partly been the result of the increased use of continuous-casting technology in Japan.[13]

Following the second oil crisis in 1979, energy-saving efforts were intensified and measures were taken to reduce the industry's dependence on oil. Blast-furnace oil injection was almost completely halted by 1981 as most blast furnaces converted to all-coke operations.[14]

Waste energy recovery has also contributed significantly to the iron and steel industry's energy-saving achievements. Recovered energy is used in downstream processes. The use of high-temperature recovery systems such as blast-furnace top-pressure recovery turbines and coke oven heat exchangers has increased steadily. The increase in the latter system's use resulted from the installation of coke-dry quenching (CDQ) facilities at many Japanese steel plants.[15] The inert gases, nitrogen, carbon dioxide, and carbon monoxide, rather than water, are used as a coolant in the CDQ process. These gases are circulated through heat exchangers to generate steam for electric power production or for by-product chemical processing. In fiscal year 1981, the total power generated from these two recovery systems was 2000 million kilowatts, amounting to about 5% of Japan's integrated steelmakers' power consumption.[14]

The all-coke blast furnace operation creates a surplus in coke oven and blast furnace gases.[12] These gases can be used in place of oil in downstream steelmaking processes.

With respect to waste heat, future emphasis will probably be placed on middle- and low-temperature heat recovery. Lower-temperature heat recovery will require the use of new technology, such as the

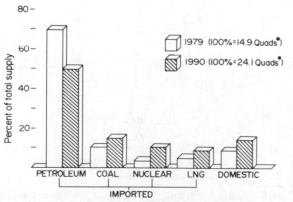

FIGURE II-2-3   Japan's energy mix, FY 1979 and 1990. (From Ref. 10.)

TABLE II-2-3   Japanese Energy Commodity Suppliers

Energy commodity	1981 suppliers		Additional future suppliers
Petroleum	Saudi Arabia	Indonesia	Canada
	United Arab Emirates	China	North Sea
	Oman	Mexico	Gabon
	Kuwait	Venezuela	
	Qatar	Ecuador	
	Iran	Nigeria	
	Iraq	Algeria	
	Bahrain	Egypt	
	Neutral Zone (Middle East)		
Natural gas (liquefied)	Indonesia		Malaysia
	Brunei		USSR
	Abu Dhabi		Canada
	United States		Australia
Coal	Australia	USSR	Indonesia (?)
	Canada	Vietnam	Colombia (?)
	United States	China	
	South Africa		
Uranium	Australia		United States (?)
	Canada		Brazil (?)
	South Africa		Argentina (?)
	Niger		Mexico (?)

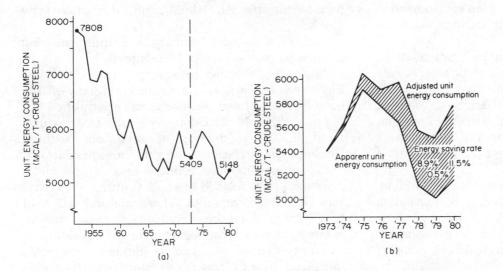

FIGURE II-2-4 (a) Annual Japanese steel industry unit energy consumption, FY 1953–1980; (b) annual Japanese steel industry unit energy consumption, FY 1973–1980; indicates unit energy savings between the observed unit energy consumption (apparent unit energy consumption) and the unit energy consumption which would have been expected had the production structure and related factors not changed since 1973 (adjusted unit energy consumption). (From Ref. 12.)

Freon turbine system in use at Sumitomo Metal Industries' Kashima steel works.[16]

Energy-saving efforts will continue to be important to the Japanese iron and steel industry. Future efforts might be expected to emphasize reductions in the blast-furnace fuel rate, increased use of hot-charging and direct-feed rolling methods, and increased recovery and use of waste heat.[12]

Cutbacks in Energy-Intensive Production Capacity. A third strategy, pursued mainly by energy-intensive mineral-processing industries, is that of a production-capacity cutback. The aluminum smelt-

ing industry is an example of one industry that has made significant capacity cutbacks. This industry encountered high energy costs in the 1970s, which resulted in it acquiring an uncompetitive position in world markets. In response, industry-wide production capacity was reduced. To satisfy domestic aluminum metal demand, Japanese producers increased the level of cheap aluminum metal imports, invested heavily in overseas production capacity, and increased the level of secondary production (Chapter II-4).

Production capacity cutbacks are also expected in the energy-intensive electric furnace steel and ferroalloy industries. Both of these industries have been designated by the government as being structurally depressed because of the high cost of energy inputs. Cutbacks have already occurred in domestic ferrosilicon capacity (Chapter II-4). However, in the early 1980s electric-furnace steelmakers had not yet cut back their production capacity, as individual companies were fighting to maintain their market share.[17]

The supply alternative pursued by the domestic ferroalloy industry has been to increase the level of ferroalloy imports. Ferroalloy imports have almost tripled since 1977, and imports of chromium and manganese ores, which make ferroalloys, have leveled off or declined slightly (Chapter II-4). Investments in overseas ferroalloy capacity have been insignificant during this time.

Stockpiling. Energy stockpiling in Japan mainly concerns oil supplies. The Japanese long-term economic and social plan (1979–1985) discussed the need for the country to strengthen its ability to respond to interruptions in imported oil supplies.[18] Two recommendations were given in this plan. The first called for increases in the national oil stockpiles. A 90-day supply target was proposed for private oil stockpiles. Second, the plan recommended the development of an oil-rationing system for emergency use.

In July 1982, the government recommendations for private oil stockpiles appeared to be met. Private stockpiles of crude oil and products were approximately 380 to 390 million barrels.[19] National stockpiles totaled 69 million barrels. Together, these represented a 121-day supply.[19] The private stockpile contribution to this total was almost a 103-day supply.

MITI's proposed budget for fiscal year 1983 included increases to cover the costs of expanding Japanese oil and LPG stockpiles.[20] Oil was to be purchased on the open market with government-subsidized loans from the Development Bank of Japan.

Non-Energy-Resource Conservation

In general, the development and use of highly efficient technology by Japan's mineral-processing industries has helped to improve product yield and thus to conserve resources. Specific recycling measures have also been significant.

Some important conservation measures are exemplified in Japan's iron and steel industry.[21] Mill wastes, such as steel scrap and iron-bearing dust, are recovered for use in the steelmaking process. Most steel plants in Japan have closed-water circulation systems. These systems treat and recirculate wastewaters for reuse. By-product slag has been used as a substitute for natural crushed stone and as fine aggregate in concrete.

Secondary aluminum metal production from scrap increased significantly during the period 1975–1981. Aside from the beneficial effect on energy supplies, the increase in secondary production will help to conserve Japan's bauxite and alumina supplies. To a lesser extent, the increase in secondary lead metal production (Chapter II-4) will help to conserve lead ore supplies.

Higher-Value Production

Emphasis on higher-value production is a strategy being pursued by many members of Japan's basic materials industry. Examples of this strategy are observed in the iron and steel, aluminum, and titanium industries.

In response to the rising competitiveness of the steel industries in the newly industrialized countries (NICs), including South Korea and Brazil, Japanese steel producers will apparently limit the growth of their annual crude steel production. In contrast, producers will probably increase their production of higher-value products, such as single-surface galvanized steel sheet, precoated steel sheet and plate, high-tensile-strength steel, and dual-phase-type steels.[22]

In the aluminum industry, aluminum smelting companies are expected to merge with aluminum rolling and processing companies in pursuit of this higher-value strategy.[23] Individual company efforts have already been initiated, as exemplified by Nippon Light Metal Company's increased involvement in the production of aluminum castings, forgings, and surface-treated aluminum products.[22]

Japanese titanium product makers and metal producers have formed business ventures with overseas titanium companies. These ventures will help to ensure that Japanese companies capture the increased revenues from higher-value production.

Domestic research and development programs in pursuit of a process for developing usable forms of high-technology commodities, such as gallium arsenide, are in agreement with this strategy. Expensive, ultrapure forms of minerals are required in these future-growth, high-technology industries.

Business Diversification

In the 1970s and early 1980s many U.S. mineral and energy companies diversified into several other fields. Similarly, in the early 1980s, Japanese mineral-processing industry members have begun to use diversification as a means to cope with their declining shipments and increasing stocks.

This strategy is observed in the Japanese iron

and steel industry. Japanese steelmakers are expected to take an active role in the development of the new basic materials industries, including carbon fiber, binary materials, and ceramics.[24] Kobe Steel and Sumitomo Metal Industries have already become involved in the production of other metals, including titanium.

The chemical industry also appears to be an attractive diversification alternative to steelmakers.[24] Initial emphasis will be given to coal chemistry, especially to the effective use of coal tar and coke oven by-product gases.

However, Japanese companies who are diversifying will often continue to make investments in their main businesses. In the iron and steel industry, capital overlays remained high during the late 1970s and the early 1980s.[14] Much of this investment has been directed toward plant and equipment rationalization, maintenance and repairs, and energy conservation (Figure II-2-5).

Industry investment will continue to be necessary in the future. The steelmaking facilities built during Japan's high-growth period in the 1960s will be due for replacement in the late 1980s.[14] If not replaced, Japanese steel producers may not be able to maintain their global competitiveness, and therefore their ability to generate profits for future investment would be curtailed.

Export of Technology

The Japanese are known worldwide for their capability of producing many basic material commodities cheaply and in an efficient manner. As a result, Japanese mineral-processing technology is in demand by those countries that are anxious to develop their own basic industries. This technology is exported in the form of processing plants and equipment, or technological cooperation.

For example, in the early 1980s Japan's iron and steel industry was involved in several overseas projects in the United States, Europe, Latin America, the Middle East, and Asia (Table II-2-4). Japan's iron and steel industry also maintained technological cooperation programs with the corresponding industries in Algeria and Italy.[14]

A 1982 agreement between Kawasaki Heavy Industries Ltd. (Japan) and Friedrich Krupp GmbH (West Germany) will enable Japan to become more involved in the growing direct-reduction (DR) iron industry.[25] After approving the agreement, Kawasaki became only the fifth company in the world to participate in the commercial production of coal-fired DR steel plants. Kawasaki already produces gas-fired DR steel plants. Exports will probably be directed toward the Southeast Asian countries.

The Japanese iron and steel industry is also active in the development and export of continuous-casting technology (Table II-2-4). Continuous casting is a process that eliminates the intermediate steps between the molten steel and semiprocessed steel stages in steelmaking. This process conserves energy, because its use reduces the number of times that materials need to be cooled and reheated. This process also results in higher steel yield, good-quality uniform products, and higher overall capital and labor productivity.[13] Japan's continuous-casting ratio (steel products produced using continuous-casting methods/total steel industry products) has increased steadily since 1973 (Figure II-2-6). In 1982, the continuous-casting ratio was 79%.

The export of aluminum smelting plants and the construction of plants overseas are other examples of technology export. As noted in Chapter II-3, Japanese smelter energy requirements per unit of metal production are lower than those of foreign counterparts. For this reason Japanese-built smelters are attractive investments to overseas buyers.

Technology export is also important in the copper metallurgy field. Japan's Mitsubishi Metals Corp. supplied the continuous copper smelting technology for Texasgulf's smelter at Timmons, Ontario, Canada, which was completed in 1981.[26] According to 1981 agreements, Sumitomo Metal Mining will provide copper smelting technology,

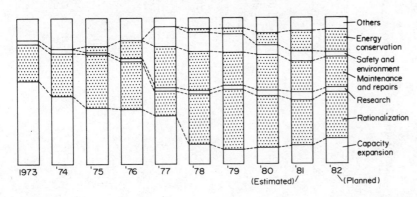

Others
Energy conservation
Safety and environment
Maintenance and repairs
Research
Rationalization
Capacity expansion

1973 '74 '75 '76 '77 '78 '79 '80 (Estimated) '81 '82 (Planned)

FIGURE II-2-5 Japanese iron and steel industry investment, 1973–1982. (From Japan Development Bank; Ref. 14.)

TABLE II-2-4 Japanese Iron and Steel Technology Export

Country	Local participants[a]	Japanese participants	Export description
Argentina[b]	SOMISA — state-owned steel company	Consortium of 50 companies: includes Nippon Steel, Mitsubishi, Export-Import Bank of Japan and OECF[c]	$130 million hot strip mill, as part of agreement in which Japan also provided $50 million investment in expansion projects[d]
Czechoslovakia[e]	NA	Japan Steel Works	Steel technology licensing agreement, new technology for existing works, engineering services, and technical training
Egypt[f]	Egyptian government	Consortium that includes Nippon Kokan, Kobe Steel, and Toyo Menka Kaisha	Install gas direct reduction integrated steel works at El Dikehila
Indonesia[g]	P.T. Krakatau Steel	Kobe Steel	Technical assistance in equipment operation and maintenance, production, and quality control, at Cilegon Steelworks
Malaysia[h]	Heavy industries Corp. of Malaysia Berhad	Nippon Steel	Construction of 600,000-metric ton/yr direct-reduction integrated steelworks at Trengganu
Mexico[i]	SICARSTA — state-owned steel company	Nippon Kokan	Construction of four 200-metric ton electric furnaces, related structures and cranes, and civil engineering and operator training
Mexico[j]	SICARSTA	Japan-Mexico Large Diameter Steel Pipe Corp.: Consortium that includes Sumitomo Metal Industries and OECF[b]	Construction of steel pipe plant in Lazaro Cardenas region
	SICARSTA	Japan-Mexico Steel Casting & Forging Co.: Consortium that includes Kobe Steel, OECF,[b] and 53 other companies	Construction of casting and forging plant in Lazaro Cardenas region
China[e]	NA	Japan Steel Works	Steel technology licensing agreement, new technology for existing works, engineering services, and technical training
China[k]	Chinese government	Nippon Steel	Second-stage construction of Baoshan steelworks, China: includes continuous cold-rolled and hot-rolled strip mills, and ingot casting unit; blast furnace, coking and sintering plants

134

Country		Japanese company	Project description
Philippines[l]	National Steel Corp. (NASCO) — state-owned steel company	Consortium that includes Marubeni Corp., Kawasaki Heavy Industries, Kobe Steel, and UBE Industries	Construction of direct-reduction furnace, ore pelletizing facility, and limestone baking furnace
Spain[m]	Industry and Energy Ministry	Kawasaki Steel	Formulate long-term reconstruction programs for three integrated blast-furnace steelmakers
Spain[e]	NA	Japan Steel Works	Steel technology licensing agreement, new technology for existing works, engineering services, and technical training
United States[n]	Armco, Inc.	Nippon Steel	Cooperation in steelmaking technology and coking coal deposit development
United States[n,o]	Republic Steel	Kawasaki Steel	Broad technological cooperation agreement that includes slab continuous casting process application at Cleveland, Ohio, plant
United States[o]	Jones & Laughlin	Consortium that includes Sumitomo Metal Industries	Construction of slab continuous caster at Indiana Harbor plant
United States[o]	Wheeling-Pittsburgh Steel Corp.	Nippon Steel	Technical assistance regarding slab continuous caster

[a]NA, not available.

[b]OECF, Overseas Economic Cooperation Fund.

[c]Japanese consortium pulled out of this project after SOMISA failed to purchase hot strip mill in 1982.

[d]"Japan Pulls Out of Argentinian Steel Project," Mining Journal 298, No. 7677 (8 October 1982).

[e]"Steel Contract with China," Asia Research Bulletin 12, No. 8, Report 5 (31 January 1983).

[f]"Japanese Reach Accord with Egyptians on Steel Plant," Japan Economic Journal 6 (October 1981).

[g]"Indonesia Needs Japanese Assistance," Ironmaking and Steelmaking 9. No. 2 (1982).

[h]"Malaysia and Japan's Nippon Steel Will Build DR Steel Plant," Engineering and Mining Journal 183 (January 1982).

[i]"News from NKK," Ironmaking and Steelmaking 9, No. 4 (1982).

[j]U.S. Department of the Interior, Bureau of Mines, "The Mineral Industry of Mexico." By O. Martinez. Minerals Yearbook 1980, Vol. III. Washington, D.C.: U.S. Government Printing Office, 1981.

[k]"China Launches Steel Complex's Second Portion," The Wall Street Journal, 9 June 1983.

[l]"Four Japanese Companies Appear Set to Build NASCO Steel Plant," Japan Economic Journal, 15 February 1983.

[m]"Kawasaki Steel Corp: Will Offer Reconstruction Aids to Spanish Steel Industry," Oriental Economist 50 (January 1982).

[n]Kenji Suzuki, "Iron and Steel: Slump of Demand Compelled Curtailment of Production in Last Half of 1980." In Industrial Review of Japan, 1981. Tokyo: Japan Economic Journal, 1981.

[o]George McManus, "A New Role for Ladle Metallurgy: High Tonnage Output," Iron Age, 7 June 1982.

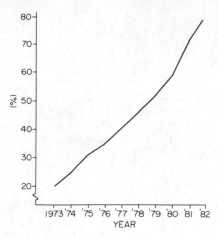

FIGURE II-2-6 Annual Japanese iron and steel industry-wide continuous casting ratio, 1973–1982. (From Refs. 12 and 14; "Steel Companies Make Steady Progress in Saving of Energy." *Japan Economic Journal*, 22 February 1983.)

and members of the Furukawa business group will provide flash smelting and electrolytic refining technologies to companies in the United States.[26] In addition, both Mitsui Mining and Smelting Co. and Sumitomo Metal Mining have been contracted to build copper smelters in the Philippines[27] and China,[28] respectively.

To maintain a large share of the technology export market, Japanese producers will have to continue to develop their processing technology. Domestic processing facilities provide a testing ground for Japanese research and development advancements. At the same time, however, the high cost of imported energy supplies may preclude the survival of some domestic processing industries.

In energy-intensive industries such as aluminum smelting, companies have already made severe capacity cutbacks. The Japanese government proposed a long-term minimum annual production capacity goal of approximately 700,000 metric tons of aluminum metal.[29] However, the annual production rate was estimated to be 260,000 metric tons, in early 1983.[30] The price of aluminum in oversupplied world markets is too low for metal production to be profitable to domestic producers, whose production cost is significantly affected by high energy costs. In another example, the high cost of energy, as well as pollution control equipment, may limit any further expansion in domestic primary copper metal production capacity.[31]

As a result, future growth in domestic metals production capacity will be limited, and in some cases all production capacity could be eliminated. As a consequence, Japanese producers may not find it profitable to continue their development of smelting technology.

SUMMARY

Japan's mineral supplies are available through its mining companies and general trading companies (GTCs). Although they lack specific knowledge in mining and related areas, GTCs can provide considerable skills in setting up essential mineral trade channels and capital for overseas development projects.

The availability of energy supplies has had an important effect on Japan's mineral industry. The OPEC oil embargoes and price hikes of the 1970s put Japan on the defensive, because of its heavy oil import dependence. Mineral industry responses to this situation involve its efforts to diversify its energy mix as well as its sources of energy supply. Conservation and energy-intensive production capacity cutbacks are two other tactics used by this industry to cope with its high energy import dependence. In addition, the Japanese maintain oil stockpiles for short-term interruptions in oil supplies.

Several interrelated strategies are employed by individual mineral companies to cope with the changing present and future economic conditions. These strategies include resource conservation, increased emphasis on higher-value production, diversification, and technology export. The first three strategies will contribute to the continued competitiveness of Japan's mineral industry with respect to those of foreign countries and will therefore provide it with the capital to continue research and development programs aimed at the development of new technology for domestic as well as export markets.

REFERENCES

1. Financial Times. *International Mining Company Yearbook 1981*. Harlow, Essex, England: Longman Group Ltd., 1981.
2. *1981–82 World Mines Register*. San Francisco: World Mining, 1981.
3. "Japan Today." *Mining Magazine* 125 (November 1971).
4. Crowson, P. C. F. "The National Mineral Policies of Germany, France and Japan." *Mining Magazine*, June 1980.
5. Lin, Kuang-Ming, and W. R. Hoskins. "Understanding Japan's International Trading Companies." *Business* 31 (September–October 1981).
6. Yoshinara, Kunio. *Sogo Shosha: The Vanguard of the Japanese Economy*. Tokyo: Oxford University Press, 1982.
7. Young, Alexander. *The Sogo Shosha: Japan's Multinational Trading Companies*. Boulder, Colo.: Westview Press, Inc., 1979.
8. "Time of Trial for General Traders: Vaunted Versatility Will Stand Them in Good Stead." *Oriental Economist* 47 (January 1979).

9. "Giant Traders Taking a High-Tech Stance." *The Oriental Economist* 50 (July 1982).

10. U.S. Department of the Interior, Bureau of Mines. "The Mineral Industry of Japan." By John Wu. *Minerals Yearbook 1980*, Vol. III. Washington, D.C.: U.S. Government Printing Office, 1981.

11. Mahler, Walter. "Japan's Adjustment to the Increased Cost of Energy." *Finance and Development* 18 (December 1981).

12. Toyoda, Shigeru. "Changes in the Use of Energy in Japanese Steel Industry—With an Emphasis on the Countermeasures Taken after the Oil Crisis," special lecture. *Transactions of the Iron and Steel Institute of Japan* 23 (January 1983).

13. Pehlke, Robert. "An Overview of Contemporary Steelmaking Processes." *Journal of Metals* 34 (May 1982).

14. Kawata, Sukeyuki, ed. *Japan's Iron and Steel Industry, 1982*. Tokyo: Kawata Publicity, Inc., 1982.

15. Howard, Al. "Dry Quench Needs Lower Cost." *American Metal Market/Metalworking News*, Steelmaking Today Supplement, 27 September 1982.

16. Horio, Koichi, Nobuyuki Kitamura, and Yutaka, Ariake. "Waste Energy Recovery at Kashima Steel Works." *Iron and Steel Engineer* 59 (July 1982).

17. Boyer, E. "How Japan Manages Declining Industries." *Fortune*, 10 January 1983.

18. Japan. Economic Planning Agency. *New Economic and Social Seven-Year Plan*, August 1979.

19. "Japan: Declining Oil Stockpile." *Asia Research Bulletin* 12, No. 7, Report 5 (31 December 1982).

20. "Basic Key Materials, Technology Top Priority for MITI Budget Plan." *Tokyo Petroleum News*, 25 August 1982.

21. "Recycling Key in Japanese Steel Success: Recovering Energy." *The Northern Miner* (Canada), 16 September 1982.

22. Ogino, Junichi. "New Basic Materials: Development of New Types Become a Matter of Increasing Interest in Industry: Metals." In *Industrial Review of Japan, 1982*. Tokyo: Japan Economic Journal, 1982.

23. Makiuchi, Iwao. "Non-ferrous Metals: Aluminum Smelters Had Worst Year in 1981; Copper, Lead and Zinc Makers Also Fared Poorly." In *Industrial Review of Japan, 1982*. Tokyo: Japan Economic Journal, 1982.

24. Suzuki, Kenji. "Iron and Steel: Booming Seamless Pipe Exports Greatly Contributed to Earning Gain in 1981." In *Industrial Review of Japan, 1982*. Tokyo: Japan Economic Journal, 1982.

25. "Kawasaki Heavy Industries: Concludes Tieup with a German Firm in Coal-Fired DR Plants." *The Oriental Economist* 50 (January 1982).

26. Suda, Satoru. "Japan." In *Mining Annual Review, 1982*. London: Mining Journal, 1982.

27. Suda, Satoru. "Japan." In *Mining Annual Review, 1981*. London: Mining Journal, 1981.

28. Suda, Satoru. "Japan." In *Mining Annual Review, 1979*. London: Mining Journal, 1979.

29. "Production Adjustments in Aluminum." *The Oriental Economist* 50 (June 1982).

30. Jamesom, Sam. "MITI Policy Irks Japan Smelters: Trade Rift over Aluminum." *Los Angeles Times*, 21 March 1983.

31. Nagano. Takeshi. "The History of Copper Smelting in Japan." *Journal of Metals* 34 (June 1982).

Infrastructure and Environmental Factors

The availability of infrastructure is an important factor in the economic consideration of a mine development or mineral-processing project. The cost of infrastructure can be so high that a promising project may not turn out to be profitable.

Similarly, the cost of environmental protection equipment can be very high. Despite this high cost, increased public opposition to industrial pollution has forced Japanese companies to begin to employ protection measures.

This chapter examines some important types of physical infrastructure, such as energy supplies, transportation, and land availability. In addition, Japan's environmental management position is examined with respect to pollution levels, applicable legislation, the system of pollution monitoring and enforcement, and industrial antipollution investment.

INFRASTRUCTURE

Infrastructure is usually divided into two types: physical, such as energy, transportation, and water; and social, such as housing, training facilities, and services. Some examples of physical infrastructure that have considerable effects on Japan's mineral industries are: energy, land availability, and transportation. Most of Japan's energy supplies are imported; therefore, a diversified mix of oil, natural gas, uranium, coal, and synthetic fuel imports reduces this country's vulnerability to an industrial shutdown due to a cutoff in the supply of any one energy commodity. Japan's mountainous terrain limits the locations available for plant siting, and high land prices can add significantly to the cost of a new project. Japan's industries have long been aware of their dependence on foreign countries as sources of raw materials and as markets for their finished products. Because of this awareness, processing plants and transportation networks have been located with this dependence in mind. These important types of infrastructure are discussed below.

Energy Supplies

As noted above, imported oil, LNG, uranium, and coal make up the majority of Japan's total energy supplies. The small domestic input includes contributions of each of these energy types as well as hydropower. It is hoped that significant future input could come from solar power, geothermal power, and nuclear fusion. The Japanese supply situation for oil, natural gas, uranium, and coal is discussed below. In addition, some information is provided regarding Japan's involvement in synthetic fuels production.

Petroleum. Production from Japan's limited domestic crude oil reserves is very small, amounting to 2,868,000 bbl in 1981.[1] This level represented about 0.2% of Japanese oil requirements in 1981. Annual domestic production has declined since 1970, when Japan reached a peak of 5,656,000 bbl (Figure II-3-1).

More than 99% of Japan's petroleum supplies are imported.[2] Annual crude petroleum imports rose steadily during the 1960s and early 1970s as rapid industrialization took place in the Japanese economy (Figure II-3-2). During the period 1973–1979, annual oil imports leveled off, following OPEC's first round of oil price increases. Since the second round of oil price increases in 1979–1980, the annual level of crude petroleum imports has declined. Annual oil imports in 1982 amounted to 1.35 billion barrels (3.7 million barrels per day), their lowest level since 1971. The decline in oil imports after 1979 is due to energy conservation successes, diversification of the energy mix, and an economic slowdown. The extent to which this decline will continue depends on the level of development of Japan's nuclear power industry, the extent of the Japanese industries' conversion to

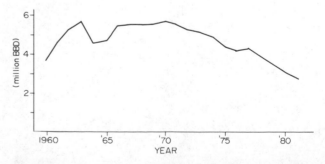

FIGURE II-3-1 Annual Japanese crude petroleum production, 1960–1981. (From U.S. Department of the Interior, Bureau of Mines. *Minerals Yearbook*.)

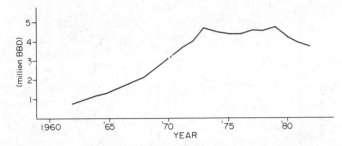

FIGURE II-3-2 Annual Japanese crude petroleum imports, 1962–1982. (From U.S. Department of the Interior, Bureau of Mines. *Minerals Yearbook*. "Oil Imports Drop to Lowest Level in 12 Years." *Japan Economic Journal*, 1 February 1983.)

non-oil-based energy, and continued conservation efforts.

Prior to the early 1970s, Japan depended on the large multinational oil companies for most of its oil supplies.[3] Japan's lone oil producer, the Arabian Oil Company, supplied only 10% of Japan's crude oil needs during the 1960s and early 1970s. Arabian Oil is 80% owned by Japanese interests and 10% each by the Saudi Arabian and Kuwaiti governments. This company produced oil from a field it discovered in the neutral zone off the coasts of Saudi Arabia and Kuwait.

Following the oil crisis in 1973–1974, Japan increased its efforts to become directly involved in the development of foreign oil fields. Japanese companies became involved in exploration and development projects in the Middle East, Africa, Southeast Asia, China, North America, and Oceania. More recently, in October 1982, Japan National Oil Corp. (JNOC) became the first company commissioned by the Chinese government to explore for onshore petroleum resources.[4]

Japanese companies have been successful in their more recent overseas endeavors. For example, information provided during development drilling by the Japan China Oil Development Co. (JCOD)

in China's Gulf of Bohai has apparently increased the original reserve estimates.[5] Production from the Gulf will probably exceed 200,000 barrels per day, once development is completed. Prior to these additional discoveries, Japan was to receive $42\frac{1}{2}\%$ of production during the period 1982–1996.[1]

In Gabon, Japan's Mitsubishi Petroleum Development Co., in joint venture with France's Elf Aquitane and the government of Gabon, had developed a production capacity of 7000 barrels per day by July 1982.[6] An additional capability of 800 barrels per day was expected once drilling was completed in late 1982. Mitsubishi was to receive $37\frac{1}{2}\%$ of the total production, which was expected to begin on a commercial level in late 1982.

A large amount of Japanese crude oil imports is still acquired through spot-market purchases and direct deals with producing country governments. In the early 1980s, Japan's general trading companies (GTCs) purchased 40% of Japan's imported oil supply requirement using these methods.[7]

Japan imports its oil supplies from several countries around the world (Table II-3-1). The Middle East was Japan's main oil supplier in 1981, and Saudi Arabia accounted for 34% of the total imported supplies.

Japan has increased its present and planned future petroleum imports from other regions, including Latin America and Africa, to lessen its dependence on the Middle East. In the early 1980s, Mexico, Nigeria, and Algeria all increased their oil exports to Japan.[8] Additional future imports will come from these countries, China, Canada,[9] and the North Sea.[10] Future imports will be important because they are expected to make up 50% of Japan's total energy supply in 1990.[11]

However, in the presence of an oil glut and discounted oil prices during 1982–1983, Japan's refiners and general trading companies increased their dependence on the spot market for crude oil supplies. In the first half of 1982, Japanese pur-

TABLE II-3-1   1981 Japanese Crude Petroleum Imports, by Source

Region	Amount (bbl)	Percent of total	Major suppliers (percent of total)
Middle East	2,737,230	69	Saudi Arabia (34); United Arab Emirates (14); Oman, Kuwait, Qatar, Iran, Iraq, Bahrain, and Neutral Zone (21, combined)
Asia	975,882	25	Indonesia (18), China (5)
Latin America	154,713	4	Mexico, Venezuela, Ecuador
Africa	95,208	2	Nigeria, Algeria, Egypt
Total	3,967,000	100	

Source: Ref. 1.

chases of Iranian oil increased significantly in response to Iran's discounted price.[12] In 1982, Japan's crude oil imports from Iran increased by 67.8%.[13] In early 1983, Asia Oil Co. cut back on its more expensive contract purchases of Saudi Arabian crude oil from two international oil companies, Mobil and Exxon.[14] Prior to this decision, Asia Oil was 85% dependent on contract purchases from these two companies. It intended to make up for this volume reduction with spot-market purchases. These examples indicate that Japanese oil importers are willing to sacrifice the security of contract oil purchases for price savings from other types of purchases.

Natural Gas. Japan's domestic reserves of natural gas are insignificant with respect to total natural gas demand. Japan's domestic production of natural gas has declined from its peak of 104 billion cubic feet in 1973 (Figure II-3-3). In 1981, Japan produced 74 billion cubic feet of natural gas.

Japan imports most of its natural gas supplies, usually on a long-term contract basis. In 1981, 16,832,000 metric tons (660 billion cubic feet) of liquefied natural gas (LNG) were imported from the following countries: Indonesia (51%), Brunei (31%), Abu Dhabi (12%), and the United States (6%).[1]

Future annual LNG imports will increase as Japan attempts to become less dependent on imported oil. In fiscal year 1979, LNG imports repre-

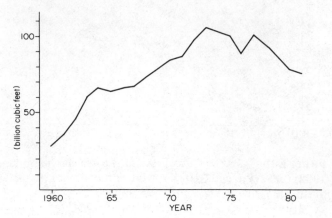

FIGURE II-3-3 Annual Japanese natural gas production, 1960–1981. (From U.S. Department of the Interior, Bureau of Mines. *Minerals Yearbook.*)

sented 4.8% of Japan's total energy supplies.[11] By fiscal year 1990, it is expected that imports will represent 9.0% of the total supplies. Additional future supplies will come from Indonesia as well as several other countries (Table II-3-2). Some supplies may also be available from Thailand. The development of the gas fields in the Gulf of Thailand was initiated with the Japanese export market in mind. However, because of production problems,[15] lower than expected reserves,[15] and a dispute regarding Thai government control over production,[16] the contribution to Japan's total LNG supply may be limited.

TABLE II-3-2   Selected Japanese Natural Gas Supply Contracts, 1982–1983

Country	Year of initial shipments	Amount of shipments (million metric tons/year)	Contract length[a] (years)
Australia (North-West Shelf)[b]	1986	6.0	NA
Canada (British Columbia and Alberta)[c]	1986	3.7	20
Malaysia (Sarawak)[d]	1983	6.0	20
		7.8 (after 1986)	
Indonesia (Badak)[b]	1983	3.2	NA
Indonesia (Arun)[b]	1984	3.3	NA
USSR (Sakhalin Island)[e]	1988?[f]	3.0e[g]	20

[a]NA, not available.
[b]Ref. 1.
[c]"Dome Petroleum Asks Canada Permission to Continue LNG Plan," *The Wall Street Journal,* 24 February 1983.
[d]"Malaysia Agrees to Make Shipments of LNG to Japanese," *The Wall Street Journal,* 28 March 1983.
[e]U.S. Congress, Office of Technology Assessment. *Technology and Soviet Energy Availability.* Washington, D.C.: U.S. Government Printing Office, 1981.
[f]"Japan: Joint Venture with Russia Resumed," *Asia Research Bulletin* 12, No. 4, Report 5 (30 September 1982).

Uranium. Japan's domestic uranium reserves are insignificant. They were estimated to be 8000 tons of uranium oxide in 1971.[17] These deposits, known as the Ningyo-Toge deposits, are located along the boundary between Okayama and Tottori prefectures in southwestern Honshu. The uranium occurs in sandstone and conglomerate channel structures located above a granite basement.

Following the government's decision to begin domestic nuclear energy development in the 1950s, Japan became an aggressive explorer for uranium ore deposits worldwide. It has been active in exploration and development projects in several countries,[18] including Australia, Canada, Brazil, Argentina, Mexico, Niger, the United States, and Colombia.[19]

In the early 1980s, Japan's imported uranium supplies were derived from deposits in Australia, Canada, South Africa, and Niger. Most of these supplies were available through long-term contracts, or develop-for-import joint ventures. Joint-venture projects usually involved several or all of the Japanese utility companies, the quasi-governmental Japanese agency known as the Overseas Uranium Resources Development Company (OURDC), and foreign mining concerns operating in the host country. It was reported in 1979 that Japan had arranged contracts for 150,000 metric tons of uranium oxide by fiscal year 1990, and 170,000 metric tons by fiscal year 1995.[20] Nuclear energy imports are expected to account for a 10.9% share of Japanese energy imports in fiscal year 1990, an increase from 4.9% in fiscal year 1979.[11]

Increased future uranium supplies will probably come from Australia, Canada, and Niger. An additional 12,250 metric tons of uranium oxide will be supplied by Australia over a 15-year period[21] beginning in late 1983.[22] The uranium will be mined from the Ranger operation in the Northern Territory, Australia. The Japan-Australia Uranium Resources Development Co. (JAURD), a Japanese consortium, holds a 10.1% equity interest in Energy Resources of Australia, the firm developing the Ranger mine. The other partners are Pan Continental and Getty Oil Company Ltd. of Australia. Japan will apparently participate in a uranium mine development project in the Dawn Lake area of Saskatchewan, Canada, where uranium reserves have been estimated to be 17,000 metric tons.[23] In Niger, OURDC discovered uranium resources estimated to be 30,000 tons of contained uranium.[20]

The future availability of uranium oxide from Australia may be in question because of the 1983 Australian Labor Party's election victory. The Labor Party, under Prime Minister Bob Hawke,

may require very strict safeguards on uranium fuel exports.[24] The Japanese government reached agreement with the previous Australian Liberal/National Country Party Coalition government regarding nuclear safeguards.[25] The agreement required that any plutonium produced from the reprocessing of Australian-origin nuclear material be stored and used according to the safeguards of the International Atomic Energy Agency. Negotiations regarding any new agreement could delay Australian uranium oxide exports to Japan.

As an alternative to dependence on foreign uranium sources, Japan has attempted to develop a process to extract uranium from seawater. The combined government and private research and development efforts resulted in a plan for a 10-kilogram/year uranium-from-seawater extraction pilot plant at Nio, Kagawa Prefecture, Shikoku Island. Construction of this plant began in 1981.[1] Capital costs for the plant totaled $11 million and operating costs have been estimated to be $7.7 million.

In 1982, 24 nuclear power plants were operating in Japan, with a total capacity of 17,177 megawatts (MW).[26] These plants ran at 70.2% capacity (12,058 MW). Full capacity is equivalent to 75%, because of regular plant shutdowns for maintenance during the year.

In 1979, the Japanese Electric Utility Council estimated that 51,000 to 53,000 MW of nuclear power generation capacity would be available by fiscal year 1990.[27] In the early 1980s, Japan's Science and Technical Agency reported that it had reduced this estimate to 46,000 MW.[28] However, because of the lead time required for gaining public agreement on plant site locations, and for completing necessary regulatory documentation, a goal of 40,000 MW will be difficult to achieve.[27] Nevertheless, the Japanese government remains strong in its commitment to expand its nuclear power industry and eventually to develop a commercial-size fast-breeder reactor.

If Japan follows a slow-paced installation program for nuclear power generating capacity (including a fast-breeder reactor), 168,000 tons of uranium oxide will be required by fiscal year 1990.[27] Based on its current efforts, Japan will be able to achieve this supply goal.

Before they can be used as fuel in domestic nuclear power plants, Japan's supplies of uranium oxide are routed through the United States and France for enrichment.[29] The United States accounts for 90% of the Japanese enrichment market.[30] Japan's long-term contracts for enriched uranium supplies from these two countries expire in 1990.[29] The plan to develop a domestic enrichment facility was postponed in early 1983, because Japanese utilities were unable to find a suitable plant location.[31] The stability in the uranium enrichment market was also cited as a reason for the delay. Therefore, despite the geographical diversity that Japan has realized in its uranium oxide sources, its sources of enriched uranium are still quite limited.

Coal. Japan's coal reserves are estimated to be 1050 million metric tons.[32] Most of the reserves are bituminous and occur on Hokkaido and Kyushu Island. This total represents both coking and steam coal reserves. Since the mid-1970s, Japanese annual coal production has fluctuated around a level of 18 million metric tons (Figure II-3-4).

In 1981, Japan produced 17.7 million metric tons of coal from 45 domestic mines.[1] The coal was produced from mines on Hokkaido (59.7%), Kyushu (39.7%), and Honshu (0.6%). The average mine depth has progressively increased as near-surface coal has been mined out.[33]

There were 20 different companies operating the domestic mines in 1981. The five major mining companies were Mitsui Coal Mining Co., Mitsubishi Mining Co., Sumitomo Coal Mining Co., Taiheiyo Coal Mining Co., and Matsushima Mining Co. A sixth company, Hokkaido Colliery and Steamship, reduced its annual production capacity by 1.2 million metric tons after shutting down its large Yubari Mine in 1982.[34]

Most of Japan's domestically produced coal in 1981 was steam coal (65%), and the remaining was coking coal (35%).[1] In the mid-1970s the production ratio for these two coal types was 50-50.[35] The increased domestic demand for steam coal led to the change in the ratio.

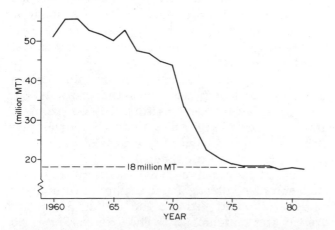

FIGURE II-3-4 Annual Japanese coal production, 1960–1981. (From U.S. Department of the Interior, Bureau of Mines. *Minerals Yearbook.*)

The Japanese government has attempted to encourage the mining and use of coal through its government coal policy. The sixth policy, adopted in 1975, had three goals:[35]

1. To maintain domestic production at 20 million metric tons annually
2. To develop overseas coal resources
3. To promote coal utilization research

The seventh policy, adopted in August 1981, emphasized domestic production.[34] The main goals were to maintain domestic production at the current levels (18 to 20 million metric tons), maintain uniform prices among domestic collieries, and to set domestic prices on the basis of the highest import prices.

The Japanese government has provided financial assistance to the domestic collieries to help them maintain production. An annual subsidy to the coal mining industry of 140 billion yen (Y) has been made available.[34] In 1981, MITI planned to provide a subsidy of $14 per metric ton to buyers of Japanese steam coal, and a subsidy of $28 per metric ton to steel mills that purchased Japanese coking coal.[36]

However, because of previously low domestic coal demand, and because of a large price differential between domestic and imported coal, Japanese coal mining companies operated with huge deficits in the early 1980s.[34] Imported coal prices were less than domestic coal prices, especially with respect to coking coal. The outlook for a reduction in the coking coal price differential was dismal in early 1983, because of the worldwide oversupply situation and the declining Japanese import price.[37]

In the early 1980s, Japan was 80% dependent on foreign sources for its coal supply requirements.[2] In 1981, Japan imported 78 million metric tons of coal, of which 11.8 million metric tons (15%) was steam coal, and 65 million metric tons (83%) was coking coal.[1] Most of the imported coal is bituminous. According to 1980 data, 98.2% of the total Japanese coal imports were bituminous, and the remainder were anthracite and lignite.[11]

The Japanese government has estimated that Japanese coal import requirements for fiscal year 1990 will be 130,000,000 metric tons.[38] If domestic production is maintained at 18 to 20 million metric tons, Japan's total coal supplies will be about 150 metric tons and imports will make up approximately 87% of total supplies.

For many years, Japan has been strongly dependent on Australian coal supplies. In 1981, almost one-half of Japan's steam and coking coal imports came from Australia (Table II-3-3). To

TABLE II-3-3 1981 Japanese Coal Imports, by Source
(Million Metric Tons)

Country	Steam coal		Coking coal	
Australia	5.4	(46)[a]	29.1	(45)
United States	2.1	(18)	21.5	(33)
China	1.2	(10)		
Canada			9.2	(14)
South Africa			2.7	(4)
Others	3.1	(26)	2.5	(4)
Total	11.8	(100)	65.0	(100)

[a]Numbers in parentheses represent percent of total imports.
Source: Ref. 1.

lessen this dependence, Japan has begun to diversify its sources of coal. Although Australia will remain an important coal supplier, several other countries will be supplying increased amounts of coal to Japan in the future. Japan hopes to eventually import its steam coal from four main sources: Australia, North America (United States and Canada), China, and South Africa, each with approximately a 25% share of total imports.[39]

Most of Japan's additional future coal imports will be available through long-term supply contracts. Japan actively pursued this strategy with Canada and Australia in the early 1980s (Tables II-3-4 and II-3-5).

Japanese institutions have also extended loans for overseas exploration and development projects (Table II-3-6). Repayment of these loans will probably involve some share of these projects' output.

Japanese companies are involved in joint-venture exploration and development projects in several countries such as Australia, China, Canada, New Zealand, the United States, and Indonesia (Table II-3-7). Of special interest in 1982 was the Japanese government's consideration of a plan to participate in Occidental's project to develop the Antaibao coal mine in Pingshuo, China.[40]

Japanese firms have also become more active in the United States. In fiscal year 1982, MITI began a Y120 million feasibility study regarding Japanese plans to import steam coal.[41] Japan's Coal Development Co., a consortium of 10 utilities, became involved in a coal development project in Colorado (Table II-3-7), and it expressed interest in another in Wyoming.[42] Japan's Kawasaki Steel Corp. discussed the sale of 50% of Amax Coal Co.'s coal assets with Amax officials.[43]

Japan had some success in its diversification plans. In 1982, Australia's share of Japan's coking coal imports dropped to 39%, from 45% in 1981, while total coking coal imports dropped only 1%.[44]

At the same time, the United States, Canada, and South Africa all posted percentage increases. Continued success is likely as Canada, China, the United States, and possibly Colombia, increase their coal exports to Japan.

Domestic demand for steam coal has been stimulated by industry efforts to reduce their dependence on oil. Several industries are cutting back on their oil consumption by converting to coal-based energy. This is especially true for the cement industry, where industry-wide conversion is almost 100% complete.[39] To differing extents, the power, steel, and nonferrous metal industries are increasing their coal consumption at the expense of oil consumption. By 1990, steam coal imports are expected to reach 53.5 million metric tons (Figure II-3-5).

Japan's utilities are mainly interested in high-Btu, low-sulfur, and low-moisture coal. Their import steam coal standards are as follows: 10,800 Btu/lb minimum, 1% sulfur maximum, 10% moisture content maximum, 15 to 20% ash content maximum, and ash fusion temperatures in a reducing atmosphere equal to 2375°F minimum.[39] Coal from several sources, including the countries mentioned above, Colombia, and the USSR, apparently meet these requirements.

Coal suppliers in Japan may have already overestimated their steaming coal requirements for the immediate future. Poor expected coal demand was indicated as the reason for the withdrawal of Japan's Electric Power Development Company from a coal mining project in Queensland, Australia (Table II-3-7). Japanese utilities have also postponed the development of a joint-venture coal mine in Colorado (Table II-3-7).

However, additional demand for steam coal might be expected as public opposition to nuclear power grows and as coal begins to represent a larger part of Japan's future energy supplies. Nuclear power was once considered to be the bridge between oil-based and other future energy sources in Japan.

Japan's demand for coking coal will depend mainly on increases in the level of domestic steel production. In 1981, 60,877,000 metric tons of imported coking coal were consumed by the iron and steel industry.[45] In other words, this industry consumed 94% of total Japanese coking coal imports for that year. As mentioned in Chapter II-2, domestic steel production is not expected to increase significantly until the late 1980s. As a result, coking coal imports will most likely remain steady or perhaps decline until this time.

Additional new technologies will apparently be incorporated in the coke-making process to ex-

TABLE II-3-4 Selected Japanese Long-Term Supply Contracts for Coal from Canadian Suppliers, 1980-1982[a]

Japanese buyer	Supplier	Mine	Coal type	Contract terms		
				Total amount (million metric tons)	Length (years)	Period
6 steel mills led by Nippon Kokan[b]	Crows Nest Resources Ltd., (subsidiary of Shell Canada Resources Ltd., Canada)	Line Creek	Coking	15	15	1983-1997
6 steel mills led by Nippon Kokan[b]	Gregg River Resources Ltd., (subsidiary of Menalta Coal Co. Ltd.)	Gregg River, Alberta	Coking	31.5	15	1983-1997
Nippon Kokan and 4 other steel companies[c]	Dennison Mines	Quintette, British Columbia	Coking	70 }18.2	14	1985-1998
Nippon Kokan and 4 other steel companies[c]	Brameda Resources (subsidiary of Teck Corp.)	Bull Moose colliery, British Columbia	Coking	1.7	15	1984-1998
9 steel mills[d]	B.C. Coal Co.	Greenhills, British Columbia	Coking	2.7	4	1983-1986
13-member consortium[e]	McIntyre Mines Ltd.	Smokey River, Alberta	Coking	1.8	2	1982-1984
NA[f]	Dennison Mines	Coalspur property, Alberta	Steam	NA	20	1982-2001

[a]NA, not available.
[b]U.S. Department of the Interior, Bureau of Mines. "The Mineral Industry of Japan." By John Wu. *Minerals Yearbook 1980*, Vol. III, table, p. 578.
[c]"Canadian Colliery Development Talks Progress; Coal Imports Up," *Japan Economic Journal*, 15 August 1982.
[d]"B.C. Coal Signs Pact to Sell from New Mine to Japanese Mills," *The Wall Street Journal*, 9 February 1983.
[e]"McIntyre Coal for Japan," *Mining Journal* 299, No. 7680 (29 October 1982).
[f]"Major Growth in Oil and Gas for Dennison," *Mining Journal* 298, No. 7642 (5 February 1982).

TABLE II-3-5 Selected Japanese Long-Term Contracts for Coal from Australian Suppliers, 1980–1982[a]

Japanese buyer	Supplier	Mine	Coal type	Contract terms Total amount (million metric tons)	Length (years)	Period
8 major steel companies led by Nippon Steel[b]	Central Queensland Coal Association	Norwick Park	Coking	9.6	8	NA
C. Itoh for a group of steel mills[b]	MIM Holdings Ltd.	Collinsville	Coking	15	15	1984–1998
Onoda Cement Co.[b]	Australian Associated Resources Ltd., (subsidiary of CSR Ltd.)	Yarrabee, Queensland	Anthracite	2	10	1981–1990
Mitsubishi Mining and Cement Corp.[b]	Warkworth Association	Warkworth, New South Wales	Steam	5.5	11	1981–1991
Unnamed steel mills[b]	Capricorn Coal Development	German Creek, Queensland	Coking	5	9	1982–1991
Unnamed steel mills[b]	Queensland State Government	Riverside, Queensland	Coking	47	14	NA
Hokkaido Electric Co.[c]	White Industries Ltd.	Ulan, New South Wales	Steam	NA	NA	1984–NA

[a]NA, not available.
[b]U.S. Department of the Interior, Bureau of Mines. "The Mineral Industry of Japan." By John Wu. *Minerals Yearbook 1980*, Vol. III, table, p. 578.
[c]"Ulan Sales Contract Signed," *Mining Journal* 300, No. 7700 (18 March 1983).

TABLE II-3-6 Selected Japanese Loans for Overseas Coal Projects

Country	Local participants	Japanese loan institutions	Loan description
Australia[a]	Mount Isa Mines Ltd.	Unnamed Japanese group that is part of a 25-member international consortium	$633 million for financing new mine at Newlands, expansion at Collinsville mine, and construction of a coal-handling port at Abbot Point
Canada[b]	Quintette Coal Ltd.	26 Japanese banks	C$950 million to finance the development of Quintette coal mine
Colombia[c]	Carbones de Colombia, S.A.	22 Japanese banks led by the Export-Import Bank of Japan	Yen credits equivalent to $14.44 billion
China[d]	Bank of China	Export-Import Bank of Japan	Y42 billion to finance coal development projects (part of a 1979 agreement)

[a]"Australian Firm to Get Loan of $633 Million for Coal Production," *The Wall Street Journal*, 1 April 1983.
[b]"Twenty-Six Japanese Banks Join Loan for Canada's Quintette Mine," *Japan Economic Journal*, 25 January 1983.
[c]"Japanese Banks to Extend Credits to Colombia Project," *The Wall Street Journal*, 27 December 1982.
[d]"Japan: Loan for Coal Projects," *Asia Research Bulletin* 12, No. 10, Report 5 (31 March 1983).

pand further the usable coal range for blast-furnace applications. This is especially important since future coking coal supplies are expected to be of poorer quality. Japanese steel mills are presently able to make good-quality coke from a blend of domestic and imported coals.[46] Designated coals are stored separately, blended according to specification, pulverized, and then injected into the blast furnace. Additional techniques that will further expand the usable coal range include the briquet-blended process, charging of preheated coal, addition of caking substances, and the form coke process.[46] These methods improve the coke strength, productivity, and fluidity of poorer-quality coking coals and noncoking coals.

Therefore, Japan's economy will become increasingly more dependent on coal. Japanese efforts are geared toward maintaining domestic coal production, increasing coal imports, and diversifying the sources of coal supply. However, many factors must be considered before Japan can significantly convert its industries to coal-based energy. These include the construction of coal transportation and handling facilities, and the installation of pollution control equipment. Large capital investments will be necessary to make this conversion. This conversion is considered essential to Japan's energy supply security. Large-scale conversion efforts would significantly reduce Japan's dependence on oil-based energy sources. A return to oil-based energy sources, such as during a time of glutted oil supply, would result in a considerable loss in capital investment and energy supply security.

Synthetic Fuels. In 1982, a Japanese consortium, Nippon Brown Coal Liquefaction Co. Ltd., received approval to construct a solvent-refined coal (SRC) liquefaction pilot plant.[47] This consortium consists of the governmental New Energy Development Organization (NEDO) and five private companies. The Japanese government was to provide Y44 billion for plant construction.[48] The Australian government was to provide the plant site in Victoria, power supply, and raw materials. The proposed plant would use brown coal from the Latrobe Valley in Victoria. The plant was expected to be operational by 1983. It was hoped that a large 5000-metric ton/day commercial plant could be developed by 1992.[47]

The Latrobe Valley brown coal was also the subject of a joint venture feasibility study conducted by Japan's Mitsui SRC Development Co. and Australia's Colonial Sugar Refineries Ltd. (CSR).[49] This study was focused on the possible construction of a commercial 6600-metric ton/day SRC plant by the late 1980s. In 1982, the study was postponed after a year of work.[50] Mitsui SRC/CSR had intended to use at its Australian plant the plant technology developed at Mitsui's joint-venture project in the United States (Morgantown, West Virginia). However, the U.S. project was suspended in 1981, before Mitsui developed the necessary plant technology. In addition, the Mitsui SRC/CSR partners were still awaiting a decision by the Victorian Brown Coal Committee regarding the designation of a mining location, and necessary developments in the SRC manufacturing technology had yet to be achieved.

Japanese concerns were involved in other synthetic fuels projects in 1982. In one example, the Japan International Cooperation Agency enlisted the help of Japanese consulting companies in order to perform a feasibility study of electricity generation from oil shale in Thailand.[51] In another example, Japan's Nippon Kokan constucted a small experimental direct hydrogeneration coal liquefaction plant at its Keihin steel works.[52] This plant was expected to be a precursor to the construction of a larger 250 to 500-metric ton/day pilot plant.

However, the glutted oil market in 1982–1983 and the resultant decline in oil prices probably reduced the economic attractiveness of the large Japanese synthetic fuels projects. Continued aggressive development efforts might only be expected when the global oil supply again becomes restricted or when oil prices rise to a price level that makes synfuel production price competitive. Even if such a change occurs, the contribution of synfuels to total future energy supplies will probably be small.

Summary. Japanese industries are attempting to diversify their geographic sources of energy supply as well as their energy mix. In the early 1980s, LGN, uranium, and coal purchases for future use were being expanded at the expense of oil purchases. Most of these future supplies will be available through long-term contracts, although joint venture overseas production will contribute to total supplies.

In pursuit of the goal of lessening Japan's dependence on oil-based energy sources, the use of LNG, uranium, and coal will become increasing important. However, the greatest emphasis may be placed on the conversion to coal-based energy for the following reasons:

Public opposition to the locations of nuclear power plants may limit the role of nuclear power in Japan's future energy supply.

The use of LNG and synfuels is expected to be small in comparison to the use of oil and coal.

The development of appropriate physical infrastructure and the installation of pollution control equipment will be necessary to facilitate a switch to coal-based sources.

Land Availability

The total land area in Japan is small, 372,313 square kilometers, and the population is quite large, 117 million people.[53] In comparison, Japan is about one-half the size of Texas, and its population is a little more than one-half that of the United States. Furthermore, Japan's mountainous terrain limits the land area available for industrial and urban development. Suitable sites for mineral-processing plants are mainly restricted to the lower-lying areas near sea level. This excludes mine-site-located refineries and smelters in Japan.

Industrialization in the 1960s and 1970s led to heavy urbanization in the Pacific Belt Zone of Japan (Figure II-3-6). This zone is an area of large cities and industrial centers along the southern coast of Honshu Island, from Tokyo to Hiroshima. Parts of 14 prefectures make up this zone, which also includes the cities of Kobe, Kyoto, Nagoya, Osaka, and Yokohama.[54] Companies favored this area because of the proximity of the major seaports, on which they depended for their imported raw material supplies. In addition, this location provided easy access to domestic and export markets for finished goods. In 1979, approximately 67% (by value) of Japan's industrial production was generated by businesses in the Pacific Belt Zone.[54]

Urbanization also created problems for the Japanese people. As discussed in a later section, severe air and water pollution resulted from the strain of the expanding residential and industrial sectors on municipal systems.

One strategy pursued by the Japanese mineral industries was to locate industrial plants on offshore islands. Mitsubishi's lead smelter and its copper refinery and smelter on Naoshima Island in the Inland Sea represent examples of this strategy (Figure II-3-7). In Tokyo Bay, Nippon Kokan K. K. constructed an offshore island and built its large Ohgishima Steelworks on top of it. Nippon Kokan also plans to build an artificial island for its Keihin Steel works.[55] Construction will require 10 years and $4 billion.

The location of metallurgical facilties on islands can present logistical and economic problems for those operations that require large amounts of labor, energy, and clean water supplies.[56] Such sites could also be vulnerable and difficult to defend during times of international conflict.

Artificial islands have also been suggested as locations for coal-fired power plants and coal storage. In 1981, MITI proposed to build two large offshore coal centers, in an effort to appease environmentalists and to avoid the high cost of land.[36]

MITI's latest plans to combat urbanization in the Pacific Belt Zone are aimed at the development of Japan's rural areas. MITI has recommended the establishment of several "technopolises," which will be centers for Japan's new industries and for research and development.[57] Nineteen medium-sized cities will be transformed in this way. These cities include Hakodate on Hokkaido Island, Akita and Aomori on Honshu Island, and Kunamoto and Miyazaki on Kyushu Island.

In the early 1980s, MITI was encouraging foreign investors to locate their industrial sites in areas outside of the Pacific Belt Zone. The incentives available to investors willing to locate in rural areas included lower taxes, lower rents, cheaper land costs, moderate wages, and lack of labor strife.[58]

TABLE II-3-7 Selected Japanese Overseas Joint-Venture Coal Projects[a]

Country	Japanese participants	Foreign participants	Coal description			Startup	Other information
			Project	Type	Production (million metric tons/year)		
Canada — British Columbia[b,c]	Mitsui Mining, 21% Tokyo Boeki, 21%	Dennison Mines, 46% Charbonnages de France, 12%	Quintette Mine	Coking steam	5 1.3	1985	Joint venture coal property development; Japan to receive all coal production for 15 years; Japanese institutions also providing loans
Australia — New South Wales[d]	Taiheiyo Coal Development Co., 25% Japan Coal Development Co., 14% Mitsui & Co., 5% C. Itoh, 5%	New South Wales Electric Commission, 51%	Birdsrock Mine	Steam	1-3	1984	Japanese utilities to receive 70% of annual output from 1984 to 1990
Australia — New South Wales[e,f]	Japan Electric Power Development Co. and Mitsui & Co., 15%	New South Wales Electric Commission, 85% (?)	Mt. Arthur	Steam	4	1984	Japanese companies were to receive 25% of annual production, and Europe and Asia the remaining 75%; Japan withdrew from project in 1982, but still intends to buy coal
Australia — Queensland[g]	Mitsui Mining Overseas Co. Ltd., 25% Showa Oil Co. Ltd., 24%	Rylance Collieries & Brickworks Pty., 51%	NA	Steam	1	NA	NA
Indonesia — Borneo[h]	Nissho Iwai, 40% (?)	Mobile (U.S.), 60%	Coal prospects in East Kalimantan Province	NA	NA	NA	Formation of an exploration and development co.; Indonesia's national coal co., PN Tambang Batubara, is entitled to 13.5% of any annual production

Location	Company	Seller/Partner	Mine	Type	Amount	Year	Comments
New Zealand — South Island[i]	Kanematsu-Gosho Ltd. Mitsui Mining Over Co. Ltd.	New Zealand Forest Products	Greymouth	Steam	2	1984	Consortium will do feasibility study of mining and export of coal
China — Shanxi Province[d]	Mitsui Mining Co.	Chinese government	Si Xiong, Kou Mine	Steam	4	1985	Joint-venture expansion and development of these properties; Mitsui also planned to form a consortium to finance and develop mines in Shanxi Province
			Xiao Kou Mine	Steam	Expansion to 1.2	1983?	
U.S. — Colorado[j,k]	Japan Coal Development Co. Sumitomo Corp.	Dorchester Gas Co.	Fruita Mine	Steam?	3.6	1986	Agreement involves sharing of mine development costs and Japanese long-term contracts for all coal production; development was postponed in 1983 as a result of waning interest from the Japanese utilities

[a]NA, not available.
[b]"Major Growth in Oil and Gas for Dennison," *Mining Journal* 298, No. 7643 (5 February 1982).
[c]"Twenty-six Japanese Banks Join Loan for Canada's Quintette Mine," *Japan Economic Journal*, 25 January 1983.
[d]Ref. 11.
[e]"Japanese to Withdraw from Mt. Arthur," *Mining Journal* 299, No. 7680 (29 October 1982).
[f]"Japanese Utility Quits Coal-Mining Project in New South Wales," *The Wall Street Journal*, 27 October 1982.
[g]"Go Ahead for Coal Project," *Mining Journal* 298, No. 7638 (8 January 1982).
[h]"Mobil and Nissho Iwai to Join in Developing Indonesian Coal Tract," *The Wall Street Journal*, 18 November 1982.
[i]"NZ Coal Consortium," *Mining Journal* 299, No. 7663 (2 July 1982).
[j]"United States: Joint Venture with Japan for Colorado Mine," *World Coal* 7 (November–December 1981).
[k]"Fruita Delayed," *Mining Journal* 300, No. 7691 (14 January 1983).

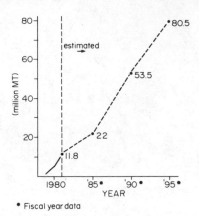

FIGURE II-3-5 Estimated annual Japanese steam coal imports for 1985, 1990, and 1995. (From U.S. Department of the Interior, Bureau of Mines. *Minerals Yearbook*; Ref. 35.)

Most of the wholly or partly foreign-owned enterprises in Japan are located in the Pacific Belt Zone. However, in late 1982, MITI reported significant increases in the number of foreign applications for industrial sites located in provincial areas.[58]

Transportation

Japan's transportation network greatly facilitates material flows to and from domestic mines and processing plants. In most cases, major mines, which are located throughout the country (Figure II-3-8), are situated near railroads, major highways, and roads. Metal-processing facilities, such as smelters and refineries (Figure II-3-7), are also well located with respect to the transportation network. Many plants are located along the coastlines, some at major seaports. Those plants that are located at a distance away from major seaports are connected by land transportation routes. Therefore, raw materials from both domestic and foreign sources can easily be transported to domestic processing plants, and processed materials can easily be transported to major distribution centers or seaports for export.

Plans for Japan's establishment of new technopolises, as described above, should include any necessary improvements in the existing transport network. If the existing network is insufficient, new roads and railways may be required to connect

FIGURE II-3-6 Map of Japan with locations of the "Pacific Belt Zone," and major seaports. (From Ref. 71; *1981-82 World Mines Register.* San Francisco: World Mining, 1981.)

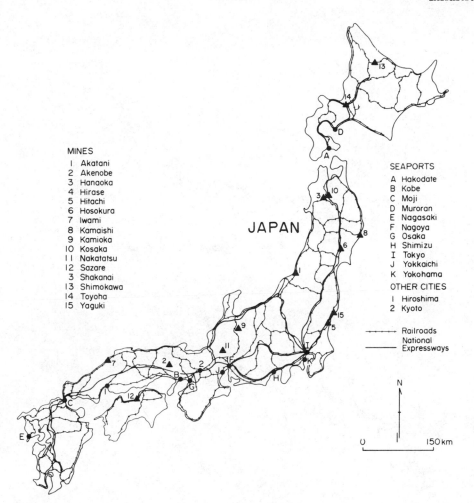

MINES
1 Akatani
2 Akenobe
3 Hanaoka
4 Hirase
5 Hitachi
6 Hosokura
7 Iwami
8 Kamaishi
9 Kamioka
10 Kosaka
11 Nakatatsu
12 Sazare
13 Shakanai
13 Shimokawa
14 Toyoha
15 Yaguki

JAPAN

SEAPORTS
A Hakodate
B Kobe
C Moji
D Muroran
E Nagasaki
F Nagoya
G Osaka
H Shimizu
I Tokyo
J Yokkaichi
K Yokohama

OTHER CITIES
1 Hiroshima
2 Kyoto

Railroads
National
Expressways

N

0   150 km

FIGURE II-3-7 Map of Japan with locations of major smelters and refineries, in relation to the domestic transportation network. (From Ref. 71; *1981–82 World Mines Register*. San Francisco: World Mining, 1981.)

these designated cities with major seaports. It has been estimated that the government will need to provide $37 billion over the next 10 years to establish the necessary infrastructure.[57]

As mentioned in Chapter II-4, imported mineral supplies will be necessary for the operations of these new industries. Good access to seaports will facilitate imports of these materials, as well as the exports of high-technology goods.

ENVIRONMENTAL FACTORS

Prior to the mid-1960s, Japanese government and industry actions to protect the environment were minimal. The environmental load increased steadily while the nation followed its policy of rapid economic growth. Air and water pollution problems resulted from the expansion of the mining and manufacturing industries; the use of advanced techniques, such as the application of insecticides and chemical fertilizers, by the agriculture, forestry, and fisheries industries; and the urbanization in the Pacific Belt Zone.[59]

However, since the mid-1960s Japan has put into effect some of the most stringent antipollu-

tion standards in the world.[60] These standards indicate a change in the government's attitude toward pollution control. Prior to this time the government promoted environmental regulations that did not interfere with rapid economic development. This change is the result of the Japanese public outcry about pollution and some important judiciary actions.

These judgments concern four pollution cases brought before the Japanese courts by pollution victims in the late 1960s and early 1970s.[60] These cases, two related to Minamata disease and one each to Itai-Itai disease and Yokkaichi asthma, were settled in favor of the plaintiffs. The implications of these suits were twofold: that a polluter would be held liable for pollution damages even if these damages were the result of negligence, without obvious intent; and that pollution control efforts would no longer have to be made in harmony with economic growth. As a result, future pollution victims would have a relatively easier time bringing suit against polluters, and companies would have to place more emphasis on environmental considerations in their developmental planning.

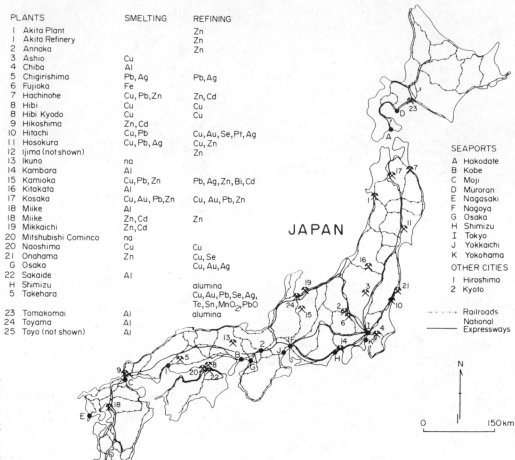

PLANTS		SMELTING	REFINING
I	Akita Plant		Zn
I	Akita Refinery		Zn
2	Annaka		Zn
3	Ashio	Cu	
4	Chiba	Al	
5	Chigirishima	Pb, Ag	Pb, Ag
6	Fujioka	Fe	
7	Hachinohe	Cu, Pb, Zn	Zn, Cd
8	Hibi	Cu	Cu
8	Hibi Kyodo	Cu	Cu
9	Hikoshima	Zn, Cd	
10	Hitachi	Cu, Pb	Cu, Au, Se, Pt, Ag
11	Hosokura	Cu, Pb, Ag	Cu, Zn
12	Ijima (not shown)		Zn
13	Ikuno	na	
14	Kambara	Al	
15	Kamioka	Cu, Pb, Zn	Pb, Ag, Zn, Bi, Cd
16	Kitakata	Al	
17	Kosaka	Cu, Au, Pb, Zn	Cu, Au, Pb, Zn
18	Miike	Al	
18	Miike	Zn, Cd	Zn
19	Mikkaichi	Zn, Cd	
20	Mitshubishi Cominco	na	
20	Naoshima	Cu	Cu
21	Onahama	Zn	Cu, Se
G	Osaka		Cu, Au, Ag
22	Sakaide	Al	
H	Shimizu		alumina
5	Takehara		Cu, Au, Pb, Se, Ag, Te, Sn, MnO_2, PbO
23	Tomakomai	Al	alumina
24	Toyama	Al	
25	Toyo (not shown)	Al	

SEAPORTS
A Hakodate
B Kobe
C Moji
D Muroran
E Nagasaki
F Nagoya
G Osaka
H Shimizu
I Tokyo
J Yokkaichi
K Yokohama

OTHER CITIES
1 Hiroshima
2 Kyoto

····· Railroads
—— National Expressways

FIGURE II-3-8 Map of Japan with locations of the major domestic mines, in relation to the domestic transportation networks. (From Ref. 71; *1981–82 World Mines Register*. San Francisco: World Mining, 1981.)

These cases also were a stimulus for later pollution control legislation.

Pollution Control Legislation

In 1967, the Basic Law for Environmental Pollution Control was passed. This law was the first government acknowledgment of its responsibility for environmental planning in economic development.[60] However, the law served only as a basis for future legislation, since it did not specify any emission standards or pollution relief measures.[61] Another weakness of the law was its inclusion of the harmony clause, which said: "Preservation of the living environment shall be carried out in harmony with the healthy development of the economy."[62] This clause was interpreted to mean that pollution control would not interfere with economic growth, and it indicated the government's continued indifference toward antipollution measures.

In 1970, several pollution measures were passed during a special session of the Japanese Diet. These measures served to strengthen and revise the Basic Law.[60] At the same time, the "harmony clause" was eliminated from the Basic Law. Local governments were given increased responsibility to estab-lish and implement environmental standards specific to their situations. Specific relief measures and a system of penalties for civil code violations were also established.[61]

The Environment Agency of Japan was created in 1971. This agency determines the ambient, emission, and related standards for air, water, and other areas,[63] that are to be met by private industry. It also coordinates environmental policies.[60]

The Environmental Agency was originally envisioned as an independent agency, responsible for the establishment and the implementation of environmental regulations. However, the regulatory power regarding pollution control remains fragmentary in Japan.[60] This power is divided among several competing government agencies and ministries, including the Environmental and Economic Planning Agencies and the Ministries of Health and Welfare, Agriculture and Fisheries, International Trade and Industry, Finance, and Justice.

Local Government Involvement

Local governments are very active in the area of pollution control in Japan. Several reasons exist to explain this situation.[60] The national government's earlier inaction toward industrial polluters was a major factor in the development of local government activism. Because of increased use of automated processes, the presence of large factories did not necessarily result in increased local employment or sufficient tax revenues to maintain industrial sites. The health of local citizens was becoming affected by industrial pollution. Local efforts were made to demonstrate a community's belief in their rights to a healthy environment. In addition, pollution control problems could also be used as a platform to voice other types of local opposition.

Because of local government activism, pollution control in Japan has become decentralized. All of Japan's prefectural governments have formulated individual pollution control ordinances.[60] In many cases, local regulations are stricter than those established by the Japanese government.

In addition to ordinances, local governments use pollution control agreements to implement local environmental standards.[60] Pollution control agreements can require companies to maintain certain environmental standards or to use the most advanced antipollution technology. Enforcement measures include stop-work orders and strict liability for damages. The number of these agreements increased steadily through the early 1970s.[60] By 1975, 8923 firms were regulated by pollution control agreements (Table II-3-8). Mining and metal-processing companies are involved in many of these agreements.

Informal guidelines for development represent another form of pollution control used by local governments. Guidelines call for environmental preservation measures, dedication of parks and schools, and building design restrictions, in return for local government approval of land development plans.[60]

Monitoring and Enforcement

Local governments are mainly responsible for monitoring industrial emission levels at the individual factory level. Various methods are used in this regard.

Local governments maintain a large network of air quality monitoring stations.[60] These stations automatically send data to computerized central stations for processing. All large factories are required to install monitoring stations. Therefore, data from individual factories can be analyzed. By 1975, 1216 monitoring stations and 74 central stations were maintained by local governments.[60] Environmental data from this type of surveillance is reported as a ratio of stations whose ambient concentrations meet the established environmental quality standards. For example, in 1978, 93.8% of

TABLE II-3-8   Number of Private Firms That Have Signed Pollution Control Agreements in Japan, by Industry, October 1, 1975

Type of industry	Number of firms
Agriculture	570
Mining	268
Construction	350
Foodstuffs	619
Clothing, textiles	383
Timber, wood products	331
Paper, pulp	353
Chemicals	942
Petroleum, coal products	183
Rubber, leather	88
Ceramics	677
Iron, steel	372
Nonferrous metals	221
Metal products	1016
Machinery	1025
Electric	177
Other	1348
Total	8923

Source: Julian Gresser, Koichiro Fujikura, and Akio Morishima, Environmental Law in Japan. Cambridge, Mass.: The MIT Press, 1981, Table 4, p. 251. Originally from: Quality of the Environment in Japan, 1976 (Environment Agency, 1976, p. 226.

the monitoring stations reported sulfur dioxide concentration levels that were in accordance with national ambient standards.[59]

The level of emissions is also monitored by pollution control managers, employed at designated factories.[60] They are responsible for the technical assessment of pollution control, such as measurement and inspection, at their factory. Pollution control managers must pass a MITI-administered national examination. Their appointment to a particular factory must be reported to the prefectural governor.

According to the Air Pollution Control Law,[64] prefectural governors are legally able to request extra data from factory owners regarding pollution abatement operations. Governors can order a shutdown at a local factory if emissions threaten human health or environmental quality. Anyone who disobeys a governor's shutdown order is subject to imprisonment and fine. Prefectural officials can also conduct periodic inspections of factories.[60]

Many Japanese pollution control laws include criminal sanctions. However, sanctions have not been used very often. Government officials prefer the use of guidance and incentives, rather than judicial trials, to control pollution.[60] Guidance is provided in the form of discussions, negotiations, and warnings. Incentives, as noted below, include financial and technical assistance for pollution control programs. The difficulty of prosecuting a pollution-related offense may also play a role in the methods used to encourage pollution control.[60]

Antipollution Investment

Japanese antipollution investments have been quite significant. In 1975, these investments totalled Y2,888,000 million, or approximately 2% of Japan's gross national product (GNP). Together, private enterprises and local governments made 90% of these investments, and the rest were made by the central government.[65] Total Japanese antipollution investments, as a percentage of GNP, doubled between 1970 and 1975 (Table II-3-9).

Five major industries accounted for 71% of all antipollution investment at this time (Table II-3-10).

TABLE II-3-9 Annual Japanese Antipollution Investment Costs as a Percentage of GNP, 1970–1975

	Antipollution[a] investment by enterprises	Antipollution[b] investment by government	Total antipollution investment
1970	0.4	0.6	1.0
1971	0.5	0.8	1.3
1972	0.5	1.0	1.5
1973	0.6	1.0	1.6
1974	0.7	1.0	1.7
1975	1.0	1.0	2.0

[a]Appendix 1.
[b]Ministry of Home Affairs; figures for 1974 and 1975 for local government were not available and have been estimated on the basis of 1970–1973 data.

Source: Environmental Policies in Japan. Paris: Organization for Economic Cooperation and Development, 1977, Table 29, p. 71.

TABLE II-3-10 Japanese Antipollution Investment, by Industry, 1974 (Percent)

	Share of total antipollution investment	Share of total investment by industry
Iron and steel	17	17
Oil	13	27
Thermal power plants	18	47
Pulp and paper	5	24
Chemicals	18	25
Other	29	7
Total	100	14

Source: Environmental Policies in Japan. Paris: Organization for Economic Cooperation and Development, 1977, Table 30, p. 71. Originally from: survey by MITI.

The iron and steel industry accounted for 17% of those investments. Antipollution investments within these five industries are also significant (Table II-3-10). In 1974, they ranged between 17% of total investment by the industry, for the iron and steel industry, and 47% for thermal power plants. These percentages may not represent long-term trends, because 1974 was a recession year and overall investments were low.

Nonetheless, antipollution investments made by the iron and steel industry remained significant in the early 1980s. Approximately 20% of the total construction cost of a recently built Nippon Kokan steelmaking plant represented the cost of pollution control equipment.[66] A Japan Steel spokesman stated that this compnay spent Y11,000 million ($14 million) on pollution control in 1980–1981.[66] Some steelmakers, unable to afford the required expensive pollution control investments, have chosen to shut down some of their plants located in large cities. In the early 1980s it appeared that industry-wide environmental investment had leveled off as pollution control objectives were met and fewer plants were built.[45]

Antipollution investments in the thermal power plant industry also continue to be significant. In 1980 it was estimated that the cost of control measures for particulates, sulfur oxides and nitrogen oxides, and for ash and wastewater treatment add 20% to the cost of the electricity generated.[67] Power plants have been located in remote areas to meet environmental standards, and thus transmission costs are higher.[68] Approximately one-half of the coal ash produced annually must be transported to the coastline to be used as landfill on land reclaimed from the ocean.[68] Pollution control equipment costs, higher transmission costs, and ash transport and disposal costs make a significant contribution to the high total cost of electric power in Japan. Additional costs could arise from time delays in plant construction, resulting from extended negotiations between industry and local governments regarding compliance with strict local environmental standards.

Financial and Technical Assistance. The "polluter pays principle" is the basis for pollution control in Japan.[69] However, government-sponsored financial assistance for pollution control investments, including reclamation work and compensatory payments to landowners, is available to private enterprises.[70] There are two types of assistance, low-interest loans and tax benefits. Subsidized loans are available from various institutions.[65] The OECD concluded that Japan's subsidy equivalent, or in other words the sum of the discounted values of the differences in interest on the loans, was similar to that provided through antipollution investment assistance schemes used in several other OECD countries in 1975.[65] Tax benefits include accelerated depreciation on pollution control equipment.

Japan's long-term economic and social plan (Economic Planning Agency, 1979–1985) calls for the continuation of government-provided financial assistance to encourage the use of pollution control equipment.[69]

Several forms of assistance are available to small and medium-sized firms for their pollution control efforts.[60] These firms often lack the technology and the capital to acquire expensive pollution control equipment.

Technical assistance is available from a network of governmental counselors, local chambers of commerce, several publicly funded national testing and research institutions, and three other groups: the Small Business Production Corporation, the Small and Medium Enterprise Agency, and the Environmental Pollution Control Service Corporation (EPCSC).

Financial assistance is available from the chambers of commerce and through loans from the Environmental Pollution Prevention Service Corporation and the EPCSC.[60] Loans can account for as much as 80% of the pollution control costs. Joint financial and technical efforts among private firms are also encouraged. Subsidies for research and development (R&D) concerning pollution control technology are available from the national testing and research institutions.[60]

Air Quality

Japanese ambient standards for some air pollutants are the strictest in the world. Table II-3-11 compares the sulfur dioxide, particulate, and nitrogen oxide objectives of several OECD countries. Environmental quality targets are set by the national government. With the exception of the nitrogen oxide standard, ambient standards are backed by a consensus in Japan.[60]

Emission standards are set by the national, prefectural, and municipal governments. The Air Pollution Prevention Law, enacted in 1968, expanded the scope of previous regulations regarding allowable sulfur dioxide, soot, and dust-particle emissions. Strict emission standards for nitrogen oxides were included in modifications to the law in 1973.[71] These standards apply to stationary sources, including metal-heating furnaces of the metal industry (excluding iron and steel) and thermal power plants. Most often, emission stan-

TABLE II-3-11 Air Quality Objectives, Japan and Other Selected OECD Countries, 1975[a,b]

Country	SO_2 (ppm)	Particulates (mg/m³)	NO_2 (ppm)
Japan	0.04	0.10	0.02
Canada	0.06	0.12	0.10[c]
Finland	0.10	0.15	0.10
Italy	0.15	0.30	NA
United States	0.14	0.26	0.13[d]
West Germany	0.06	NA	0.15[e]
France	0.38	0.35	NA
Sweden	0.25	NA	NA

[a]All figures are average daily values.
[b]NA, not available.
[c]The figure is for Ontario; the figure for Saskatchewan is much lower: 0.01.
[d]For the United States, the NO_2 objective is set in terms of average yearly value (0.05 ppm); the figure given is therefore an equivalent open to criticism.
[e]The West German standard is 0.05 ppm for "long-term exposure" and 0.15 ppm for "short-term exposure."

Source: Environmental Policies in Japan. Paris: Organization for Economic Cooperation and Development, 1977, Table 9, p. 25. Originally from: for SO_2 and for particulates — Werner-Martin and Arthur C. Stern, *The Collection, Tabulation, Codification and Analysis of the World's Air Quality Management Standards*, School of Public Health, University of North Carolina at Chapel Hill, N.C., October 1974; for NO_2 — R. Kiyoura, *International Comparison and Critical Analysis of NO_2 Air Quality Standards*, paper presented at the 69th Annual Meeting of the Air Pollution Control Association, Portland, Ore., 27 June–1 July 1976.

dards are expressed as concentrations, not as absolute quantities.[65] However, for sulfur and nitrogen oxides, these standards are also based on quantity and the speed and temperature of the exhaust.[60] Emission standards for air pollutants differ geographically.

By 1975, Japan achieved significant reductions in ambient concentration levels of sulfur dioxide and carbon monoxide.[65] However, the average levels of nitrogen oxides and oxidents generally indicated a slow increase from 1970 to 1975.[65]

By the early 1980s, significant reductions had been achieved in photochemical smog and particulate concentrations.[61] Also, the increase in average nitrogen oxide concentrations had slowed.

An example of the metal industry in which pollution control efforts have been successful with respect to sulfur dioxide emissions is that of the copper smelting industry. Japanese copper smelters capture 99% of their sulfur input, through the treatment of all of the furnace offgases, acid plant tail gases, and other fugitive gases.[72]

The Environmental Agency was expected to establish new nitrogen oxide emission standards for the iron and steel industry. By the early 1980s, this industry had apparently not yet developed the technology necessary to prevent nitrogen oxide emissions.[71] However, by the mid- to late 1980s it was expected that this industry would be able to reduce its nitrogen oxide emissions by 90%.

Water Quality

According to a 1977 OECD report,[65] some improvements in Japanese water quality were observed during the period 1970–1974. During this time, the nationwide levels of several harmful substances, such as cadmium, lead, arsenic, and total mercury were reduced by 87%, 86%, 73%, and 99%, respectively.

In the early 1980s, water quality remained high, although the level of organic pollution gave reason for some concern. Excluding the Inland Sea, most of Japan's waterways were clear according to international standards.[61]

Outlook: Improvements in Environmental Quality

Japan has achieved significant successes in its pollution abatement efforts of the 1970s and 1980s. However, pollution abatement is only part of the Japanese effort toward achievement of a better quality of life.

Japanese officials are interested in developing broader-based environmental policies to eliminate environmental discontent.[69] Future policies emphasize acquisition and maintenance of several hard-to-measure amenities, such as quietness, beauty, privacy, and social relations.

The Environmental Agency will continue to play an important role by applying even stricter air and water quality standards.[61] Some areas that will receive additional emphasis include noise, vibrations, industrial odors, soil pollution from mine and mineral processing wastes, and recycling of wastes.[69] To meet these standards, the mining and mineral-processing industries will need to continue and perhaps increase their antipollution investments. In some cases the cost of pollution control equipment may be too high, and some heavy polluters will have to be eliminated.

Additional attempts may be made to pass legislation that will require the completion of environmental impact statements on new development projects. Previous attempts by the Environmental Agency have failed because of opposition from several industries and government ministries.[61] The government could use impact statements as one of its tools to implement future broad-based environmental policies.

SUMMARY

Japan's future energy supply mix will be richer in coal, nuclear, and LNG portions compared to that of the 1970s and 1980s. However, imported coal will provide the most important alternative to dependence on imported oil if public opposition to nuclear power grows and the LNG contribution to total supply remains small. Although its use will be cut back, oil will still represent the largest contribution to the total energy supply in fiscal year 1990.

Other types of physical infrastructure that are important to the mineral industry are transportation and land availability. In general, Japan's mines and processing plants are connected by a nationwide series of railroads and roadways. The location of processing plants in port cities or at remote locations connected by the national transportation network reflects Japan's dependence on imported raw materials and energy supplies, as well as export markets for basic materials and manufactured goods. One innovative approach to Japan's problem of limited land availability is its use of both natural and artificial islands as industrial sites.

Since the early 1970s, environmental concerns have become increasingly important to the mineral industry. National legislation, local government action, and public opposition have resulted in increased antipollution investments by this industry. Continued, even broader future concerns will require that this industry, as well as the supporting power industry, maintain their efforts to control pollution.

REFERENCES

1. U.S. Department of the Interior, Bureau of Mines. "The Mineral Industry of Japan." By John Wu. *Minerals Yearbook 1981*, Vol. III, preprint. Washington, D.C.: U.S. Government Printing Office, 1982.
2. Japan. Prime Minister's Office, Statistics Bureau. *Japan Statistical Yearbook 1982*. Tokyo: Japan Statistical Association and the Mainichi Newspapers, 1982.
3. Yoshino, M. Y. *Japan's Multinational Enterprises*. Cambridge, Mass.: Harvard University Press, 1976.
4. "Joint Oil Exploration with China." *Asia Research Bulletin* 12, No. 2, Report 5 (31 July 1982).
5. "Oil and Gas Are Produced Successfully from Fifth Test Well in Gulf of Bohai." *Japan Economic Journal*, 22 February 1983.
6. "Drilling Operations Completed." *Asia Research Bulletin* 12, No. 2, Report 5 (31 July 1982).
7. "Survey: Japan: New International Strategies." *World Business Weekly*, 6 October 1980.
8. Yoshii, Hideo. "Petroleum: Japan Widens Its Purchases of Crude Oil to Such New Places as Mexico and Africa." In *Industrial Review of Japan, 1981*. Tokyo: Japan Economic Journal, 1981.

9. "Canada Agency Clears Exports of Crude Oil to Japan, South Korea." *The Wall Street Journal*, 27 April 1983, p. 56.

10. "A First Long-Term Deal to Buy North Sea Oil." *Asia Research Bulletin* 12, No. 4, Report 5 (30 September 1982).

11. U.S. Department of the Interior, Bureau of Mines. "The Mineral Industry of Japan." By John Wu. *Minerals Yearbook 1980*, Vol. III. Washington, D.C.: U.S. Government Printing Office, 1981.

12. Ibrahim, Youssef. "Iranians Propel Petroleum Sales with Cut Rates." *The Wall Street Journal*, 16 April 1982.

13. "Japan: Oil Imports Lowest in 12 Years." *Asia Research Bulletin* 12, No. 11, Report 5 (30 April 1982).

14. "Asia Oil Will Cut Purchases from Mobil and Exxon to Half." *Japan Economic Journal*, 11 January 1983.

15. Borsuk, Richard. "Union Oil Faces Production Problems at Gulf of Thailand Natural Gas Field." *The Wall Street Journal*, 28 July 1982.

16. "Seagram Unit Fights Thailand Government to Control Gas Project." *The Wall Street Journal*, 1 November 1982.

17. Ridge, John. *Annotated Bibliographies of Mineral Deposits in Africa, Asia (exclusive of the USSR) and Australasia*. Oxford: Pergamon Press, Ltd., 1976.

18. U.S. Department of the Interior, Bureau of Mines, "The Mineral Industry of Japan." *Minerals Yearbook*, Vol. III, various years. Washington, D.C.: U.S. Government Printing Office, 1963-1982.

19. "Colombian Profile." *Mining Journal* 298, No. 7638 (8 January 1982).

20. U.S. Department of the Interior, Bureau of Mines. "The Mineral Industry of Japan." By E. Chin. *Minerals Yearbook 1978-1979*, Vol. III. Washington, D.C.: U.S. Government Printing Office, 1979.

21. "Australian/Japan U Trade." *Mining Magazine*, March 1982.

22. "Government Okays Jabiluka U Mine." *World Mining* 35 (October 1982).

23. "Power Reactor Corp. Enters Talks for Saskatchewan U Mining." *Japan Economic Journal*, 25 January 1983.

24. "Australia Looks to Labor." *Mining Journal* 300, No. 7689 (11 March 1983).

25. "Australia/Japan Nuclear Agreement." *Energy World*, No. 95 (August-September 1982).

26. "N-Power Plants Operate at 70.2% Capacity." *Japan Economic Journal*, 1 February 1983.

27. "Japan's Nuclear Industry at a Turning Point," Part I. *Japan Finance and Industry*, Quarterly Survey, No. 50 (January-June 1982).

28. "Japan: U.S. $27b Nuclear Energy Scheme." *Asia Research Bulletin* 12, No. 3, Report 5 (31 August 1982).

29. Japan. Science and Technology Agency. "Atomic Energy." *The White Papers of Japan 1979-1980*, annual abstracts of the Japanese government. Tokyo: Japan Institute of International Affairs Editorial Section of English Annual, 1981.

30. "U.S. Gov't Suggests New Formula for Accepting Uranium Enriching." *Japan Economic Journal*, 11 January 1983.

31. "Delay in Uranium Group Formation." *Japan Economic Journal*, 18 January 1983.

32. "Japan: Coal Mine Reopened in Northern Japan." *World Coal* 7 (July-August 1981).

33. Suda, Satoru. "Japan." In *Mining Annual Review, 1980*. London: Mining Journal, 1980.

34. Matsumoto, Toshio. "Coal: Demand Is Indicating Increasing Trend, but Domestic Miners Suffer from Deficits." In *Industrial Review of Japan, 1982*. Tokyo: Japan Economic Journal, 1982.

35. Matsumoto, Yoshio. "Coal: Some Miners Appear to Have Broken Even from Industries Shifting Away from Oil." In *Industrial Review of Japan, 1981*. Tokyo: Japan Economic Journal, 1981.

36. "Japan: Energy Plans Swing Back to Coal." *World Business Weekly*, 15 June 1981.

37. "Australia: Japan Drives Down Coal Prices." *Mining Journal* 300, No. 7697 (25 February 1983).

38. "Japan: World's Largest Importer Is Now Facing Problems." *World Coal* 8 (November-December 1982).

39. Miller, R. E. "Export Opportunities for Western Coal." *Mining Congress Journal* 68 (February 1982).

40. "Japan: Occidental Talks about Pungshuo in China." *World Coal* 8 (November-December 1982).

41. "Japan Eyes U.S. West Coast Coal." *Japan Economic Review*, 15 February 1982.

42. "Japanese Interest in Coal Canyon Mine." *Mining Journal* 298, No. 7641 (29 January 1982).

43. "Amax to Slash 1982 Moly Output." *Mining Journal* 298, No. 7652 (16 April 1982).

44. "Japan Drives Down Coal Prices." *Mining Journal* 300, No. 7697 (25 February 1983).

45. Kawata, Sukeyuki, ed. *Japan's Iron and Steel Industry—1982*. Tokyo: Kawata Publicity, Inc., 1982.

46. Nakamura, N., Y. Togino, and T. Adachi. "Philosophy of Blending Coals and Coke-Making Technology in Japan." *Ironmaking and Steelmaking* 5, No. 2 (1978).

47. "Australia/Japan Coal Liquefaction Plant." *Mining Magazine*, May 1982.

48. "Japan and Asutralia Launch Joint Brown Coal Liquefaction Project." *The Oriental Economist* 50 (April, 1982).

49. "Oil from Brown Coal Planned for Victoria, Australia." *Mining Engineering* 31 (June 1982).

50. "Mitsui Postpones SRC Plant." *Mining Journal* 298, No. 7661 (18 June 1982).

51. "Thai Shale to Be Studied." *Mining Journal* 299, No. 7687 (17 December 1982).

52. "Japan: Coal Liquefaction." *Energy World*, Bulletin of the Institute of Energy of the United Kingdom, No. 97 (November 1982).

53. *Atlas of the World*, 5th ed. Washington, D.C.: National Geographic Society, 1981.

54. Lehner, Urban. "Japan's Kyushu Promoted as 'Sun Belt' in Hopes of Luring Factories to the Sticks." *The Wall Street Journal*, 16 June 1982.

55. U.S. Department of the Interior, Bureau of Mines. "Japan: Steelmakers Hurt by World Recession." *Minerals and Materials: A Bimonthly Survey*, October-November 1982. Washington D.C.: U.S. Goverment Printing Office, 1982.

56. Alexander, W. O., and E. G. West. "Metallurgical Oper-

ations on Islands." *The Metallurgist and Materials Technologist* 14 (April 1982).

57. "Developing Technopolises for the 21st Century." *Asia Research Bulletin* 12, No. 8 (31 January 1983).

58. "News from Japanese Industry." *Bulletin of American-Japanese Economic Institute*, No. 234 (November-December 1982).

59. Japan. Environment Agency. "Environmental Pollution." *The White Papers of Japan 1979–1980*, annual abstracts of official reports and statistics of the Japanese government. Tokyo: Japan Institute of International Affairs Editorial Section of English Annual, 1981.

60. Gresser, Julian, Koichiro Fujikura, and Akio Morishima. *Environmental Law in Japan.* Cambridge, Mass.: The MIT Press, 1981.

61. Pempel, T. *Policy and Politics in Japan: Creative Conservation.* Philadelphia: Temple University Press, 1982.

62. Japan. *Basic Law for Environmental Pollution*, 1967. Cited in T. Pempel, *Policy and Politics in Japan: Creative Conservation.* Philadelphia: Temple University Press, 1982.

63. Japan. *Environmental Agency Establishment Law*, Law No. 88, 1981. Cited in Julian Gresser, Koichiro Fujikura, and Akio Morishima, *Environmental Law in Japan.* Cambridge, Mass.: The MIT Press, 1981.

64. Japan. *Air Pollution Control Law*, 1968. Cited in Julian Gresser, Koichiro Fujikura, and Akio Morishima, *Environmental Law in Japan.* Cambridge, Mass.: The MIT Press, 1981.

65. *Environmental Policies in Japan.* Paris: Organization for Economic Cooperation and Development, 1977.

66. "Japan's Steel Industry: Trouble Hits the East." *Iron and Steel International* 55 (October 1982).

67. Wilson, Caroll, Program Director. *World Coal Study*, Vols. I and II. Cited in T. A. Siddiqi, and D. James "Coal Use in Asia and the Pacific: Some Environmental Considerations." *Energy* 7 (March 1982).

68. Siddiqi, T. A., and D. James. "Coal Use in Asia and the Pacific: Some Environmental Considerations." *Energy* 7 (March 1982).

69. Japan Economic Planning Agency. *New Economic and Social Seven-Year Plan*, August 1979.

70. Crowson, P. C. F. "The National Mineral Policies of Germany, France and Japan." *Mining Magazine*, June 1980.

71. *Exporter's Encyclopedia*, 77th ed. New York: Dun & Bradstreet International Ltd., March 1982.

72. Nagano, Takeshi. "The History of Copper Smelting in Japan." *Journal of Metals* 34 (June 1982).

Commodity Profiles of Strategic Minerals

Mineral commodities have a wide variety of applications in Japan. Some of these applications, such as alloy steel, aluminum metal, copper wire, and zinc coatings, are old and well established. Others, such as semiconductors, superalloys, and photovoltaic cells, are newer and have led to increased demand for their appropriate raw materials. Mineral raw materials were required for the development of almost every manufacturing industry in Japan.

Most of Japan's strategic mineral supplies are obtained through trade. Domestic mineral reserves are limited, and mine production provides only a small contribution to domestic ore and concentrate requirements.

Japanese mineral supplies are imported from several different countries. One Japanese mineral policy goal has been to develop a diversified set of supply sources in order to lessen the dependence on any one country's exports.

These mineral supplies are imported in various forms. In several examples, semiprocessed and fully processed forms have become favored over once-traditional ore and concentrate imports. Increased energy and raw material costs, a changing national economic base, and the development of local processing capacity in mineral-producing countries are reasons for this trend. Japanese banks and companies have also provided financing, or have become joint-venture partners, in the development of local processing facilities.

This chapter provides specific information regarding Japan's strategic mineral commodities. Where possible, the domestic sources, foreign sources, domestic processing, domestic consumption, exports, and significant trends are examined. Special emphasis is placed on the form of mineral imports, the effects of energy on supplies and consumption, and the level of Japanese participation in overseas resource development and mineral processing projects.

STRATEGIC MINERALS

A strategic mineral can be defined as one that is, at the time of designation, essential for the operation of a key industrial or military sector of a country, available only in limited quantities around the world and not available to a large extent domestically. A fourth characteristic could be included: that the mineral-producing country, or countries, are politically unstable. An interruption in the supply of a strategic mineral could result in higher prices for that mineral, a disruption in the operation of the consuming country's industries that are dependent on that mineral, and could limit the country's ability to build military hardware required to defend itself in a time of war.

Japan has several strategic mineral commodities. Domestic reserves are available for several of these, but only in limited quantities (Table II-4-1). As a result, Japan maintains a high net import dependence for its strategic mineral commodities (Table II-4-2).

It is difficult to classify Japanese strategic mineral commodities according to a traditional scheme that is based on usage, such as the one shown in Table II-4-3. Many of these commodities have wide-ranging applications, and could be designated in more than one category simultaneously. For example, the ferrous metals, cobalt, chromium, manganese, nickel, molybdenum, and vanadium could also be considered as nonferrous metals because of their use in nonferrous and superalloys. Several other metals, including beryllium, cobalt, gallium, germanium, platinum, rare earth metals, silicon, strontium, tantalum, titanium, and zirconium, could be designated in a high-technology category, because of their knowledge-intensive industrial applications. Therefore, the classification scheme is provided in Table II-4-3 as a matter of convenience.

Because of their strategic nature, many of these commodities are included in the national and private stockpiles (Table II-4-3). The national stockpile mainly includes the ferrous metals, and some nonferrous and precious metals with high-technology applications. Nonferrous metals are the basis of the private stockpiles.

ALUMINUM

Japan does not have any domestic bauxite resources. In 1980, 5.708 million metric tons of bauxite were imported, mostly from Australia (65%), Indonesia (21%), and Malaysia (11%).[1]

TABLE II-4-1   Japanese Domestic Mine Production, Percentage Contribution to Domestic Ore and Concentrate Requirements, and Reserves[a]

Commodity	1981 mine production[b]: gross weight/metal content (metric tons)	1981 contribution to domestic ore requirements[b] (%)	Exploitable crude ore reserves[c] Tonnage: gross weight/metal content (metric tons)	Year of survey
Chromium	10,959/3625	2	381,000/74,000	1979
Copper	NA/51,459	4	60,600,000/775,000	1980
Fluorspar	NA	NA	88,000/NA	1963
Gold	NA/3	7	5,705,000/24	1980
Iron	442,000/275,000	0.4	5,678,000/1,885,000	1980
Iron sand			77,449,000/11,839,000	
Lead	NA/44,932	22	61,886,000 /549,000	1980
Zinc	NA/242,042	36	/3,089,000	
Manganese (metallurgical)	87,208/21,134	4	3,193,000/529,000	1979
Molybdenum	NA/80 (concentrate)	NA	155,000/1000	1975
Silver	NA/279	21	196,000/4	1976
Tin	NA/562	NA	1,528,000/11,000	1976
Tungsten	NA/667	20-25	4,551,000/21,000	1979

[a]NA, not available.
[b]Ref. 1.
[c]Japan. Prime Minister's Office, Statistics Bureau. *Japan Statistical Yearbook.*

TABLE II-4-2   Japanese Net Import Reliance for a Group of Selected Strategic Mineral Commodities, 1980 (Percent)

Columbium	100	Iron ore	99
Bauxite	100	Molybdenum	99
Cobalt	100	Platinum group	98
Tantalum	100	Manganese	97
Fluorspar	100	Tin	96
Nickel	100	Gold	94
Antimony	100	Copper	87
Titanium	100	Lead	75
Vanadium	100	Tungsten	75
Phosphate rock	100	Silver	73
Chromium	99	Zinc	59

Source: U.S. Department of the Interior, Bureau of Mines. Cited by Paul Portney, ed., *Current Issues in Natural Resource Policy.* Baltimore, Md: The Johns Hopkins University Press, 1982.

Annual bauxite imports have leveled off since 1973, after exhibiting rapid growth in the 1960s and early 1970s (Figure II-4-1). Future bauxite imports might be expected to decline if domestic refining, as well as smelting, capacity is reduced. In 1982, it was reported that Australian bauxite exports to Japan had been significantly reduced.[2]

Japan's Ministry of International Trade and Industry (MITI) is investigating the use of domestic resources of aluminum-rich clay as a substitute for bauxite.[3] However, problems in the production process will probably not be solved for several years, and commercialization would require at least an additional 10 years.[4]

Japan also imports aluminum oxide, hydroxide, and metal. In 1980, aluminum oxide and hydroxide imports totalled 735,000 metric tons, or approximately 28% of total supplies of alumina, oxides, and hydroxides combined.[1] Most of these imports came from Australia. A record high of 1,129,322 metric tons of aluminum and aluminum alloy metal supplies were imported in 1981, mostly from the United States (24%), Venezuela (14%), Canada (10%), New Zealand (10%), Australia (6%), the USSR (6%), and Argentina (5%).[1] Aluminum scrap imports totaled 275,696 metric tons in 1980, and most were imported from the United States (82%).[1] Oxide and hydroxide, and metal imports increased during the period 1963–1981 (Figure II-4-1). The increase in metal imports was most pronounced in the late 1970s and early 1980s, as domestic smelting capacity was reduced.

Japan produces most of the alumina it requires as a feedstock for primary aluminum metal production. In 1980, domestic alumina production amounted to 72% of the total aluminum oxide and hydroxide, and alumina suppplies. The history of annual alumina production (Figure II-4-2), mimics that of annual aluminum metal production (Figure II-4-3).

TABLE II-4-3   Selected Japanese Strategic Mineral Commodities

Ferrous	Nonferrous	Precious	Industrial
Chromium[a]	Aluminum[b]	Gold	Fertilizer minerals
Cobalt[a]	Antimony[a]	Platinum-group metals[a]	Fluorspar
Columbium[a]	Beryllium	Silver	
Ferroalloys[c]	Copper[b]		
Iron and steel	Gallium		
Manganese[a]	Germanium		
Molybdenum[a]	Lead and zinc		
Nickel[a]	Rare earth metals		
Tungsten[a]	Silicon		
Vanadium[a]	Strontium[a]		
	Tantalum[a]		
	Tin		
	Titanium		
	Zirconium		
	Zinc[b]		

[a]Included in the proposed national stockpile.
[b]Included in the private stockpiles.
[c]Ferroalloys will probably be included in the national stockpile under headings of other ferrous metals.

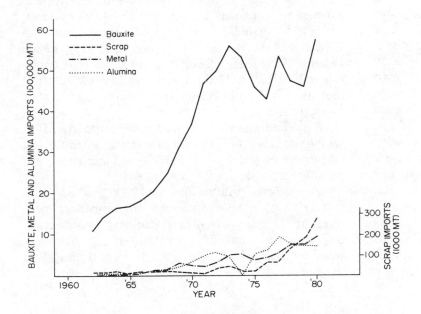

FIGURE II-4-1   Annual Japanese aluminum imports, 1962-1980. (From U.S. Department of the Interior, Bureau of Mines. *Minerals Yearbook*.)

Japan's aluminum industry thrived in the 1960s and early 1970s. During this time, Japan underwent rapid industrial and urban growth, a situation that created a large demand for aluminum. Both bauxite and cheap oil were available on world markets. Annual primary production increased sevenfold, from 171,000 metric tons in 1962 to 1,205,000 metric tons in 1973 (Figure II-4-3).

Following the oil price hikes of 1974, the Japanese aluminum industry became less competitive on international markets. Japanese producers, who depend on oil-generated electricity, were unable to compete with foreign producers who had access to cheaper energy supplies. The total production cost of Japanese aluminum eventually reached $1.08 per pound in 1980, and approximately one-half of that amount represented power costs.[5]

Annual production declined from 1973 to 1976 because of the energy cost increases and the global economic recession. Primary production increased in 1977, but then began a general decline that continued into 1982. Primary production in 1981 was 770,600 metric tons, a 64% drop from the produc-

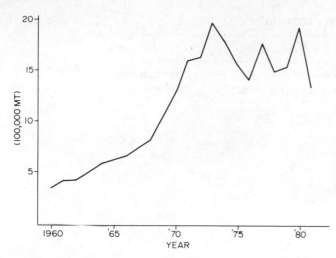

FIGURE II-4-2 Annual Japanese alumina production, 1960–1981. (From U.S. Department of the Interior, Bureau of Mines. *Minerals Yearbook*.)

tion peak in 1973. Production in 1982 was estimated to be even lower, totalling 400,000 metric tons.[6]

In 1978, the Japanese aluminum industry was declared structurally depressed by the Japanese government (Chapter II-5). Producers shut down inefficient plants, and annual production capacity dropped from 1.64 million metric tons in 1973 to 1.1 million metric tons in 1981 (Figure II-4-3). MITI proposed a further cutback to 712,000 metric tons by the end of fiscal year 1983.[7] MITI did not recommend elimination of all domestic capacity because it wanted to maintain bargaining power with foreign suppliers, refining technology, and emergency capacity in case of any future metal supply cutoffs.

One producer, Nippon Light Metal, sold some potlines from one of its smelters to Alusaf (South Africa), as an alternative to scrapping its facilities.[5] Additional future sales may occur, as China has expressed interest in purchasing some idle smelting facilities.[8] Japanese smelting facilities are attractive investments because their energy consumption per unit of production is lower than that of foreign facilities.

Japanese smelters also used the strategy of converting from oil-generated to non-oil-generated electricity.[4] Mitsui Aluminum's Miike smelter already uses coal to generate electricity. Sumitomo Aluminum and Kokuriku Power Co. will jointly build a coal-fired power plant to power Sumitomo's smelter at Kikumoto.[1] Sumitomo was also reported to be considering the use of geothermal power at its operations.[9] Nippon Light Metal's Kambara smelter runs on energy generated by hydropower.[1]

MITI's goal of 712,000 metric tons of domestic capacity represents approximately one-third of Japan's expected aluminum needs for 1985.[7] The rest of Japan's supplies will be derived from metal imports and secondary production. Additional future imports will come from several countries.

Japan made significant investments in overseas processing capacity in Indonesia, New Zealand, Brazil, Venezuela, Australia, Canada, and the United States in the 1970s and the early 1980s (Figure II-4-4). By 1985, it is expected that about one-half of Japan's aluminum imports of 756,000 metric tons will come from Japanese-invested foreign capacity.[10]

To encourage increased metal imports, MITI will eliminate the 9% import tariff on imported aluminum metal, for a volume of imports comparable to the domestic loss in capacity.[7]

Secondary production in 1981 was 814,000 metric tons, a higher volume than primary production (Figure II-4-3). Secondary production requires less than 20% of the energy normally required for primary production,[11] and therefore lowers production costs. As might be expected,

FIGURE II-4-3 Annual Japanese aluminum production and capacity, 1960–1982. [From *production statistics:* U.S. Department of the Interior, Bureau of Mines. *Minerals Yearbook. Japanese Aluminum Federation News*, No. 4 (1982). "Japanese Al Predictions." *Mining Journal* 300, No. 7691 (14 January 1983). *Capacity statistics:* American Bureau of Metal Statistics. *Nonferrous Metal Data.* "Six Aluminum Refiners Announce Production Curtailment Plans." *The Oriental Economist* 50 (January, 1982).

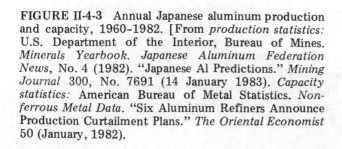

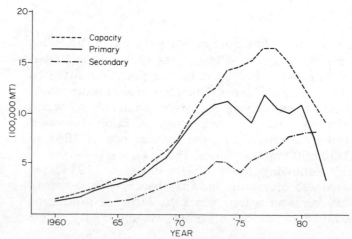

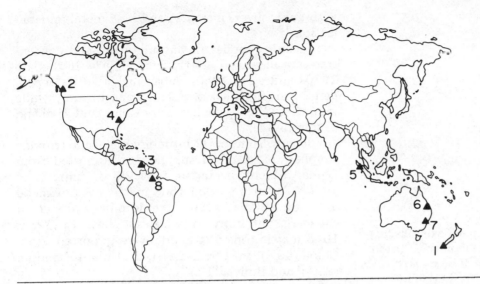

FIGURE II-4-4   Japanese-invested overseas aluminum processing capacity. [From Ref. 1; "First A1 Exports from Indonesia." *Mining Journal* 299, No. 7680 (29 October 1982). "Queensland Has Integrated Aluminum Industry." *World Mining* 55 (November 1982). "Japanese Kission to Discuss Albras." *Mining Journal* 299, No. 7682 (12 November 1982). "Nippon Amason Aluminum Co. . . ." *Mining Engineering* 35 (March 1983).]

Country: project	Annual capacity (thousand metric tons)	Japanese output share (thousand metric tons)	Status
1 New Zealand: Bluff	152	75	Completed in 1977
2 Canada: Kitimat	90	45	Completed in 1977
3 Venezuela: Venalum	280	160	Completed in 1979
4 US: Alumax	180	45	Completed in 1980
5 Indonesia: Asahan	225	170	Scheduled for completion in 1984 (first phase completed in 1982
6 Australia: Gladstone (Boync)	206	101	Scheduled for completion in 1984 (partially completed in 1982)
7 Australia: Hunter Valley	232	(a)	Scheduled for completion in 1984
8 Brazil: Albras	320	160	Originally scheduled for partial completion in 1984, but projects delays will probably delay startup by several years

aluminum scrap imports increased in 1975–1980, paralleling the increase in secondary production (Figures II-4-1 and II-4-3).

There are five aluminum producers in Japan. They are Mitsubishi Light Metal Industries, Ltd., Mitsui Aluminum Co. Ltd., Showa Denko K. K., and Sumitomo Aluminum Smelting Co. Ltd. A sixth producer, Sumikei Aluminum Industries, shut down its operations in 1982.[12] According to an announcement in 1982, a 50% stake in Showa Denko is to be acquired by CRA, the Australian mining group of Rio Tinto Zinc Corp.[13] The agreement commits Showa Denko to a long-term aluminum purchase contract with Comalco, which is a subsidiary of CRA.

The main domestic applications for aluminum are in the construction, transportation, fabricated metals, and communication machinery sectors. Economic downturns in the first two sectors in the early 1980s lessened aluminum demand and increased stocks. Increased substitution of aluminum for steel in the transportation sector has not

occurred because of aluminum's high price in domestic markets.[14] Future increases in domestic aluminum demand depend on the price of imported metal and the degree of economic recovery.

Japan exports aluminum mainly in the form of metal, and oxide and hydroxide. In 1980, Japan exported about 400,000 metric tons of aluminum, an amount equivalent to 15% of total oxide and hydroxide and alumina supplies combined, mainly to Canada (38%), Egypt (23%), and South Korea (15%).[1] In 1980, the major importers of Japanese aluminum and aluminum alloy metal included Australia and the Philippines.[1] The United States and Asian countries, including China, Hong Kong, and Taiwan, accepted many of Japan's exports of semimanufactured aluminum.[1]

Since 1978, aluminum metal exports have declined and alumina exports have increased (Figure II-4-5). These patterns reflect the decline in domestic primary metal production, for which alumina is a feedstock, and continued domestic aluminum metal demand.

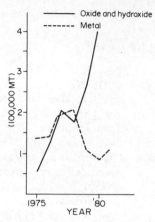

FIGURE II-4-5 Annual Japanese aluminum oxide, hydroxide, and metal exports, 1975–1981. [From U.S. Department of the Interior, Bureau of Mines. *Minerals Yearbook. Japanese Aluminum Federation News*, No. 4 (1982).]

ANTIMONY

Commercial domestic mine production of antimony stopped in 1971 (Figure II-4-6). Japanese antimony deposits are vein-type, and are sometimes associated

with gold, silver, mercury, and base-metal mineralization.[15]

Japan's supplies of antimony ore and concentrate are imported mostly from Bolivia. In the late 1970s and early 1980s, Japan increased its dependence on Bolivian ores and concentrates, while importing fewer supplies from China and Thailand (Table II-4-4).

Japan also imports antimony metal and trioxide supplies. Most metal supplies are imported from China, which accounted for 92% of imports in 1980.[1] Japan imported 1563 metric tons of metal in 1980. China is also the leading supplier of Japanese antimony trioxide supplies. In 1981, 1048 metric tons of trioxide were imported, from China (49%), the USSR (28%), the United Kingdom (22%), and Bolivia (1%).[1]

Since the early 1970s antimony ore and concentrate imports have generally declined, while metal imports have increased (Figure II-4-7). The availability of cheap imports, mainly from China, resulted in this overall change in imported forms. In addition, antimony trioxide imports, used to

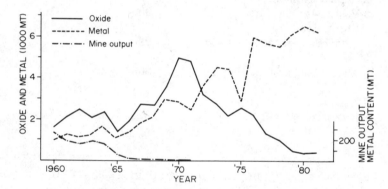

FIGURE II-4-6 Annual Japanese antimony mine output, and oxide and metal production, 1960–1981. (From U.S. Department of the Interior, Bureau of Mines. *Minerals Yearbook.*)

TABLE II-4-4 Annual Japanese Antimony Ore and Concentrate Imports, by Source, 1970–1980[a,b] (Metric Tons)

| Year | Total imports | Source | | | | |
		Bolivia	South Africa	Thailand	China	Other
1970	17,344 (100)	10,630 (61)	NA	1745 (10)	NA	4969 (29)
1971	10,197	5,272 (52)	NA	1036 (10)	3038 (30)	806 (8)
1972	13,312	6,504 (49)	1485 (11)	NA	2182 (16)	3141 (24)
1973	13,959	7,345 (53)	NA	1526 (11)	3014 (22)	2074 (15)
1974	10,857	6,216 (57)	NA	907 (8)	1994 (18)	1740 (16)
1975	9,012	4,402 (49)	NA	NA	2437 (27)	2173 (24)
1976	12,703	8,444 (66)	NA	NA	819 (6)	3440 (27)
1977	7,501	NA	NA	NA	NA	NA
1978	6,553	4,699 (72)	NA	1032 (16)	NA	822 (13)
1979	6,702	5,503 (82)	536 (8)	335 (5)	NA	328 (5)
1980	6,966	6,055 (87)	446 (6)	NA	NA	465 (7)

[a]NA, not available.
[b]Numbers in parentheses represent percent of total imports.
Source: U.S. Department of the Interior, Bureau of Mines, *Minerals Yearbook.*

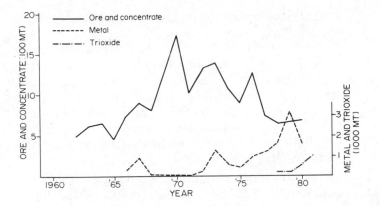

FIGURE II-4-7   Annual Japanese antimony ore and concentrate, and metal imports, 1962–1981.

make metal and flame-retardant products, were increasing in the late 1970s and early 1980s.

The recent increase in antimony trioxide imports has created alarm among domestic antimony producers. These producers have already realized significant metal production declines since the middle 1970s, while metal imports increased (Figures II-4-6 and II-4-7). Despite the overall rise in oxide production that has taken place since the early 1960s (Figure II-4-6), domestic producers fear that increased trioxide import duties will be required to prevent cheaper imports from meeting a larger portion of domestic demand.[16] Japanese producers have requested an increase in the import duty from 10.6% to 12%. However, according to GATT terms, this duty is to be reduced to 7.2% by 1987. Increased imports of trioxide and metal may be unavoidable, since future world antimony trade will probably involve more metal and oxide forms from Bolivia, China, and South Africa.[17]

Antimony has many uses.[18] In metallic form, its major applications are as an alloying agent in lead-based alloys for storage batteries, power transmission and communications equipment, and solder. In nonmetallic form, its major applications include flame-retardant chemicals, ceramics, plastics and rubber products. The increase in trioxide imports noted above is the result of increased demand from flame-retardant producers.[16] Although the demand for lead-antimony alloy storage batteries is strong,[19] the increased use of maintenance-free lead-tin-calcium alloy batteries, and increased recycling of used lead-antimony batteries will lead to decreased demand for antimony in battery applications.

BERYLLIUM

Japan imports all its beryllium supplies, mostly in the oxide form. In 1980, Japan imported 85 metric tons of beryllium oxide and 1611 kg (1.6 metric tons) of beryllium metal.[1] All of the oxide and most of the metal was imported from the United States. A small amount, 7 kg (0.007 metric ton) of metal was imported from the USSR.

Beryllium's main applications are in the nuclear power, aerospace, and electrical industries. Beryllium is mostly used in alloy with copper, but it is also used in oxide and metal forms. Small amounts of beryllium are required in these applications. In its structural applications, such as aircraft frames and satellites and space vehicles, beryllium metal has several substitutes, including steel, titanium, and boron or graphite composites.[20] Graphite can substitute for beryllium metal in its nuclear applications. Substitution is also possible for beryllium-copper alloys and beryllium oxide forms, but at a performance loss in some use sectors.[20] Japan's supplies of all of these substitutes, excluding steel, must all be imported.

CHROMIUM

Japan produced 10,959 metric tons of chromite from domestic mines in 1981.[1] This ore was produced at two mines,[5] Hirose and Wakamatsu, located in the Chugoku district along the boundaries of the Tottori, Okayama, and Hiroshima prefectures.[15] The alumina-rich chromite is low grade, 30 to 35% Cr_2O_3, and it commonly occurs as disseminated ore in ultrabasic rocks.

Domestic mine production is insignificant with respect to total chromite ore demand (Table II-4-1), and mine production has declined since the early 1960s (Figure II-5-4). According to a 1979 survey, Japan has 381,000 metric tons of exploitable chromium ore reserves, with a metal content of 74,000 metric tons (Table II-4-1).

Japan imports most of its chromium supplies. Annual chromium ore and concentrate imports generally parallel annual Japanese steel production (Figures II-4-8 and II-4-22). The declines in imports in 1963, 1965, 1972, and 1978 are the result of similar declines in steel production. Three countries, South Africa, the Philippines, and India, account for the bulk of imported supplies (Table

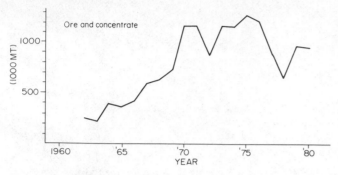

FIGURE II-4-8 Annual Japanese chromium ore and concentrate imports, 1962–1980. (From U.S. Department of the Interior, Bureau of Mines. *Minerals Yearbook*.)

II-4-5). Chromium ore and concentrate purchases from South Africa represent almost one-half of Japan's imports. Increased purchases from other suppliers, such as India and the Philippines, would not only lessen Japan's dependence on South African supplies, but would increase the average grade of Japanese imports. Chromium ore, with grades as high as 56% Cr_2O_3, can be imported from these two Asian countries.[21]

Japanese involvement in overseas chromium mine development is limited, although Japan was actively involved in overseas exploration during the 1970s. Japan's Kawasaki Steel Corp. is a joint-venture partner with Amax, Inc. (U.S.) and Philchrome Mining Corp. (Indonesia) in a chromite mine development project on Palawan Island, Indonesia.[22] Mine production, which began in 1981, is expected to reach 20,000 metric tons per year of refractory-grade chromite.

Japan was involved in the development of the Serjana chromite mine in Brazil, but withdrew its support in 1980.[21] The consortium of five Japanese ferrochromium producers sold their 49% share in the project to their Brazilian partner, Ferroligas da Bahia, S.A. This deposit was estimated to have 100 million metric tons of chromium ore, and

annual mine production was expected to be 100,000 to 200,000 metric tons of chromium ore.[23] The Japanese pulled out because of the project's large financial losses,[21] the anticipated high ore transport costs,[1] and the expected reduction in Japanese ferroalloy production capacity (Chapter II-5). Another reason for the withdrawal might be that the initial reserve estimates were overly optimistic. In 1979, reserves were estimated to be only 10 million metric tons.[24]

Japan imports significant amounts of ferrochrome. In 1981, approximately 50% of the Japanese ferrochromium requirements were imported.[22] South Africa is the source of most of Japan's imported ferrochrome supplies. However, South Africa's share of the Japanese import market dropped to 66% (127,000 metric tons) in 1981, from 72% in 1980.[25] This decline was apparently due to South Africa's price increase request and Japan's ability to increase imports from other sources. If Japanese ferrochrome production capacity is decreased, future ferrochrome imports would be expected to remain steady or increase, whereas chromium ore and concentrate imports would be expected to decrease.

The main consumers of chromium in Japan are the ferrochrome and steel industries. Chromium is an essential element in stainless steel and there is no substitute for it in this application. The history of annual ferrochrome production parallels that of annual steel production (Figures II-4-16 and II-4-22).

Ferrochromium production is energy intensive and domestic producers have had to cope with the high costs of imported energy supplies. Because of the ferroalloy industry's inclusion in the list of industries covered under Japan's depressed industry laws (Chapter II-5), ferrochrome production capacity will probably decline in the future. Producers will merge their operations and eliminate excess and uneconomic capacity. Higher-grade

TABLE II-4-5   Annual Japanese Chromium Ore and Concentrate Imports, by Source, 1978–1981[a] (Metric Tons)

Year	Total imports	Source			
		South Africa	India	Philippines	Other
1978	670 (100)	350 (52)	73 (11)	74 (11)	173 (26)[b]
1979	962	443 (46)	156 (16)	184 (19)	179 (19)
1980	950	407 (43)	170 (18)	208 (22)	165 (17)
1981	744	350 (47)	141 (19)	149 (20)	104 (14)

[a]Numbers in parentheses represent percent of total imports.
[b]Other suppliers in 1981 were the USSR (6%), Albania (4%), and China, Vietnam, Pakistan, Turkey, Cuba, and Finland (combined 4%).

Source: U.S. Department of the Interior, Bureau of Mines, *Minerals Yearbook*.

Indian ores were also being used to reduce the energy requirements for processing.

Imports of charge chrome, or prereduced chromite, might be expected to increase in the future, because the use of charge chrome in ferroalloy production results in energy savings.[26] Increased imports may be the goal of those producers involved in the development of overseas charge chrome production capacity.[27] At Byrapura, India, the Karnataka government's Mysore Minerals, Ltd. is expected to receive Japanese financial and technical assistance for its charge chrome plant construction project. Also in India, two Japanese companies will supply the equipment and machinery required to build a charge chrome plant in Orissa state.

Japanese producers have also tried to reduce their excess inventories in the early 1980s. Exports of Japanese ferrochrome were being promoted in foreign markets, such as Taiwan, Brazil, and Yugoslavia.[21]

Chromium metal production in Japan increased steadily from the early 1960s to 1981 (Figure II-4-9). In 1981, 3625 metric tons were produced for use in superalloys, nonferrous alloys and welding rods. Increased demand for special alloys in the 1970s contributed to this production increase. Two Japanese chromium metal producers, Nippon Denko Ltd. and Toyo Soda Manufacturing Co., underwent significant production capacity expansions in 1980-1981.[21]

Small amounts of chromite ore are exported annually. In 1980, 5678 metric tons were exported, mostly to North Korea (77.9%) and South Korea (21.7%).[1] Chromite exports in 1980 were more than twice as large as 1979 exports, reflecting the large domestic inventories of ferrochrome, and possibly the future reductions in ferrochrome capacity.

COBALT

Japan imports all its cobalt supplies. The Philippines and Australia supply Japan with mixed nickel-cobalt sulfides, from their nickel refining and mining operations, for cobalt refining in Japan.[1]

Japan also imports cobalt oxides and hydroxides, mostly from Belgium. Cobalt metal and alloy imports amounted to 886 metric tons in 1981, mostly from Belgium (28%), Zaire (24%), the United States (16%), and Finland (11%).[1]

The history of annual Japanese cobalt imports has been erratic, with increases and decreases that reflect the country's general economic conditions and cobalt's main uses in growth industries (Figure II-4-10). During slowdowns in economic growth, including those in 1971, 1975, the late 1970s, and the early 1980s, cobalt imports declined. The higher price of cobalt in 1978-1980 and the interruption in supplies from Zaire, probably contributed to the low level of cobalt metal imports during that time.[28]

Cobalt metal is produced by two companies in Japan, Sumitomo Metal and Nippon Mining. Their combined annual production capacity is 2800 metric tons.[1] Annual cobalt metal production grew significantly, from 10 metric tons in 1974 to 2867 metric tons in 1980 (Figure II-4-11). In 1977, Japan surpassed Finland and became the world's fourth largest cobalt metal producer, behind Zaire, the USSR, and Zambia.[29]

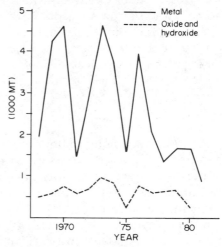

FIGURE II-4-10 Annual Japanese cobalt metal, and oxide and hydroxide imports, 1968-1981. (From U.S. Department of the Interior, Bureau of Mines. *Minerals Yearbook*.)

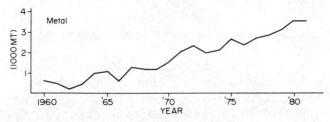

FIGURE II-4-9 Annual Japanese chromium metal production, 1960-1981. (From U.S. Department of the Interior, Bureau of Mines. *Minerals Yearbook*.)

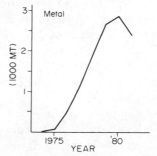

FIGURE II-4-11 Annual Japanese cobalt metal production, 1974-1980. (From U.S. Department of the Interior, Bureau of Mines. *Minerals Yearbook*.)

The main uses for cobalt in Japan are in advanced or high-technology industries, such as the manufacture of magnetic alloys, specialty steel, heat-resisting alloys, ultrahard alloys, and catalysts.[5] Magnetic alloys make up 30% of domestic consumption. Domestic consumption in 1980 totaled 1914 metric tons.[5]

Future Japanese cobalt demand will depend on several factors. The planned development and expansion of Japan's aerospace industry, as called for in the country's future growth plans (Chapter II-5), will probably increase the demand for cobalt in nickel-based superalloys. At the same time, increases in the price of cobalt metal would stimulate research and development activities for more efficient usage of cobalt and for cobalt substitutes. Technological advances made in the production process for combined rare-earth and cobalt alloy magnets have led to the miniaturization and a reduction in cobalt requirements in the Japanese consumer electronics industry.[30] In the United States, NASA researchers have been able to substitute nickel for some of the cobalt in at least one alloy used for aerospace applications.[31]

Japan exports toll-refined cobalt and small amounts of cobalt oxides and hydroxides. In 1980, toll-refined cobalt exports represented 33% of total domestic production.[5] Since cobalt is produced as either a by-product or co-product of copper and nickel refining, future supplies will be affected by the production levels of the latter two metals.

COLUMBIUM

Japan has no domestic columbium resources. Its supplies are derived mostly from imports of ore and concentrate from pyrochlore deposits in Canada and Brazil. In 1980, 1764 metric tons of ore and concentrate were imported.[1] Canada supplied 52% of these imports and Brazil, 40%. Japan may become more dependent on Brazilian imports as a result of the shutdown of the lone Canadian columbium mine in 1982.[32]

Most of the imported columbium ore and concentrate is used by the Japanese iron and steel industry. Columbium is added in small amounts to stainless and high-strength low-alloy (HSLA) steels. Domestic ferrocolumbium production dropped to 825 metric tons in 1981, a 21% drop from the 1980 level,[1] in response to the decline in steel production.

Columbium can also be used in nickel, iron, and cobalt-based superalloys. Future demand for columbium imports will depend on the demand for steel as well as special alloys.

COPPER

Japan's exploitable crude copper ore reserves are estimated to be 60,600,000 metric tons with a metal content of 775,000 metric tons (Table II-4-1). Several types of copper deposits have had economic importance in Japan. The most important types are the copper-lead-zinc-silver-sulfate Kuroko-type massive sulfide deposits (black ores) in northern Honshu (Figure II-5-6). Kosaka and Hanaoka are two large mines that produce ores of this type.[33] Japan's continuing exploration program for domestic Kuroko-type deposits has resulted in several discoveries (Chapter II-5). Copper is also produced from copper-pyrite Besshi-type massive sulfide deposits.[34] Typical Besshi-type deposits occur on Shikoku Island, but the Hitachi mine on northeastern Honshu and the Shimokawa mine on Hokkaido occur in similar metamorphic rocks.[35] Other important deposits include the epithermeral vein-type, such as the Ikuno and Akenobe mines on southwestern Honshu[36] and the Hosokura mine in northeastern Honshu;[35] and skarns, such as the Kamaishi iron-copper mine on northeastern Honshu.[35]

Domestic copper ore and concentrate production has declined steadily from a high of 138,000 metric tons (metal content) in 1972 to 51,449 metric tons (metal content) in 1981 (Figure II-5-2). Ore grade has declined and reserves have been exhausted at many mines. As a recent example, Shimokawa Mining Company was expected to close its Shimokawa copper mine (Table II-2-1) at the end of 1982, because of depleted reserves.

The bulk of Japanese copper supplies are imported from several foreign countries, mostly in the ore and concentrate form. In 1981, Japan imported 3,338,326 metric tons of ores and concentrates.[1] The breakdown of suppliers was as follows: the Philippines 30%, Canada 23%, the United States 15%, Papua New Guinea 9%, Chile 6%, Indonesia 4%, Malaysia 4%, Australia 4%, Zaire 3%, and others 6%. Japan also imported 241,146 metric tons of refined copper in 1981, mostly from Zambia (15%), Chile (15%), and Peru (14½%).[1] Japan also imports moderate amounts of copper scrap and semimanufactured metal. In 1980, 49,929 metric tons of scrap and 3330 metric tons of semimanufactures were imported.

Japan's annual imports of copper ore and concentrate have generally increased since the early 1960s (Figure II-4-12). Domestic demand was strong through this period. Annual copper imports continued to increase in the early 1980s. The

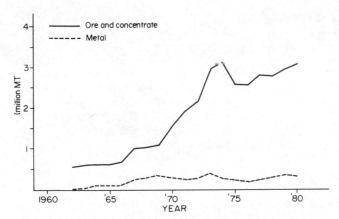

FIGURE II-4-12 Annual Japanese copper ore and concentrate, and metal imports, 1962–1981. (From U.S. Department of the Interior, Bureau of Mines. *Minerals Yearbook*.)

annual import level in 1981 was even higher than the pre-1975 recession high (1974) (Table II-4-6), another characteristic not observed for many of Japan's other ore concentrate imports.

As mentioned in Chapter II-2, Japanese mineral policy has emphasized the development and acquisition of overseas copper ore supply sources. The involvement of Japanese companies in several overseas copper mines (Table II-2-2) demonstrates this emphasis.

In the early 1980s, Japanese producers were involved or interested in several other overseas copper mining and development projects. Japan's Marubeni Corp. increased its equity in the Philippines CDCP Mining Corp. to 35%.[37] These two companies maintained a joint-venture exploration agreement which led to the discovery of additional copper-molybdenum reserves in 1981. In China, Japan's Mineral Mining Agency will spend Y2000 million over four years for detailed exploration

and development at the Anguin copper deposit.[38] Japan's Sumitomo Metal Mining Co. is a member of the consortium developing the Kutcho copper deposit in Canada. Japan has also provided a $475,800 development loan for this project.[1] In 1982, Japan's MMAJ was considering a request from the Panamanian government for partial funding of an exploration and delimiting program at the Petaquilla porphyry copper deposit.[39]

Following the closing of Anaconda's (U.S.) copper smelter at Anaconda, Montana, seven Japanese smelters signed a multiyear toll-refining agreement to process 400,000 to 500,000 metric tons of copper concentrates from Anaconda's mines in Montana, Arizona, and Utah.[5] The Japanese smelters agreed to purchase 50% of the refined copper output. However, in mid-1983, Anaconda planned to close its Butte, Montana, copper mine temporarily because of poor ore and concentrate market conditions.[40] As a result, Japan's supplies from Anaconda will probably be cut back to some extent.

Japanese producers have also become directly involved in several overseas smelting operations (Table II-4-7). Unlike the aluminum example described earlier, Japanese domestic smelters are apparently trying to maintain their own market share. As noted above, they have continued their efforts to acquire imported concentrate supplies. However, aggressive Japanese efforts in the Philippines could lead to strained relations with that country. Japanese smelters offer custom processing rates to Philippine concentrate suppliers that are 10 to 20% cheaper than those offered by the Philippine's PASAR smelter, which is partly Japanese-owned (Table II-4-7).[41] The Philippines government considers Japanese rates to be sub-

TABLE II-4-6 Annual Japanese Copper Imports, by Form, 1974–1981[a]
(Thousand Metric Tons)

| Year | Ore and concentrate | Metal | | | |
		Scrap	Unwrought	Semimanufactures	Total
1974	3124	35	303	17	335
1975	2605	39	184	2	225
1976	2587	34	233	1	268
1977	2823	41	272	1	314
1978	2818	55	333	1	389
1979	2969	50	395	3	449
1980	3104	50	290	3	344
1981	3338	NA	NA	NA	NA

[a]NA, not available.

Source: U.S. Department of the Interior, Bureau of Mines, *Minerals Yearbook*.

TABLE II-4-7   Japanese Involvement in Overseas Copper Processing Capacity[a]

Country	Local participants	Japanese participants	
Philippines[b,c]	PASAR (29%) — Consortium of Philippine ore and concentrate suppliers Philippine government (34%) — National Development Co. International Finance Corp. (5%)	Marubeni Corp. (16%) Sumitomo Corp. (10%) C. Itoh & Co. (6%)	PASAR will supply copper ore and concentrates. Japanese will provide technology and receive 105,000 metric tons/yr of refined copper (76% of total production capacity of 138,000 metric tons/yr). Startup: 1983
United States[d]	Kennecott Minerals Co. — Chino Mines Division	Mitsubishi Corp.	Mitsubishi will provide $116 million for multiyear smelter modernization and expansion program, in return for $33\frac{1}{3}\%$ ownership of Chino Mines. Mitsubishi to receive rights to one-third of output. Startup: 1985
Mexico[e,f]	Mexicana de Cobre, S.A.	Japan Export-Import Bank	Japan will provide $96 million loan for construction of a 180,000 metric ton/yr electrolytic copper refinery. Marubeni Corp. (Japan) and a Mitsui Smelting and Refining Co. subsidiary will supply equipment and engineering services, respectively. Startup: 1985
Peru[g]	NA	NA	Japan will provide $130 million loan for equipment purchased to develop Peru's Cerro Verde copper complex.

[a]NA, not available.

[b]"PASAR Copper Smelter Heads to 1983 Opening," *World Mining* 35 (June 1982).

[c]The Philippines government apparently increased its share in the project to over 60%, after the withdrawal of some of the local companies. "Philippines: The Philippine Government Has Increased . . .", *Engineering and Mining Journal* 183 (December 1982).

[d]Ref. 5.

[e]"La Caridad Financing," *Mining Journal* 299, No. 7686 (10 December 1982).

[f]"Marubeni Awarded Mexican Copper Refinery Contract," *Mining Journal* 299, No. 7677 (8 October 1982).

[g]"Japanese Loan for Peruvian Development," *Mining Journal* 299, No. 7679 (22 October 1982).

economic and accuses the Japanese government of attempting to discourage the proposed future expansion in PASAR's smelting capacity. Japan appeared to have a controlling role in this situation during the depressed market conditions in the early 1980s. Japanese suppliers have contracted for most of PASAR's initial output (Table II-4-7). In 1982, the Philippines government was very interested in beginning negotiations with Japan for $120 million in advance ore and concentrate payments to help its depressed economy. Japanese smelters received approximately 74% of the Philippines copper ore and concentrate exports in 1980.[42]

Annual domestic copper metal production increased steadily through Japan's industrialization period in the 1960s and early 1970s (Figure II-4-13). Increases in domestic smelting and refining capacity were necessary to keep up with domes-

tic metal demand. Following the 1975 recession, annual production continued to increase, a characteristic not observed in the production histories of other nonferrous metals such as aluminum, lead, and zinc (Figures II-4-3, II-4-25, and II-4-27). The overall increase in annual production since the early 1970s can be attributed to increased demand from the telecommunications, electrical machinery, electric power, electrical equipment, and automobile industries.

Domestic production reached an all-time high of 1.08 million metric tons in 1982 (Figure II-4-13). The increased level of custom copper concentrate imports from the U.S. contributed significantly to this high rate of production. However, the growth in demand leveled off, and producer inventories increased to 114,800 metric tons of refined copper at the end of 1982.[43] This level was

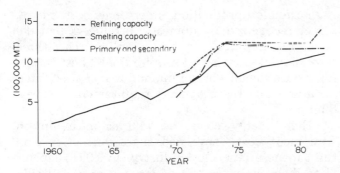

FIGURE II-4-13 Annual Japanese copper metal production, 1960–1982 and production capacities, 1970–1982. (From *production statistics*: U.S. Department of the Interior, Bureau of Mines. *Minerals Yearbook*; Ref. 43. *capacity statistics*: American Bureau of Metal Statistics. *Non-ferrous Metal Data*.)

twice that observed at the beginning of the same year.

Japanese domestic refining and smelting capacities leveled off after 1974 (Figure II-4-13). Capacity expansion at several refineries led to an increase in refining capacity in 1982, to 1,396,710 metric tons per year. Annual smelting capacity in 1981 was 1,167,120 metric tons. Production capacities may increase further if Sumitomo Metal follows through on a plan to increase capacity at its Toyo Smelter by 12,000 metric tons per year and at its Niihama refinery by 21,000 metric tons per year.[1]

Japanese copper metal producers have initiated energy-saving and fuel conversion measures to reduce the consumption of expensive heavy oil imports. These measures include the capture and use of waste heat and the use of more oxygen in the smelting process.[44] At the Onahama smelter, computer-controlled blending and charging of copper concentrate and optimized blower and compressor operation have contributed to energy savings.[45] Fuel substitution of pulverized coal at the Onahama[45] and Naoshima[46] smelters, and waste tires at Onahama,[45] have been successful in reducing heavy oil consumption.

Technology development and export are additional strategies used by domestic copper producers. For example, Mitsubishi Metal developed the world's only commercial process for continuous copper smelting.[44] In addition, Japanese companies will construct copper refineries in China,[47] at Guiyuan, Guixi Province, and in Mexico, at Enpalme, Sonora state (Table II-4-7).

Annual copper metal and copper metal alloy exports, including waste, scrap, unwrought, and semimanufactures, have increased significantly since the early 1960s (Figure II-4-14). Total metal exports in 1980 reached a record high of 409,358

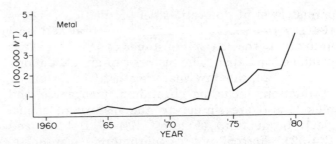

FIGURE II-4-14 Annual Japanese copper metal exports, 1962–1980. (From U.S. Department of the Interior, Bureau of Mines. *Minerals Yearbook*.)

metric tons, including 205,728 metric tons of refined copper metal.[5] In 1981, refined copper exports dropped by 81% to 38,301 metric tons,[1] and a significant decline probably occurred in total metal exports.

The United States and Far East Asian nations accepted most of Japan's total copper metal exports in 1980. The United States imported 152,288 metric tons (37%) and Taiwan, 79,923 metric tons (20%).[1] Hong Kong, Singapore, China, and South Korea also imported significant amounts of copper metal supplies from Japan.

Several other factors will affect the future expansion of the Japanese copper industry. Two of these are the future level of fiber-optics substitution for copper wire in the telecommunications industry, and the demand for by-product acid and gypsum.

Increased substitution of fiber optics for copper in the telecommunications industry, especially where greater density transmission is required and space in underground transmission-line conduits is limited,[48] could reduce the future copper requirements of this industry. However, the potential for fiber-optics substitution is controversial, because of improvements in the transmission density capability of copper wire. As noted in Chapter II-5, the fiber-optics industry is one of those slated for future growth in Japan.

Second, the domestic markets for by-product acid and gypsum are no longer expanding and export markets have become more significant to domestic producers. Further copper metal production capacity expansion will depend on the level to which revenues can be generated from these by-products.

FERROALLOYS

Most of Japan's supplies of ferroalloys are produced domestically from imported raw materials. Through the 1960s and 1970s, Japan's annual ferroalloy production reflected changes in the annual level of domestic steel production (Figures II-4-15, II-4-16, and II-4-22). The iron and steel industry is the major consumer of the ferroalloys produced in Japan. The histories of annual imports of many of the raw materials used in ferroalloy production, including chromium, manganese, and nickel, also are similar to the history of annual steel production (Figures II-4-8, II-4-28, and II-4-32). Ferroalloys of chromium, manganese, silicon, and nickel represented the majority of ferroalloy production during this period (Figure II-4-16).

Ferroalloy production is very energy intensive, a fact recognized by domestic producers that depend on imported energy supplies. The three ferroalloying metals that require the highest energy inputs are silicon, chromium, and manganese (listed in order of decreasing energy requirements for production).[49]

High energy costs, as well as antipollution equipment costs, have had significant effects on the Japanese ferroalloy industry. In 1978, Japan's ferrosilicon industry was designated as structurally depressed, and later in 1983, the entire ferroalloy industry was to receive this designation (Chapter II-5). These designated industries are eligible to receive special governmental and private financial assistance and to form capacity-reducing cartels. Production capacity cutbacks have already been observed in the ferrosilicon and ferronickel industries, and further cutbacks might be expected industry-wide.

To satisfy the continued high ferroalloy demand from the domestic iron and steel industry, Japanese imports of ferroalloys have increased dramatically since 1975 (Figure II-4-17). Ferroalloy imports in 1981 totaled 503,000 metric tons, almost three times that of 1976 and 5.5 times that of 1975. South Africa was the major supplier of Japanese ferroalloy imports during the 1970s and early 1980s (Table II-4-8). Other supplies in 1979–1980 included Norway, Canada, Brazil, and the United States.[50] Increased ferroalloy imports might be expected in the future if Japan's steel production remains high and if domestic ferroalloy production is drastically reduced, as in the aluminum example.

Involvement in joint venture overseas ferroalloy plants is another strategy available to domestic producers. This strategy has already been pursued for ferrosilicon.

Although Japan's ferroalloy production is quite large, little is exported (Table II-4-9). Most is con-

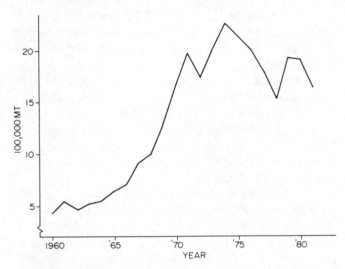

FIGURE II-4-15 Annual Japanese total ferroalloy production, 1960–1981. (From U.S. Department of the Interior, Bureau of Mines. *Minerals Yearbook*.)

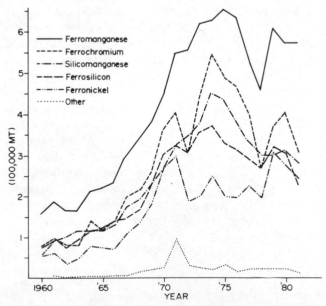

FIGURE II-4-16 Annual Japanese ferroalloy production, by type, 1960–1981. (From U.S. Department of the Interior, Bureau of Mines. *Minerals Yearbook*.)

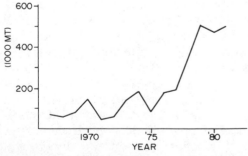

FIGURE II-4-17 Annual Japanese ferroalloy imports, 1967–1981. (From U.S. Department of the Interior, Bureau of Mines. *Minerals Yearbook*.)

TABLE II-4-8   Annual Japanese Ferroalloy Imports, by Source, 1970–1980

Year	Total imports (thousand metric tons)	Major suppliers[a,b] (% of total imports)
1970	148.9	RSA (36), IDA (35)
1971	48.0	RSA (69), IDA (4), US (8)
1972	59.7	RSA (57), TWN (8), ROK (7)
1973	142.9	RSA (36), IDA (15), NOR (10)
1974	186.7	RSA (40), NOR (11), FRN (12)
1975	90.2	RSA (51), TWN (17)
1976	179.1	RSA (56), TWN (14)
1977	191.0	NA
1978	361.2	RSA (42), NOR (8), BZL (12)
1979	503.8	RSA (50), NOR (8), BZL (9), US (1)
1980	477.3	RSA (50), NOR (6), US (4), CAN (4)

[a]NA, not available.

[b]RSA, South Africa; IDA, India; US, United States; TWN, Taiwan; ROK, Korea; NOR, Norway; FRN, France; BZL, Brazil; CAN, Canada.

Source: U.S. Department of the Interior, Bureau of Mines, Minerals Yearbook.

TABLE II-4-9   Annual Japanese Ferroalloy Production and Exports, 1970–1980 (Thousand Metric Tons)

Year	Total production	Total exports	Ferromanganese exports[a]
1970	1667	16	8 (50)
1971	1965	68	19 (28)
1972	1729	106	50 (47)
1973	2015	50	26 (52)
1974	2255	177	54 (31)
1975	2130	260	131 (50)
1976	2008	211	122 (58)
1977	1808	75	49 (65)
1978	1531	70	37 (53)
1979	1926	131	93 (70)
1980	1907	71	40 (57)

[a]Numbers in parentheses represent percent of total ferroalloy exports.

Source: U.S. Department of the Interior, Bureau of Mines, Minerals Yearbook.

sumed domestically by the iron and steel industry. Ferromanganese makes up the bulk of ferroalloy exports. Export destinations include the United States, the Netherlands, Middle Eastern countries, and eastern and southeastern Asian countries.[50] Australia has also imported significant quantities of Japanese ferroalloys in the late 1970s and early 1980s.

The specific mineral commodities used in the ferroalloy industry are discussed in other sections of this chapter.

FERTILIZER MINERALS

Japanese fertilizer consumption is larger than 200 pounds per acre of cropped land, a level that is comparable to that of other small, highly populated countries, including East and West Germany, France, Denmark, the United Kingdom, Poland, and Hungary.[51] Japan satisfies its demand for fertilizer supplies through imports and domestic production.

Potassium

All of Japan's potash supplies are imported in manufactured form. In 1980, Japan imported 1.533 million metric tons of manufactured fertilizer materials, mostly from Canada (42%) and the USSR (19%).[1] The United States was also a major supplier, accounting for 13% of total 1980 imports. Annual manufactured potassic fertilizer material imports have grown very gradually (Figure II-4-18). In 1980, these imports were 43% larger than those observed in 1963.

During the 1970s, Japan exported small amounts of potassic fertilizer materials, mainly to East and Southeast Asian countries, which included South Korea, Thailand, and the Philippines.[50] Exports dropped to only 6 metric tons in 1980, from 2006 metric tons in 1979.[1]

Phosphorus

Japan imports its phosphate supplies in ore and manufactured forms. In 1981, 2.256 million metric tons of phosphate ore were imported, mostly from the United States (58%), Morocco (27%), and

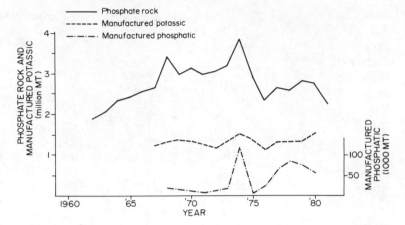

FIGURE II-4-18 Annual Japanese phosphate and potassic fertilizer imports, 1962–1981. (From U.S. Department of the Interior, Bureau of Mines. *Minerals Yearbook. Japan.* Prime Minister's Office, Statistics Bureau. *Japan Statistical Yearbook.*)

Jordan (10%).[1] In addition, 58,924 metric tons of manufactured phosphate fertilizer materials were imported, mostly from South Korea (54%) and the United States (45.4%).[1]

Between 1975 and 1980, it appeared that increased domestic phosphate demand was being met by increased imports of manufactured phosphate materials (Figure II-4-18) rather than imports of phosphate ore. Future increases in phosphatic materials demand will probably be met with increased imports of manufactured materials because of expected cutbacks in domestic fertilizer production capacity (Chapter II-5).

In early 1983, Japan's Zen-Noh Phosphate Corp., which is based in Florida, was reported to be purchasing a joint venture interest in Estech's (U.S.) Watson phosphate rock mine in Fort Meade, Florida.[52] Zeh-Noh is jointly owned by the Japanese National Federation of Agricultural Cooperatives, and Mitsubishi Chemical Industries Ltd. (Japan).

Japanese annual production of superphosphates declined steadily during the 1960s and early 1970s (Figure II-4-19). From 1960 to 1977, annual production dropped by 72% to 607 metric tons. By fiscal year 1981, production declined to 498 metric tons.[1] This production decline reflects the decreased demand for superphosphate because of its low

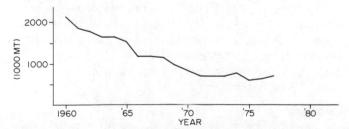

FIGURE II-4-19 Annual Japanese superphosphate production, 1960–1977. (From U.S. Department of the Interior, Bureau of Mines. *Minerals Yearbook.*)

phosphorus content,[53] as well as the chemical fertilizer industry efforts to cut back excess production capacity. Wet-process phosphatic fertilizer production capacity, which includes superphosphate production capacity, was reduced by 17% during 1979 alone.[5] MITI's Industry Structure Council previously recommended that capacity be cut by 20%. Mergers, such as that involving the four fertilizer producers, Nitto Chemical Industry, San Kagaku, Rasa Industries, and Tohoku Hiryo K.K.,[54] were also proposed to help solve the industry's economic problems.

Japanese exports of manufactured phosphatic fertilizer materials have been directed to East and Southeast Asian and Oceanic area markets, which include Burma, Taiwan, Fiji, and Indonesia.[50] In 1980, 23,477 metric tons of manufactured phosphatic fertilizer materials were exported, mainly to Burma (43%) and Taiwan (25%).[1] Exports reached a peak of 87,000 metric tons in 1972.[50]

Nitrogen

Most of Japan's nitrogen supplies are produced domestically using a process that fixes nitrogen from the atmosphere. Japan is a major producer and exporter of nitrogenous fertilizers, such as urea and ammonia.

However, Japan's fertilizer industry is in decline because of the increased cost of raw materials, such as naphtha and liquefied petroleum gas;[54] the development of production capacity in China, which is a major export market; and the availability of cheap Middle Eastern exports in traditional Japanese export markets.[1] This industry was included in the initial list of those designated as structurally depressed in the Japanese government's depressed industries law of 1978 (Chapter II-5).

As in the example of phosphatic fertilizer production, production capacity has been reduced in the ammonia and urea production industries. By

June 1980, ammonia capacity dropped 26% to 3.3 million metric tons per year, and urea capacity dropped 42% to 2.2 million metric tons per year.[54] Further capacity reductions might be expected since urea export demand has not remained as high as originally anticipated. Japan's chemical fertilizer industries were also included for coverage under the 1983 depressed industries law.

Japan imported 3000 metric tons of Chilean nitrates in 1980.[1] In addition, 34,937 metric tons of manufactured nitrogenous fertilizer materials were imported, mostly from Chile (63%), Norway (13%), and South Korea (13%).[1] If nitrogenous fertilizer material production declines to a level that is less than domestic consumption, imports of manufactured materials will most likely be increased to meet domestic demand.

Mixed Fertilizer Materials

Japan also imports large amounts of mixed manufactured fertilizer materials. In 1980, 224,089 metric tons were imported, mostly from the United States (92%), South Korea (4%), and Canada (2%).[1]

FLUORSPAR

Japan's reserves of fluorspar were estimated to be 88,000 metric tons in a 1963 survey (Table II-4-1). These reserves are not of a high enough quality for use in making hydrofluoric acid, which is used by the aluminum industry.[15] Production, if any, of these reserves is insignificant.

Japan imports its supplies of fluorspar mainly from three countries: China, South Africa, and Thailand. In 1980, 487,455 metric tons of fluorspar were imported by Japan and these three countries accounted by 46%, 27%, and 26%, respectively, of total imported supply.[1]

Fluorspar is an important raw material for the steel and aluminum industries. In steelmaking, it is used as a metallurgical flux. In the aluminum industry, hydrofluoric acid, which is produced from the fluorspar, is used to produce additives for the electrolyte from which aluminum is produced during electrolysis.[55]

Japanese fluorspar demand would be expected to reflect that of steel and aluminum. Assuming a consumption rate of 170 pounds of fluorspar per metric ton of aluminum,[55] the reduction in aluminum production from 1,010,000 metric tons in 1979 to 851,000 metric tons (estimated) in 1982 would decrease the aluminum industry's fluorspar requirement by 51,125 metric tons. This figure represents only an 11% decrease in the 1979 level of fluorspar imports, indicating the significance of steelmaking applications on fluorspar use. There-

fore, the future import level of fluorspar will depend mainly on the level of steel production in Japan.

GALLIUM

Scant information is available regarding Japan's sources of gallium. Worldwide, most gallium is recovered as a by-product in the extraction of alumina from bauxite. A second important source is zinc processing wastes. The concentration of gallium in bauxite and zinc minerals is low, averaging 0.005%.[56] Only a small amount of gallium is economically recoverable. Gallium can also be recovered from coal ash and fly ash. The by-product recovery from domestic bauxite and zinc processing is probably a major component in Japanese gallium supply.

The cutback in Japanese domestic aluminum production capacity may affect future Japanese gallium supply. If alumina production capacity is also reduced, Japan will become more dependent on foreign sources of supply, such as Switzerland. In the late 1970s and early 1980s, domestic alumina production and bauxite imports remained high (Figures II-4-2 and II-4-1).

Gallium's main applications include light-emitting diodes (LEDs), photoelectric materials, semiconductors, and solid-state devices.[56] Gallium is also considered important for use in emitters and receivers in optical fibers.[57] Researchers have shown that gallium can be a better element to use in terms of the desired output of many applications, including semiconductors and solar cells.[58] The high cost and difficulty of producing high-purity gallium crystals has limited gallium's use. However, research is being conducted in many countries, including Japan,[59] to develop an economic production process.

GERMANIUM

In 1981, Japan imported 19 metric tons of semi-processed germanium dioxide, mostly from Belgium and West Germany.[1] A small amount of metal, 134 kg (0.1 metric ton) of metal was also imported, mostly from China.

Most germanium is obtained as a by-product of copper-zinc and zinc mineral refining.[60] Japan probably receives some of its germanium supplies as a by-product of its large-scale domestic nonferrous-metal-refining operations.

Japan's production of germanium metal and oxide has been erratic, declining during periods of economic stagnation and increasing during periods of economic growth. In 1981, the estimated production of germanium oxide was 12 metric tons, and germanium metal, 11 metric tons.[1]

Germanium's main application is in the cores of optical fibers used in telecommunications.[60] Because of the inclusion of the fiber-optics industry among those designated for future growth in Japan, domestic germanium requirements will probably increase. Additional requirements will most likely be met by increased imports. Other uses for germanium include semiconductors, solar photovoltaics, and military thermal imaging systems.[60] In electronic and energy applications, silicon can substitute for germanium. Tellurium, selenium, indium, and gallium bimetals can also be used as substitutes in some electronic applications.[6] Zinc selenide can substitute for germanium metal in thermal imaging systems.

GOLD

The most important gold and silver deposits in Japan occur in Tertiary epithermal veins that crossout volcanic and sedimentary sequences.[15] Productive mines are located on Hokkaido, Kyushu, and Honshu Islands. Annual domestic gold production has been declining since the late 1960s (Figure II-5-5).

The Mineral Mining Agency of Japan (MMAJ) maintains an active exploration program for precious metals, and this agency's efforts have led to the discoveries of new gold and silver deposits. One promising deposit discovered by the agency is the proposed Hishikari gold mine in Kagoshima Prefecture on Kyushu Island.[61] A Sumitomo Metal Mining Co. subsidiary was expected to begin mine production in the fall of 1984 at a daily rate of 200 metric tons of ore. The average ore grade is 80 g/metric ton (2.6 oz/ton), and the estimated reserves are 120 metric tons of gold metal content. Eventually, this mine is expected to produce more gold than Japan's largest gold mine, Konomai Gold Mine in Hokkaido.[62] Konomai had already produced 73 metric tons of gold by 1982.

Japan's major suppliers of gold metal are the United Kingdom, Switzerland, and the USSR. In 1981, these countries supplied 39%, 35%, and 22% of the total Japanese imports, respectively.[1] Gold imports reached a record high total of 5.3 million troy ounces in 1981 (Figure II-4-20). The reason for this increase was the government-proposed tax on certain savings accounts.[63] Many Japanese, who are known for their propensity to save,[64] turned to gold purchases as an alternative tax haven. Gold purchases are not as closely monitored by Japanese tax officials. Gold's lower price in the second half

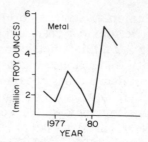

FIGURE II-4-20 Annual Japanese gold metal imports, 1976–1982. (From U.S. Department of the Interior, Bureau of Mine. *Minerals Yearbook*; Ref. 65.)

of 1981 probably also contributed to the surge in sales and imports.[65] The high Japanese savings rate, the imposition of a tax on savings accounts, and the availability of gold through over-the-counter public sales have all been suggested as reasons for expanded future Japanese gold demand, and therefore, continued high levels of gold imports.[66] In 1982, gold imports dropped to 4.49 million troy ounces, but remained higher than late-1970s imports (Figure II-4-20).

Japan's domestic consumption of gold in fiscal year 1979 was 4.6 million troy ounces, mainly for jewelry (32%) and private hoarding (24%).[5] Other significant uses were plating (14%), electronic communication (10%), and dentistry (8%). In 1981, a 280% increase over 1980 levels was observed in gold purchases earmarked for private hoarding.[1] The use of substitutes for gold in plating and electronic communication applications might be expected to increase if the price of gold continues to rise during the 1980s.

IRON AND STEEL

According to a 1980 survey, Japan had 5,678,000 metric tons of exploitable crude iron ore reserves, with a metal content of 1,885,000 metric tons (Table II-4-1). The most important deposits are magnetite-bearing skarns, such as the Kamaishi[35] ore deposits in northern Honshu.

Japan has 77,449,000 metric tons of iron sand reserves with a metal content of 11,839,000 metric tons (Table II-4-1). Many deposits are located along shorelines and in the foothills of coastal ranges in Hokkaido, Iwate, Amori, and Chiba prefectures.[15]

Domestic mine production from iron ore and iron sand deposits is insignificant. In 1981, domestic ore and concentrate supplied less than 1% of the Japanese iron and steel industry consumption (Table II-4-1). Domestic mine production declined significantly during the period 1960–1981 (Figure II-5-3). Iron sand production actually ceased in 1980.[50]

In 1980, Japan was almost 100% import dependent for its iron ore supplies. Suppliers include Australia, Brazil, India, Chile, South Africa, Canada, the Philippines, and Peru (Table II-4-10). Australia is by far the most important supplier.

Japan's dependence on Australian iron ore dates back to 1964, when Japanese steelmakers signed long-term purchase contracts with Australian producers.[67] These contracts provided guaranteed markets and were responsible for the initial development of Australia's iron ore mines. In 1974, Japan was 70% dependent on Australia for its total import supplies. This dependence was reduced in the following years through increased imports of iron ore and concentrate from Brazil, South Africa, and the Philippines (Table II-4-10).

In 1982, Japan became involved in the huge Grande Carajas project, in northeastern Brazil.[68] The project involves the eventual development of iron and aluminum ore reserves and possibly the development of manganese, copper, tin, and nickel deposits. Large infrastructure investments will be required prior to the exploitation of these resources. The total project cost is expected to be $60 billion.[68]

A consortium of Japanese banks has provided a $500 million loan for the development of the iron ore reserves at Carajas.[68] The total cost of the iron ore mine and related infrastructure is expected to reach $3.6 billion. About one-half of the financing is in the form of loans from the World Bank, the European Community, Japan, and the United States. The remaining funds will be supplied by Brazilian interests. Production will probably begin in 1985, at an initial rate of 10 million metric tons per year, eventually increasing to 35 million metric

tons per year by 1988. According to a World Bank report, full production at Carajas is expected to coincide with a global increase in iron ore demand during the second half of the 1980s.[69]

Japan has signed a long-term contract to receive 10 million metric tons per year of iron ore from Carajas.[70] The contract covers the period 1985–1999. Steel producers are also considering the development of other iron ore mines in South America and Australia.

In 1982, the Japanese-Brazilian joint venture Capanema iron ore mine was opened.[71] Approximately 1.5 to 1.7 million metric tons of iron ore was expected to be produced in 1982, all of it destined for the Japanese market. Starting in 1983, approximately 8.5 to 9.5 million metric tons of iron ore, 81 to 83% of total production, was to be available for export. The remaining 2 million metric tons would be sold to Tubarao Steel Works, a joint venture Brazilian-Japanese-Italian company in Brazil. The Capanema mine is operated by Minas de Serra Geral, S.A., a joint venture between CVRD (51%) and eight Japanese companies (49%).

By increasing the level of iron ore shipments from Brazil, Japan will further reduce its dependence on Australian iron ore. However, this reduction could lead to a decline in ore quality. Australian mines require high-quantity production to maintain ore grade.[72]

During the 1960s and early 1970s, annual Japanese iron ore imports increased steadily while the domestic steel industry expanded (Figure II-4-21). Annual imports leveled off in the 1970s as global steel demand stagnated.

Most of Japan's supplies of iron and steel scrap, an important raw material for steel and foundry

TABLE II-4-10   Annual Japanese Iron Ore and Concentrate Imports, by Source, 1974-1981[a,b]

| Year | Total imports | Sources | | | | | |
		Australia	India	Brazil	South Africa	Philippines	Other
1974	141,816 (100)	99,436 (70)	17,407 (12)	19,523 (14)	NA	1118 (1)	4,332 (3)[c]
1975	131,657	63,253 (48)	16,790 (13)	23,460 (18)	1632 (1)	1513 (1)	25,009 (19)
1976	133,727	64,094 (48)	NA	25,380 (19)	NA	NA	44,253 (33)[d]
1977	132,571	63,096 (48)	17,878 (13)	23,743 (18)	5556 (4)	2042 (1)	20,256 (15)
1978	114,465	52,626 (46)	14,355 (13)	20,815 (18)	5831 (5)	3620 (3)	17,398 (15)
1979	130,277	55,297 (42)	17,095 (13)	26,136 (20)	7197 (6)	4045 (3)	20,507 (16)
1980	133,721	60,040 (45)	16,507 (12)	28,523 (21)	6279 (5)	4060 (3)	18,312 (14)
1981	123,362	54,861 (44)	15,636 (13)	27,165 (22)	5765 (5)	3639 (3)	16,296 (13)

[a]NA, not available.
[b]Numbers in parentheses represent percent of total imports.
[c]Includes South Africa.
[d]Includes India, Brazil, and South Africa.

Sources: Ref. 70; U.S. Department of the Interior, Bureau of Mines, Minerals Yearbook.

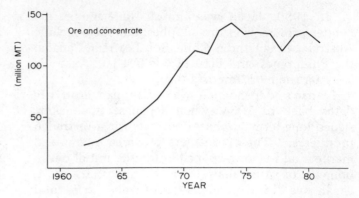

FIGURE II-4-21 Annual Japanese iron ore and concentrate imports, 1962–1981. (From U.S. Department of the Interior, Bureau of Mines. *Minerals Yearbook*.)

production, is available domestically. Imports account for less than 10% of the scrap consumed each year (Table II-4-11). In 1981, only 6% of the iron and steel scrap consumed by the steel industry was imported. Most of the scrap imports comes from the United States.

Japan's imports of iron and steel metal amounted to 3.15 million metric tons in 1981 (including pig iron, ferroalloys, ordinary and special rolled steel products, semifinished products, and

secondary products).[70] This amount is insignificant in relation to total domestic iron and steel metal production.

The main consumer of iron ore in Japan is the iron and steel industry, which annually produces about 80 million metric tons of pig iron and 100 million metric tons of crude steel. Japan ranks with the United States and the USSR as one of the top three steel producers in the world (Table II-4-12a).

There are six major steelmakers in Japan:

TABLE II-4-11   Annual Japanese Iron and Steel Scrap Imports, by Source, 1978–1981 (Metric Tons)

			Source[b]			
Year	Total imports[a]	Total domestic consumption[a]	US	Australia	USSR	Other
1978	3229 (10)	32,700 (100)	2691 (83)	172 (5)	165 (5)	201 (6)
1979	3346 (9)	38,242	2727 (82)	314 (9)	132 (4)	173 (5)
1980	2986 (8)	36,305	2581 (86)	184 (6)	119 (4)	102 (3)
1981	1791 (6)	32,560	1132 (63)	264 (15)	155 (9)	240 (13)

[a]Numbers in parentheses represent percent of total domestic consumption.
[b]Numbers in parentheses represent percent of total imports.
Source: Ref. 70.

TABLE II-4-12   Steel Production (Million Metric Tons)

(a) Major steel-producing countries, 1980 and 1981

	1981		1980	
Country	Rank	Tonnage	Rank	Tonnage
USSR	1	149.0	1	147.9
US	2	108.8	3	101.5
Japan	3	101.7	2	111.4
West Germany	4	41.6	4	43.8
China	5	35.6	5	37.1
Italy	6	24.8	6	26.5
France	7	21.3	7	23.2
Poland	8	15.6	8	19.5
United Kingdom	9	15.6	15	11.3
Czechoslovakia	10	15.2	11	14.8

Source: "World Steel Statistics," *Iron and Steel Engineer* 59 (September 1982), table, p. Ch-87.

(b) Steel production of Japan's six major steel producers

Company	1981 Tonnage
Nippon Steel	29.6
Nippon Kokan	12.6
Kawasaki	11.4
Sumitomo	11.4
Kobe Steel	6.7
Nisshin Steel	2.6
Total	74.3
Total Japanese production	101.7
	73% of total Japanese production

Source: "World Steel Statistics," *Iron and Steel Engineer* 59 (September 1982).

(c) Major steel-producing companies of the world, 1980 and 1981

Company[a]	1981 Rank	1981 Tonnage	1980 Rank	1980 Tonnage
Nippon Steel	1	29.6	1	32.9
US Steel	2	21.2	2	21.1
Bethlehem	3	15.2	5	13.6
Finsider	4	13.9	4	13.7
British Steel	5	13.2	11	8.4
Nippon Kokan	6	12.6	3	14.0
Thyssen Aktiengesellschaft	7	11.6	8	12.4
Kawasaki	8	11.4	7	12.7
Sumitomo	9	11.4	6	12.7
Uswor[b]	10	10.6	9	9.2
Jones & Laughlin	11	9.9	10	8.8
Republic Steel	12	8.6	12	7.7
Pohang Iron & Steel	13	8.2	19	5.9
Broken Hill	14	7.5	13	7.6
Armco	15	7.4	17	6.6
National Steel	16	7.4	16	6.9
Inland	17	7.3	18	6.4
Iscor	18	6.9	15	7.0
Kobe Steel	19	6.7	14	7.4
Steel Authority of India	20	6.6	21	5.6

[a]International Iron and Steel Institute members.
[b]1980 figure noncomparable.

Source: "World Steel Statistics," *Iron and Steel Engineer* 59 (September 1982), table, p. Ch-88.

Nippon Steel, Nippon Kokan, Sumitomo Metal Industries, Kawasaki Steel, Kobe Steel, and Nisshin Steel (listed in decreasing order of annual crude steel production). These six accounted for 73% of Japan's total steel production in 1981 (Table II-4-12b). Nippon Steel is the world's largest steelmaker. Nippon Kokan, Kawasaki, and Sumitomo are also large producers, ranking sixth, eighth, and ninth in 1981 world production, respectively (Table II-4-12c).

Japanese annual crude steel production reached its peak in 1973, totaling 119 million metric tons (Figure II-4-22). Annual steel production increased rapidly in the 1960s in response to increased public and private sector demands. However, production leveled off through the 1970s and the early 1980s, as the growth in steel demand declined. The extended worldwide economic recession of the late 1970s contributed to this decline. Annual crude steel production declined to 99.5 million metric

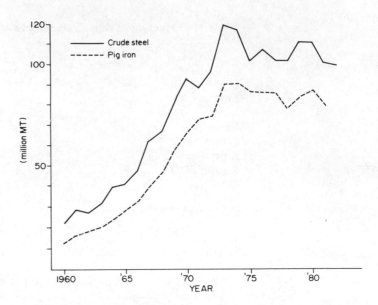

FIGURE II-4-22 Annual Japanese crude steel and pig iron production, 1960–1982. U.S. Department of the Interior, Bureau of Mines. *Minerals Yearbook.* "Steel Output Dips below 100 mil. Tons." *Japan Economic Journal*, 1 February 1983.)

tons in 1982, the lowest annual level in the previous 10 years.

According to a 1980 estimate made by the Japan Iron and Steel Federation, annual crude steel production was expected to increase steadily during the 1980s, and be in the range 118 to 132 million metric tons by 1990.[70] The extended economic recession has already lowered production levels below the most pessimistic predictions made by the Federation for the early 1980s. The future level of production will depend on the extent of global economic growth in the 1980s, protectionist measures taken by other industrialized countries to limit steel imports, the level of steel production in the developing countries of Asia and Latin America, and the use of substitute materials for steel.

Japan exports significant amounts of iron and steel products (Table II-4-13 and Figure II-4-23). Annual exports peaked in 1976 and then leveled off during the late 1970s and early 1980s. Annual exports dropped below 30 million metric tons in 1981 and 1982, and were expected to remain below this mark in 1983.[73] The decline in annual export growth is the result of the worldwide recession, the establishment of voluntary export limits with industrialized countries, and increased competition from steel producers in developing countries.

Japan exports its iron and steel products mainly to North America, Latin America, the USSR, Asia, and the Middle East. Annual exports to China were expected to increase in the early 1980s, while this

TABLE II-4-13   Annual Japanese Iron and Steel Product Exports, by Product, 1976–1981 (Thousand Metric Tons)

Year	Total exports	Pig iron	Ferroalloys	Semifinished products[a]	Ordinary rolled steel products[b]	Special rolled steel products	Cast-iron pipe and secondary products[c]
1976	37,035	156	211	1046	32,504	1731	1370
1977	34,983	575	76	419	30,602	1805	1494
1978	31,554	43	70	106	28,226	1761	1348
1979	31,496	61	132	124	28,245	1679	1256
1980	30,327	15	71	184	27,305	1591	1162
1981	29,134	12	64	115	26,110	1626	1208

[a]Semifinished steel products include ordinary and special semifinished steel ingots.
[b]Ordinary rolled steel products include coated and clad products.
[c]Secondary steel products include iron wires, welding electrode cores, wire nails, wood screws, galvanized hard steel wires, coated and clad wires, barbed wires, fencing wires, wire nettings, welded wire nettings and ropes, strand wires, welded wires, cold finished bars, nuts and bolts, rivets, cold finished hoop, shovels, scoops, picks, cans, and high-pressure containers.

Source: Sukeyuki Kawata, ed., *Japan's Iron and Steel Industry 1982.* Tokyo: Kawata Publicity, Inc., 1982, "Exports: Summary" table, p. 207.

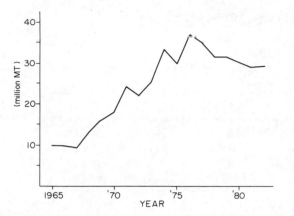

FIGURE II-4-23 Annual Japanese iron and steel product exports, 1965–1981. (From Refs. 70 and 73.)

country took advantage of its improved foreign exchange position, and pursued its development plans. In 1983, China increased its purchases of ordinary rolled steel products to 6 million metric tons.[74] Annual exports to Iran were also expected to increase.[73] Annual exports to the United States and Europe have declined since the early 1980s. Pressure from the United States will probably force Japanese producers to continue reducing their level of United States exports.[75] Exports to Europe will probably remain low as long as depressed market conditions exist.[76] A maximum limit of 1.49 million metric tons has been established for Japanese iron and steel exports to Europe, but export levels have been far below that limit. Japanese steel exports to Europe totaled 620,000 metric tons in 1980, 300,000 metric tons in 1981, and an estimated 500,000 metric tons in 1982.[76]

Japan exports very small quantities of iron ore and concentrate. Hong Kong and Taiwan received most of these supplies in the early 1980s.[1]

LEAD AND ZINC

Japan's exploitable crude lead and zinc ore reserves are estimated to be 61,886,000 metric tons, with a lead content of 549,000 metric tons and a zinc content of 3,089,000 metric tons (Table II-4-1). There are three economically important types of lead zinc deposits in Japan. These include skarns,[35] such as at the Kamioka and Nakatatsu mines; epithermal base metal-mineralized veins, such as at the Hosokura[35] and Toyoha[77] mines; and Kuroko-type massive sulfides,[33] such as the Kosaka, Hanaoka, and Shakanai mines. These major domestic lead-zinc mines are listed in Table II-2-1. In 1981, domestic mines produced 44,392 metric tons of lead and 242,042 metric tons of zinc, accounting for 22% and 36% of the raw material requirements used in domestic metal production, respectively

(Table II-4-1). In recent years, Japan's Metal Mining Agency, as well as private companies, have made several lead and zinc discoveries mainly in the vicinity of preexisting mines (Figure II-5-7 and Table II-5-1). These discoveries will help extend domestic reserves and maintain production levels.

Annual domestic lead mine production during the period 1960–1981 was fairly steady, averaging about 50,000 metric tons (metal content) (Figure II-5-2). Annual domestic zinc production peaked at 294,000 metric tons in 1971 and has fluctuated in a downward direction since that time (Figure II-5-2). During the period 1979–1981, zinc production leveled off at about 241,000 metric tons.

Japanese private and public interests are involved in lead and zinc resource development activities overseas. For example, Japan's Mitsui and Co. is part owner of the Huanzala lead/zinc mine in Peru, which supplies lead-zinc ore to Japan (Table II-2-2). Japan's Mineral Mining Agency has maintained an active exploration program in the vicinity of the Huanzala mine which has resulted in the confirmation of lead and zinc mineralization.[46] Also, Japan's Mitsubishi Corp. and the Cia Minera Milpo (Peru) will jointly develop the San Hilarion mine in central Peru.[78] The reserves at San Hilarion were estimated to be 10 million metric tons of ore at 6% zinc, 2% lead, and 70 g/metric ton of silver.

Canada and Peru are major sources of Japanese lead ore and concentrate imports (Table II-4-14). In 1981, U.S. shipments amounted to 10% of total imported supplies. During the same year, the Bunker Hill lead/zinc smelter (U.S) was closed because of economic and environmental protection reasons.[19] Increased future U.S. ore and concentrate shipments might be expected if more U.S. smelters are closed down.

Japan also imported 90,712 metric tons of unwrought lead metal in 1981. Metal imports came from several countries, including North Korea, Mexico, and Peru (Table II-4-14). Both ore and concentrate and metal imports have been on the rise in the late 1970s and early 1980s (Figure II-4-24). By 1980, the level of annual imports exceeded the pre-1975 highs of 243,000 metric tons of ore and concentrate and 64,483 metric tons of metal set in 1973.

In 1981, the main uses of primary lead in Japan were in the manufacture of storage batteries (45%), inorganic chemicals (25%), cable sheathing (9%), and pipe and sheet (7%). The use of lead in storage batteries has remained strong despite declines in other industrialized countries.[19] Increased use of lead in cable sheathing also contributed to the high consumption levels.[79]

TABLE II-4-14   Annual Japanese Lead Imports, by Source, 1978-1981[a,b]
(Thousand Metric Tons)

(a) Ore and concentrate

		Source				
Year	Total imports	Canada	Peru	US	Australia	Other
1978	222 (100)	154 (69)	NA	NA	NA	68 (31)
1979	219	144 (66)	44 (20)	0	15 (7)	16 (7)
1980	259	165 (64)	43 (17)	0	NA	50 (19)
1981	256	115 (45)	54 (21)	26 (10)	10 (4)	61 (24)

(b) Metal (unwrought)

		Source				
Year	Total imports	North Korea	Mexico	Peru	US	Other
1978	59 (100)	19 (32)	10 (17)	NA	NA	30 (51)
1979	63	11 (17)	13 (21)	13 (20)	0.2 (1)	26 (42)
1980	91	22 (25)	17 (19)	14 (15)	6 (7)	31 (34)

[a]Numbers in parentheses represent percent of total imports.
[b]NA, not available.

Source: U.S. Department of the Interior, Bureau of Mines, Minerals Yearbook.

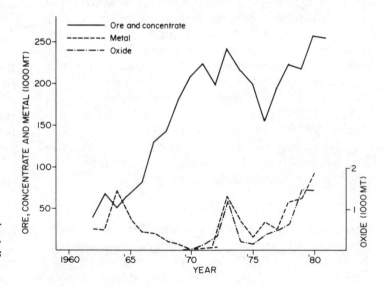

FIGURE II-4-24   Annual Japanese lead ore and concentrate, metal, and oxide imports, 1962-1981. (From U.S. Department of the Interior, Bureau of Mines. Minerals Yearbook.)

Annual primary lead production was stable in the late 1970s and early 1980s, following its rapid growth in the 1960s and early 1970s (Figure II-4-25a). Domestic lead consumption was on the rise in the late 1970s and early 1980s, exceeding production levels (Figure II-4-25b). The difference was made up with increased metal imports (Figure II-4-24) and producer stocks. Secondary lead production also exceeded metal imports in 1981 (Table II-4-15). Japan's imports of zinc oxide were also increasing (Figure II-4-26). In 1980, Japan imported 4544 metric tons of zinc oxide, an amount

that was almost 12 times larger than the level of 1970 imports. Most of Japan's zinc oxide supplies are imported from Taiwan (31%), South Korea (29%), and Singapore (23%).[1]

According to 1981 consumption levels, steel sheet galvanizing was the major application for zinc in Japan.[1] Other major uses included diecasting; wire, tube, and general galvanizing; brass and manufactured rolled zinc. The largest use for zinc oxide is as an accelerator in the vulcanizing process for rubber tires. Many of zinc's applications are in the construction, automobile and

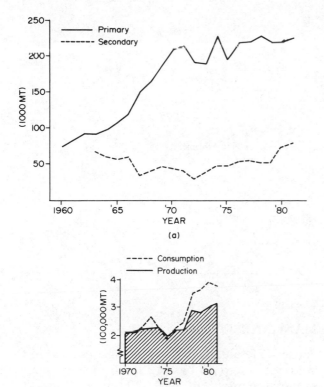

FIGURE II-4-25   (a) Annual Japanese lead metal production, 1960-1981; (b) annual Japanese refined lead metal
production and consumption, 1970-1981. [(a) from U.S.
Department of the Interior, Bureau of Mines, *Minerals
Yearbook*; (b) from *Mining Annual Review*.]

consumer-goods industries, and therefore, zinc
demand is greatly affected by the general economic
cycle.

Annual zinc metal production increased rapidly
in the 1960s and early 1970s, and then leveled off
in the middle-1970s (Figure II-4-27a). In the late
1970s and early 1980s, production declined.
Throughout the 1970s, production exceeded domestic consumption (Figure II-4-27b). In the early
1980s production declined because of high zinc ore
and electric power costs, and low demand for
galvanized steel sheet, the major zinc application
in Japan.[9]

Because of the high domestic power costs and
power requirements for zinc smelting operations,
Japanese smelters might be expected to become
involved in the development of overseas smelting
capacity. As described earlier, domestic aluminum
smelting companies have already pursued this
strategy. At least one company is involved in
developing overseas zinc capacity. Japan's Mitsui
Engineering & Construction, Co. owns a very small
(3%) interest in Thailand's Padaeng Industry Co.
Ltd., which is planning to build a zinc smelter at
Tak, Thailand.[80] This smelter will process zinc
silicate ores from Thailand's Mae Sot ore body, as
early as 1985.

Zinc metal and oxide exports are directed to

TABLE II-4-15   Annual Japanese Zinc Imports, by Source, 1978-1981[a,b]
(Thousand Metric Tons)

(a) Ore and concentrate

| Year | Total imports | Source | | | |
		Canada	Australia	Peru	Other
1978	937 (100)	366 (39)	228 (24)	288 (31)	55 (6)
1979	959	359 (37)	255 (26)	250 (26)	96 (10)
1980	805	259 (32)	272 (34)	197 (24)	77 (10)
1981	888	276 (31)	265 (30)	256 (29)	91 (10)

(b) Metal

| Year | Total imports | Source | | | |
		North Korea	South Korea	Peru	Other
1978	34 (100)	26 (76)	NA	NA	8 (24)
1979	41	29 (71)	4 (10)	4 (9)	4 (9)
1980	46	32 (70)	7 (15)	NA	7 (15)

[a]Numbers in parentheses represent percent of total imports.
[b]NA, not available.
Source: U.S. Department of the Interior, Bureau of Mines, *Minerals Yearbook*.

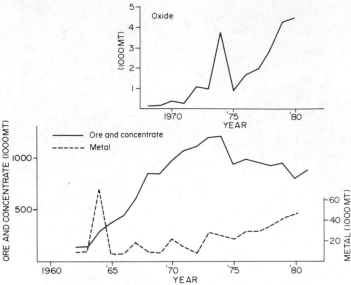

FIGURE II-4-26 Annual Japanese zinc ore and concentrate, metal and oxide imports, 1962–1981. (From U.S. Department of the Interior, Bureau of Mines. *Minerals Yearbook*.)

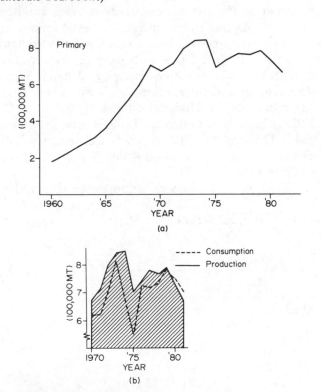

FIGURE II-4-27 (a) Annual Japanese zinc metal production, 1960–1981; (b) annual Japanese refined zinc metal production and consumption, 1970–1981. [(a) from U.S. Department of the Interior, Bureau of Mines, *Minerals Yearbook*; (b) from *Mining Annual Review*.]

east and southeast Asian nations. In 1980, 45,157 metric tons of zinc metal (all forms) was exported mostly to Taiwan (20%), South Korea (19%) and the Philippines (15%).[1] In the same year 455 metric tons of zinc oxide were imported mostly to South Korea (30%), Thailand (18%), and Taiwan (12%).[1]

MANGANESE

Japan's metalurgical grade manganese ore reserves were estimated to be 3,193,000 metric tons, with a metal content of 529,000 metric tons (Table II-4-1). Most of the domestic production is derived from hydrothermal rhodocrosite vein deposits in southwestern Hokkaido.[15] In the early 1980s, manganese deposits were discovered at Ohe mine and at Kaminokuni on Hokkaido Island (Figure II-5-7 and Table II-5-1). However, domestic mine production is insignificant and has declined steadily since the late 1960s (Figure II-5-4 and Table II-4-1).

Most of Japan's manganese ore supplies are imported. In 1981, Japan imported large quantities of manganese ore, ferruginous manganese ore, and manganese dioxide ore and concentrate (Table II-4-16). South Africa, India, and Australia are Japan's major ore and concentrate suppliers. Japanese imports of semiprocessed manganese oxides totaled 762 metric tons, mostly from Belgium.[1] The iron and steel industry is the biggest consumer of manganese ore in Japan. Manganese is an essential alloying ingredient in steelmaking. In 1980, the iron and steel industry consumed 90% of the manganese ore that was consumed in Japan.[5] As might be expected, the history of annual manganese ore and concentrate imports closely parallels that of annual crude steel and ferromanganese production (Figures II-4-28, II-4-22, and II-4-16). In a recent example, Japan reduced its 1982 purchases of high-grade manganese ore by 300,000 metric tons, because of the expected slump in steel production.[81]

Because of the expected reduction in ferroalloy

TABLE II-4-16 Japanese Manganese Ore and Concentrate Imports, by Source, 1981[a] (Thousand Metric Tons)

Source	Amount	
Manganese ore and concentrate		
South Africa	820	(55)
Australia	423	(28)
Brazil	110	(7)
Mexico	66	(4)
Other	81	(5)
Total	1,500	(100)
Ferromanganese ore and concentrate		
South Africa	532	(55)
India	391	(41)
Other	37	(4)
Total	960	(100)
Manganese dioxide ore and concentrate		
Australia	4.443	(41)
Gabon	4.200	(39)
China	2.229	(20)
Total	10.872	(100)

[a]Numbers in parentheses represent percent of total imports.

Source: U.S. Department of the Interior, Bureau of Mines, *Minerals Yearbook.*

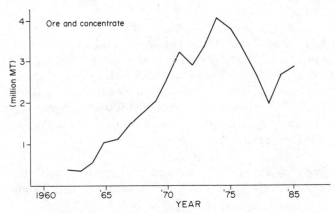

FIGURE II-4-28 Annual Japanese manganese ore and concentrate imports, 1962–1980. (From U.S. Department of the Interior, Bureau of Mines. *Minerals Yearbook.*)

production capacity, manganese ore and concentrate imports will probably never again reach their early 1970s peak of 4.04 million metric tons (Figure II-4-28). Ore and concentrate imports might be expected to decline to lower levels, as ferroalloy imports increase.

In addition, Japanese firms are planning to become involved in the overseas production of ferromanganese. Two Japanese companies, Nippon Kokan and Marubeni, have expressed interest in producing ferromanganese at Brazil's Blue Creek

manganese deposit, which is a part of the hugh Carajas development project.[87] Several other companies, from the United States, Germany, and Brazil, have expressed a similar interest.

Increased use of the electric furnace steelmaking process could lead to reduced consumption of manganese per unit of steel production, because of the furnace's efficient operation.[83] In 1981, 25% of Japanese steel was made in electric furnaces.[70] However, the electric-furnace steelmaking industry is energy intensive, and the Japanese government has designated it as structurally depressed. MITI and the steel industry association have recommended that electric furnace steel producers cut back their production capacity (Chapter II-5). A cutback could eliminate any reduction in manganese consumption already obtained.

The annual Japanese production of manganese dioxide, used to make dry cell batteries, has grown steadily since 1977 (Figure II-4-29). In 1981, production was estimated to be 44,296 metric tons. Producers have expanded their production capacities to keep up with higher expected future demand (Table II-4-17).

Japanese manganese metal production leveled off at about 4200 metric tons, following a steady decline from a peak of 12,657 metric tons in 1972 (Figure II-4-29).

Japan exports manganese in several forms. Small amounts of manganese ore and concentrate are exported, mostly to East and Southeast Asian countries.[50] Exports of processed manganese oxide totaled 22,944 metric tons in 1980, an amount equivalent to 58% of domestic production.[1] The United States and Indonesia are major destinations

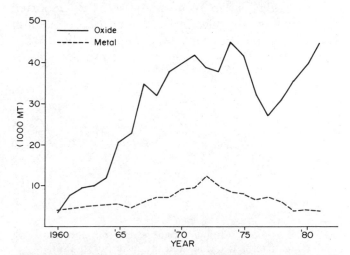

FIGURE II-4-29 Annual Japanese manganese oxide and metal production, 1960–1981. (From U.S. Department of the Interior, Bureau of Mines. *Minerals Yearbook.*)

TABLE II-4-17   Annual Japanese Manganese Dioxide Production Capacity, 1980–1983

Company	Plant/location[a]	Annual production capacity (thousand metric tons)		
		1980	1981	1983e[b]
Toyo Soda Manufacturing Co.	Hinata, Shizuoka Prefecture	18	24	24
	Tekkosha Hellas A.B.E., Greece	10	10	10
	Domestic plant to be built in 1983	—	—	12
Mitsui Mining & Smelting	Takehara, Hiroshima Prefecture	19	25	25
	Mitsui Denham, Ireland	12	12	12
Daiichi Carbon Co.	NA	4	4	4
Japan Metals & Chemicals Co.	Takaoka, Miyazaki Prefecture	—	6	6
Total domestic/overseas capacity		41/22	59/22	71/22

[a]NA, not available.
[b]e, estimated.
Source: Ref. 5.

for Japanese manganese oxide exports. The United States was also the largest consumer of Japanese ferromanganese exports in the 1970s and early 1980s.[50] In 1980, 40,271 metric tons of ferromanganese were exported, and 11,786 metric tons were shipped to the United States.[1]

MOLYBDENUM

Japan's exploitable crude molybdenum ore reserves are estimated to be 155,000 metric tons, with a metal content of 1000 metric tons (Table II-4-1). Most domestic mine production is from pegmatitic to hydrothermal vein deposits,[15] such as at Hirase mine (Table II-2-1). Domestic production is insignificant with respect to total ore and concentrate demand (Table II-4-1).

In 1981, Japan received most of its imported molybdenum supplies in the form of molybdenum oxide (roasted concentrate). Most of the 16,276 metric tons of molybdenum oxide imports came from the United States (52%), Canada (29%), and Chile (15%).[1] Imports of unroasted ore and concentrate totaled 14 metric tons, a decline of 86% from 1980. Japan's imports of molybdenum trioxide, 660 metric tons in 1981,[1] were acquired mostly from the United States. In addition, Japan imports small amounts of molybdenum metal, mainly from the United States and West Germany.[50]

Except for downturns in 1972, 1975, and late 1970s, and the early 1980s, annual molybdenum ore and concentrate imports grew during the period 1962–1980 (Figure II-4-30).

Most of the molybdenum consumed in Japan is used by the domestic steel industry. Some supplies are used in the production of superalloys and nonferrous alloys.

Annual molybdenum metal production has been cyclic, paralleling the history of annual Japanese crude steel production (Figures II-4-22 and II-4-31). However, production generally con-

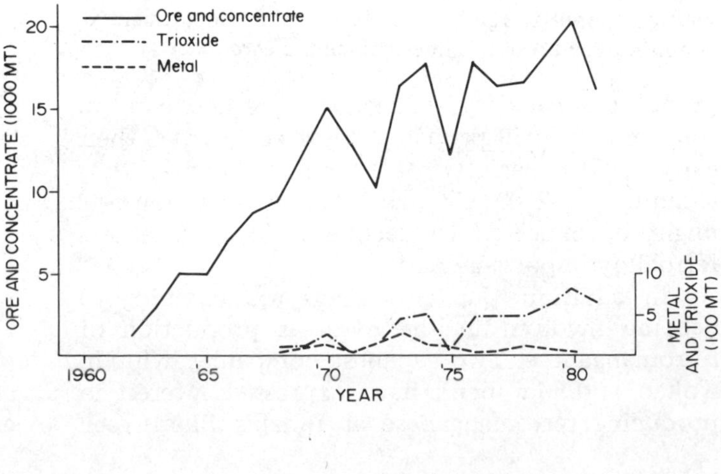

FIGURE II-4-30 Annual Japanese molydbenum ore and concentrate, trioxide, and metal imports, 1962–1981. (From U.S. Department of the Interior, Bureau of Mines. *Minerals Yearbook.*)

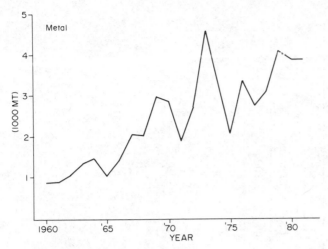

FIGURE II-4-31 Annual Japanese molybdenum metal production, 1960–1981. (From U.S. Department of the Interior, Bureau of Mines. *Minerals Yearbook*.)

tinued to increase during the period 1960–1981. The availability of molybdenum, its versatility, and the increased interest in the production of specialty steel and special alloys have all contributed to this growth.

Japan exported only 10% of its molybdenum metal production in 1980.[1] However, exports have been increasing since 1975. The main export markets for Japanese molybdenum metal in 1980 included Hungary, Taiwan, and the United States.

NICKEL

Japan imports all its nickel supplies. The main suppliers are New Caledonia, Indonesia, Australia, and the Philippines (Table II-4-18). Japanese companies are joint venture partners in nickel mining projects in both the Philippines and Indonesia (Table II-2-2).

Since 1972, annual ore and concentrate imports have fluctuated between 3 and 4.0 million metric tons, a level less than the 1971 high of 4.9 million metric tons (Figure II-4-32). At the same time, matte and metal imports have increased. Although still fluctuating with the level of economic activity, annual matte imports are generally at a higher level than pre-1972 levels. Over the long term, 1972–1981, matte imports appeared to be on the rise, whereas ore and concentrate imports had leveled off. Annual metal imports have increased much more gradually since 1963.

Higher energy costs since 1973–1974 have resulted in increased operating costs for domestic nickel ore and concentrate processors, especially for those companies that process nickel laterite ores. Fuel oil is usually used in the drying and smelting processes for wet laterite ores.[84] Producers have coped with higher energy costs by increasing matte, and probably metal, imports since 1973–1974.

The three main nickel-consuming industries in Japan are the specialty steel (38%), plating (20%), and nonferrous alloy (17%) industries.[1] Nickel was also used to make magnetic materials, fabricated products, storage batteries, and catalysts. Increased demand from the specialty steel and nonferrous alloys industries, and possibly the electronics and microprocessing industries,[85] might be expected in the future because these industries will be important in Japan's future economy (Chapter II-5). Similar to the histories of annual imports of other ferroalloy raw materials, annual nickel ore and matte imports reflect the history of annual crude steel production. Domestic ferronickel production increased significantly in 1979–1980, coinciding with an increase in demand for specialty steels (Figure II-4-16). Because of the expected decline in ferroalloy production capacity, future ferronickel demand from Japan's specialty steel industry will probably be satisfied by increased imports. In the early 1980s, ferronickel producers were reducing overseas ore purchases,[86] reducing production levels,[87] cutting back production capacity,[88] and employing energy conservation measures.[89]

Refined nickel metal production increased gradually during the period 1960–1981 (Figure II-4-33). The increased demand for nonferrous nickel-bearing alloys contributed to this increase. In 1981, refined nickel production was 23,790 metric tons and consumption was 34,114 metric tons.[1] The difference was made up by imports and producer stocks.

The two domestic refined metal producers, Nippon Mining and Sumitomo Metal Mining Companies, have a combined production capacity of 26,880 metric tons.[84]

Nickel metal exports have declined from their peak of 6083 metric tons in 1977 (Table II-4-19). Annual metal exports may be expected to stay low even after an economic recovery in the early 1980s, if domestic demand is strong and if domestic ore-processing capacity is not expanded. Nickel metal export destinations in the early 1980s were mainly the United States, Taiwan, South Korea, and Indonesia.[50]

TABLE II-4-18   Japanese Nickel Imports, by Source, 1981[a] (Thousand Metric Tons)

Import form	Total imports	Source			
		New Caledonia	Indonesia	Philippines	Australia
Ore and concentrate	3463 (100)	1662 (48)	1177 (34)	623 (18)	—
Matte and speiss	36.6 (100)	2.6 (7)	18.6 (51)	—	15.4 (42)

		Source					
		Canada	USSR	Australia	Philippines	US	Zimbabwe	Other
Refined metal	17,732 (100)	5142 (29)	3192 (18)	2837 (16)	1950 (11)	1507 (9)	1064 (6)	1950 (11)

[a]Numbers in parentheses represent percent of total imports.

Source: U.S. Department of the Interior, Bureau of Mines, *Minerals Yearbook.*

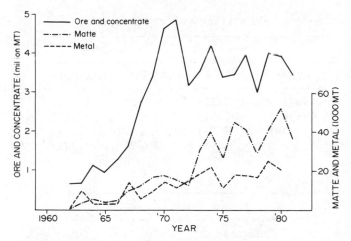

FIGURE II-4-32 Annual Japanese nickel ore and concentrate, matte, and metal imports, 1962–1981. (From U.S. Department of the Interior, Bureau of Mines. *Minerals Yearbook*.)

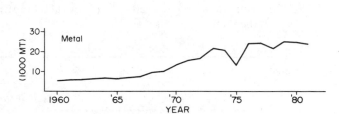

FIGURE II-4-33 Annual Japanese nickel metal production, 1960–1981. (From U.S. Department of the Interior, Bureau of Mines. *Minerals Yearbook*.)

TABLE II-4-19   Annual Japanese Refined Nickel Metal Exports, by Destination, 1979–1980[a,b] (Metric Tons)

| Year | Total exports | Export destination | | | | |
		US	Taiwan	South Korea	Indonesia	Other
1979	2623 (100)	1232 (47)	205 (8)	162 (6)	NA	1024 (39)
1980	2031	756 (37)	230 (11)	139 (7)	225 (11)	681 (34)

[a]Numbers in parentheses represent percent of total exports.
[b]NA, not available.

Source: U.S. Department of the Interior, Bureau of Mines, *Minerals Yearbook*.

PLATINUM-GROUP METALS

Japan's domestic resources of platinum-group metals (PGMs) are insignificant. However, some production did occur from domestic placer deposits of iridosmine, the native iridium-osmium alloy, on Hokkaido Island during World War II.[15]

Japan is 100% dependent on foreign PGM sources. Almost all PGMs are imported in metal form. The major suppliers of PGMs are the USSR, South Africa, the United States, and the United Kingdom (Table II-4-20). West Germany and Switzerland supply the majority of PGM alloy imports.

This history of Japanese platinum and palladium imports (Figure II-4-34) generally reflects the individual metals' uses. For example, the increases in platinum metal imports during the commodity boom of the early 1970s, and the economic downturns of 1975 and 1977–1978, demonstrate the demand for platinum as a store of wealth. In this example, the high demand for platinum is similar to that for gold during times of world crisis and economic uncertainty. However, during the above two economic downturns, palladium imports declined, reflecting its main usage in growth industries such as automobiles, chemicals, and electronics.

The main applications for PGMs in Japan are the manufacture of jewelry, catalysts for automobile exhaust systems, and electrical and electronic devices. Demand for platinum jewelry is strong and continued growth is probable. Approximately 70% of the platinum consumed in Japan is used to make jewelry.[90] Rhodium, iridium, and ruthenium also have applications in the jewelry sector.

Increased demand for PGM might be expected from Japan's future growth industries. Platinum-rhodium alloys used to make glass-fiber manufacturing equipment,[90] have applications in the fiber-optics industry. PGMs will also be in demand because of their many applications in the electrical and electronics sectors.

Japan produces small amounts of platinum and palladium metal. During the period 1972–1980, annual platinum metal production increased more

TABLE II-4-20   Japanese Platinum-Group Metal Imports, by Source, 1980[a,b] (Thousand Troy Ounces)

Metal	Total imports[c]	Source				
		South Africa	USSR	US	UK	Other
Platinum	1010 (100)	589 (58)	160 (16)	120 (12)	126 (12)	15 (1)
Palladium	767	105 (14)	491 (64)	87 (11)	NA	84 (11)
Rhodium	26	7 (27)	10 (38)	8 (31)	NA	1 (4)
Iridium, osmium, and ruthenium	28	12 (42)	NA	4 (14)	11 (39)	1 (4)

		Source			
		West Germany	Switzerland	US	Other
Alloys	23 (100)	15 (63)	5 (19)	2 (9)	2 (8)

[a]Numbers in parentheses represent percent of total imports.
[b]NA, not available.
[c]Numbers may not add up to totals because of rounding.
Source: U.S. Department of the Interior, Bureau of Mines, *Minerals Yearbook.*

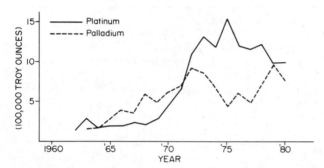

FIGURE II-4-34   Annual Japanese platinum and palladium metal imports, 1962–1981. (From U.S. Department of the Interior, Bureau of Mines. *Minerals Yearbook.*)

than threefold to 12,366 troy ounces (Figure II-4-35). During the same period palladium production increased fivefold to 28,968 troy ounces.

In 1980, Japanese exports of PGMs totaled 189,000 troy ounces, an amount equivalent to 10% of total PGM imports.[1] Taiwan was the major im-

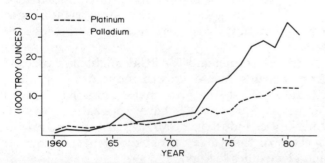

FIGURE II-4-35   Annual Japanese platinum and palladium metal production, 1960–1981. (From U.S. Department of the Interior, Bureau of Mines. *Minerals Yearbook.*)

porter, receiving almost half of the total exports. Other export destinations in 1980–1981 included the United Kingdom, Switzerland, the Netherlands, and South Korea.[50]

RARE EARTH METALS

In 1980, Japan imported 3333 metric tons of rare earth metal oxides and chlorides, mostly from China (61%).[1] Chinese imports have been increasing because of a shortage of available exports from India, another Japanese supplier.[91] The United States supplied 18% of Japan's total imports. Japan also imported 42 metric tons of rare earth metals in metal form, mainly from Brazil (76%).[1] Annual rare earth metal oxide and chloride imports have generally increased during the period 1975–1980 (Figure II-4-36). Annual metal imports in the late 1970s and early 1980s were significantly higher than in the early 1970s (Figure II-4-36).

Japan's Mitsui Mining and Smelting Co. was to provide technical aid to the PRC by building two rare earth plants in Baotou, Inner Mongolia.[92] One plant would process and the other would smelt the ores from China's Bayan Obo rare earth metal mine. However, the Chinese have apparently decided not to go ahead with these development plans.[91]

Japan's annual production of processed rare earth metal elements, mainly lanthanum and cerium oxides, generally increased during the period 1970–1981 (Figure II-4-37). By mid-1982, Mitsubishi Chemical Industries was to complete a rare earth metal separating and refining plant, to pro-

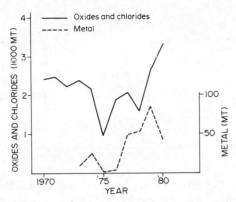

FIGURE II-4-36 Annual Japanese rare earth metal imports, in metal, oxide, and chloride forms, 1970-1981. (From U.S. Department of the Interior, Bureau of Mines. *Minerals Yearbook*.)

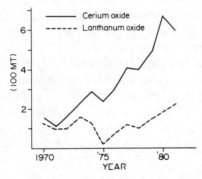

FIGURE II-4-37 Annual Japanese rare earth metal production, 1970-1981. (From U.S. Department of the Interior, Bureau of Mines. *Minerals Yearbook*.)

duce yttrium, europium, samarium, gadolinium, and terbium oxides.[93] The increased domestic demand for these materials has been significant. Domestic demand rose 8 to 9% from 1980 to 1981, a lower growth rate than that observed in previous years.[93]

Most rare earth metals are used in a compound form, but the demand for separated, high-purity rare earth metals in specific end uses is expected to increase steadily.[94] Some uses of separated rare earth metals include capacitors (lanthanum or neodymium),[119] and permanent magnets (samarium).[95]

Japan's emphasis on the development of new energy sources, information processing, and new basic material industries could lead to increased demand for rare earth metals, especially those in the lanthanum series. For example, gadolinium in alloy with cobalt appears to have important industrial applications, such as the lining of heat pumps and storage tanks.[95] The use of samarium in the construction of small, lightweight, permanent magnets has helped to make miniaturization possi-

ble in the consumer electronics industry.[30] Research and development regarding gadolinium oxide use in pressurized-water reactors was being undertaken in the early 1980s.[93] Gadolinium oxide use, in combination with uranium oxide in boiling-water reactors, has already resulted in more efficient use of uranium. Future demand for rare earth metals will depend on continued research and development successes.

Japan's future requirements for rare earth metals will probably be satisfied by imports. The PRC is thought to have the largest potential resources of rare earth minerals in the world.[95] Brazil, India, Australia, and the United States are other important sources. Japan's already strong presence in the mining industries of China, Brazil, and Australia will probably aid in the procurement of its future rare earth metal supplies.

SILICON

Japan imports all its supplies of silicon. Silicon is available from many sources worldwide, and Japan's main suppliers include Spain, South Africa, France, and the United States.[1] World resources will probably be sufficient to supply the ferrosilicon and silicon metal industries in future years.[96]

There was a 20-fold increase in annual Japanese imports of elemental silicon over the period 1968-1979 (Figure II-4-38). Downturns in 1971 and 1975 reflect depressed economic conditions and reduced steel output. Increased demand for polycrystalline silicon, the feedstock for semiconductor applications, probably contributed to the rapid increase in elemental silicon imports since 1975.

Metallurgical-grade silicon has important applications in the Japanese steel industry, where it is used to make ferrosilicon. It is also used in alloys with aluminum, copper, and nickel.[96]

Ferrosilicon production declined from its peak of 369,000 metric tons in 1974 to 234,524 metric

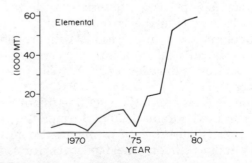

FIGURE II-4-38 Annual Japanese elemental silicon imports, 1968-1980. (From U.S. Department of the Interior, Bureau of Mines. *Minerals Yearbook*.)

tons in 1981, as a result of the reduction in steel production and increased energy costs (Figure II-4-16). Ferrosilicon and silicon metal production are both energy intensive and involve the use of electric furnaces. The oil price hikes of the 1970s have increased production costs and have made Japanese-produced ferrosilicon and silicon metal uncompetitive in world markets.

MITI designated the ferrosilicon industry as structurally depressed in 1978, making it eligible for financial assistance and special treatment under the antimonopoly laws (Chapter II-5). In response, members of the Ferro-Alloy Association of Japan agreed to make a 20% or 100,000-metric ton cutback in ferrosilicon production capacity.[97] Since that time, several companies have eliminated capacity, and some have withdrawn from ferrosilicon production entirely.[98]

Japanese imports of ferrosilicon increased in the late 1970s and early 1980s, as domestic production capacity declined. Both the Philippines[98] and China[99] increased their exports to Japan during this time.

Some domestic producers have become involved in the development of joint venture overseas ferrosilicon capacity. In 1982, discussions were held regarding the following projects in British Columbia, Canada:[100] a Cominco (Canada)-Mitsui (Japan) joint venture in Kimberly, a Mitsubishi-Nippon Kokan (both Japan) joint venture in the Kamloops Area, and a Sumitomo (Japan)-SKW (Canada) joint venture also in the Kamloops area. The output from these plants would probably be destined for the Japanese market. In addition, Pacific Metals Co. Ltd. (Japan) and PT Aneka Tambang (Indonesia) planned to build a joint-venture, 15,000 to 20,000-metric ton/yr ferrosilicon plant in Celebes, Indonesia, by 1985.[101]

Ultrahigh-purity silicon is used as an important raw material in the electronics and energy industries. Silicon is used to make semiconductors, rectifiers, and transistors in the former, and is the most cost-competitive material used to make photovoltaic cells in the latter industry.[96] Demand for silicon in these applications will probably increase as Japan's information processing and new energy source industries expand their operations (Chapter II-5).

Annual Japanese polycrystalline silicon production increased dramatically during the period 1968–1981 (Figure II-4-39). Increased demand for polycrystalline silicon, the feedstock for semiconductors, probably came from Japan's expanding consumer electronics and information processing industries. Future production levels will eventually

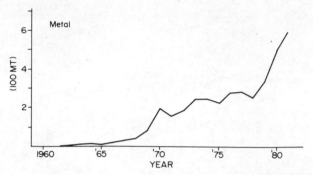

FIGURE II-4-39 Annual Japanese polycrystalline silicon metal production, 1960–1981. (From U.S. Department of the Interior, Bureau of Mines. *Minerals Yearbook.*)

be dependent on the level of silicon conservation that results from the miniaturization of electronic components, the advancement of production techniques, and the substitution by materials such as gallium arsenide.

Future demand for silicon will probably remain high as Japan continues its development of advanced computers and new energy sources. Because imported energy costs will probably continue to be significant, larger imports of ferrosilicon and silicon metal might be expected to meet domestic demand.

SILVER

Domestic silver mines supplied 21% of the raw material used in refined silver metal production in 1980–1981 (Table II-4-1). Silver is produced from several epithermal vein-type deposits such as at the Kushikino mine, Kagoshima Prefecture, the Sado mine, Niigata Prefecture, and the Chitose mine, Hokkaido.[35] In addition, silver is produced as a by-product in the processing of base-metal ores, such as those from skarn deposits at the Kamioka and Nakatatsu mines,[35] Kuroko-type deposits at Hanaoka mine,[33] and vein-type deposits at Hosokura mine[15] (Table II-2-1).

Most of Japan's silver supplies are imported. In 1981, 14.3 million troy ounces of silver metal were imported, mostly from Mexico (63%), Peru (17%), and Australia (11%).[1] In 1980–1981 Mexico became Japan's major source of silver metal, accounting for more than 60% of total metal imports (Table II-4-21). The silver price increase in 1979–1980 led to the expansion of existing mines and the mining of previously uneconomic deposits in Mexico. The extent to which Mexico will remain the leading supplier of Japanese silver, as well as the world's leading producer, will depend on the future price of silver and Mexico's ability to generate investment funds for the development of its mines.

TABLE II-4-21   Annual Japanese Silver Metal Imports, by Source, 1972-1981[a,b] (Million Troy Ounces)

Year	Total imports	Source		
		Mexico	Peru	Others
1972	18.6 (100)	3.6 (19)	7.4 (40)	7.6 (41)
1973	34.3	18.3 (53)	5.5 (16)	10.5 (31)
1974	30.9	9.9 (32)	12.1 (39)	8.9 (29)
1975	17.1	3.4 (20)	9.7 (57)	4.0 (23)
1976	20.5	3.9 (19)	9.8 (48)	6.8 (33)
1977	17.9	NA	NA	NA
1978	17.2	4.4 (26)	6.4 (37)	6.4 (37)
1979	23.5	8.8 (37)	9.2 (39)	5.5 (24)
1980	19.1	11.7 (61)	4.1 (21)	3.3 (18)
1981	14.3	9.0 (63)	2.4 (17)	2.9 (20)

[a]Numbers in parentheses represent percent of total imports.
[b]NA, not available.
Source: U.S. Department of the Interior, Bureau of Mines, Minerals Yearbook.

Japanese silver metal imports declined between 1979 and 1981 (Table II-4-21). This decline reflects Japan's mainly industrial uses of silver. The increase in silver prices in 1979-1980 also contributed to this decline and stimulated increased secondary recovery of silver. In fiscal year 1979, 4.1 million troy ounces were recovered secondarily.[1]

Japan also imports small amounts of silver ore. In 1980, 3400 metric tons of silver ore and concentrate were imported, all from South Korea.[1]

Japan's refined silver production has increased steadily since the early 1960s (Figure II-4-40). In 1981, a record high 40.25 million troy ounces was produced. In fiscal year 1979, domestic consumption of silver was 21 million troy ounces, mostly for use in silver nitrate sensitive film (48%), other

silver nitrate (10%), point connectors (13%), rolled sheet (7%), and solder (7%).[1]

STRONTIUM

No information was available regarding Japanese strontium supplies. Japan most likely receives most of its strontium through world trade. Mexico, Spain, Turkey, Algeria, and Iran are major strontium producers.[1]

Strontium has been included in the list of commodities designated for inclusion in the national stockpile (Chapter II-5). Strontium's major applications are in color-television tubes, ferrite-ceramic permanent magnets, high-purity electrolytic zinc preparation, and pyrotechnics. Increased demand from the first two use sectors has been largely responsible for the ninefold increase in world strontium production that occurred during the period 1960-1981.[102] A steady growth in demand from these five sectors is expected until at least 1990. The consumer electronics industry is probably the major consumer of strontium in Japan.

TANTALUM

Japan imports all its tantalum supplies. Ore and concentrate imports increased fivefold during the second half of the 1970s from 62 metric tons in 1975 to 309 metric tons in 1980 (Figure II-4-41). Thailand was the source of 46% of imports in 1980 and Malaysia, 42%.[1] Japan also imports tantalum metal, mostly from the United States, West Germany, and Taiwan.[1] In 1980, 46 metric tons of tantalum metal were imported.

The major application for tantalum is in electronic components, especially capacitors.[103] Other applications include metalworking machinery, chemical process equipment, and superalloys for the transportation and aerospace industries. Increased demand from all these applications might

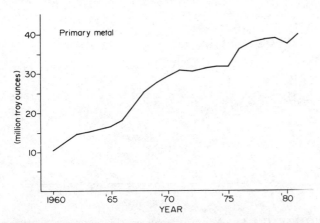

FIGURE II-4-40   Annual Japanese silver metal production, 1960-1981. (From U.S. Department of the Interior, Bureau of Mines. Minerals Yearbook.)

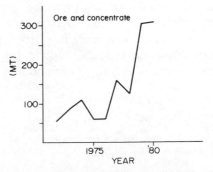

FIGURE II-4-41   Annual Japanese tantalum ore and concentrate imports, 1973-1980. (From U.S. Department of the Interior, Bureau of Mines. Minerals Yearbook.)

be expected as Japan develops its high-technology industries. Substitutes for tantalum are available in all of these applications. However, most of these substitutes, which include aluminum, silicon, germanium, zirconium, selenium, titanium, mischmetal, columbium, hafnium, glass, platinum, molybdenum, and tungsten, are available mainly through imports.

Excluding a decline during economic slowdowns in 1975 and 1978, Japanese tantalum metal production remained steady with little increase from 1974 to 1981 (Figure II-4-42). In 1981, 55 metric tons of tantalum metal were produced in Japan.[1]

Japan exports a large amount of tantalum metal compared to its level of metal production. In 1980, 40 metric tons were exported to Western Europe and the United States.[1]

TIN

One of the most important types of Japanese tin deposits is that observed at the Akenobe copper-gold-tin-tungsten-zinc mine in Hyogo Prefecture. At this mine, tin-bearing minerals occur in high-temperature hydrothermal, or xenothermal, veins, which are believed to be associated with subsurface granites.[35]

However, Japanese tin mine production is small and reserves are limited (Table II-4-1). In 1981, ore with a tin metal content of 562 metric tons was produced at domestic mines. Domestic mine production has declined very gradually since 1960 when production was 855 metric tons (Figure II-5-5). Production reached a peak of 1185 metric tons in 1967. According to a 1976 survey, Japanese exploitable crude tin ore reserves totaled 1,528,000 metric tons of ore with a metal content of 11,000 metric tons (Table II-4-1). Recent tin discoveries at the Akenobe mine (Figure II-5-7 and Table II-5-1) will probably help maintain, rather than increase, the level of domestic mine production. Most of Japan's tin supplies are imported in metal form from Southeast Asian countries. In 1980, 31,155 metric tons were imported, and shipments from Malaysia (18,455 metric tons), Indonesia (6324 metric tons), and Thailand (6166 metric

tons), accounted for more than 99% of total imports.[1] Annual tin metal imports have leveled off since their decline in 1975 (Figure II-4-43).

Small amounts of tin metal are produced annually from domestic ores, imported tin oxides, and probably, scrap tin. Metal production increased in the 1960s but has since declined to a level that is about the same as that of the early 1960s (Figure II-4-44). In 1980, tin metal production was 1319 metric tons. This amount is equivalent to 4% of the amount of imported tin metal supplies.

Future tin production and consumption will depend on the extent of substitution of alternative materials for tin, such as tin-free steel and aluminum for tinplate in the container industry. The substitution of tin-free steel for tinplate was stimulated by the increase in tin prices that occurred during 1980–1982.[104]

Japanese exports of tin metal amounted to only 1450 metric tons in 1980.[1] In 1979–1980, Burma, Taiwan, Singapore, Tanzania, and Mozambique received the majority of Japanese tin metal exports.

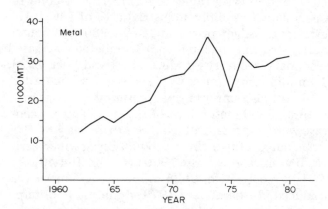

FIGURE II-4-43 Annual Japanese tin metal imports, 1962–1980. (From U.S. Department of the Interior, Bureau of Mines. *Minerals Yearbook*.)

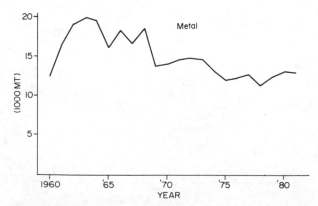

FIGURE II-4-44 Annual Japanese tin metal production, 1960–1981. (From U.S. Department of the Interior, Bureau of Mines. *Minerals Yearbook*.)

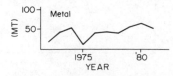

FIGURE II-4-42 Annual Japanese tantalum metal production, 1973–1981. (From U.S. Department of the Interior, Bureau of Mines. *Minerals Yearbook*.)

TITANIUM

Japan imports all its titanium ore and concentrate (ilmenite and rutile) supplies. The main suppliers are Malaysia, Australia, India, and Sri Lanka (Table II-4-22). Japan's dependence on Malaysian ore and concentrates increased significantly in 1977–1978, following the development of an ilmenite ore deposit of 2 to 3 million metric tons in northeastern Malaysia[105] (Table II—4-22). This project involved a Japanese-German-Malaysian joint venture company. Japanese ore and concentrate imports have not risen above their pre-energy-crisis peak of 681,000 metric tons (Figure II-4-45). In the late 1970s and early 1980s, ore and concentrate imports appeared to be in a long-term decline.

Japan also imports titanium slag from Canada and South Africa (Table II-4-22). Japanese imports of titanium slag increased almost fivefold during the period 1976–1980 (Figure II-4-45).

Japan is the world's second largest titanium

sponge producer. Japanese sponge production has been cyclic (Figure II-4-46), because of changes in export demand, which resulted from economic cycles in the aerospace industry, and changes in domestic demand.

In 1981, Japan produced a record high 25,000 metric tons of sponge metal (Figure II-4-46). In 1982, production dropped again as a result of the downturn in the aerospace industry.

The Kroll reduction process, used to produce titanium sponge metal, is energy intensive.[106] By way of conservation measures, the two largest Japanese producers, Toho Titanium and Osaka Titanium, reduced their electricity consumption per ton of produced sponge by 17% in 1980.[5]

However, the high cost of imported Japanese energy supplies was apparently not expected to alter the capacity expansion plans of Japanese sponge producers. Annual sponge capacity at the end of 1982 was expected to reach 32,400 metric tons, an increase of 18% over the 1980 level and 54% over the 1979 level.[1]

Japan's three sponge producers, Osaka Titanium Co., Toho Titanium Co. Ltd., and New Metals Industries (a subsidiary of Nippon Soda Co. Ltd.), anticipate increased future demand for titanium metal in industrial markets, such as chemical equipment, power generation, and desalinization.[106] Japan mainly produces unalloyed titanium metal, the form most suitable in industrial markets.

Ishizuka Research Institute was expected to produce titanium sponge, by way of a modified magnesium process, at an annual capacity of 3 million pounds (1361 metric tons) in 1982.[106] Ishizuka Research Institute is also the majority owner of International Titanium (U.S.). During 1981, Inter-

TABLE II-4-22 Annual Japanese Titanium Ore and Concentrate Imports, by Source, 1975-1981[a] (Thousand Metric Tons)

Year	Total imports	Source		
		Malaysia	Australia	Other[b]
1975	444 (100)	105 (24)	116 (26)	223 (50)
1976	497	124 (25)	203 (41)	170 (34)
1977	496	130 (26)	150 (30)	216 (44)
1978	399	162 (41)	113 (28)	124 (31)
1979	420	187 (45)	71 (17)	162 (38)
1980	409	173 (42)	111 (27)	125 (31)
1981	323	155 (48)	71 (22)	97 (30)

[a]Numbers in parentheses represent percent of total imports.
[b]Other suppliers include India, Sri Lanka, and Canada.
Source: U.S. Department of the Interior, Brueau of Mines, *Minerals Yearbook*.

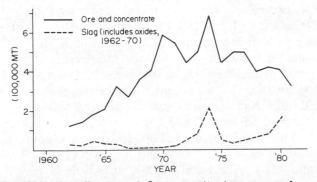

FIGURE II-4-45 Annual Japanese titanium ore and concentrate, and slag imports, 1962-1981. (From U.S. Department of the Interior, Bureau of Mines. *Minerals Yearbook*.)

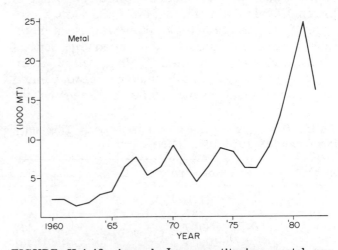

FIGURE II-4-46 Annual Japanese titanium metal production, 1960-1982. (From U.S. Department of the Interior, Bureau of Mines. *Minerals Yearbook*.)

national Titanium began construction in Oregon of a sponge plant with a capacity of 10 million pounds per year (4536 metric tons per year). Initial production was expected to begin in 1982.

A 1981 report indicated that Mitsubishi Metal was planning to become a titanium sponge producer.[107] If so, it will become Japan's first integrated titanium producer. The other producers depend on the Japanese GTCs to obtain titanium raw material supplies, whereas Mitsubishi Metal would provide its own.

Domestic demand for titanium sponge grew significantly during the period 1972–1981. The increase in domestic titanium demand came mainly from the industrial markets.[50] The export demand is still significant, accounting for 40% of total production in 1981. In the late 1970s and the early 1980s, the United States was the main destination for Japanese titanium metal exports (Table II-4-23).

Increased demand from Japan's new aerospace industries might be expected in the future. This industry is one of those expected to contribute to Japan's future economic growth (Chapter II-5).

Recent agreements between Japanese, U.S., and British firms will probably increase Japanese access to foreign export markets and technology. In 1983, Sumitomo Metal Industries (Japan) and Allegheny Ludlum Steel Corp. (U.S.) formed a joint venture to process Sumitomo's semifinished products for sale in North American industrial markets.[108] Also in 1983, Kobe Steel (Japan) agreed to process RMI's (U.S.) semifinished titanium, and to sell the finished products in the United States.[109] This agreement will allow Kobe to enter the U.S. market without having to pay the high U.S. duty levied on titanium sponge and mill products. In 1981, Mitsui & Co. obtained an equity interest in International Titanium, a new U.S. sponge producer.[109] International Titanium will use a modified magnesium process developed by Japan's Ishizuka Research Institute.[107] In addition, Sumitomo and titanium product-maker IMI Ltd. (United Kingdom) entered into a technology agreement.[108]

TABLE II-4-23   U.S. Share of Annual Japanese Titanium Metal Exports, 1978–1981[a] (Metric Tons)

Year	Total exports	U.S. share
1978	4176 (100)	843 (20)
1979	4887	2164 (44)
1980	8070	3332 (41)
1981	10005	4291 (43)

[a]Numbers in parentheses represent percent of total exports.

Source: U.S. Department of the Interior, Bureau of Mines, Minerals Yearbook.

Japan is a leading producer of titanium dioxide pigment. Annual production capacity was estimated to be 227,000 metric tons in 1982.[6] Major uses for titanium dioxide pigment include surface coatings, paper coatings and fillers, and plastics.[106]

Most of Japan's oxide production is consumed domestically or exported indirectly. Japan exported 16,768 metric tons of titanium oxides in 1980, mainly to the United States (38%), Taiwan (19%), China (13%), and South Korea (8%).[1]

TUNGSTEN

Domestic tungsten mine production is from pegmatitic and hydrothermal vein deposits, such as at the Kaneuchi mine, Kyoto Prefecture; tungsten-bearing skarns, such as at the Fujigatani mine, Yamaguchi Prefecture; and from accessory minerals in polymetallic xenothermal vein deposits, such as at the Akenobe mine, Niigata Prefecture.[35]

During the early 1980s several tungsten discoveries were made in southwestern Honshu island. Two tungsten vein deposits were discovered in 1980, with ore grades as high as 3.7% in the Nishigawa district, Yamaguchi Prefecture, in southwestern Honshu.[5] These deposits are located in the area of the Kuga and Kichita mines. Also in 1980, an ore deposit of 120,000 metric tons of gold, silver, copper, lead, zinc, and tungsten was discovered at the Tsumo mine in Tottori Prefecture, Honshu (Figure II-5-7 and Table II-5-1). Tungsten recovery will be emphasized when this new deposit is developed. In 1982, the Metal Mining Agency reported the discovery of another tungsten-bearing vein in Yamaguchi Prefecture.[110] The average tungsten ore grade is as high as 0.49%. This vein also contains tin, copper, and zinc mineralization. Tungsten deposits in southwestern Honshu are skarn-type with associated quartz-scheelite veins.[15]

According to a 1979 survey, Japan had 4,551,000 metric tons of exploitable crude tungsten ore reserves, with a metal content of 21,000 metric tons (Table II-4-1). Domestic mine production contributes as much as one-fourth of total Japanese ore and concentrate requirements. Domestic production has been in decline since 1974 (Figure II-5-4). The recent discoveries noted above will help to maintain annual production levels.

Japan imported 3480 metric tons of tungsten ore and concentrate in 1980, from several sources, including South Korea (21%), Canada (19%), Australia (16%), Portugal (10%), and Bolivia (9%).[5] Other suppliers included Thailand, China, and the United States. Imports of ore and concentrate dropped to 2256 metric tons in 1981. Tungsten ore and concentrate imports have been very cyclic, reflecting tungsten's use in growth industries (Figure II-4-47). Japan also imported 135 metric tons of tungsten metal in 1980, mostly from South Korea (83%).[1] In an effort to develop new sources, Japan and China are involved in a joint exploration venture for tungsten and other rare metals in both countries.[111]

In Japan, tungsten ore and concentrate is used to make tungsten metal, calcium tungsten acid, and ferrotungsten. Its use is in growth industries such as metalworking, steel, superalloys, and chemicals,[1] and therefore, tungsten production usually reflects the general economic conditions (Figure II-4-48). The economic slowdowns in Japan during 1965, 1968–1969, 1971–1972, 1975, 1978, and 1981 can easily be correlated with the downturns in metal production, as well as tungsten ore imports.

Japan exports small amounts of tungsten metal. In 1980, it exported an amount equivalent to approximately 11% of production.[1] Export destinations include the United States, the USSR, and West Germany.

VANADIUM

Japan has no domestic sources of vanadium, and it relies on foreign sources for most of its supplies. In 1980, 3404 metric tons of semiprocessed vanadium pentoxide were imported.[1] South Africa was the source of 90% of imported supplies. Less than 5% was imported from the United States. Japan also imports some ferrovanadium, mostly from Austria, West Germany, and the United States.[112]

Japan is capable of producing vanadium pentoxide from imported ore. Japanese vanadium pentoxide production capacity in 1980 was 1361 metric tons.[113] This capacity level is expected to remain stable through 1986. Some vanadium is also produced domestically through secondary recovery.[112]

Japanese 1981 consumption of vanadium was estimated to be 5625 metric tons.[113] Annual consumption is expected to grow to 7258 by 1986. Most of the vanadium consumed in Japan is used by the iron and steel industry. However, Japanese

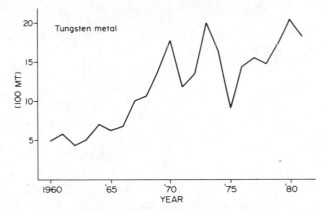

FIGURE II-4-48 Annual Japanese tungsten metal production, 1960–1981. (From U.S. Department of the Interior, Bureau of Mines. *Minerals Yearbook*.)

steelmakers are expected to cut back on the amount of ferrovanadium used in large-diameter pipe.[57] Vanadium is also used to make catalysts and nonferrous alloys.

In comparison with the steel industries of the United States and Western European countries, the steel industry of Japan consumes smaller amounts of vanadium per ton of steel.[113] This is apparently due to Japanese emphasis on the production of plain carbon ship plate steel, rather than low-alloy carbon steel, and to the use of a sophisticated hot steel rolling process that improves the strength and hardness qualities of steel plate.

Annual Japanese vanadium imports generally parallel annual steel and ferroalloy production levels, as well as annual imports of other ferroalloy raw materials such as chrome and manganese (Figures II-4-49, II-4-22, II-4-15, II-4-16, II-4-18, and II-4-28). However, the history of Japanese pentoxide imports in the late 1970s reflects a change in this trend. Following the decline in the middle to late 1970s, Japanese annual vanadium pentoxide imports increased well above the previous mid-1970s peak (Figure II-4-49). During the same period, chromium and manganese imports did not match their previous high levels (Figures II-4-8 and II-4-28). The higher levels of Japanese pentoxide

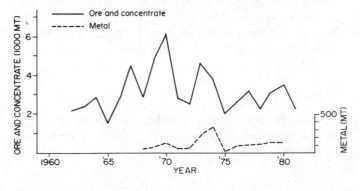

FIGURE II-4-47 Annual Japanese tungsten ore and concentrate, and metal imports, 1962–1981. (From U.S. Department of the Interior, Bureau of Mines. *Minerals Yearbook*.)

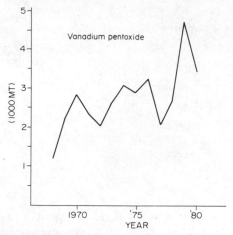

FIGURE II-4-49 Annual Japanese vanadium pentoxide imports, 1968-1981. (From U.S. Department of the Interior, Bureau of Mines. *Minerals Yearbook*.)

imports probably reflect the use of vanadium-bearing steel in line-pipe production. The demand for line pipe in energy pipeline applications increased following the energy crises in the 1970s. However, in the early 1980s line-pipe production declined, as did the level of Japanese vanadium pentoxide imports.

Japanese demand for vanadium in nonferrous metal alloy applications might be expected to increase in the future. Vanadium is an essential component in titanium alloys,[113] and its use has become increasingly more important in alloys destined for jet engine and aircraft applications.[112] The aerospace industry is one that has been targeted for future growth in Japan.

Future Japanese demand for vanadium will reflect the demand for steel as well as nonferrous alloys. Increased vanadium pentoxide import supplies from other sources, such as Australia, New Zealand, and China,[114] will be necessary to satisfy future demands. In 1982, it was reported that Chinese exports to Japan were already expanding rapidly.[57]

ZIRCONIUM

Japan imports all its zirconium supplies. In 1980, Japanese imports of zirconium ore and concentrate (zircon) totaled 190,109 metric tons.[1] Australia supplies 86% of these supplies, and South Africa, 13%. Australia's share represented 33% of its total domestic production.[115] Japan also imported 103 metric tons of zirconium metal, mostly from France (35%), the United States (31%), and Canada (17%).[1] Japanese imports of ore and concentrate increased steadily during the period 1977-1980 (Figure II-4-50).

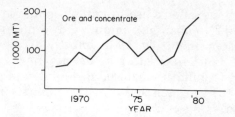

FIGURE II-4-50 Annual Japanese zirconium ore and concentrate imports, 1968-1980. (From U.S. Department of the Interior, Bureau of Mines. *Minerals Yearbook*.)

Japan's usage of zircon is as follows: refractories (70%), foundries (16%), and ceramics, abrasives, metal sponge, and alloys (14%).[116] The use of zircon for steel industry refractory applications is favored over other materials because of its long life and ability to withstand higher temperatures.[116]

Zirconium metal alloys are used mainly in nuclear reactors. This metal also has corrosion-resistant and superalloy applications.

Japan's three zirconium metal producers were all planning production capacity increases in the early 1980s. Nippon Mining was to increase its production capacity to 500 metric tons per year in 1981.[116] Mitsubishi Metal was to increase its production capacity for zirconium alloy to about 60 metric tons per year in 1981.[117] Zirconium Industry, a joint venture of Teledyne Wah Chang Albany (U.S.), Ishizuka Research Institute Ltd. (Japan), and Mitsui and Co. (Japan), was to begin production in late 1982 of a zirconium production plant with a capacity of 1000 metric tons per year.[118] This company will use a Ishizuka-developed distillation process, which is expected to result in halfnium-free zirconium at a cost savings of about 50% over the conventional three-step method.[119]

The future level of metal use will depend mainly on the rate of development of the nuclear power industry in Japan and worldwide, because Japan is a supplier of zirconium metal to the U.S. nuclear power industry.[118] Annual domestic metal demand is expected to be 500 to 600 metric tons in 1985-1986.[119]

Japan was tied with West Germany as the third largest producer of zirconium dioxide (zirconia, range: 94% ZrO_2) in 1980. It ranked behind South Africa and the United States.[116] Zirconia has applications in the electronics, ceramics, and refractories industries.[118]

CONCLUSION

In general, three recurring themes can be observed in the profiles of Japanese strategic mineral commodities:

1. Many processed mineral commodity imports have become favored over traditional ore and concentrate imports, especially for those commodities whose processing is energy intensive.
2. Many Japanese mineral processors and consumers have become directly involved in the development of overseas mineral processing capacity, and are rationalizing or eliminating domestic capacity.
3. The development and expansion of Japanese knowledge-intensive industries has stimulated demand for specific mineral commodities in high-technology applications.

Several factors have affected the form of Japanese mineral commodity imports. These include increased raw material and energy prices, a changing national economic base, and the development of local processing in mineral-producing countries. These factors have contributed to the two basic trends observed in the history of Japanese strategic mineral commodity imports (Figure II-4-51).

The first trend, I, represents rapidly increasing import levels in the 1960s and early 1970s, and then a decline after the oil price hikes in 1973–1974,

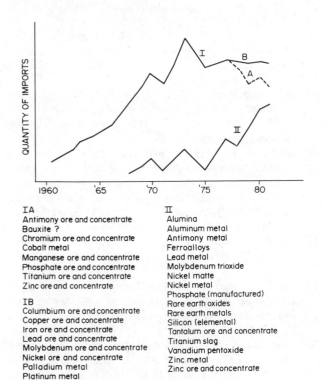

IA
Antimony ore and concentrate
Bauxite ?
Chromium ore and concentrate
Cobalt metal
Manganese ore and concentrate
Phosphate ore and concentrate
Titanium ore and concentrate
Zinc ore and concentrate

IB
Columbium ore and concentrate
Copper ore and concentrate
Iron ore and concentrate
Lead ore and concentrate
Molybdenum ore and concentrate
Nickel ore and concentrate
Palladium metal
Platinum metal
Silver metal
Tin metal
Tungsten ore and concentrate

II
Alumina
Aluminum metal
Antimony metal
Ferroalloys
Lead metal
Molybdenum trioxide
Nickel matte
Nickel metal
Phosphate (manufactured)
Rare earth oxides
Rare earth metals
Silicon (elemental)
Tantalum ore and concentrate
Titanium slag
Vanadium pentoxide
Zinc metal
Zinc ore and concentrate

FIGURE II-4-51 Generalized Japanese stratigic mineral commodity annual import trends, 1962–1981. (From U.S. Department of the Interior, Bureau of Mines. *Minerals Yearbook.*)

continuing until the global economic recession in 1975. In the late 1970s and early 1980s, two alternative pathways exist. The first, IA, is an eventual decline in imports, and the second, IB, is a leveling off of imports.

Mineral commodities in the I category are generally in the ore and concentrate form. Most were the raw materials required for the industrialization of the Japanese state in the 1960s and early 1970s. Their use sectors include the basic materials, chemicals, construction, and automotive industries, which have long been important in the Japanese economy. For some of these, category IB, their future level of imports depends on the general economic conditions, because of their application in growth industries, and the costs of processing. For others, category IA, future import levels might decline. Domestically processed forms of the ores and concentrates in this group have become uncompetitive in world markets.

The second trend, II, represents a gradually increasing import level through most of the 1960s and 1970s. In the late 1970s and early 1980s import levels increased more rapidly. Mineral commodities in the II category are mainly in semiprocessed or processed forms. Several of these, including aluminum metal, antimony metal, ferroalloys, and zinc metal, are the processed forms of those commodities listed in category IA. Silicon, tantalum, titanium, vanadium, and zirconium imports exhibit this trend because of increased demands from Japan's future growth industries.

Since the 1970s and early 1980s, Japanese companies have become directly involved in overseas mineral processing. They have become part-owners in some examples or have provided technical and financial assistance. The reasons for this trend include the lack of cheap domestic energy sources required for competitive processing costs, public concerns about processing-related pollution, the desire of mineral-producing countries to capture the value added in processing, and Japan's interest in maintaining secure sources of processed mineral supplies for its manufacturing and other future growth industries. Some of the best examples of overseas involvement are observed in the aluminum, ferrosilicon, copper, and steel industries. Further involvement might also be expected in the zinc, ferromanganese, ferrochromium, and ferronickel industries, all of whose processing sectors require significant energy inputs.

As mentioned above, Japan's changing national economic base, which emphasizes knowledge-intensive industries such as information processing, has led to increased demand for specific commodities:

silicon, tantalum, rare earth metals, and germanium. In addition, increased demand has been observed for several multiuse commodities in new, expanding use sectors, such as aerospace, superalloys, nonferrous alloys, chemicals, and new basic materials. These commodities include cobalt, manganese, titanium, vanadium, nickel, antimony, and copper.

REFERENCES

1. U.S. Department of the Interior, Bureau of Mines. "The Mineral Industry of Japan." By John Wu. *Minerals Yearbook 1981*, Vol. III, preprint. Washington, D.C.: U.S. Government Printing Office, 1982.
2. "Comalco Assesses Aluminum's Future at Smelter Opening." *Engineering and Mining Journal* 183 (October 1982).
3. "New Japanese Process to Smelt Aluminum from Clay Tested." *Engineering and Mining Journal* 182 (August 1981).
4. Jameson, Sam. "MITI Policy Irks Japan Smelters." *Los Angeles Times*, 21 March 1983.
5. U.S. Department of the Interior, Bureau of Mines. "The Mineral Industry of Japan." By John Wu. *Minerals Yearbook 1980*, Vol. III. Washington, D.C.: U.S. Government Printing Office, 1981.
6. U.S. Department of the Interior, Bureau of Mines. *Mineral Commodity Summaries 1983*. Washington, D.C.: U.S. Government Printing Office, 1983.
7. "Production Adjustments in Aluminum." *The Oriental Economist* 50 (June 1982).
8. "Chinese Envisage Big Project for Production of Aluminum." *Japan Economic Journal*, 25 January 1983.
9. Makiuchi, Iwao. "Non-ferrous Metals: Aluminum Smelters Had Worst Year in 1981; Copper, Lead, Zinc Makers Also Fared Poorly." In *Industrial Review of Japan, 1982*. Tokyo: Japan Economic Journal, 1982.
10. Chiba, Atsuko. "Japan's Aluminum Woes." *Forbes*, 21 July 1980.
11. Fitzgerald, M. D., and G. Pollio. "Aluminum: The Next Twenty Years." *Journal of Metals* 34 (December 1982).
12. "Sumikei Aluminum Formally Decides on Ending Business." *Japan Economic Journal*, 8 June 1982.
13. "CRA Ltd. to Acquire 50% of Japan Firm; RTZ Earnings Drop." *The Wall Street Journal*, 16 April 1982.
14. Inaba, M. "Japan Shuns Metal Substitution." *American Metal Market/Metalworking News*, 20 September 1982.
15. Japan. Geological Survey. *Geology and Mineral Resources of Japan*, 2nd ed. Edited by M. Saito, K. Hashimoto, H. Sawata, and Y. Shimazaki. Japan: Geological Survey of Japan, 1960.
16. "Japan: Antimony Import Problems. . . ." *Mining Journal* 299, No. 7670 (20 August 1982).
17. Way, H. J. R. "Antimony." In *Mining Annual Review, 1981*. London: Mining Journal, 1981.
18. U.S. Department of the Interior, Bureau of Mines. "Antimony." By John Rathjen. *Mineral Facts and Problems, 1980 Edition*. Washington, D.C.: U.S. Government Printing Office, 1980.
19. Stubbs, R. L. "Lead and Zinc." In *Mining Annual Review, 1982*. London: Mining Journal, 1982.
20. U.S. Department of the Interior, Bureau of Mines. "Beryllium." By Benjamin Petkof. *Mineral Facts and Problems, 1980 Edition*. Washington, D.C.: U.S. Government Printing Office, 1980.
21. U.S. Department of the Interior, Bureau of Mines. "Chromium." By E. C. Peterson. *Minerals Yearbook 1980*, Vol. I. Washington, D.C.: U.S. Government Printing Office, 1981.
22. U.S. Department of the Interior, Bureau of Mines. "Chromium." By John Papp. *Minerals Yearbook 1981*, Vol. I, preprint. Washington, D.C.: U.S. Government Printing Office, 1982.
23. U.S. Department of the Interior, Bureau of Mines. "The Mineral Industry of Japan." By K. P. Wang. *Minerals Yearbook 1972*, Vol. III. Washington, D.C.: U.S. Government Printing Office, 1972.
24. Thompson, A. G. "Chromite." In *Mining Annual Review, 1979*. London: Mining Journal, 1979.
25. "Less SA Ferrochrome for Japan." *Mining Journal* 298, No. 7648 (19 March 1982).
26. U.S. Department of the Interior, Bureau of Mines. "Chromium." By J. L. Morning, N. A. Matthews, and E. C. Peterson. *Mineral Facts and Problems, 1980 Edition*. Washington, D.C.: U.S. Government Printing Office, 1980.
27. "Ferrochrome Supply and Demand." *Mining Journal* 298, No. 7639 (15 January 1982).
28. U.S. Department of the Interior, Bureau of Mines. "Cobalt." By Scott Sibley. *Mineral Facts and Problems, 1980 Edition*. Washington, D.C.: U.S. Government Printing Office, 1980.
29. U.S. Department of the Interior, Bureau of Mines. "Cobalt." *Minerals Yearbook*, various years 1978–1981. Washington, D.C.: U.S. Government Printing Office, 1980–1982.
30. Jones, S. L. "Samarium–Cobalt Magnets Spur Electronic Advances." *American Metal Market/Metalworking News*, 28 June 1982.
31. Vernyi, Bruce. "Cutting Down on Cobalt." *American Metal Market/Metalworking News*, Aerospace Metals and Machines Supplement, 1983.
32. "Niobec to Idle Colombium Mines." *American Metal Market/Metalworking News* 91, No. 55, (21 March 1983).
33. Lambert, Ian, and Takeo Sato. "The Kuroko and Associated Ore Deposits of Japan: A Review of Their Features and Metallogenesis." *Economic Geology* 69 (December 1974).
34. Franklin, J. M., D. M. Sangster, and J. W. Lydon. "Volcanic-Associated Massive Sulfide Deposits." *Economic Geology, Seventy-Fifth Anniversary Volume, 1905–1980*. El Paso, Tex.: Economic Geology Publishing Co., 1981.
35. Imai, Hideki. *Geological Studies of the Mineral Deposits in Japan And East Asia*. Tokyo: University of Tokyo Press, 1978.
36. Ridge, John Drew. *Annotated Bibliographies of Mineral Deposits in Africa, Asia (exclusive of the USSR) and Australasia*. Oxford: Pergamon Press Ltd., 1976.

37. "Philippines: Under a $2 Million Joint Exploration Agreement. . . ." *Engineering and Mining Journal* 182 (November 1981).

38. "Anguin Agreement." *Mining Journal* 298, No. 7658 (28 May 1982).

39. "Japanese Interested in Developing Petaquilla." *World Mining* 35 (June 1982).

40. "Arco Unit to End Copper Mining in Butte, Mont." *The Wall Street Journal*, 10 January 1983.

41. "Philippines Subsidizes Its Copper Production to Keep Mines Running." *The Wall Street Journal*, 6 July 1982.

42. U.S. Department of the Interior, Bureau of Mines. "The Mineral Industry of the Philippines." By John Wu. *Minerals Yearbook 1981*, Vol. III. Washington, D.C.: U.S. Government Printing Office, 1982.

43. "Japan: Record Copper Output." *Mining Journal* 300, No. 7694 (4 February 1983).

44. Nagano, Takeshi. "The History of Copper Smelting in Japan." *Journal of Metals* 34 (June 1982).

45. Kohno, H., H. Asao, and T. Amano. "Use of Alternative Fuel at Onahama." *Journal of Metals* 34 (July 1982).

46. Suda, Satoru. "Japan." In *Mining Annual Review, 1982*. London: Mining Journal, 1982.

47. "Sumitomo Finalizes Chinese Copper Refinery Agreement." *Mining Journal* 299, No. 7677 (8 October 1982).

48. Brown, Stuart. "Staving Off Glass Inroads: Gains in Electronics Enable Copper Wires to Transmit More Data." *American Metal Market/Metalworking News*, Copper Supplement, 22 November 1982.

49. U.S. Department of the Interior, Bureau of Mines. "Ferroalloys." By Frederick Schottman. *Mineral Facts and Problems, 1980 Edition*. Washington, D.C.: U.S. Government Printing Office, 1980.

50. U.S. Department of the Interior, Bureau of Mines. "The Mineral Industry of Japan." *Minerals Yearbook*, Vol. III, various years. Washington, D.C.: U.S. Government Printing Office, 1963–1982.

51. U.S. Department of the Interior, Bureau of Mines. "Potash." By James Searles. *Mineral Facts and Problems, 1980 Edition*. Washington, D.C.: U.S. Government Printing Office, 1980.

52. "Esmark Unit, Japanese Firm to Set Up Mining Venture." *The Wall Street Journal*, 29 April 1983.

53. U.S. Department of the Interior, Bureau of Mines. "Phosphate Rock." By W. F. Stowasser. *Mineral Facts and Problems, 1980 Edition*. Washington, D.C.: U.S. Government Printing Office, 1980.

54. Ninose, Tomoko. "Chemicals: Companies in 1981 Made Rush to Advance into Biotechnology as New Field of Growth." In *Industrial Review of Japan, 1982*. Tokyo: Japan Economic Journal, 1982.

55. U.S. Department of the Interior, Bureau of Mines. "Fluorspar." By David Morse. *Mineral Facts and Problems, 1980 Edition*. Washington, D.C.: U.S. Government Printing Office, 1980.

56. U.S. Department of the Interior, Bureau of Mines. "Gallium." By Benjamin Petkof. *Mineral Facts and Problems, 1980 Edition*. Washington, D.C.: U.S. Government Printing Office, 1980.

57. "Mixed Fortunes for Minor Metals." *Mining Journal* 298, No. 7650 (2 April 1982).

58. Post, Charles. "Gallium Arsenide: Silicon's New Competition." *Iron Age*, 1 November 1982.

59. "New Process Is Devised for Gallium Arsenide Crystals." *Japan Economic Journal*, 1 February 1983.

60. U.S. Department of the Interior, Bureau of Mines. "Germanium." By John Lucas. *Mineral Facts and Problems, 1980 Edition*. Washington, D.C.: U.S. Government Printing Office, 1980.

61. "Sumitomo Metal Mining. . . ." *Engineering and Mining Journal* 183 (October 1982).

62. "Sumitomo Metal Mining Reveals a New Gold Vein." *The Oriental Economist* 50 (May 1982).

63. "Rise in Japanese Gold Imports." *Mining Journal* 298, No. 7646 (5 March 1982).

64. Lehner, Urban. "The Outlook: Japan's High Savings: Boon or Burden?" *The Wall Street Journal*, 25 October 1982.

65. "Japan: Gold Imports Totalled 167.3 Tons in 1981." *Asia Research Bulletin* 11, No. 12, Report 5 (30 April 1982).

66. "Japan Demand for Gold Likely to Recover." *Asia Research Bulletin* 12, No. 11, Report 5 (30 April 1983.

67. Rodrik, Dani. "Managing Resource Dependency: The United States and Japan in the Markets for Copper, Iron Ore and Bauxite." *World Development* 10 (July 1982).

68. Ridley, R. S., and A. F. Kaba. "Brazil Battles the Jungle to Mine Carajas Minerals." *World Mining* 36 (January 1983).

69. Ulman, N. "Brazil Pursues Amazon's Riches with a Project on a Huge Scale Befitting the Region's Vastness." *The Wall Street Journal*, 28 October 1982.

70. Kawata, Sukeyuki, ed. *Japan's Iron and Steel Industry, 1982*. Tokyo: Kawata Publicity, Inc., 1982.

71. "Brazil: An Opening Ceremony Was Held August 5. . . ." *Engineering and Mining Journal* 183 (September 1982).

72. "Australia Losing Some Japanese Iron Orders to Brazilian Producers." *The Wall Street Journal*, 30 September 1982.

73. "JISEA Figures Exports of Steel Will Drop in 1983." *Japan Economic Journal*, 8 February 1983.

74. "China Seems Due to Purchase Three Million Tons of Steel." *Japan Economic Journal* 21, No. 1046 (1 March 1983).

75. "Steel Industry's Case against Japan Is Rejected by U.S." *The Wall Street Journal*, 28 February 1983.

76. Peters, Hans-Jurgen. "Japan Steel to U.S., EEC: Market Conditions Dictate Export Flow." *American Metal Market/Metalworking News*, 24 January 1983.

77. Shikazono, Naotatsu. "Mineralization and Chemical Environment of the Toyoha Lead–Zinc Vein-Type Deposits, Hokkaido, Japan." *Economic Geology* 70 (June–July 1975).

78. Soldi, C. G. "Peru." In *Mining Annual Review, 1982*. London: Mining Journal, 1982.

79. Stubbs, R. L. "Lead And Zinc." In *Mining Annual Review, 1981*. London: Mining Journal, 1981.

80. "Work on Thailand's Zinc Smelter Moves toward 1984 Completion." *Engineering and Mining Journal* 183 (December 1982).

81. Ellison, T. D. "Manganese." In *Mining Annual Review, 1982*. London: Mining Journal, 1982.

82. "CVRD Obtains Mining Concession for Carajas Manganese Reserves." *Engineering and Mining Journal* 183 (November 1982).

83. Thompson, A. G. "Manganese." In *Mining Annual Review, 1981*. London: Mining Journal, 1981.

84. U.S. Department of the Interior, Bureau of Mines. "Nickel." By Norman Matthews and Scott Sibley. *Mineral Facts and Problems, 1980 Edition*. Washington, D.C.: U.S. Government Printing Office, 1980.

85. Japanese Mineral Demand May Boost Pacific Northwest and Alaska." *Mining Engineering* 35 (February 1983).

86. "Nickel Imports Thinned." *Mining Journal* 298, No. 7662 (25 June 1982).

87. "Nippon to Increase Copper Output." *Mining Journal* 299, No. 7678 (15 October 1982).

88. "Showa Halts Nickel Output." *Mining Journal* 299, No. 7683 (19 November 1982).

89. U.S. Department of the Interior, Bureau of Mines. "Nickel." By Scott Sibley. *Minerals Yearbook 1981*, Vol. I. Washington, D.C.: U.S. Government Printing Office, 1982.

90. U.S. Department of the Interior, Bureau of Mines. "Platinum Group Metals." By J. H. Jolly. *Mineral Facts and Problems, 1980 Edition*. Washington, D.C.: U.S. Government Printing Office, 1980.

91. Chegwidden, Judith. "Rare Earths." In *Mining Annual Review, 1981*. London: Mining Journal, 1981.

92. U.S. Department of the Interior, Bureau of Mines. "Rare Earth Minerals and Metals." By James Hendrick. *Minerals Yearbook 1981*, Vol. I. Washington, D.C.: U.S. Government Printing Office, 1982.

93. Farr, P. J. "Rare Earths." In *Mining Annual Review, 1982*. London: Mining Journal, 1982.

94. U.S. Department of the Interior, Bureau of Mines. "Rare Earth Elements and Yttrium." By Christine Moore. *Mineral Facts and Problems, 1980 Edition*. Washington, D.C.: U.S. Government Printing Office, 1980.

95. Jones, S. L. "Rare Earth's Development Said to Be in Its Infancy." *American Metal Market/Metalworking News*, 9 August 1982.

96. U.S. Department of the Interior, Bureau of Mines. "Silicon." By Peter Kuck. *Mineral Facts and Problems, 1980 Edition*. Washington, D.C.: U.S. Government Printing Office, 1980.

97. Shiota, S. "Balancing Ferro-Silicon Supply and Demand." *Metal Bulletin*, No. 107 (November 1979). Cited in U.S. Department of the Interior, Bureau of Mines. "Ferroalloys." By Frederick Schottman. *Minerals Yearbook 1978-79*. Vol. I. Washington, D.C.: U.S. Government Printing Office, 1979.

98. U.S. Department of the Interior, Bureau of Mines. "Silicon." *Minerals Yearbook*, Vol. I, various years, 1978-1981. Washington, D.C.: U.S. Government Printing Office, 1979-1982.

99. U.S. Department of the Interior, Bureau of Mines. "Ferroalloys." By Raymond Brown. *Minerals Yearbook 1981*, Vol. I. Washington, D.C.: U.S. Government Printing Office, 1982.

100. Canada. Department of External Affairs. *Canada's Export Development Plan for Japan*, August 1982.

101. "Indonesia: Negotiations Are Under Way. . . ." *Engineering and Mining Journal* 182 (July 1981).

102. "New Technology Boosts Demand for Strontium." *Mining Journal* 299, No. 7688 (24 December 1982).

103. U.S. Department of the Interior, Bureau of Mines. "Tantalum." By Thomas Jones. *Mineral Facts and Problems, 1980 Edition*. Washington, D.C.: U.S. Government Printing Office, 1980.

104. Suzuki, Kenji. "Iron and Steel: Slump of Demand Compelled Curtailment of Production in Last Half of 1980." In *Industrial Review of Japan, 1981*. Tokyo: Japan Economic Journal, 1981.

105. Gadsen, P. H. "Titanium." In *Mining Annual Review, 1975*. London: Mining Journal, 1975.

106. U.S. Department of the Interior, Bureau of Mines. "Titanium." By Langtry Lynd. *Mineral Facts and Problems, 1980 Edition*. Washington, D.C.: U.S. Government Printing Office, 1980.

107. "Titanium: No More Boom-and-Bust Cycles?" *Chemical Week*, 26 August 1981.

108. "Sumitomo Firms Will Tie-Up with Alleghany Ludlum Steel Corp." *Japan Economic Journal*, 18 January 1983.

109. Kingston, John. "RMI/Kobe Steel Reach Titanium Accord." *American Metal Market/Metalworking News* 91, No. 40 (28 February 1983).

110. "Japanese Tungsten Discovery." *Mining Journal* 299, No. 7688 (24 December 1982).

111. "Japan: International." *Bulletin of Anglo-Japan Economic Institute*, No. 234 (November–December 1982).

112. U.S. Department of the Interior, Bureau of Mines. "Vanadium." By George Morgan. *Mineral Facts and Problems, 1980 Edition*. Washington, D.C.: U.S. Government Printing Office, 1980.

113. Korchynsky, M. "Vanadium–Present and Future." *CIM Bulletin* 75, No. 843 (July 1982).

114. Sage, A. M. "Vanadium." In *Mining Annual Review, 1982*. London: Mining Journal, 1982.

115. U.S. Department of the Interior, Bureau of Mines. "The Mineral Industry of Australia." By Charlie Wyche. *Minerals Yearbook 1981*, Vol. III. Washington, D.C.: U.S. Government Printing Office, 1982.

116. Fuller, Ronald. "Zirconium and Hafnium." In *Mining Annual Review, 1981*. London: Mining Journal, 1981.

117. "Minor Precious Metals." *Metal Bulletin*, No. 6527 (30 September 1980). Cited by U.S. Department of the Interior, Bureau of Mines. "Zirconium and Hafnium." By William Kirk. *Minerals Yearbook 1980*, Vol. I. Washington, D.C.: U.S. Government Printing Office, 1981.

118. U.S. Department of the Interior, Bureau of Mines.

"Zirconium and Hafnium." By William Kirk. *Minerals Yearbook 1981*, Vol. I. Washington, D.C.: U.S. Government Printing Office, 1982.

119. "New Technology for Zirconium Sponge Production under Test by Japanese," *Engineering and Mining Journal* 181 (November 1980).

CHAPTER II-5

Mineral Supply Strategies

Japanese strategies used to maintain the security of their mineral supplies have changed many times during the development of the modern Japanese State. Methods have included those involving normal trade relations, such as spot-market purchases, and short- and long-term purchase contracts. Forceful acquisition methods have also been used, as in World War II. In part, the reasons for Japanese expansionism in East Asia (Figure II-5-1) during this time were to satisfy their increasing need for raw materials and living space.[1] This expansion resulted in Japanese control of supplies of oil, coal, iron ore, cobalt, nickel, copper, tungsten, chromium, bauxite, tin, and rubber, which were so vital to their war economy.

This chapter examines present-day Japanese strategies used to maintain the security of mineral supplies. These strategies include the following:

1. Development of domestic resources
2. Development of foreign resources (with emphasis on overseas processing capacity)
3. Stockpiling
4. Deep-seabed mining
5. Change in national economic base

These strategies are implemented through the use of various mineral policy elements. These elements are discussed where they affect the different strategies, and they are summarized afterward. Many of the elements of Japanese mineral policy discussed below are taken from a study made by Crowson.[2]

FIGURE II-5-1 The boundaries of the Japanese state before, during, and after World War II. (From John Whitney Hall, *Japan: From Prehistory to Modern Times*. New York: Dell Publishing Company, 1979, Table, p. 335.)

.... The Japanese Empire in 1931

— — — Farthest extent of Japanese expansion, 1942-1943

■ Post-surrender limits of the Japanese State, 1945

The success of these strategies depends on the relationship between mineral and foreign policies, which is well defined in the Japanese example. The unique level of cooperation between the Japanese government and the mineral industries is also a major element in mineral supply security.

DEVELOPMENT OF DOMESTIC RESOURCES

The importance of Japanese domestic strategic mineral mine production to processed metal production is small. Domestic mine production for several commodities, such as copper, lead, zinc, manganese, chromium, tungsten, molybdenum, gold, silver, tin, and iron, has remained steady or has declined during the period 1960–1981 (Figures II-5-2 to II-5-5). Except in a few cases, such as lead, zinc, and silver, domestic mine production is

insignificant in relation to total Japanese demand (Table II-4-1).

In an effort to aid domestic mining companies and to continue the level of domestic production, the Japanese government established an Emergency

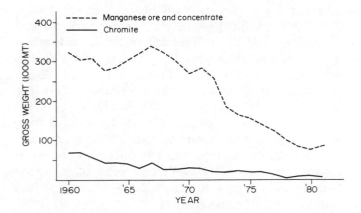

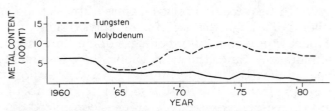

FIGURE II-5-4 Annual Japanese domestic mine production of manganese, chromium, tungsten, and molybdenum, 1960–1981. (From U.S. Department of the Interior, Bureau of Mines. *Minerals Yearbook*.)

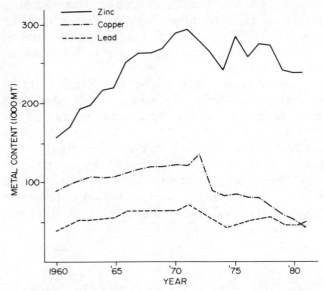

FIGURE II-5-2 Annual Japanese domestic mine production of copper, lead, and zinc, 1960–1981. (From U.S. Department of the Interior, Bureau of Mines., *Minerals Yearbook*.)

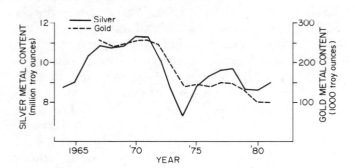

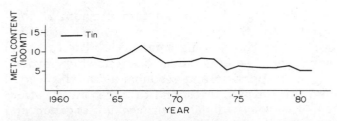

FIGURE II-5-5 Annual Japanese domestic mine production of gold, 1967–1981; silver, 1964–1981; and tin, 1960–1981. (From U.S. Department of the Interior, Bureau of Mines. *Minerals Yearbook*.)

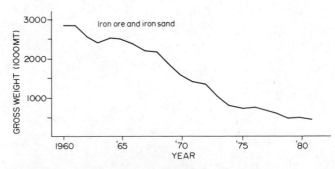

FIGURE II-5-3 Annual Japanese domestic mine production of iron, 1960–1981. (From U.S. Department of the Interior, Bureau of Mines. *Minerals Yearbook*.)

Finance Fund in late 1978.[3] This fund was capitalized at Y25 billion. The incentive for this action was the decline in domestic zinc and copper mine production resulting from weakened metal prices[2] (Figure II-5-2). Mining companies are able to receive loans from the fund at very low interest rates, when market prices fall below certain levels: Y361,000 ($1523) per metric ton for copper and Y196,000 ($827) per metric ton for zinc. If metal prices increase to higher levels: Y500,000 ($2110) per metric ton for copper, and Y210,000 ($886) per metric ton for zinc, mining companies are in a position to contribute to the fund. The government's financial responsibility to the fund is twofold: to guarantee loans to the MMAJ from commercial banks, and to pay the difference between the rate at which MMAJ borrows the loan money and the lower rate at which it makes loans to the domestic mining companies.

Despite the limited potential for future Japanese mine production, exploration programs continue to be pursued in several areas of the Japanese islands. Through the use of loans and subsidies, the Japanese government encourages domestic mineral exploration. Since the early 1960s, the government has supported a broad, three-level national exploration program.[2] The first level corresponds to regional geologic assessment, which is contracted out by the MMAJ and is totally funded by government funds. Second-level exploration surveys are performed in the most promising areas. Concessions are granted to mining companies for these areas. The national and prefectural governments supply 80% of the necessary funds, and mining companies, 20%. At the third level, mining companies perform more detailed exploration work in selected areas. Government encouragement is provided through subsidized loans to large companies, and direct subsidies to smaller companies. Government funding can only be used by Japanese nationals or Japanese mining companies, in exploration for any of 13 metallic mineral commodities. These commodities include copper, lead, zinc, manganese, gold and silver.

This three-level exploration program is comparable to the typical exploration activities performed by U.S. mining companies.[4] The first level corresponds to target identification, which involves regional appraisal and reconnaisance. The second and third levels correspond to target investigation, which involves detailed surface investigation and three-dimensional physical sampling of the target area. However, there is no nonemergency U.S. program to provide government loans and subsidies.

In 1980, the MMAJ performed general geologi-cal structure surveys in 14 areas and intensified surveys in eight areas.[24] A total of 17,027 meters of core drilling and 1645 meters of prospect drilling were also completed by the agency. In 1981, the number of general surveys was 22, and detailed surveys, 21.[5] The area of Kuroko-type massive sulfide deposits in the vicinity of Akita Prefecture, northeastern Honshu Island (Figure II-5-6), has been explored extensively since the 1960s. Kuroko-type deposits contain complex assembleges of copper, zinc, lead, silver, and sulfate minerals (black ores). Discoveries made in this area during the 1960s significantly extended Japan's copper, zinc, and lead reserves.[6] By 1971, most of the copper-in-concentrates, lead-in-concentrates, and zinc-in-concentrates produced domestically came from Kuroko-type deposits.[7] Exploration work in the late 1970s and early 1980s, resulted in the extension of copper, lead, and zinc reserves at several mines, such as Hanaoka Fukazawa, Hanaoka Matsumine, and Ezuri mines in Akita Prefecture (Figure II-5-7). The potential for additional mineral production from future discoveries of Kuroko-type deposits appears to be favorable.

The original reserves of the Ezuri mine in Ohdate City, Akita Prefecture, were discovered through the continuing exploration program of the governmental Metal Mining Agency of Japan (MMAJ).[5] Ezuri's ore reserves of copper (0.89%), lead (3.5%), zinc (10.1%), gold (1.3 g/metric ton), and silver (180 g/metric ton) are estimated to be 3 million metric tons.[8] Ore production at this mine began in 1980 at a monthly level of 6000 metric

FIGURE II-5-6 Area of Kuroko-type deposits on northeastern Honshu Island, Japan. (From Ref. 7.)

FIGURE II-5-7 Locations of selected discoveries at existing mines in Japan, 1979–1982; descriptions provided in Table II-5-1. (From *Mining Annual Review*; *Mining Journal*.)

tons. Production was expected eventually to reach a monthly level of 10,000 metric tons.

Another promising area for exploration is the southern part of Kagoshima Prefecture on Shikoku Island (Figure II-5-8). MMAJ's exploration surveys in this area have led to the discovery of several gold and silver deposits.[5] A core drilling program has already begun at one gold discovery, Hishikari, in Kagoshima Prefecture.

Continued private and government exploration

FIGURE II-5-8 Location of Kagoshima Prefecture, Kyushu Island, Japan.

efforts have been successful in extending the resource base at several preexisting mines (Figure II-5-7). Table II-5-1 lists several of the discoveries made in 1979–1981. In general, these new discoveries will probably not lead to increased domestic production overall, but rather to the extension of domestic production into the future.

Future exploration efforts will be aided by the proposed launch of an exploration satellite in 1986.[5] This satellite is part of a 20-year research and development program that Japan's Ministry of International Trade and Industry initiated in 1981. The goal of the program is the development of technology for future industries. Also included within the program is the establishment of an Institute and Analysis Center for Natural Resources.

DEVELOPMENT OF FOREIGN RESOURCES

The Japanese long-term economic plan (1979–1985) recognizes the vulnerability of the domestic economy to changes in prices of natural resources, and in policies of resource-exporting nations.[9] It does not directly address the country's vulnerability to imported resource supply cutoffs. However, it recommends the following options to maintain a steady supply of resources:

Promotion of overseas resources
Diversification of resource sources and supply routes
Increased involvement in overseas processing of raw materials

These options are available to most industrialized resource-poor countries. The first two have been pursued by the Japanese since the early 1900s, the latter since the early 1970s.

It is important to note the comprehensive nature of these first two options. They indicate Japan's recognition of its limited resource potential, and the intention of the government to play a role in dealing with this situation. Promotion of overseas resources is an obvious alternative to a resource-poor country with intentions of developing a world-class economy. But to recommend geographical diversification of resources and supply routes is to understand the vulnerability of a country's dependency on foreign raw material supplies. An importer's supply of minerals can be easily interrupted by changes in the exporter's internal policies, or by external factors that affect either party.

Several methods have been used to promote and diversify foreign sources of mineral supplies. These include direct involvement, in the form of equity investments, loans, yen credits, technological cooperation, and technology transfer. In return for

TABLE II-5-1   Description of Selected New Discoveries at Existing Domestic Mines, 1979–1982

Mine (prefecture)[a]	Commodities	Description	Year
Ohe	Pb, Zn, Ag, Mn	Four veins, 1.6–3.9 m wide, Pb: 1.0–6.0%, Zn: 4.6–10.4%, Ag: 27–51 g/MT, Mn: 7.2–11.8%	1980–1981
Toyoha	Pb, Zn, Ag	1.5 m deposit, Ag: 200–300 g/Mt, combined Pb and Zn: 10–15%	1981
Adjacent area	Pb, Zn, Ag	Two deposits, 3 m and 9 m wide, at depths of 400–500 m	1981
Kaminokuni	Pb, Zn, Au, Ag, Mn	Pb: 3.51%, Zn: 6.02%, Au: 3.9 g/MT, Ag: 1079 g/MT, Mn: 8.07%	1982
Hanaoka Fukazawa	?	Two new orebodies	1980
Hanaoka Matsumine	Cu, Pb, Zn	Cu: 4%, Pb: 5%, Zn: 7%	1981
Ezuri	Cu?, Pb?, Zn?	Total deposit: 120,000 MT ore	1980
Kamioka (Gifu)			
Tochibara pit, Maruyama pit, Mozumi pit	Pb, Zn	100,000 MT ore (at deeper levels)	1979–1981
320-m-level	Ag, Pb, Zn	Two deposits, Ag: 334 g/MT, Pb: 1.76%, Zn: 1.2%	1980
Adjacent area	Pb, Zn, Au	Au values of isolated samples: 92 g/MT, 27 g/MT	1981
Nakatatsu	Ag, Pb, Zn	Ag: 113 g/MT, Pb: 0.16%, Zn: 1.2%	1980
Adjacent area	Pb, Zn	Indication of mineralized zone	1979–1980
Akenobe	Cu, Zn, Sn, Au, Ag	Confirmation of 60,000-MT reserve, Two veins: (1) Au: 1.6 g/MT, Ag: 364 g/MT (2) Au: 0.78 g/MT, Ag: 1560 g/MT	1980
Tsumo	Au, Ag, Cu, Pb, Zn, W	Au: 0.04 g/MT, Ag: 3 g/MT, Cu: 0.1%, Pb: 0.16%, Zn: 0.07%, WO_3: 0.4%	1979
Fujigatani	W	Vein deposit, 1.0–3.8% WO_3	1981

[a]Locations shown on Figure II-5-7.
Sources: Mining Annual Review; Mining Journal.

this involvement, the Japanese participants usually receive a share of the final output of the project, which could be mined ore and concentrate, or processed metals. An alternative method, having a longer history of use, is that of less direct involvement through long-term purchase contracts or spot market purchases.

Long-Term Contracts

In general, long-term purchase contracts (LTCs) assure the producer of a guaranteed market for its production, and the consumer of guaranteed mineral supplies for its industries. For this reason, LTCs can be thought of as being similar to vertical integration.[10]

The average length of a contract period is 10 to 12 years, but it can be as long as 25 years.[11] Either party can usually cancel an agreement with 30 to 60 days' advance notice and the payment of a penalty fee. Purchase prices are usually not fixed, and can either be indexed to a standard reference, such as the London Metal Exchange (LME) price index, or be renegotiated at various times if market conditions vary.[12] Indirectly, the indexing of prices to a standard reference may indicate that the consumer's main interest is the security of supply, rather than price stability.[10] LTCs are often tied to initial financing arrangements for the specific resource project. Develop-for-import projects are examples of this type of arrangement.[12]

LTCs have been used extensively by the Japanese to maintain their supplies of raw materials. Rodrik[13] states that prior to the 1970s, most of Japan's supplies of iron ore, bauxite, and copper came from LTCs with independent mines. Japan's relaxed antitrust regulations allowed consuming companies to form import-purchase cartels. As a cartel, these companies could maintain a bargaining advantage with ore-producing companies. Compared to the U.S. strategy of vertical integration and direct control of overseas mines, the Japanese strategy of LTC use resulted in lower bauxite and iron ore prices for domestic consumers. Over the long term, 1960–1976, Japan obtained bauxite and iron ore supplies at discounts of 20 to 50% below U.S. prices.[13]

Japanese LTCs have been subject to frequent renegotiation. For example, Japan has renegotiated Australian iron ore contracts many times since 1972, following the decline in the growth of annual Japanese steel production.[13] Renegotiation mainly concerned contracted tonnage, not price.

Negotiations in 1982–1983 concerned price and tonnage. In late 1982, Japan called for price and tonnage reductions with its main suppliers, Australia, Brazil, and India.[14] Earlier in the year these suppliers had obtained price increases of 17%, 13.6%, and 15.5%, respectively. Similarly, Japan has requested price and tonnage reductions regarding its coking coal contracts with its main suppliers, Australia, United States, Canada, and South Africa.[15] The continued slump in steel demand has led to both of these requests.

Direct Involvement

Japanese direct involvement methods have seen increased use since the early 1970s. The reasons for this change are at least twofold. First, LTCs were less secure, following the rise in commodity prices in the early 1970s, as producing nations canceled contracts and hoarded their supplies.[12] Second,

Japan considered overseas investment as a way to reduce the controversy over its large balance-of-payments surpluses that were generated during the 1960s.[13]

Walsh has observed an increased level of Japanese direct involvement in overseas mineral-related projects in a study concerning France, West Germany, and Japan.[12] He discussed the "develop-for-import" (DFI) strategy in which foreign involvement, in the form of production-sharing joint ventures, export credits, and technical aid, is repaid with a share of the project's output. For several selected commodities, Walsh noted a distinct growth in the amount of metals committed under DFI arrangements in 1984 compared to 1978 (Table II-5-2). With respect to Japan, growth was seen in ore and concentrate commitments for chromite, copper, iron ore, iron pellets, manganese, and nickel.

The annual level of Japanese direct foreign investment has been on the rise since the early 1970s (Figure II-5-9). In 1971, the Japanese government began to encourage direct foreign investment by raising the ceiling on allowable investment levels.[16] This action was meant to reduce the huge balance-of-payment surpluses built up by Japan's export-oriented industries during the 1960s.

In 1981, direct foreign investment reached a record high of $8906 million,[17] a 190% increase

TABLE II-5-2 Japanese Mineral Commodity Supply Commitments from DFI Projects Scheduled to Come On-Stream in 1978 and 1984, Compared to Actual Imports in 1980[a] (Thousand Metric Tons)

	Committed 1978	Committed 1984	Actual 1980
Ores and concentrates			
Bauxite	—	—	5,708
Chromite	30	130	950
Copper	290	506	3,104
Iron ore and pellets	11,800	54,400	116,404
Manganese	—	383	1,820
Molybdenum	—	—	20
Nickel	385	3,075	3,950
Tungsten (metric tons)	90	90	3,480
Metals and alloys			
Alumina	—	490	NA
Aluminum	—	620	910
Copper matte/blister	—	235	NA
Nickel matte	—	24	52

[a]—, none; NA not readily available.

Source: J. I. Walsh, "The Growth of Develop-for-Import Projects," *Resources Policy* 8 (December 1982), after Table 4, p. 280. Originally from: actual imports, 1980 — National Import Reports; projected commitments, 1984 — press reports.

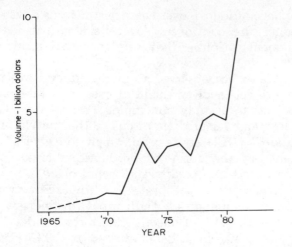

FIGURE II-5-9 Annual Japanese direct foreign investment, 1965 and 1968–1981. (From Japan. Prime Minister's Office, Statistics Bureau. *Japan Statistical Yearbook*.)

from the 1980 level. This figure corresponds to foreign investments in all industrial sectors, including resource development and overseas mineral processing.

The Japanese government encourages overseas exploration subsidies through the use of loans and investment.[2] Mining companies can obtain subsidies for general survey activities, and low-interest loans for more detailed exploration. Loan repayment periods can be 10 or more years' long. Loans are available to cover as much as 50% of the costs of approved exploration projects for copper, lead, zinc, manganese, nickel, bauxite, and chromite ores. In some cases, a loan can represent as much as 70% of a project's cost.

The government allocated Y800 million ($3.0 million) each year during fiscal years 1981 and 1982, for loans and investments to companies operating abroad.[18] However, during this time, low metal prices and the prolonged recession curtailed the interest in overseas resource development.

Several government agencies and programs are involved in the development of overseas mineral resources. These include the Metal Mining Agency of Japan (MMAJ), the Overseas economic Cooperation Fund (OECF), the official development assistance (ODA), and the quasi-governmental Overseas Mineral Resources Development Company (OMRD).

The Metal Mining Agency of Japan (MMAJ) is responsible for most of the minerals activities of the Ministry of International Trade and Industry (MITI).[2] The MMAJ serves as a funding agency for overseas mineral exploration and development. MMAJ can guarantee mining company loans up to a maximum of 80%. Guarantees are provided to those companies who are involved in the develop-

ment of overseas copper, lead, zinc, manganese, uranium, nickel, bauxite, or chromite mines. In 1978, MMAJ held approximately Y34 billion ($144.7 million) in outstanding loan guarantees, for seven development projects around the world.[2] Five of these projects have resulted in the development of mines, whose production is shipped in part to Japan (Table II-2-2). Two of these five are copper mines, Musoshi in Zaire and Mamut in Chile; two are nickel mines, Soroaco in Indonesia and Rio Tuba in the Philippines; and one is a lead and zinc mine, Huanzala in Peru. In 1978, the maximum guarantee limit of the fund was Y52.3 billion.

MMAJ also maintains a political risk guarantee fund for investments in designated "safe" countries.[2] This type of fund is important since the overseas operations of Japan's basic materials industries have already been affected by political changes.[19] For example, construction of the joint-venture Japan-Iran petrochemical complex in Iran has been postponed twice, because of the Iranian revolution in the late 1970s and the early 1980s, and the later Iran-Iraq war. The second-phase construction of the Baoshan Steelworks (China), in which Japan's Nippon Steel is involved, was postponed because of changes in the Chinese government's economic plans. In addition, Japanese businesses in Central America have also been affected to some extent by the political instability in that region.

MMAJ's future involvement in overseas development and mining projects will be expanded.[20] Emphasis will be placed on production feasibility studies of undeveloped mineral resources in Africa.[21] In the same vein, Japan and China will jointly explore for tungsten, tin, iron, and other deposits in China.[22] However, the extent of MMAJ's expansion will depend on the budgetary outlays of the Ministry of Finance, which was planning budgetary cutbacks in early 1983.

The Overseas Economic Cooperation Fund (OECF) was originally set up to distribute foreign aid. However, within Japan, the OECF was considered to be an export promotion agency.[23] Foreign aid was often provided by the OECF in exchange for purchases of Japanese plants or equipment ("tied loans"). In the early 1980s, the Japanese government was attempting to increase the number of untied loans in the ODA.[24]

The OECF is one of the operational arms of Japan's official development assistance (ODA). The ODA includes technical assistance, which is used frequently in the development of overseas resource projects. In 1980, Y60.9 billion ($278 million) was distributed in technical assistance.[25]

The Japanese government has set a goal of doubling the aggregate level of ODA distributed from 1976 to 1980, during the period 1981-1985.[25] Aggregate ODA disbursement from 1976 to 1980 was Y2500 billion ($10.7 billion).

Geographic areas targeted for increased ODA expenditure include African and Latin American countries. Both regions are important suppliers of Japanese strategic minerals. Japan budgeted Y43.7 billion ($186 million) for African countries and Y39 billion ($166 million) for Latin American countries in FY 1980.[24] The total budgeted ODA for FY 1980 was Y840.2 billion ($3.6 billion).

The Japanese government's plan to increase the level of the ODA is part of the Ministry of Foreign Affairs' strategy to protect Japan's national, or economic, security.[24] Some officials expect that a high level of ODA will offset Japan's limited level of defense spending. However, as with the expansion of MMAJ activities, and the increases in defense spending, the planned ODA increases will depend on a consensus among the Ministries of Finance, Foreign Affairs, and International Trade and Industry.

The Overseas Mineral Resources Development Company (OMRD) is also involved in the promotion of overseas mineral projects.[2] Together with the Overseas Economic Cooperation Fund (OECF), the MMAJ holds one-half of the interest in the Overseas Mineral Resources Development Company (OMRD). Private mining and smelting companies hold the other half of the interest in OMRD. Funding is provided by all the participants. The authorized capital for the OMRD is Y7 billion.[2] By 1978, Y5.8 billion had been issued in loans.

OMRD has performed exploration surveys in many countries around the world. Locations of past surveys include Indonesia, Chile, Ecuador, and Mexico.[2] In 1981, exploration surveys were continued in the vicinity of the Frieda River deposit, Papua New Guinea; the Kutcho copper deposit, Canada; the Huanzala lead-zinc mine, Peru; the Katanga copper mine, Zaire, and in southern and western Australia; northern Minas Gerais, Brazil; and Saskatchewan, Canada.[5]

Increased Emphasis on Overseas Processing Capacity

The third option, increased involvement in overseas processing, has been pursued since the late 1970s. Japan's involvement in overseas resource development projects dates back to the early 1970s. The inclusion of this option demonstrates Japan's recognition of the trend toward the development of local processing facilities in mineral-producing nations. The authors of the 1980 Brandt Report noted that mineral-producing, developing nations have not been receiving the "value added" through processing of their raw material commodities.[26] The report cites research performed by the U.N. Secretariat regarding the affects of local semiprocessing on 10 commodities. Using 1975 trade figures, it was estimated that local semiprocessing could add $27 billion annually in gross additional export earnings to the developing nations.

Indeed, the trend among developing countries is to capture this "value added." In mineral-rich countries where raw materials, energy supplies, and labor are abundant, this is a wise strategy. At the same time, developing nations can incorporate the most sophisticated technologies in their new processing plants. This is similar to the industrial development that occurred in Japan following the destruction of that country's industries in World War II.[26] The Japanese were able to take advantage of the newest technology in the construction of their mineral-processing plants. They were in a much better position than the Western nations, whose intact industrial plants would have had to be retooled to incorporate the newest technology. The combination of abundant resources, inexpensive labor, and sophisticated technology will probably result in mineral processing costs that are much lower than those of the developed nations.

In response to this trend, Japan has taken the initiative and has begun investing in overseas processing capacity. The data collected by Walsh[12] indicate the growth in this type of investment (Table II-5-2). The growth in Japan's DFI metal commitments included increased amounts in the semiprocessed form. Whereas Japan did not use this strategy to commit itself to semiprocessed imports in 1978, it expected to be doing so, to a large extent, in 1984.

These data are in agreement with the recent trend in Japanese direct foreign investment. As mentioned earlier, direct foreign investment has increased significantly since the early 1970s. A change in the type of investment has been noted during this time (Figure II-5-10). The composition of direct foreign investment in fiscal years 1978-1980 compared to that for fiscal years 1972-1977 indicates increased investment in the manufacturing and service sectors and a decline in the resource development sector. In other words, more money is being spent in the development of overseas processing centers, such as for primary aluminum (Figure II-4-4), and less is being spent on overseas mine development. Similar overseas processing involvement can be observed in the copper, chromium, silicon, and steel industries (Chapter II-4).

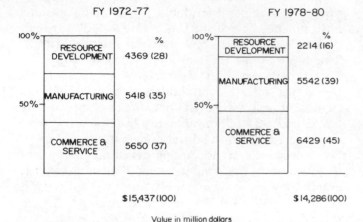

FIGURE II-5-10 Comparison of the composition of Japanese direct foreign investment between FY 1972–1977 and FY 1978–1980. (From Hisao Saida, "High-Level Overseas Investment Becomes Common among Japanese Enterprises." In *Industrial Review of Japan, 1982.* Tokyo: Japan Economic Journal, 1982.)

As a result, these new suppliers will be meeting some of the Japanese demand for steel and nonferrous metals that is currently being met by Japanese companies. While Japan may be facing a decline in some of its major processing industries as a result of entry of new, lower-cost suppliers into the market, it is ensuring that it will still be able to obtain the necessary processed mineral supplies to run other industries that will be responsible for its future growth. The alternative, taking an observer's position, could result in an unexpected loss of resource supplies, and an abrupt decline in the Japanese processing industries.

Japan receives other benefits through its direct involvement in overseas mineral resource development and processing projects. Japanese producers who are involved in overseas projects will be able to maintain their awareness of technological and economic changes in the industry worldwide. At the same time, these producers will be able to maintain their strong bargaining position in negotiations for raw material and processed mineral supplies. In addition, they can increase their processing capacity at overseas plants without increasing Japan's environmental load.

STOCKPILING

Strategic mineral stockpiling in Japan has taken place for a long time, but mostly through the efforts of private interests. Prior to the 1970s, several Japanese steel companies developed private stockpiles for nickel, cobalt, and tungsten.[2] Two additional metals, molybdenum and chromium,

were included in these stockpile plans, but were never stockpiled.[2]

Other private groups have been active in strategic mineral stockpiling. The Metallic Mineral Stockpiling Association maintains stockpiles of copper and zinc metals. In September 1982, the Association held 4780 metric tons of copper metal and 91,570 metric tons of zinc metal.[18] The Light Metal Association maintains stocks of aluminum metal, and it held about 30,000 metric tons in September 1982.[18] In fiscal year 1982, both of these groups were to receive a total of Y20.5 billion ($79 million) in government-guaranteed loans to maintain mineral stockpiles.[18]

Aluminum metal stockpile purchases increased in 1982–1983. The Light Metal Association purchased 7300 metric tons of aluminum metal in early 1982 from local smelters.[27] In addition, Japan's Ministry of International Trade and Industry (MITI) agreed in early 1983 to arrange for a government corporation to purchase 100,000 metric tons of aluminum ingots.[28] The ministry will subsidize the interest payments on the loans the corporation will use to make these stockpile purchases. These purchases are part of the efforts to help the domestic aluminum industry, which was faced with decreased demand and high production costs. At the same time, Japan's security of aluminum metal supply was strengthened, since domestic aluminum smelting capacity had been substantially reduced.

In the early 1980s plans were made to establish a national mineral stockpile. The plan evolved in several steps (Figure II-5-11).

Designated Mineral Commodities	1981 Proposal	1982 Plan	1983 Plan	1983 News Report
Chromium	X	X	X	X
Cobalt	X	X	X	X
Molybdenum	X	X	X	X
Nickel	X	X	X	X
Tungsten	X	X	X	X
Manganese			X	X
Vanadium	X		X	X
Columbium	X		X	
Antimony			X	
Beryllium	X			
Bismuth	X			
Gallium	X			
Germanium	X			
Lithium	X			
Palladium			X	
Platinum			X	
Rare earths	X			
Selenium	X			
Silicon	X			
Strontium			X	
Tantalum	X		X	
Tellurium	X			
Titanium	X			
Zirconium	X			

FIGURE II-5-11 Evolution of the Japanese national stockpile plan, 1981–early 1983.

In 1981, the Primary Product Committee (PPC), an advisory board for the Trade Bureau of MITI, recommended a broad national stockpiling program.[29] The committee's goal was to stockpile critical minerals that had important applications to both the metallurgical and the high-technology industries. Its list of recommended commodities included the following: beryllium, bismuth, gallium, germanium, lithium, columbium, rare earths, selenium, silicon, tantalum, tellurium, titanium, vanadium, zirconium/hafnium, chromium, cobalt, molybdenum, nickel, and tungsten. Japan is strongly dependent on foreign supplies for all these commodities.

The intent of the committee's proposal is similar to that of one made by MITI in 1972. MITI's proposal recommended the stockpiling of several metals, including beryllium, columbium, rare earths, tantalum, and zirconium.[30] The reason given at that time was that the electronics and related high-technology industries depended heavily on imports of these metals. Ferroalloys were excluded from MITI's 1972 list because a private group—the Japan Rare Metals Stockpile—had already been established to stockpile nickel, tungsten, cobalt, and molybdenum. These early stockpile goals were farsighted since Japan was already considering proposals to shift away from its heavy industrial economic base.

Following the PPC's recommendation in 1981, MITI responded by continuing to hold private industry responsible for maintaining a strategic minerals stockpile. A private group, the Strategic Metals Stockpile Association, was established to administer a stockpile of the following five metals: chromium, cobalt, molybdenum, nickel, and tungsten.[31] This association was to receive both private and governmental support. The funds necessary to make the stockpile acquisitions were to be provided through a Y11.5 billion ($44 million) government-guaranteed loan.[32] Other government support includes a subsidy of Y493.2 million ($1.97 million) to pay two-thirds of the interest payment on this loan.[32,33] This stockpile was to be a temporary one and would be replaced by a national stockpile in the following year. Stockpiling acquisition was originally expected to begin in April 1982, but was postponed until December because of preparation delays. Stockpile goals were to be completed by March 1983.

The temporary stockpile goals and acquisitions are given in Table II-5-3. At the end of December 1982, approximately $27.9 million had been spent. For chromium and nickel, the stockpile goals include the ferroalloy forms rather than unprocessed ore. These goals would be consistent with a Japanese plan to reduce domestic ferroalloy production capacity.

A more permanent national stockpile was proposed for 1983, by the Japanese Natural Resources and Energy Agency.[34] This proposal recommended the stockpiling of the following 13 commodities: antimony, chromium, cobalt, columbium, manganese, molybdenum, nickel, palladium, platinum, strontium, tantalum, tungsten, and vanadium. Stockpiling was expected to begin as early as July 1983. Stockpile goals were set to meet a 60-day supply (Table II-5-4). These goals were to be achieved over a five-year period, fiscal years 1983–1987, with the accumulation of an additional 12-day supply each year.[32] The stockpile would be administered by Japan's Metal Mining Agency (MMAJ). The total estimated cost of the stockpile was Y120 billion ($470 million). The cost to meet

TABLE II-5-3 Supply Goals and Acquisitions of the Temporary Japanese National Stockpile, Fiscal Year 1982[a]

Commodity	Form	Quantity (metric tons)	Acquisitions[b] (thousand dollars)	Sources
Chromium	Ferrochrome (at 25% chromium)	13,857	8,800	Philippines, South Africa
Cobalt	NA	45	650	Philippines, Australia
Molybdenum	Concentrates	358	c	NA
Nickel	Metal	1,042		
	Ferronickel	1,470	18,400	New Caledonia, Indonesia, Philippines
	Nickel oxide sinter	159	NA	NA
Tungsten	Concentrates	43	c	NA

[a]NA, not available.
[b]Purchases made in December 1982.
[c]Purchases of tungsten and molybdenum combined were expected to be $2.4 to $2.8 million in January 1983.
Source: Ref. 33.

TABLE II-5-4 Goals of the Japanese National Mineral Stockpile[a] (Metric Tons)

Antimony	1,677	Palladium	7
Chromium	105,104	Platinum	NA
Cobalt	284	Strontium	7,047
Columbium	536	Tantalum	NA
Manganese	101,580	Tungsten	1,021
Molybdenum	3,402	Vanadium	1,070
Nickel	56,744		

[a]Supply goal: 60 days; cost: 120 billion yen ($470 million). NA, not available.

Source: U.S. Department of the Interior, Bureau of Mines, "Japan: Metals Stockpile," Metals and Materials: A Bimonthly Survey, August–September 1982.

the fiscal year 1983 goals was estimated to be Y12.8 billion ($50 million). Funds were to be generated from the sale of 8¼% interest government bonds.

The national stockpile proposal is more comprehensive than earlier stockpile plans. Japanese policymakers recognized the reliance of high-technology industries on imported supplies of cobalt, palladium, platinum, strontium, tantalum, and vanadium. However, it failed to meet the earlier recommendations made by the PPC.

The Japanese stockpile is quite small compared to the U.S. stockpile (Table II-5-5). As a result, completing the goals of this stockpile will be relatively easy, and purchases will have only small effects on world mineral markets. But in the event of a supply interruption longer than two months, Japanese consumers would be in an unfavorable supply position.

The national stockpile plan makes the Japanese government financially and administratively responsible for stockpile purchases. Funding will be subject to the constraints of the government budget. The Liberal Democratic Party (LDP) government's desire to reduce government spending[35] may have already altered these national stockpile plans. In early 1983, the NREA announced a three-month

TABLE II-5-5 Comparison of the National Mineral Stockpiles of the United States and Japan

	Japan	United States
Mineral family categories	13	61
Supply goals	60 days	3 years
Value of stockpile	$470 million	$12.5 billion

Source: U.S. White House Report. National Materials and Minerals Program Plan and Report to Congress, No. 1982-505-002/60. Washington, D.C.: U.S. Government Printing Office, April 1982.

postponement in the startup of the program.[36] In addition, the announcement highlighted only cobalt, nickel, chromium, tungsten, and vanadium as being included in the stockpiling plan. This may indicate that stockpiling of the remaining six metals will be postponed as well.

Therefore, with respect to stockpiling, Japan will rely on new government programs as well as continued private programs. The extent of government stockpiling will depend on the health of the Japanese economy and the elimination of constraints on government spending. However, the level of Japanese stockpiling will limit the use of this strategy to very short term supply interruptions.

DEEP-SEABED MINING

Japanese activity in this area of resource development dates back at least to the early 1970s. At that time, the Geological Survey of Japan, and the National Research Institute for Pollution and Resources (NRIPR) conducted marine mineral exploration and sampling in the central Pacific Ocean, south of Japan and Hawaii[37] (Figure II-5-12).

Another group, the Deep Ocean Minerals Association (DOMA), conducted similar surveys for Japan's Metal Mining Agency. DOMA is a private group formed by 39 companies.[38] Their goal is to develop deep-seabed mineral deposits.

In late 1981, MITI began a manganese module mining research project as part of the National Research and Development Program.[5,39] The goal of this project is to develop a one-third-scale hydraulic mining system by 1989. The government allocated Y20 billion ($85 million) to this program over a period of nine years.

In 1982, a new company was formed to explore and develop deep-seabed mineral resources.[22] This company was capitalized by Y10 million ($42.6 million) and will be supported by the government and private industry. The Metal Mining Agency's manganese module exploration program in the Pacific Ocean, south of Hawaii, will be taken over by this new company (Figure II-5-12). The formation of this company follows the enactment of the Deep Seabed Mining Provisional law, which allows the government and private industries to form joint companies to develop deep-seabed resources.[38]

At one time, the Japanese government was interested in devising a legal framework to assign marine mining rights to interested prospectors. It appeared to favor a bilateral or multilateral agreement, similar to that established between the United States and West Germany (FRG).[40] The U.S.-FRG agreement prohibits commercial mining

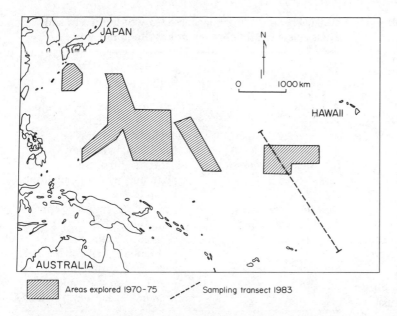

Areas explored 1970-75   Sampling transect 1983

FIGURE II-5-12   Locations of Japanese manganese nodule exploration in the north Pacific Ocean. (From Refs. 37 and 39.)

until 1988, but allows for noncommercial mining necessary for prospecting and testing activities. Two trading groups, Sumitomo and Mitsubishi, are interested in developing deep seabed deposits, once the appropriate legislation is enacted.

Although Japanese representatives did not oppose the adoption of the Law of the Sea Treaty (UNCLOS) in April 1982,[41] they did not sign the treaty during formal ceremonies in December 1982.[42] However, Japan announced that it intended to sign the treaty in the future. Japan eventually signed the treaty in February 1983.[43] The Japanese government apparently believed that an international treaty would be necessary to secure deep-sea mining rights.

The Japanese foreign policy of being friendly with all nations is easily observed in this example. By delaying its approval of the UNCLOS treaty, Japan avoided offending the United States and other industrialized nations who opposed the treaty. At the same time, developing nations, who have depended on Japanese financial and technological assistance in the past, could still anticipate Japan's future approval of the treaty.

CHANGE IN NATIONAL ECONOMIC BASE

New Knowledge- and Service-Intensive Industries

A fifth strategy used by Japan to cope with its high level of foreign resource dependency is that of a change in its national economic base. This change is a switch away from energy- and resource-intensive industries to knowledge- and service-intensive industries.[9]

During the 1960s and 1970s, Japan became a leading importer of mineral ores and concentrates, and leading producer of processed metals and metal products (Figure II-5-13). In the future, Japan will most likely import more of its mineral commodity needs in the form of processed metals, to produce higher-technology goods and services.

The Japanese industries that have been emphasized for future growth include the following five:[44,45,46]

1. *Aerospace:* including artificial satellites and efficient transport aircraft
2. *New energy sources:* nonoil sources, including nuclear and solar
3. *Information processing:* development of new

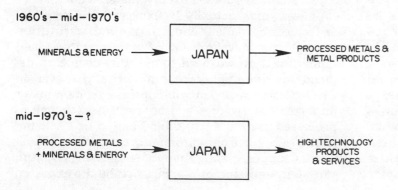

FIGURE II-5-13   Simple "black-box" model of the Japanese economy.

semiconductors, supercomputers, and fiber-optics communication

4. *New basic materials:* ceramics, amorphous metals, carbon and silicon carbide fibers, and probably superalloys

5. *Biotechnology:* gene recombination, cell fusion, and other genetical and biochemical research in medicine, pharmocology, agriculture, and environmental protection

A sixth industry that may be emphasized is that of specialty steels.[47] This may or may not involve a shift away from the production of bulk steel. Capital investment in the specialty steel sector was expected to be Y144.5 billion in fiscal year 1982.[48] This amount is more than double that of the previous year. However, at the same time, Japanese producers may also be interested in supplying semifinished steel to American producers for finishing in the United States.[49] This strategy could involve a supply agreement similar to the proposed 1983 agreement between British Steel and U.S. Steel. According to this agreement, U.S. Steel was to receive several million tons of semifinished British steel for finishing at U.S. Steel's Fairless Steel plant in the United States.[49] If Japanese steelmakers become suppliers of semifinished steel to U.S. producers, Japanese crude steel production may not decline.

In general, a switch to knowledge- and service-intensive industries is meant to satisfy the Japanese government's new industrial policy. The goal of this policy is to target those industries that are highly productive, resource conservative, and whose output captures higher added values.[50]

Effects on Energy and Mineral Requirements

While a switch in economic base may be regarded as a means to conserve both energy and mineral resources, Japan's proposed switch will probably have its greatest initial effects on energy needs. The movement toward knowledge- and service-intensive industries will probably reduce Japanese energy supply requirements.

Japanese industries will continue to be dependent on several mineral commodities, many in the ore and concentrate forms. Selected mineral commodity requirements are listed in Table II-5-6. As indicated, many of these commodities will have to be obtained from foreign sources. In many cases, such as for cobalt, PGM, manganese, chromium, rare earth metals, and nickel, these foreign sources are geographically limited or are politically unstable.

The heavy industries, such as steel and nonferrous metals, will continue to be part of the Japanese economy. The new Japanese industries will require

TABLE II-5-6 Future Japanese Growth Industries and Their Selected Mineral Commodity Requirements

Industry	Selected Mineral Commodity Requirements
Aerospace	Titanium,[a] cobalt,[a] beryllium,[a] nickel,[a] vanadium[a]
New energy sources	Silicon,[a] zirconium,[a] beryllium,[a] uranium[a]
Information processing	Silicon,[a] gallium,[a] germanium,[a] tantalum,[a] copper,[a] platinum-group metals[a]
New basic materials	Clays, petrochemicals,[a] cobalt,[a] manganese,[a] chromium,[a] vanadium,[a] titanium,[a] rare earth metals,[a] silicon[a]
Speciality steels	Manganese,[a] chromium,[a] molybdenum,[a] vanadium,[a] titanium,[a] tungsten,[a] nickel[a]

[a]Indicates continued dependence on foreign sources of supply.

materials produced by these industries, that are still competitive in international markets. For example, Japanese steelmakers have some of the most technologically advanced and economical facilities in the world. This is evidenced by Japan's exports of its steel technology. Although the annual level of Japanese steel production may not increase significantly in the future, because of increased competition from newly industrialized countries, Japan will still require huge amounts of imported iron ore. Other reasons for a gradual switch away from the heavy industrial sector include:

The strong desire to avoid immediate high levels of unemployment

The physical logistics of a switch in economic base are slow

The large export demand for some of Japan's processed mineral products, such as steel and titanium sponge

Therefore, producer-country concerns regarding Japan's decreased need for raw material and energy supplies[51] may actually be exaggerated. Certainly, any successes made in conservation and substitution through the development of new materials, and miniaturization will lead to eventual decreased demand for some mineral commodities. However, in the near future Japan will continue to be a major importer of raw ores and concentrates, as well as processed metals. A continued period of static import levels for iron, chromium, manganese, aluminum, copper, lead, molybdenum, nickel, tin, tungsten, and zinc ores appears to be the expected future outlook. Static import levels have been observed since the early 1970s, when Japan first considered a switch away from its heavy industrial base.

Research and Development

An essential factor in developing new industries is research and development (R&D). Japanese combined public and private R&D spending quadrupled during fiscal years 1971–1981.[52] The net increase in spending from fiscal year 1980 to 1981 was 9.7%, the highest annual net increase in 10 years.

In the past, Japanese R&D goals have emphasized improvements in basic R&D (applied and developmental research) rather than new basic research.[46] Of the total Y5982.4 billion ($25.5 billion) spent on R&D in fiscal year 1981, 14.6% was assigned to basic research, 25.6% to applied research, and 59.8% to developmental research.[52] This total represented 2.3% of Japan's GNP in fiscal year 1981. Future R&D goals will apparently emphasize basic research.[47] The following examples illustrate the government's efforts to meet this new goal:

New energy sources: Y588 billion ($2.5 billion) annually for energy research, about half of which goes to nuclear power research[53]

Information processing: Y10.3 billion ($44 million) grant over three years to the Institute for the New Generation Computer Technology (ICOT), for the development of a fifth-generation computer[54]

New basic materials: Y105.8 billion ($450 million) grant over 10 years for research on high-performance ceramics and plastics[53]

However, the largest amount of R&D spending is made by private companies (Figure II-5-14). For example, private companies accounted for 57 to 60% of total R&D spending during 1975–1980. In 1981, the share of private company R&D spending was higher, reaching 70% of the total R&D spending. Therefore, advances in new industrial technology will depend on increased private company spending for basic research. Government spending in these areas will be beneficial, but may not contribute as much to future technology as could private company spending.

GOVERNMENT–INDUSTRY INTERACTION: MINERAL POLICY

In general, Japanese policymakers play an important role in encouraging the acquisition of strategic mineral supplies for their nation's economy. As noted above, this encouragement is provided through the use of government-sponsored incentives

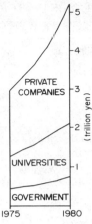

FY	Total	Private Companies	Government Institutions	Universities
1975	2974.6 (100%)	1684.8 (57%)	449.9 (15%)	839.8 (28%)
1976	3320.7 (100%)	1882.2 (57%)	504.4 (15%)	934.0 (28%)
1977	3651.3 (100%)	2109.5 (58%)	529.5 (15%)	1012.3 (28%)
1978	4045.9 (100%)	2291.0 (57%)	603.8 (15%)	1151.1 (28%)
1979	4607.8 (100%)	2664.9 (58%)	660.4 (14%)	1282.5 (28%)
1980	5246.2 (100%)	3142.3 (60%)	763.9 (15%)	1340.1 (26%)
1981	5982.4 (100%)	4187.7 (70%)	NA	NA

NA- not available

FIGURE II-5-14 Japanese expenditures for R&D projects, by originator, 1975–1981. (From Japan. Prime Minister's Office. Cited by: Kohno Teruyuki. "Priority Will Go to New Materials, Information Processing and Biotechnology." In *Industrial Review of Japan, 1982.* Tokyo: Japan Economic Journal, 1982, Table, p. 36.)

and programs. These incentives and programs are its elements of mineral policy.

Several policy elements are discussed above and will not be repeated here. These include loans, loan guarantees, subsidies, government investment in overseas projects, a national stockpiling plan, R&D spending, and the manganese nodule program. Further clarification is provided below regarding other policy elements, such as tax incentives, tariffs, and government aid for depressed basic materials industries.

Tax Policy

Japanese mining companies pay normal corporate tax rates.[2] The Japanese corporate tax rate depends on the method by which income is distributed.[55] For nondividend income of companies whose paid-in capital exceeds Y100,000,000 ($425,000), the tax rate is 42%. For companies with less paid-in capital, a lower tax rate, 32%, is available on their income of less than Y8,000,000 ($34,040). For income distributed as dividends, the tax rate is as follows: 24% on income that does not exceed Y8,000,000, and 32% on income in excess of Y8,000,000.

A summary of Japanese tax measures for fiscal year 1979, and a list of depreciation rates are provided in Tables II-5-7 and II-5-8. A few important

TABLE II-5-7 Summary of Japanese Taxation Measures Affecting the Mineral-Processing Industry, 1980

Tax rate	Estimated aggregate effective tax rate for the year ended 31 March 1980 for corporate, prefectural, municipal, and enterprise tax was 49.47% (Note: Mining companies are exempt from enterprise tax, which was imposed at effective rate of 10.71%)
Deductions	
Investment allowance/ development expenditure	Nil
Relevant infrastructure expenditure	No special deductions
Accelerated depreciation	In addition to normal depreciation, increased initial depreciation allowance of 10–27% in first year of use is available for pollution control and energy-saving equipment, and for facilities and equipment in underdeveloped or industry development areas
Other deductions[a]	No special deductions
Is processing taxed under general tax system or mining tax system?	General
Tax holidays	Nil
Tax credits/exemptions	Industries ascribed "permanently depressed" receive investment tax credit of 10% of cost of new equipment subject to credit ceiling of 20% of tax due (unused credit can be carried over for three years)
Processing incentives	Nil
Regional tax concessions	Nil
Withholding tax	20%

[a]Corporate net income (for taxation purposes) coincides with corporate profit before tax, as shown in the corporation's profit/loss statement.

Source: Australian Trade Development Council. *Minerals Processing.* Canberra: Australian Government Publishing Service, 1980, table, p. 183.

TABLE II-5-8 Description of Japanese Tax Depreciation Rates, 1979

Item	Amount (%)	Method[a]
Buildings (factories)	1.5–4	SL
Plant and equipment	7–17	SL
Vehicles	25	SL

[a]Application of the reduced-balanced method is not clear. SL, straight-line method.

Source: Japan: Ministry of Finance, Study group on Structural Adjustment, *An Outline of Japanese Taxes,* Appendix 7.6, 1979.

deductions are available under the Japanese tax system, such as increased initial depreciation for pollution control and energy-saving investments, and investment tax credits for equipment investments made by "permanently depressed" industries. As mentioned below, these industries have been structurally depressed as a result of the energy cost increases that took place in the 1970s.

Another tax incentive is available to mining companies exploring for metallic minerals, oil, coal, and limestone. This incentive allows a company to reduce its tax burden on a particular year's high profits by earmarking these profits for the next three years of exploration.[2]

Tariffs

Through the period of rapid industrialization during the 1960s and 1970s, Japan's tariff policy was used to protect domestic mineral-processing industries. Ore and concentrate imports were favored over more processed forms of imports, by the presence of duties on the latter.

However, since future imports will feature increased levels of processed mineral forms, the tariff policy will have to be readjusted. Readjustments have already taken place to some extent. The specific reduction in tariffs on aluminum metal imports, as described in Chapter II-4 is an example of the readjustment: However, the quota of unrestricted imports may have to be increased, since aluminum production has apparently declined to a lower level than originally expected by MITI.[28] Higher increases in aluminum metal imports will be necessary to meet domestic demand.

Readjustments in tariff policy will probably be needed with respect to ferroalloy imports. Because of its designation as a structurally depressed industry, ferroalloy production capacity will most likely be cut back. As discussed in Chapter II-4, cutbacks have already occurred in ferrosilicon capacity. Annual imports of ferroalloys have increased since the late 1970s. Any future increases in imports that are necessary to satisfy domestic demand will need to be compensated for in Japanese tariff policy.

Government Aid for Depressed Basic Materials Industries

Several Japanese basic material industries became structurally depressed following the first round of energy price hikes in 1973–1974. High energy prices made these industries uncompetitive in world markets. Many of these industries were already in a state of overcapacity following the rapid industrial growth in Japan in the 1960s and early 1970s. In 1978, specific aid measures were made available through the enactment of the Temporary Law for Stabilization of Specific Depressed Industries.[56] The law originally designated eight industries, aluminum smelting, cardboard, cotton and woolspinning, electric furnace and open-hearth steelmaking, ferrosilicon, fertilizers, shipbuilding, and synthetic fibers, as being structurally depressed. Eventually, 14 industries received this designation.[57]

The law's main objective was to promote the disposal of excess production capacity in those designated industries.[56] Exemptions were granted from Japanese antimonopoly laws to allow for the formation of industry-specific cartels, whose role was to supervise the orderly reduction in excess capacities. These became known as "recession cartels." Many firms sought to eliminate uneconomical capacity, and merged their operations. Tax exemptions and monetary aid were also made available.

Financial aid for the most severely affected members of these industries was supplied by other less affected members, industrial associations, and banks.[58] Government participation in financial assistance schemes was mainly through the provision of loan guarantees. The government's Depressed Industries Guaranty Fund was created to provide loan guarantees up to a total of $440 million.[58] The majority of this fund was held by the Japan Development Bank (80%), and the remainder by private banks (20%). Loan guarantees were beneficial because they could be used as collateral by severely affected companies, in the absence of plants and equipment that had been scrapped. Banks also made other aid available in the form of low-interest loans, loan-repayment moratoriums, and elimination or reduction of the 30% compensating balance normally required for bank loans.[58]

Industry restructuring also affected the employees of these depressed industries. Several of Japan's large and medium-sized firms follow the practice of lifetime employment for their personnel, and these firms strive to avoid layoffs during economic slowdowns. These companies attempt to find alternative employment for their affected employees in subsidiaries, other divisions, affiliates, and even other companies in unrelated industries.[58] For example, during a three-month period in early 1983, the recession-plagued Kawasaki Steel Corp. transferred 80 workers to new jobs at Suzuki Motor Co.[59] If intercompany transfers result in lower salary levels, the original company will make up the wage difference. In the presence of prolonged recession in the early 1980s, the number of intercompany transfers decreased as more companies were affected by the recession.[59]

Companies offer retraining programs to affected employees designated for new jobs. The Japanese government's employment insurance system provides financial support to the designated depressed industries for retraining purposes.[59]

Other measures are used by companies to help maintain their employment levels. These measures include the reduction in overtime and part-time work, and the reduction in salaries, bonuses, and hiring quotas.[58] In addition, older workers are encouraged to take voluntary retirement, often with higher than normal severance pay and job placement assistance.

The 1978 law was scheduled to expire in June 1983. In early 1983, a new law, entitled the Temporary Law for Stabilization of Specific Depressed Industries, was formulated.[56] This law was expected to take effect in July of the same year and to expire in 1988. Seven industries were designated for inclusion: aluminum smelting, chemical fertilizers, chemical fibers, electric furnace steelmaking, ferroalloys, paper and paperboard, and petrochemicals. As with the previous law, this new law affects the energy-intensive basic materials industries.

The new law stresses the structural improvement of these industries.[56] The previous measure involving the use of cartels to eliminate excess capacity is again emphasized.

However, specific emphasis is placed on the concentration of business operations.[56] Business mergers or tie-ups will require the approvals of the Minister of International Trade and Industry and the chairman of the Fair Trade Commission. It might be expected that the elimination of excess capacity will be more selective, being based on the economic position and abilities of the individual companies.

Criticism of MITI's depressed industries laws has come from several parties. Japan's Fair Trade Commission and private-industry members opposed the protection of declining companies with inefficient operations. The FTC also felt that MITI was sidestepping the country's antimonopoly laws by encouraging oligopolistic structure in the basic materials industries through mergers and tie-ups.[60] Opposition from private industry concerned company fears that the government would have too much control over private industry's decisions.[61] Company officials in some industries believed that their capacity reductions and mergers were as much a result of their industry's international uncompetitiveness as they were the result of the government's measures.[56] Opposition from foreign countries, especially the United States, centered on their belief that the laws result in import restrictions.[61] Foreign officials felt that MITI's policy would not be consistent with the OECD-endorsed positive adjustment policy (PAP). PAP is a policy that calls for the elimination of inefficient industrial sectors in a country, and the redirection of efforts toward efficient sectors.

There is still a large amount of controversy regarding the eventual effects of these laws. Private industry and the FTC question whether these laws will strengthen the international competitiveness of the basic material industries. For example, the continued decline of the aluminum smelting industry, despite its designation under the first law and its two stages of capacity cutbacks (Chapter II-4), gives rise to such questions. Private industry may be most interested in ways to lower the cost of power and raw materials in Japan. MITI is apparently considering such action.[57]

At least, these laws are short-term measures to ease the inevitable decline of some of Japan's energy-intensive basic materials industries. In acknowledgment of this goal, Japan's FTC chairman, Osamu Hashiguchi, suggested that it was better to allow the formation of recession cartels so that they could be subject to FTC supervision rather than to have unsupervised illegal cartels.[60]

Perhaps the most significant long-term action taken by the Japanese government is its encouragement of a change in the country's national economic base, as described in the section above. The switch to faster-growing knowledge- and service-intensive industries is intended to help maintain Japan's world-class economy. These proposals indicate that the government is concerned about the decline of its basic materials sector; and that because this sector is the important basis for any national economic transition, its survival is essential.

Analysis

The various elements of Japanese mineral policy are summarized below:

Loans, loan guarantees, and subsidies for exploration, development, environmentally related investments, stockpiling, and industry restructuring

Establishment of tariffs on imported processed minerals and materials

Governmental financial investment in overseas resource-related projects

Tax allowances

Government-level advisory council for minerals-related decisions

National stockpiling plan

Research and development spending

Manganese module development program

These different policy elements can be analyzed for their effects on four basic mineral policy options available to the Japanese (Figure II-5-15). These four options are modified after those that are available to the U.S.:[62]

1. Increased domestic production
2. Substitution, recycling, and conservation
3. Stockpiling
4. Diversification of foreign sources of supply

In the Japanese analysis, maintenance of domestic production is used rather than increased domestic production, because of Japan's limited potential for strategic mineral development.

Maintenance of Domestic Production. Loans, subsidies, and tax incentives for private exploration, as well as the exploration surveys performed by the MMAJ, help to maintain domestic production. Continued exploration has led to new discoveries and to the development of new mines. The availability of low-interest loans as provided by the Emergency Finance Fund has also played a part in the continued production from existing mines.

Substitution, Recycling, and Conservation. Government R&D spending on new basic materials and new energy sources are the policy elements that contribute the most to this option. New sub-

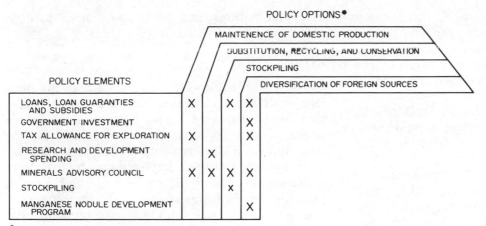

POLICY OPTIONS *

MAINTENENCE OF DOMESTIC PRODUCTION

SUBSTITUTION, RECYCLING, AND CONSERVATION

STOCKPILING

DIVERSIFICATION OF FOREIGN SOURCES

POLICY ELEMENTS				
LOANS, LOAN GUARANTIES AND SUBSIDIES	X		X	X
GOVERNMENT INVESTMENT				X
TAX ALLOWANCE FOR EXPLORATION	X			X
RESEARCH AND DEVELOPMENT SPENDING		X		
MINERALS ADVISORY COUNCIL	X	X	X	X
STOCKPILING			x	
MANGANESE NODULE DEVELOPMENT PROGRAM				X

* Size of X indicates significance of effect on policy option

FIGURE II-5-15   Analysis of Japanese mineral policy.

stitute materials could eventually displace steel and nonferrous metals in some of their present applications. Government spending on energy research will lead to increased conservation of energy minerals. The availability of tax benefits for energy-saving investments also help to increase the level of energy conservation.

However, since government spending represents only a small portion of total R&D spending, the effect on this option may be limited.

Stockpiling. The stockpiling option will be affected slightly by the completion of the government-sponsored national mineral stockpile. Because of its limited size the stockpile will serve only as an alternative to very short term supply interruptions.

The government assistance available to private stockpiling associations for mineral acquisition and maintenance encourages private stockpiling. Private stockpiles are very important to the Japanese economy, since supplies of copper, lead, zinc, and aluminum are held only in these stockpiles. These commodities have not been earmarked for the proposed national stockpile.

Diversification of Foreign Sources of Supply. The availability of loans, investment funds, subsidies, and limited risk guarantees has encouraged Japanese mineral industry involvement in the development of foreign sources of mineral supplies. Japanese mining, manufacturing, and trading companies have participated in ventures located in several regions of the world. Diversification has been stressed in Japanese policy statements, and the results can be seen to some extent in the many import sources of Japanese mineral supplies (Chapter II-4). The activities of the MMAJ and the OMRD have indicated to the mining industry that the government is interested in overseas mineral development.

Japanese tariff policy has also been an incentive to foreign resource development. Further readjust-

ments of this policy may be necessary to encourage the development of overseas processing.

Summary. Japanese mineral policy affects all four of these options. Most notable is its effects on the diversification of foreign sources of supply. This is the most important option to Japanese strategists, because of Japan's limited resource potential. Yet at the same time, the maintenance of domestic production option is also encouraged. Domestic production is important for its beneficial affects on the Japanese economy and technology.

Japanese mineral policy has less significant effects on the substitution, recycling and conservation, and stockpiling options. The lack of encouragement of these options indicates the government's reliance on the private sector in these areas.

FOREIGN-POLICY LINKAGE

Japan's long-time recognition of its resource vulnerability has had a strong effect on its foreign policy. Foreign-policy initiatives have been considered for their impact on natural resources, prior to their initiation. Japan's high import dependence is partly responsible for this country's international position of "being defenseless on all sides,"[63] as described by Saburo Okita, the former Foreign Minister of Japan. Okita notes that Japan's most basic foreign policy option is to be friendly with all nations.

In this manner, Japan maintains a resource-sensitive foreign policy, otherwise known as resource diplomacy. Terutomo Ozawa divides Japanese resource diplomacy into three different stages.[64] These stages could be labeled initiation, promotion, and commitment. They can be used to describe the pattern of Japanese direct involvement in foreign resource projects.

The initiation stage involves a large Japanese firm, or a consortia of firms, and the host government. These parties work to make a preliminary study of the proposed resource project. At the same time, the firm or firms develop a good working relationship with the host government. The firm will emphasize the "economic development" nature of the project. Applicable projects are usually large, and require infrastructure development, such as power or shipping facilities. Overall, each can contribute significantly to the development of the host country.

The second stage, promotion, involves the Japanese firm, the Japanese government, and the host government. The host government promotes the project and seeks financial support from the Japanese government. This promotion involves meetings between high-ranking officials from both governments. Meanwhile, the initiating firm investigates the availability of financing from financial agencies associated with the Japanese government, as well as from other Japanese firms. The second stage usually ends with an unclear commitment from the Japanese government to take part in the project.

Stage three, commitment, involves the initiating firm, co-investors, Japanese government, host government, and probably a government-owned firm in the host country. The host government, unable to obtain a firm commitment from the Japanese government, will discuss the project with other industrialized nations. The Japanese government, which may be delaying its support of the project for any of several reasons, will be pressured by the interested Japanese firms and will eventually provide a firm commitment to the host government.

Two examples cited by Ozawa are the Asahan aluminum project in Indonesia and Mitsui's Bandar petrochemical complex in Iran.[64] Both of these projects started as private ventures before they became national projects.

Another example is observed in Japan's involvement in Panama's Petaquilla porphyry copper deposit.[65] In 1973, a private consortium, composed of Mitsui Mining and Smelting, Dowa Mining, and Mitsubishi Metal Companies, began an exploration program in the area of this deposit. In 1982, this project appeared to be in Ozawa's second, or promotion, stage of development. The Panamanian government approached Japan's MMAJ with a request for partial financing of a $3 million detailed exploration program. Other groups, including Japan's OMRD, Sumitomo Metal Mining, and Furukawa Mining, were also apparently interested in providing funds for the project.

Therefore, Japan's use of resource diplomacy indicates the consistency between its mineral policy and foreign policy. Japanese policymakers apparently recognize their nation's economic dependency on other nations, and have taken steps to cope with that dependence.

CONCLUSION

Japan has been successful in maintaining the security of its strategic mineral supplies. Several strategies have been used to meet its mineral supply needs. The most important one is that of the development of foreign sources of mineral supplies. This strategy has been implemented by long-term contracts, as well as more direct methods such as joint venture investments, loans, and technological assistance. Domestic mineral development also plays a role in Japanese mineral supply security. Continued exploration has led to mineral discoveries which will extend the producing life of many domestic mines. Both private and public stockpiling are important strategies, but only for very short term supply interruptions. The mining of deep-seabed minerals is a strategy that may have significant future use. The Japanese are continuing their research and development in this area. The switch away from energy-intensive industries will reduce Japan's energy supply requirements. However, Japan will continue to be dependent on foreign sources of mineral supplies, many of which are necessary raw materials for its future industries. In the long term, technological advances, especially in the new basic materials industries, could increase the potential for substitution in many mineral-use sectors.

Japan's ability to maintain the security of its mineral supplies has depended on close cooperation between government and private industry. Japanese mineral policy helps to promote this cooperation through the provision of loans, subsidies, tax incentives, and government-financed programs. Continued cooperation will be necessary during the transition to a knowledge- and service-intensive economy, all the while maintaining the security of strategic mineral supplies.

The consistent relationship between Japan's foreign policy and mineral policy, has been a key element in Japan's mineral supply security. Japan's poor domestic mineral potential, and its economy's large mineral supply requirements have helped to mold its resource-sensitive foreign policy. This type of policy will continue to be necessary as Japan develops its future national economy.

REFERENCES

1. Eckes, Alfred E. *The United States and the Global Struggle for Minerals*. Austin, Tex.: University of Texas Press, 1979.
2. Crowson, P. C. F. "The National Mineral Policies of Germany, France and Japan. *Mining Magazine*, June 1980.

3. Suda, Satoru. "Japan." In *Mining Annual Review, 1979.* London: Mining Journal, 1979.

4. U.S. Congress, Office of Technology Assessment. *Management of Fuel and Non-Fuel Minerals in Federal Land.* Washington, D.C.: U.S. Government Printing Office, April 1979.

5. Suda, Satoru. "Japan." In *Mining Annual Review, 1982.* London: Mining Journal, 1982.

6. U.S. Department of the Interior, Bureau of Mines. "The Mineral Industry of Japan." *Minerals Yearbook,* Vol. III, various years.

7. Ridge, John D. *Annotated Bibliographies of Mineral Deposits in Africa, Asia (Exclusive of the USSR) and Australasia.* Oxford: Pergamon Press Ltd., 1976.

8. Suda, Satoru. "Japan." In *Mining Annual Review, 1980,* London: Mining Journal, 1980.

9. Japan. Economic Planning Agency. *New Economic and Social Seven-Year Plan,* August 1979.

10. Cobbe, James. *Governments and Mining Companies in Developing Countries.* Boulder, Colo.: Westview Press, Inc., 1979.

11. Smith, B. "Scarcity and Stability in Mineral Markets: The Role of Long Term Contracts." *The World Economy,* January 1979. Cited by James Walsh. "The Growth of Develop-for-Import Projects." *Resources Policy* 8 (December 1982).

12. Walsh, James. "The Growth of Develop-for-Import Projects." *Resources Policy* 8 (December 1982).

13. Rodrik, Dani. "Managing Resource Dependency: The United States and Japan in the Markets for Copper, Iron Ore and Bauxite." *World Development* 10 (July 1982).

14. "Japan Seeks Price Reduction." *Mining Journal* 299, No. 7688 (24 December 1982).

15. Bayless, Alan. "Japan Wants Coal Producers to Cut Prices and Deliveries As Steel Demand Declines." *The Wall Street Journal,* 21 January 1983.

16. Tsurumi, Yoshi. *The Japanese Are Coming.* Cambridge, Mass.: Ballinger Publishing Co., 1976.

17. "Direct Foreign Investment Hit Record $8.9 Billion." *Japan Economic Journal,* 15 February 1983.

18. U.S. Department of State. Telegram 18,018, Tokyo, October 1982.

19. Ueda, Katsumi. "Overseas Strategy: 'Country Risk' Becomes Serious Problem with International Tensions and Troubles." In *Industrial Review of Japan, 1981.* Tokyo: Japan Economic Journal, 1981.

20. "Main Points of Mining Policies for FY '83." *Japan Metal Journal* 12, No. 31 (2 August 1982).

21. ". . . and Rare Metal Plans." *Mining Journal* 299, No. 7670 (20 August 1982).

22. "Japan: International." *Bulletin of Anglo-Japan Economic Institute,* The Economic and Trade Picture, No. 234 (November–December 1982).

23. Johnson, Chalmers. *MITI and the Japanese Miracle.* Stanford, Calif.: Stanford University Press, 1982.

24. Tamura, Hideo. "Foreign Aid: Japan's ODA Ratio to GNP in 1980 Reached Figure of 0.31 Per Cent." In *Industrial Review of Japan, 1981.* Tokyo: Japan Economic Journal, 1981.

25. Atsumi, Keiko. "Foreign Aid: New Departure in Policy in 1981 Was to Place Emphasis on Doubling ODA." In *Industrial Review of Japan, 1982.* Tokyo: Japan Economic Journal, 1982.

26. Independent Commission on Development Issues. *North-South: A Programme for Survival.* Cambridge, Mass.: The MIT Press, 1980.

27. "Japan to Stockpile Aluminum." *Mining Journal* 298, No. 7651 (9 April 1982).

28. Jameson, Sam. "MITI Policy Irks Japan Smelters: The Rift over Aluminum." *Los Angeles Times,* 21 March 1983.

29. "Stockpile Program." *Mining Journal* 298, No. 7637 (1 January 1982).

30. U.S. Department of the Interior, Bureau of Mines. "The Mineral Industry of Japan." By K. P. Wang. *Minerals Yearbook 1972,* Vol. III. Washington, D.C.: U.S. Government Printing Office, 1972.

31. "Japan Was Scheduled to Begin Stockpiling." *Engineering and Mining Journal* 183 (May 1982).

32. "MITI Decides to Make National Stockpile of 11 Rare Metals from Next July." *Japan Metal Journal,* 30 August 1982.

33. U.S. Department of the Interior, Bureau of Mines. "Japan: Japanese Private Metal Stockpile." *Minerals and Materials: A Bimonthly Survey,* December 1982–January 1983. Washington, D.C.: U.S. Government Printing Office, 1983.

34. "Second Japanese Stockpile Plan." *Metal Bulletin,* No. 6700 (29 June 1982).

35. "Japanese Prime Minister Aims at Easing of Monetary Policy as Way to Aid Growth." *The Wall Street Journal,* 3 January 1983.

36. "Japanese Stockpiling." *Mining Journal* 300, No. 7690.

37. Itoh, F., and T. Hirota. "On Development and Researches of Marine Mineral Resources in Japan." In *World Mining and Metals Technology,* Proceedings of the Joint MMIJ–AIME Meeting, Denver, Colo.: September 1976, Vol. 1. Edited by A. Weiss. New York: American Institute of Mining, Metallurgical, and Petroleum Engineers, 1976.

38. "Japan Approves Deep Sea Law." *Mining Journal* 299, No. 7665 (16 July 1982).

39. Padan, John, U.S. Department of Commerce, National Oceanic and Atmospheric Administration, Ocean Minerals and Energy Division, Washington, D.C. Personal correspondence, 27 April 1983.

40. "Sea-Bed Mining: Japan Proposes Legislation." *Mining Journal* 296, No. 7591 (13 February 1981).

41. "U.N. Approves Sea Law Treaty." *Austin-American Statesman,* 1 May 1982.

42. "Sea-Law Pact Is Signed by 118 Nations." *The Wall Street Journal,* 13 December 1982.

43. "Japan: Government Signs Sea Law Code." *Asia Research Bulletin* 12, No. 10, Report 5 (31 March 1983).

44. Kohno, Teruyuki. "Science and Technology: Priority Will Go to New Materials, Information Processing and Biotechnology." In *Industrial Review of Japan, 1982.* Tokyo: Japan Economic Journal, 1982.

45. Matsumoto, Toshio. "New Basic Materials: Development of New Types Become Matter of Increasing Interest in Industry." In *Industrial Review of Japan, 1982.* Tokyo: Japan Economic Journal, 1982.

46. Shishido, Toshio. "Japanese Industrial Development and Policies for Science and Technology." *Science* 219, No. 4582 (21 January 1983).

47. Starrels, John. "Steel's Stiff Competition." *The Wall Street Journal*, 9 July 1982.

48. Furukawa, Tsukasa. "Japanese Tend to Bigness." *American Metal Market/Metalworking News*, Specialty Steels Supplement, 2 August 1982.

49. Boyle, Thomas. "Forging Ahead: Laid Low by Recession, Big Steel Companies Consider Major Change." *The Wall Street Journal*, 27 May 1983.

50. Schwartz, Lloyd. "Japan Shifts Stress in Industrial Policy." *American Metal Market/Metalworking News*, 20 September 1982.

51. Ward, Peter. "The Onward March of Japan That Threatens to Trample Australia." *The Australian*, 11 March 1983.

52. "Science and Technology Expenditure in FY 1981 Reached Y5.9 trillion." *Japan Economic Journal*, 18 January 1983.

53. Ramsey, Douglas, Kim Willenson, Ayako Doi, and Frank Gibney. "Japan's High-Tech Challenge." *Newsweek*, 9 August 1982.

54. Inaba, Minoru. "In Japan, Computer Development Continues." *American Metal Market/Metalworking News* 91, No. 70 (11 April 1983).

55. Japan. Ministry of Finance, Minister's Secretarial, Research and Planning Division. *Quarterly Bulletin of Financial Statistics*, September 1982.

56. Ishizuka, M. "Industrialists Are Not Convinced Whether It Will Become a Panacea: Law for 'Structural Improvement' of Industries." *Japan Economic Journal*, 22 February 1983.

57. "MITI Studies Ways to Aid Basic Materials Industries: Bogged Down in Serious Situation." *Japan Economic Journal*, 1 September 1981.

58. Boyer, E. "How Japan Manages Declining Industries." *Fortune*, 10 January 1983.

59. Kanabayashi, Masayoshi. "Japan's Recession-Hit Companies Make Complex Arrangements to Avoid Layoffs." *The Wall Street Journal*, 17 February 1983.

60. Murray, Alan. "'In Japan, FTC Is Challenged by Not Only Business But Bureaucrats': Hashiguchi Notes Growth of FTC since 1977." *Japan Economic Journal*, 15 June 1982.

61. Okabe, Naoki, and Hisao Saida. "Rescue Plan for Troubled Industries Draws Suspicion at Home and Abroad: Special Law for Structurally Depressed Industries." *Japan Economic Journal*, 25 January 1983.

62. Calaway, Lee, and W. C. J. van Rensburg. "U.S. Strategic Minerals: Policy Options." *Resources Policy*, 8 (June 1982).

63. Okita, Saburo. "Natural Resource Dependency and Japanese Foreign Policy." *Foreign Affairs* 52 (July 1974).

64. Ozawa, Terutomo. "Japan's New Resource Diplomacy: Government-Backed Group Investment." *Journal of World Trade Law* 14 (January–February, 1980).

65. "Japanese Interested in Developing Petaquilla." *World Mining* 35 (June 1982).

External Factors Affecting the Security of Mineral Supply and Some Policy Responses

External factors are those that usually exist outside the realm of mineral policy. They can affect the security or access of essential strategic mineral supplies. In response to these factors, mineral policy can be modified or can be made consistent with other policy areas, such as trade and foreign policy.

Japanese policymakers have developed policy responses to some of these external factors (Table II-6-1). In response to producer-country efforts to improve their bargaining strength, the Japanese have diversified their sources of mineral supplies. With respect to the existence of trade restrictions and surpluses, Japan has made small improvements in trade relations and has increased its level of foreign investment in industrialized nations. Japan has committed itself to protecting its nearby sea-lanes to guarantee safe transport of its mineral supplies.

These factors and responses are discussed below. In addition, some recommendations are given to improve the responses.

PRODUCER-COUNTRY EFFORTS TO IMPROVE THEIR BARGAINING POSITIONS

Dependence on one foreign country's mineral reserves for the imported supply of any one strategic mineral commodity is not a wise strategy for a mineral-poor nation. If the shipment of imported supplies from that foreign country is interrupted, the importer could face the possible collapse of its domestic industries that require those supplies. Of course, where geologic and geographic factors limit the global distribution of a mineral ore, such as with cobalt and chromium, an importer is limited in its efforts to diversify its sources of supply. However, when a strategic mineral can be found and produced at several localities in the world, an importer should attempt to derive its supplies from as many of these sources as possible. If these attempts are successful, the importer would be less affected by the cutoff of supplies from any one exporter. This is simply an importer's argument for diversification of its foreign sources of mineral supplies. From an exporter's point of view, diversification of export markets is also a desirable goal.

An exporter, whose revenues are derived mainly from the sale of its exported mineral commodities, would rather not depend on any one export market alone. If this dependence exists, an economic recession in the importer's economy can lead to a recession in the exporter's economy. For example, if a bauxite importer cannot sell its processed supplies of aluminum, it will import less bauxite. To protect itself from this situation, an exporter could choose to develop other sectors of its economy to lessen its dependence on its minerals sector. Alternatively, it could attempt to diversify the export markets for its mineral commodities. Diversification may not be possible if there are a limited number of export markets, or if there are numerous sources of supply, for a particular mineral commodity. For the latter example, that of a competi-

TABLE II-6-1 External Factors Affecting the Security of Japanese Mineral Supply, and Some Policy Responses

External factors	Policy responses
Producer country efforts to improve their bargaining positions	Diversification of foreign sources of mineral supplies
	Establishment of fair contract agreements
Trade restrictions and surpluses	Removal of market barriers
	Maintenance of good foreign relations
Mineral transport through sea-lanes	Increase defense spending

tive market, an exporter's best strategy may be to seek cooperation with the other mineral-supplying nations. The type of cooperation considered here is that of the exchange of information regarding export markets. This is different from that of a producers' cartel, whose goal is to control price, production, or marketing of the mineral commodity. With the exclusion of OPEC, producers' cartels in the energy and minerals industries have had limited success in the past.

One example of an information exchange association is that of the Association of Iron Ore Exporting Countries (AIOEC).[1] This association was set up in 1975, and is composed of 11 nations, who together control 75% of the noncommunist iron ore exporting countries. AIOEC's goal is to consider the interests of both the producing and consuming nations. In the words of an Australian delegate, the purpose of AIOEC is to serve as a "forum for consultation and a clearing house for the exchange of information."[1] This could be an especially advantageous association for mineral-exporting nations who trade with Japan.

In mineral trade, the Japanese are usually represented by their large general trading companies (GTCs). The 10 largest GTCs have controlled more than 50% of Japan's external trade.[2] They occupy a strategic position in the Japanese economy, having taken on the responsibility of acquiring imported raw material supplies for their nation. GTCs are quite competitive in the international markets. Several of them often act as a consortium in securing long-term mineral supply contracts, or in making foreign mineral development investments. The business acumen of Japanese companies is well illustrated by Rodrik, in his description of their acquisition of iron ore supplies.[3] In 1964, a consortium of Japan's 10 largest steel companies secured iron ore supplies at a cost that was approximately 20% lower than the costs of imports from other sources.

An association of exporters may eventually have some success in trading with such an aggressive competitor as Japan. Australia has, in fact, sent a trade mission to India and has plans for a future one to Brazil.[4] Part of their discussions probably concern iron ore trade. As mentioned in Chapter II-4, these three countries are the major suppliers of iron ore to Japan. A successful iron ore exporters' association could be in a position that would allow it to delay the shipment of iron ore to Japan until acceptable supply agreements were reached.

However, Japan may be taking the initiative to limit the development of any significant bargaining strength among mineral suppliers. For example, it appears to be playing its major iron ore suppliers against one another. Australian officials have recently expressed concern over declining iron ore exports to Japan. In a study that compared trade figures from fiscal years 1977, 1978, and 1979 to those for 1981, Australia's share in the supply of Japan's iron ore needs dropped from 48% to 45%.[5] Because of the large volume of trade, a small percentage drop can mean a large drop in profits. This is an unfortunate situation for Australia, whose export trade represents a significant portion of its national economy. At the same time, Brazil's share increased from 19% to 23%. These changes could create an unfavorable climate for multilateral iron ore information exchange among producers.

Therefore, in response to any efforts made by mineral exporters to improve their bargaining position, Japan is diversifying its sources of foreign mineral supplies. At the same time, it is reducing its dependence on any one nation for its mineral supplies.

TRADE RESTRICTIONS AND SURPLUSES

Trade restrictions within the Japanese market have led to friction between Japan and its trading partners. The trade surpluses that Japan achieved in the 1970s and 1980s have also contributed to this friction. Since most of Japan's mineral supplies are available only through foreign trade, good trade relations are essential. If efforts are not successful in reducing this trade friction, the Japanese might encounter future interruptions in their resource supplies.

The controversy that erupted in the early 1980s over the presence of nontariff barriers in the Japanese market, concerned the trade of products that are not directly related to energy and mineral trade.[6] However, this controversy has had an indirect effect on energy and mineral trade, because of its bearing on overall trade relations. The following example illustrates this point.

Japan is currently involved in an oil and gas joint-venture exploration project with the USSR, offshore of Sakhalin Island (Figure II-6-1). The Japanese are represented by a government-backed consortium of companies, called the Sakhalin Oil Development Cooperation Company (SODECO).[7] Beginning in 1975, SODECO agreed to provide $100 to $150 million in Japanese credits for exploration equipment. In addition, SODECO agreed to provide some exploration equipment available only from U.S. suppliers.[8] In exchange, SODECO will receive 50% of any crude oil or gas produced offshore for 10 years.

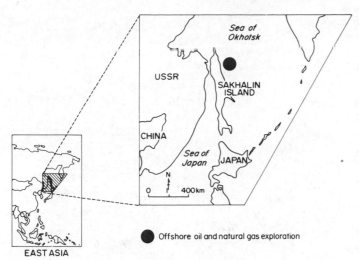

FIGURE II-6 1 Location of Sakhalin Island oil and gas exploration project. (From Ref. 7.)

In 1982, this project was delayed because of the Reagan administration's ban on the use of U.S. technology on Soviet energy projects. SODECO had to proceed with a Soviet-built drilling rig rather than the expected U.S.-equipped Japanese rig.[9] The Soviet rig has a smaller drilling capacity than the original Japanese rig. The 1982 drilling plan for five wells had to be modified to include only two wells. The project is expected to be delayed by at least one year.[10] In early 1983, the Reagan administration eliminated the trade sanctions, and Japan was again able to purchase U.S. oil and gas exploration technology.[11]

President Reagan's ban was initially meant to affect the use of U.S. technology in the development of the Soviet natural gas pipeline to Western Europe. It was later broadened to affect any Soviet oil and gas projects. The Japanese request for an exemption from the ban was not honored. Instead, U.S. government officials insisted on some progress in the U.S.-Japan trade dispute before they would consider an exemption.[12]

Fortunately for Japan, this project is not very significant with regard to Japan's total energy needs. SODECO expects to receive 3 to 3.5 million tons per year of liquefied natural gas (LNG) from the Sakhalin Island projects.[7] This amount represents 12% of the expected Japanese LNG imports in 1985, but less than 1% of Japan's total energy imports in the same year. However, this example illustrates the possible effects of trade friction on the Japanese supply of natural resources.

A more significant problem could arise with Japanese-Brazilian relations. The Japanese have established a strong presence in Brazil. In fiscal year 1980, Japanese direct foreign investment in Brazil amounted to $588 million.[13] Investments have been made in state-owned companies in sev-

eral industries.[14] As mentioned in Chapter II-4, a consortium of Japanese banks provided a multi-million dollar loan for the development of Brazil's huge Carajas iron ore deposits. Japanese plans to diversify their foreign sources of iron ore supplies include imports of ore produced at Carajas.

Japanese trade policies may be affecting their relations with Brazil. Reportedly, some Brazilians believe that Japanese trade barriers have restricted Brazilian exports of oranges and orange juice to Japan.[14] If this belief prevails among government officials, shipments of iron ore supplies could be used as a bargaining wedge to gain better treatment in the Japanese market. The loss of iron ore supplies would result in increased dependence on Japan's other suppliers. This would be a setback to Japan's strategy of diversifying its foreign sources of mineral supplies, and could lead to a short-term interruption in supplies.

Because of the potential indirect effects of trade friction on energy and mineral trade, Japan must eliminate any protectionist barriers that exist in its domestic markets. U.S.-Japan trade negotiations have had little success in addressing this problem. In the early 1980s, the mood in the United States favored the establishment of protectionist measures in the United States to retaliate against the Japanese policy. However, in late 1982, Japanese Prime Minister Yasuhiro Nakasone indicated his interest in improving trade relations.[15,16] The extent of his efforts will probably not be known until later in his administration.

Japan is pursuing a different course of action to counter this trade friction by increasing its investments in the Western industrialized nations. In 1974, Japanese investment in North America (the United States and Canada) and Europe amounted to 22% and 8% of its total foreign investment, respectively.[17] In 1980, Japanese foreign investment in North America rose to 34%, and in Europe, 12%, of total foreign investment. These Western nations are the ones that are affected the most by trade barriers in Japanese domestic markets. Through increased foreign investment in the West, Japan hopes to reduce some of the tension in its foreign trade relations.

For example, in 1982–1983, some Japanese steel producers, including Nippon Kokan,[18] Kobe Steel,[19] and Sumitomo Metal Industries,[19] demonstrated considerable interest in the possible purchases of steelworks in the United States. By investing in U.S. plants, Japanese steelmakers felt that they could limit the effects of any future U.S. steel import quotas, and perhaps even increase their U.S. market share. Sumitomo Metal Industries eventually purchased Tube Turns, Inc., a maker of welded steel pipe joints, from Allegheny International, Inc.[20] However, by mid-1983, Japanese interest appeared to be waning. In one example, Nippon Kokan terminated its negotiations with Ford Motor Co. (U.S.) regarding the purchase of a controlling interest in Ford's Rouge Steel division.[21] It was suggested that the Japanese and the labor union negotiators were unable to agree on a contract that was acceptable to both parties. Another reason given for Nippon Kokan's withdrawal was a possible change in strategy—that it become a supplier of semifinished steel to U.S. producers (Chapter II-4).

MINERAL TRANSPORT THROUGH SEA-LANES

Virtually all of the energy and mineral resources imported by Japan arrive by ship at Japan's many seaports. Transport ships must travel by way of various shipping routes on their way to Japan. Figure II-6-2 illustrates the main Eastern Hemisphere shipping routes in the Pacific and Indian Oceans. These routes are important for the transport of Japanese energy and mineral supplies from North and South America, Australia, Africa, and Asia. Several narrow sea-lanes exist within these shipping routes. A blockage in one or more of these sea-lanes could delay, and in some cases prevent, the delivery of essential energy and mineral supplies to Japan.

One very important sea-lane is the Strait of Malacca, which is located between Malaysia and Indonesia (Figure II-6-3). Significant amounts of Japan's imported supplies of oil, coal, iron, manganese, chromium, cobalt, and vanadium must be transported through this strait. Other sea-lanes important to mineral transport include the Formosa

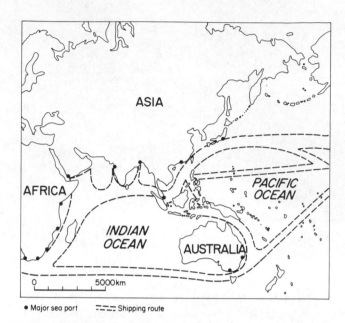

FIGURE II-6-2 Location of Eastern Hemisphere main shipping routes. (From Republic of South Africa, Department of Mining and the Environment. *Atlas of Marine Resources*, 1974.)

and Luzon straits. Open passage through all of these sea-lanes is essential to the security of Japanese energy and mineral supplies.

In response to this factor, Japan could develop a military force capable of protecting its nearby sea-lanes. The development of such a force is extremely controversial. Article 9 of the Japanese Constitution prohibits the development of a military force capable of threatening war as a means of settling international disputes.[22] The Constitution was written at the end of World War II, following the destruction of the Japanese war machine. However, the wording in Article 9 has been interpreted to allow for the development of a self-defense force.[22]

The Japanese Self Defense Force (JSDF) consists of approximately 240,000 personnel,[23] in the three services: the Ground, Maritime, and Air Self Defense Forces (GSDF, MSDF, and ASDF, respectively). In 1981, approximately 160,000 personnel made up the infantry and mechanized divisions of the GSDF. The MSDF consisted of 90 surface ships, 14 submarines, and 100 aircraft. There were 450 aircraft and five ground-to-air missile groups in the ASDF.[24]

The strength of the JSDF has been supplemented in the past by the presence of U.S. military forces in the western Pacific Ocean region. However, U.S. military strength in this region has declined drastically in recent years, and at the same time, the military position of the USSR has been

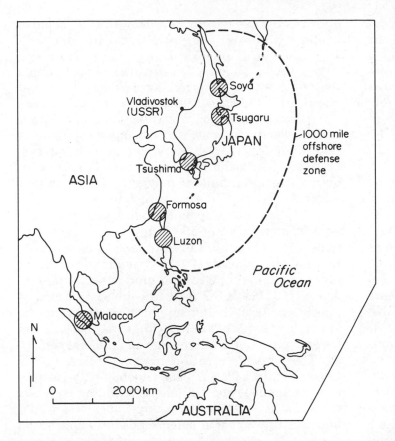

FIGURE II-6-3 Location of western Pacific sea-lanes in the vicinity of Japan; and the Japanese 1000-mile offshore defense zone.

strengthened. Retired U.S. Rear Admiral R. J. Hanks[25] describes the Soviet position as one in which

> Moscow can now operate its submarines in such a way as to menace all shipping lanes from the Strait of Malacca to the Sea of Japan, as well as to govern the maritime approaches to Australia and New Zealand.

The Soviet Pacific fleet consists of 700 ships.[26] In addition, the Soviets have based more than 300 airplanes and 151 submarines at installations along the western Pacific Ocean.[27] In contrast, the U.S. Pacific fleet consists of 350 ships.[26]

The United States has attempted to be the prime counterforce to potential Soviet aggression in the western Pacific. Actions in the Middle East (1980–1981 Iranian hostage crisis), and South Asia (1981 Soviet invasion of Afghanistan) have forced the U.S. military to increase its presence in the Indian Ocean region. It has done this by diverting its naval vessels from the western Pacific Ocean.[25]

Current bilateral agreements bind the United States to continued defense of Japan during times of military conflict.[24] However, the United States may no longer be able to meet this defense goal and still protect nearby sea-lanes. The weakened U.S. military position and the increased Soviet military stance in the western Pacific have led to the increased American pressure on Japan to carry a greater share of the defense burden in the Far East.

The ongoing international discussion about Japanese defense came to a head in 1981–1982, with the United States calling on Japan to improve its military defense position. U.S. pressure stems from its concern about the potential for Soviet aggression in the western Pacific region.

Because of its geographic position, Japan is strategically important to the Western world. The islands of Japan form an arclike chain, enclosing the Sea of Japan and, most important, the Soviet military bases at the port of Vladivostok in the eastern USSR (Figure II-6-3). The defense of Japan has long been recognized by the United States as an important element in the defense of western interests in the western Pacific Ocean region. The Tsushima strait, adjacent to the Asian mainland, and the Soya and Tsugaru straits, between the Japanese islands, are the main sea-lanes that Soviet military vessels would use to travel to and from Vladivostok during a military conflict. Straits between the Kurile Islands, to the north, could also be used for this purpose.

Japan has proposed to improve its military strength by increasing national defense spending. In 1981, Japan spent 0.9% of its GNP for defense purposes. In contrast, the United States spent 6.1% of its GNP on defense.[26] In July 1982, Japan set an upper limit of 7.3% for an increase in defense spending from fiscal year 1982 to fiscal year 1983.[28] However, in early 1983, the Japanese government approved only a 6.5% increase in defense spending.[29] This increase will limit Japanese defense spending to a level that is still less than 1% of its GNP.

In late 1982, Japanese Prime Minister Nakasone was asked about the 1976 Japanese Cabinet's decision to limit defense spending to an amount less than 1% of GNP. Nakasone replied that he would like to respect the past Cabinet's decision, but that it would depend on the performance of the Japanese economy.[23] The prime minister's vague comment might be interpreted to mean that defense spending could represent a larger portion of GNP if the economy does not grow as significantly in fiscal years 1982 and 1983 as it has in previous years.

Former Japanese Prime Minister Zenko Suzuki stated that his country's defense goal was to protect Japanese sea-lanes and the overlying airspace to a distance of 1000 miles from the Japanese shoreline[23] (Figure II-6-3). Even if Japanese military strength is increased to meet this goal, the Japanese will still be unable to protect other important sea-lanes, such as the Strait of Malacca. Japan will continue to depend on a strong U.S. military presence in the western Pacific to protect its vital sea-lanes.

Increased Japanese military spending has been opposed by several groups. Many of the Japanese people, who well remember the Japanese military power of World War II, are in opposition to a large military buildup. The Japanese people purged their government of its military leaders after World War II, and have since maintained pacifist attitudes.

Increased spending would be opposed by other western Pacific nations, whose peoples also fear the return of Japan's military power. Philippines' President Ferdinand Marcos has expressed concern over a Japanese military buildup.[30] Other nations might be expected to express such fears.

It is important to note that Japan imports a significant portion of its energy and mineral supplies from the nations of Southeast Asia. In 1981, this region supplied 22% (347 million barrels) of the crude oil and petroleum products imported by Japan.[31] Table II-6-2 indicates the significance of Southeast Asia for the supply of several strategic minerals. Good relations with these nations are necessary for continued energy and mineral trade.

The present situation in the western Pacific Ocean region calls for Japan to build up its military strength. Proposed defense spending increases may not be large enough. Larger increases would improve the Japanese defense position, and would allow it to achieve its goal of protecting vital sea-lanes within a distance of 1000 miles of its shorelines.

To reduce the fear of a major Japanese military buildup, discussions should be initiated with other western Pacific nations. The goal of improving the defense of all of East Asia should be stressed at such meetings.

TABLE II-6-2 Amounts and Sources of Selected Strategic Mineral Commodities Exported to Japan by Southeast Asian Countries, 1980

Commodity	Amount[a] (metric tons)	Source
Aluminum ore and concentrate	1,875,000 (32)	Indonesia, Malaysia
Chromium ore and concentrate	208,376 (22)	Philippines
Copper ore and concentrate	919,000 (30)	Philippines
Fluorspar	126,512 (26)	Thailand
Nickel ore and concentrate	1,310,000 (33)	Indonesia
Nickel matte	26,763 (52)	Indonesia
Tantalum ore and concentrate	272,000 (88)	Malaysia, Thailand
Tin metal	30,945 (99)	Malaysia, Indonesia, Thailand
Titanium ore and concentrate	173,380 (42)	Malaysia
Sponge iron	30,253 (71)	Indonesia

[a]Numbers in parentheses represent percent of total imports.

Source: U.S. Department of the Interior, Bureau of Mines, "The Mineral Industry of Japan." By John Wu. Minerals Yearbook 1981, Vol. III, preprint. Washington, D.C.: U.S. Government Printing Office, 1982.

If a sufficient military build-up takes place, a part of the U.S. military force could be freed from the burden of defending Japan. The United States could move its naval patrols to other areas. These areas include the Indian Ocean, as well as the sea-lanes beyond the 1000-mile limit, which are so important to the security of Japan's mineral supplies. Without an improved Japanese defense posture, and with the increased size of the U.S. defense burden, vital sea-lane defense may not be possible.

CONCLUSION

In conclusion, several external factors can affect the security of Japanese mineral supply. Preventive measures to reduce the call for improvements in collective bargaining among mineral exporters include the diversification of sources of foreign mineral supplies and the establishment of fair contract agreements. The development of solutions to problems of trade restrictions and surpluses that affect mineral trade relations should be given a high priority. Steps should be taken to improve and to maintain good foreign trade relations. The existence of potential mineral transport chokepoints along various sea-lanes calls for a defensive military buildup.

REFERENCES

1. van Rensburg, W. C. J., and Susan Bambrick. *The Economics of the World's Minerals Industries*. New York: McGraw-Hill Book Company, 1978.
2. Lin, Kuang-Ming, and W. R. Hoskins. "Understanding Japan's International Trading Companies." *Business* 31 (September–October 1981).
3. Rodrik, Doni. "Managing Resource Dependency: The United States and Japan in the Markets for Copper, Iron Ore and Bauxite." *World Development* 10 (July 1982).
4. Bambrick, Susan. National Energy Research and Development Council, Canberra, Australia. Interview, 29 November 1982.
5. "Australia Losing Some Japanese Iron Orders to Brazilian Producers." *The Wall Street Journal*, 30 September 1982.
6. Agress, Philip. "U.S. Calls on Japan to Eliminate Import Barriers." *Business America* 5, No. 6 (22 March 1982).
7. U.S. Congress, Office of Technology Assessment. *Technology and Soviet Energy Availability*. Washington, D.C.: U.S. Government Printing Office, 1981.
8. Lehner, U. C., and Masayoshi Kanabayashi. "Japan Minimizes New U.S. Curbs on Soviet Trade." *The Wall Street Journal*, 23 June 1982.
9. "Sakhalin Project Uses Soviet Drilling Rig." *The Wall Street Journal*, 29 July 1982.
10. "Japan to Continue Soviet Energy Project despite Ban by U.S." *The Wall Street Journal*, 7 July 1982.
11. "Sanctions: And When They Are Only Halfway Up. . . ." *The Economist*, 22 January 1983.
12. "Trade Showdown with Japan." *Newsweek*, 15 May 1982.
13. Saida, Hisao. "High-Level Overseas Investment Becomes Common among Japanese Enterprises." In *Industrial Review of Japan, 1982*. Tokyo: Japan Economic Journal, 1982.
14. Ulman, N. "Brazil's Japanese Population Thrives, Helps Tokyo Forge Ties with Brasilia." *The Wall Street Journal*, 24 September 1982.
15. "Japan May Open Market Further to U.S., but Steps Might Not Satisfy White House." *The Wall Street Journal*, 6 December 1982.
16. "Nakasone Tells Cabinet to Map Out Steps to Further Open Japan to Foreign Goods." *The Wall Street Journal*, 8 December 1982.
17. Beal, Tim. "Economics: Japan: Japanese Bids to Outwit Protectionists." *South*, No. 17 (March 1982).
18. Asano, J. "Nippon Kokan Makes Major Strategy Change: From Defensive to Preemptive Attack." *The Oriental Economist* 50 (November 1982).
19. Nag, Amal. "Japanese Steel Concerns, Fearing Quotas, Weigh Buying Stakes in U.S. Counterparts." *The Wall Street Journal*, 4 June 1982.
20. "Sumitomo Metal Will Take Over U.S. Pipe Joint Firm, Tube Turns." *Japan Economic Journal*, 18 January 1983.
21. Boyle, Thomas. "Forging Ahead: Laid Low by Recession Big Steel Companies Consider Major Change." *The Wall Street Journal*, 27 May 1983.
22. Emmerson, J. K. *Arms, Yen and Power: The Japanese Dilemma*. New York: Dunellen, 1977.
23. Lohr, S. "Japanese Premier Urges an Increase in Armed Strength." *The New York Times*, 28 November 1982.
24. Tsurutani, Taketsugu. *Japanese Policy and East Asian Security*. New York: Praeger Publishers, 1981.
25. Hanks, R. J. *The Pacific Far East: Endangered American Strategic Position*, Special Report. Cambridge, Mass.: Institute for Foreign Policy Analysis, Inc., November 1981.
26. Ulman, N. "U.S. and Japan: Defense Tensions." *The Wall Street Journal*, 1 December 1982.
27. Ulman, N. "U.S. Runs Risk Pressuring Japanese." *The Wall Street Journal*, 8 December 1982.
28. "Japan Plans to Lift Defense Outlays 7.3% in Year Starting in April." *The Wall Street Journal*, 12 July 1982.
29. "Military Spending Still Less than 1% GNP." *Asia Research Bulletin* 12, No. 8 (31 January 1983).
30. Keatley, Robert. "The Far East Worries As U.S. Pressures Japan to Rearm." *The Wall Street Journal*, 29 November 1982.
31. Scanlon, A. F. G., ed. *BP Statistical Review of World Energy 1981*. London: The British Petroleum Company, PLC, 1981.

Comparison with U.S. Policy

An analysis of Japan's strategic mineral dependence has significance for U.S. mineral policy because one can draw parallels between the mineral position of each country. First, both nations are major consumers of mineral and energy supplies. Second, both depend on foreign sources for significant amounts of these supplies. Japan is much more dependent on foreign sources and it has fashioned its foreign and domestic policies with reference to its national resource supply vulnerability.

In contrast, the U.S. policy arena is less sensitive to the country's natural resource supply needs. The Reagan administration's proposed mineral policy does recognize many of the previously neglected minerals-related issues. Its proposals would have their greatest effect on increasing domestic production and on stockpiling. But the policy is lacking in its encouragement of other policy options. Few incentives are provided to industry to develop and diversify foreign sources of mineral supplies. In addition, the relationship between foreign policy and mineral policy goals requires some clarification. The success of U.S. mineral policy depends on such clarification, as well as the cooperation between the government and the minerals industry.

Plans to increase this level of cooperation were reemphasized in 1983, during the initial stages of the 1984 U.S. presidential election campaigns. However, the application and use of such plans remains controversial.

MINERAL POLICY COMPARISON

Both the Japanese and the U.S. mineral policies can be analyzed for their effects on four basic policy options. These options[1] are increased domestic production; substitution, recycling, and conservation; stockpiling; and diversification of foreign sources of supply.

An analysis of Japanese mineral policy, based on a modified version of these policy options, was made in Chapter II-5. In summary, the effects of government incentives and programs provided in the Japanese mineral policy are mainly to encourage the diversification of foreign supply sources (Figure II-7-1). The policy elements also have a strong effect on the maintenance of the domestic production option.

The proposed 1982 U.S. mineral policy[2] has been analyzed in a similar manner in a previous study.[3] The results of that study are summarized in Figure II-7-1. In contrast to the Japanese policy, the elements of the U.S. policy affect mainly the increased domestic production and stockpiling options.

Increased domestic production is encouraged through:

Increased land availability, by way of accelerated land withdrawal reviews, the opening of public lands closed to exploration and mining because of executive or secretarial order, modifications

FIGURE II-7-1 Comparison of the effects of U.S. and Japanese mineral policy elements on basic policy options. [From Priscilla McLeroy, Thomas Braschayko, and W. C. J. van Rensburg. Resources Policy: The 1982 Report to Congress." *Resources Policy* 8 (December 1982).]

POLICY OPTIONS *

POLICY ELEMENTS	INCREASED DOMESTIC PRODUCTION	SUBSTITUTION, RECYCLING, AND CONSERVATION	STOCKPILING	DIVERSIFICATION OF FOREIGN SOURCES
INCREASED LAND AVAILABILITY	X			
LAW OF THE SEA REVISIONS	x			x
IMPROVED MINERALS DATA	x			x
LONG TERM RESEARCH AND DEVELOPMENT SPENDING	X	x		x
REGULATORY REFORM	X			
STOCKPILING	X		X	
CABINET-LEVEL COUNCIL	X	X	X	X

*Size of X indicates significance of effect on policy option

of the Wilderness Act of 1964 and the requirements of critical and strategic mineral impact analyses on lands proposed for withdrawal

Possible improvements in minerals data collection

Long-term research and development spending

Regulatory reform

The possible use of the amended Defense Production Act (DPA) of 1950, Title III provisions, as an alternative to stockpiling

The establishment of a U.S. executive cabinet-level council to coordinate national mineral activities

The stockpiling option is affected to some extent by the Reagan administration's proposals to acquire strategic mineral commodities, although the methods described for planning and acquisition may present some difficulties.[3]

Diversification of Foreign Sources of Supply

The proposed U.S. policy has very little effect on the diversification of foreign supply sources.[3] (Figure II-7-1). The plans to continue the involvement of the International Development Cooperation Agency/Trade and Development Program (IDCA/TDP) in overseas resources and minerals-related infrastructure and transportation studies, and to improve the minerals-related training of the State Department's Regional Resource Officers (RRO), will have only a small effect on this option.

The lack of encouragement for this option is a major weakness of the proposed U.S. mineral policy. The United States maintains a strong dependence on foreign sources for several mineral commodities that are very important to its economy (Table II-7-1). For some of these, such as manganese, cobalt, chromium, and the platinum-group metals, the sources of supply are not politically stable. Encouragement of the diversification

TABLE II-7-1   U.S. Net Import Reliance for Selected Mineral Commodities, 1982

Strontium	100	Ilmenite	72
Columbium	100	Tin	72
Manganese	99	Potash	71
Bauxite and alumina	97	Gallium	61
Cobalt	91	Silver	59
Tantalum	90	Zinc	53
Chromium	88	Tungsten	48
Fluorspar	87	Gold	43
Platinum-group metals	85	Iron ore	36
Nickel	75		

Source: U.S. Department of the Interior, Bureau of Mines, *Mineral Commodity Summaries 1983.*

of supply option should be one of the major goals of U.S. mineral policy.

Export trade legislation passed by the 97th Congress in 1982[4] may be a forerunner of the encouragement needed to promote foreign supply diversification. The new law, known as the Export Trade Act, makes possible the formation of export trading companies (ETCs) in the United States. Specific provisions of the act include:[5]

Antitrust exemptions for ETCs and export trade associations

Tax and financing incentives for ETC formation

Improved guaranteed loan programs from the U.S. Export-Import Bank

Free flow of information from the U.S. Department of Commerce to U.S. manufacturers

Allowance of limited investment in ETCs by financial institutions.

The inspiration for this law may have been the success of the large Japanese general trading companies (GTCs). The GTCs have taken on part of the responsibility of acquiring Japan's necessary strategic mineral supplies.[6] To be as successful in the acquisition of minerals, American ETCs would have to meet other conditions.[5,7] The objective of an ETC would have to be modified to emphasize a balance of imports and exports, rather than a one-sided emphasis on exports. Industrial companies and a central bank would have to be combined together with a modified-ETC to form an organization similar to a Japanese business group. The formation of such a business group may still be restricted by the U.S. antitrust laws. At the same time, U.S. banks and stockholders would have to accept the expected low profit margins of ETCs. In addition, mineral distribution systems and trade channels would all have to be established.

Another factor that might affect the success of American ETCs is historical. One of the reasons why GTCs developed in Japan was the existence of a cultural gap between this country and other foreign countries.[7] Because of Japan's isolation from the rest of the world prior to the Mejii Restoration in the late nineteenth century, an organization was needed to overcome the nation's barriers to outside interaction. The GTCs and their predecessors have filled this gap. A cultural gap between most of the industrialized countries, as well as some of the developing countries, and the United States may not be as wide as in the past Japanese example.

If successful, a modified ETC could be an important asset to link (barter) trade with raw materials producers. The proposed Reagan mineral policy included the consideration of using barter trade to acquire future supplies of strategic minerals.[2]

These shortcomings indicate that an American ETC may not play a significant role in purchasing and/or developing foreign supplies of mineral resources or processed minerals. Changes in the law are necessary if ETCs are to become more involved in acquiring mineral imports.

Foreign Policy

Japan's all-important need to develop foreign sources of strategic mineral and energy supplies has enabled that country to develop a resource-sensitive foreign policy. In the 1960s and early 1970s, Japan maintained an independent position in world affairs. This position allowed it to maintain friendly relations with the many countries of the world that supply it with essential raw materials and energy supplies. As a result, the inward flow of resource supplies was guaranteed.

However, U.S. mineral and foreign-policy goals may not be consistent. Several questions, with respect to foreign policy, remain to be answered. These involve U.S. policy toward strategic mineral-supplying countries with poor human rights records. Other questions involve the development of military strength to protect sea-lanes necessary for mineral transport,[1] and to protect foreign producing and processing facilities during military conflict.[8] Definitive positions regarding U.S. foreign policy for minerals-related issues need to be established.

Increased Domestic Production

With respect to domestic production, the use of the DPA, Title III provisions should be strongly considered. These provisions,[3] such as certificates of necessity for rapid tax amortization, guaranteed purchase levels at guaranteed prices, loans at below-market interest rates, loan guarantees, and research and development grants, are mineral development incentives. Some of these are similar to the elements of Japanese mineral policy already in use. Some types of incentives may be necessary to induce companies to increase their domestic production of strategic minerals, such as cobalt in Missouri and Idaho.[9]

Despite its limited effect on total Japanese strategic mineral supplies, domestic mine production is encouraged by Japanese mineral policy. Japanese production of several strategic minerals such as copper, lead, zinc, manganese, gold, and tungsten has continued for many years. Government-sponsored incentives and programs have been partly re-

sponsible for this continued production. Declining ore grade and declining reserve base are the factors that will eventually halt domestic mine production. Excluding the production of zinc, tungsten, lead, and silver, domestic mine production makes only a small contribution to total Japanese mineral supply requirements. It might be expected that U.S. domestic production of strategic minerals could be stimulated by similar government-sponsored incentives, and to a greater extent than in Japan because of the richer mineral potential of U.S. lands.

GOVERNMENT AND INDUSTRY COOPERATION

Part of the success of Japanese mineral-procurement strategies has been due to the cooperative relationship between government and private industry. The question could be asked whether such cooperation could ever be achieved between U.S. government and private industry. The U.S. history of laissez-faire government, and the differences between Japanese and U.S. culture, would appear to preclude this possibility. The suggestion of government intervention or assistance in industrial activity is considered by many Americans to threaten the established free-market system.

The Reagan administration policy does not favor strong government intervention in private industry. Secretary of Commerce Malcolm Baldridge stated in early 1983 that the private sector is responsible for targeting new industries.[10] He described the government's role as one to create a business environment that is conducive to private firms, in order to develop new technologies.

To encourage joint R&D ventures, the Reagan administration has proposed changes in the antitrust law.[11] These changes would relax the restrictions that normally prohibit the formation of such ventures. An example of a new U.S. R&D partnership is the Microelectronics and Computer Technology Corporation (MCC),[12] which was formed in 1982. MCC is composed of 10 computer and electronics firms, whose goal is to pool their research efforts to respond to foreign competition (mainly Japan) in the information-processing industry. Although no government organizations are among its members, MCC can expect government cooperation in its activities. The U.S. Department of Justice indicated that MCC's formation apparently does not violate the Sherman Antitrust Act. The establishment of MCC is unusual for the U.S. industry, whose members have labeled similar organizations within Japan as detrimental to overall product competition.[13] Also, because of strong competition within the United States, American companies have attempted to maintain the confidentiality of their research data.

Calls for greater U.S. government-industry cooperation were heard frequently in early 1983. Many of the Democratic candidates for the 1984 U.S. presidential election indicated their general agreement with some form of governmental industrial policy. An industrial policy could involve government planning, trade restrictions, tax concessions, loan guarantees, and subsidies to promote U.S. industry. However, industrial policy is still a controversial issue in the American political scene. The debate centers around the question of whether or not the government could pick a winning industry to support.

A poll of high level U.S. executives taken in early 1983 indicated that a consensus for government intervention does not exist.[14] A high proportion of executives indicated that they would let uneconomic industries, perhaps including steel and mineral processing, decline even if this action led to continued structural unemployment and higher prices for basic materials and products. Most of the executives were strongly opposed to the use of subsidies, direct federal loans, and higher tariffs on foreign manufactured goods to save failing industries. Tax incentives and increased access to the tax-exempt debt market were the favored methods to aid declining industries.

In general, it appears that strong government involvement in U.S. industrial planning will not be seen in the near future. Cooperation may be increased through relaxed antitrust limitations on cooperative R&D ventures, and the use of tax breaks for failing industries. However, the level of cooperation between government and industry within Japan will probably not be achieved in the United States.

CONCLUSION

U.S. mineral policy should encourage greater use of the diversification of foreign supply options. To this end, the development of a consistent foreign policy, and the use of foreign mineral development incentives may be helpful. Strong consideration should be given to the use of government-sponsored incentives, such as those provided by the DPA,

Title III provisions, to stimulate domestic strategic mineral production. Cooperation between U.S. government and industry in meeting national mineral security goals may best be achieved through the use of tax incentives and the relaxation of antitrust regulations. However, the level of cooperation between government and industry in Japan may not be obtainable in the United States.

REFERENCES

1. Calaway, Lee, and W. C. J. van Rensburg. "U.S. Strategic Minerals: Policy Options." *Resources Policy* 8 (June 1982).
2. U.S. White House Report. *National Materials and Minerals Program Plan and Report to Congress*, No. 1982-505-002/06. Washington, D.C.: U.S. Government Printing Office, April 1982.
3. McLeroy, Priscilla, Thomas Braschayko, and W. C. J. van Rensburg. "U.S. Resources Policy: The 1982 Report to Congress." *Resources Policy* 8 (December 1982).
4. *Export Trading Company Act of 1982: Export Trading Companies*. U.S. Code, No. 9 (November 1982).
5. Lin, Kuang-Ming, and W. R. Hoskins. "Understanding Japan's International Trading Companies." *Business* 31 (September–October 1981).
6. Young, Alexander. *The Sogo Shosha: Japan's Multinational Trading Companies*. Boulder, Colo.: Westview Press, Inc., 1979.
7. Yoshinara, Kunio. *Sogo Shosha: The Vanguard of the Japanese Economy*. Tokyo: Oxford University Press, 1982.
8. Miller, J. A. "Reagan Administration's Materials and Minerals Policy: A Tremendous Step in the Right Direction." *Alert Letter on the Availability of Raw Materials*, No. 12 (May 1982).
9. "Anschutz Seeks Federal Aid in Developing Cobalt Mine." *Mining Engineering* 34 (November 1982).
10. Baldridge, Malcolm. "Limited R&D Partnerships Can Spur High Technology." *American Metal Market/Metalworking News* 91, No. 50 (14 March 1983).
11. Brooks, Rosanne. "Anti-trust Law Changes Could Ease Cooperative R&D Efforts." *American Metal Market/Metalworking News* 91, No. 65 (4 April 1983).
12. Warsh, David. "The MCC Gamble: Consortium Bets Research Effort Will Triumph Over Knot of Hazards." *Austin-American Statesman*, 17 April 1983.
13. "U.S. Electronics Firms Form Venture to Stem Challenge by Japanese." *The Wall Street Journal*, 26 August 1982.
14. "Business Week/Harris Poll: Executives Split on Saving Smokestack Industries." *Business Week*, 4 April 1983.

CHAPTER II-8

Conclusions

Several conclusions can be made about Japan's strategic mineral position:

1. Japan's mineral industry employs four basic strategies in its attempts to remain internationally competitive:
   a. Resource conservation
   b. Emphasis on higher-value production
   c. Business diversification
   d. Technology export
2. Japanese mineral-processing companies are almost wholly dependent on foreign sources for energy and mineral supplies.
3. The Japanese strategies used to deal with their energy import dependence include:
   a. Diversification of energy sources and energy types, with emphasis on increased imports of coal, natural gas, and uranium
   b. Conservation
   c. Cutbacks in energy-intensive production
   d. Stockpiling
4. The Japanese employ five strategies to manage their high mineral import dependence:
   a. Development of domestic resources
   b. Development of foreign resources, with an emphasis on overseas processing capacity
   c. Strategic mineral stockpiling
   d. Deep-seabed mining
   e. Change from energy-intensive to knowledge- and service-intensive national economy
5. In Japan, there is a generally increasing preference for processed mineral imports. This is especially true for those commodities, whose processing is energy intensive, such as aluminum, ferroalloys, zinc, and nickel.
6. Japanese mineral policy emphasizes two main policy options:
   a. Diversification of foreign sources of strategic minerals
   b. Development of domestic resources
7. Japan's mineral policy is linked strongly to its foreign policy, and its success requires government-industry cooperation.
8. The Japanese mineral processing industry will have to continue to meet existing, and perhaps even stricter future, environmental quality regulations, in order to receive the support of the local citizenry.
9. Foreign diversification of imported strategic mineral supplies, and the employment of fair supply contract agreements with foreign mineral producers, are the most important means of maintaining a secure and steady supply of strategic minerals.
10. Japan must eliminate the barriers within its domestic markets that make foreign entry difficult; good trade relations are necessary for Japan to maintain its mineral export and import markets.
11. Japanese military strength should be built up to a level where it can help protect Far Eastern sea-lanes, which are vital to mineral transport.
12. U.S. mineral policy should promote the increased use of the foreign diversification option. Successful development of foreign sources of supply, especially overseas processing capacity, will require a consistent foreign policy and increased government-industry cooperation.

ECONOMIC INCENTIVE OR SECURITY INCENTIVE?

This chapter has examined the Japanese strategies used to maintain the security of strategic mineral supplies. Inherent in this analysis was a premise that the Japanese are more concerned about maintaining the security of their mineral supplies than with the price competitiveness of their finished products on world markets. However, this may be a false premise.

The latest Japanese long term (1979–1985) social and economic plan makes no mention of the country's vulnerability to imported resource supply cutoffs. Instead, emphasis is placed on the effects of unstable resource prices and foreign country policies on the Japanese economy. This emphasis appears to give priority to the continued international competitiveness of Japanese exports, and considers access to raw materials supplies as a given.

Past Japanese investment in overseas resource projects may have resulted mainly from the fear of import barriers placed upon Japanese exports. In the early 1980s, the same pattern was observed. Japanese investments have increased in the industrialized countries, where trade protectionist sentiments are the strongest, at the expense of developing nations.

Crowson[1] notes that the MMAJ, the Japanese mineral activities arm, was originally formed as an indirect result of outside pressure for reductions in Japanese import tariffs. Policymakers were concerned that domestic products would not be competitive with cheap, duty-free processed metal imports. It was hoped that the agency could develop new sources of raw material supplies that would give domestic producers a price advantage over foreign producers.

In the presence of an oil glut in the early 1980s, Japanese suppliers, who purchase oil from foreign sources, broke expensive long-term supply contracts with established, politically stable foreign sources. Cheap spot-market and short-term contract purchases from politically unstable sources, Iran and Iraq, were increased, at the same time. The reason for this change was a desire on the part of Japanese suppliers to maximize short-term profits. Japan's heavy dependence on imported oil did not appear to prevent this change.

These examples indicate the priority that Japan has placed on the economic attractiveness of its export products. It could be argued that Japan's mineral supply strategies have not grown out of the fear of a mineral crisis or embargo, equivalent to the 1973-1974 OPEC oil embargo, but rather the fear of destabilized economic conditions which would reduce the competitiveness of Japanese exports.

FUTURE CONSIDERATIONS

Over the long term, the increased involvement of Japanese companies in knowledge- and service-intensive industries, such as information processing and new basic materials, and reduced involvement in the capital- and energy-intensive heavy industries, will probably result in reduced domestic raw material requirements. The demand for processed materials will continue to be significant. The latter materials are the building blocks for Japan's future growth industries.

Unit material reductions will probably occur as a result of miniaturization in the information processing and consumer electronics industries, and substitution with new basic materials.

Japanese participation in the development of overseas processing capacity is a good strategy for maintaining future processed strategic material supply. This action will allow Japanese companies to keep abreast of technological advancements in processing, as well as obtain processed material supplies in return for financing and services.

It is especially important for Japan to eliminate any nontariff barriers in its domestic markets that prevent foreign entry. Increasingly, future mineral imports will be in higher-value processed forms. If domestic processing capacity for these commodities is reduced or eliminated, imports will compose the major portion of domestic supplies. Good trade relations will be required to maintain a secure and steady flow of processed material supplies to Japan.

The proposed national mineral stockpile is limited in size and mineral type. In the event of an extensive supply interruption that lasts longer than two months, some Japanese industries could encounter very restrictive materials supply conditions. A larger stockpile which includes other commodities required by Japan's high-technology industries, might be a more responsive measure to maintain the security of mineral supply.

Because of their changing mineral import needs, Japanese national stockpiling plans should make special consideration of the form of the minerals being stockpiled. Form was apparently considered when initial stockpile purchases were made in late 1982. For example, most of the nickel and chromium supplies were purchased in ferroalloy or processed metal forms rather than in raw material forms (Table II-4-8). Private stockpiles already contain aluminum, copper, and zinc supplies in processed metal forms. Processed forms will continue to be required in the national stockpile, especially for those commodities whose domestic processing is limited or nonexistent.

Finally, a Japanese military buildup will remain controversial. Economic and foreign relations considerations, as well as domestic public opposition, will postpone any developments in this area. A military buildup might be successful only if a consensus is reached among nations in East and Southeast Asia that a common defense is needed. Future transport of mineral supplies to that region will require unrestricted sea-lanes.

REFERENCES

1. Crowson. P. C. F. "The National Mineral Policies of Germany, France and Japan." *Mining Magazine*, June 1980.

PART III WESTERN EUROPE

AN EXAMINATION OF THE ENERGY AND NONFUEL MINERAL POLICY OPTIONS FOR WESTERN EUROPE

Western Europe, and more particularly the European Economic Community (EEC),* represents one of the principal concentrations of political and economic power that is highly dependent on imported raw materials to meet consumption demands. In comparison with the United States and Japan, it occupies a middle ground, being more reliant on foreign supplies than the former but less import dependent than the latter. Although dependence of this magnitude on foreign energy and minerals has been viewed in the United States as a serious matter, promoting vulnerability of both industrial and defense capabilities, this has not necessarily been similarly regarded in Western Europe. Trade has historically held a more prominent position in European economies.[1] Prior to the oil embargo of 1973, therefore, heavy reliance on imports raised no undue alarm.

The energy crisis proved a critical factor in persuading Western Europeans that their national security was not necessarily limited to a direct attack on their shores. Rather, dependence on external sources of vital raw materials provided a more subtle, but nonetheless substantial, potential for severe consequences of both a political and economic nature should an interruption in the flow of these supplies occur. Western European countries maintain a key role, not only as NATO's front line of defense, but as major producers and consumers in international markets as well. The responses of the various nations of Western Europe, most particularly the Big 3 (France, West Germany, and the United Kingdom) to the issue of reliability of supply are therefore a matter of some significance. Although energy needs have been most actively addressed, policies, often quite divergent, have been implemented to assure secure sources of nonfuel minerals as well.

ENERGY POLICY ALTERNATIVES

During the period from 1950 to the first oil crisis of 1973–1974, Western Europe underwent a drastic and rapid change in the structure of its energy

usage. Coal's share of total consumption dropped sharply, and oil became the dominant fuel. This substitution can be attributed to many factors, especially comparative prices, as well as ease of extraction, transport, storage, and utilization. Shifts in fuel consumption had several effects. First, the decreasing emphasis on coal was reflected by drops in domestic coal production and in the decline of the European coal industry generally. Second, as a result of increasing oil usage, Western Europe's position altered from relative self-sufficiency in energy to rising dependence on imports, predominantly imported crude oil. Whereas in the 1950s, only 10% of Western Europe's primary energy consumed was imported, this figure was 60% in 1975 and 56% in 1980.[2] A third notable aspect of this period of rapid substitution was the increasing prominence of natural gas, beginning in the 1960s.

The oil price hikes and their subsequent harsh impacts in the Western European nations drove home the need for a reduction in oil imports. Dependence on imported energy supplies, particularly from OPEC, made these supplies vulnerable to manipulation. Certainly, a more secure energy base was needed. To achieve this goal, the nations of Western Europe examined the options of conservation, structural changes in supplies (such as occurred in the switch from coal to oil), diversification of foreign sources of supply, and increasing production from indigenous reserves. It should be added, however, that 10 years after the first oil shock rumbled through the European countries, no clear community-wide approach to the problem of establishing a secure energy supply yet exists. Rather, energy policies remain ad hoc, formulated in response to the demands of individual national needs. This dearth of a cooperative strategy is a major flaw in the composition of the EEC. Nor could it be said that the EEC or other European countries have proven particularly adept at reducing overall import dependence, as illustrated in Table III-1. It appears that the import reliance of some nations has actually increased.

Potential for Increased Domestic Production

Changes in the structure of fuel use since 1960 for the EEC, and for the Big 3, are shown in Figures III-1 to III-4. In each case, oil usage peaked in 1973. The relative proportions of the mix comprising total consumption vary considerably from country to country. Note, for example, that oil represents a larger share of consumption in France

---

*Belgium, Denmark, France, Greece, Ireland, Italy, Luxembourg, the Netherlands, the United Kingdom, and West Germany.

239

TABLE III-1 Selected Western European Import Reliance
for Primary Energy Needs (Percent)

	1973	1980
EEC	61	59
France	81	78
Italy	89	85
Netherlands	26	12
Norway[a]	32	Exporter
Portugal[a]	70	78
United Kingdom	51	6
West Germany	56	58

[a]Non-EEC.
Source: Ref. 2.

than it does in the United Kingdom (UK), West
Germany (FRG), or for the EEC as a whole, whereas coal is more prominent in the UK and the FRG.
These variations are largely in response to differing
national endowments of the various types of fuels.
Western Europe's physical infrastructure is shown
in Figure III-5.

One of the main strategies for securing a reliable energy framework is through increased domestic production. This is often the most emphasized
component of any energy policy, domestic production being normally perceived as "secure." Typically, the primary obstacle is a dearth of sufficient
economically exploitable deposits. It is unusual
that in Western Europe, although it is believed that

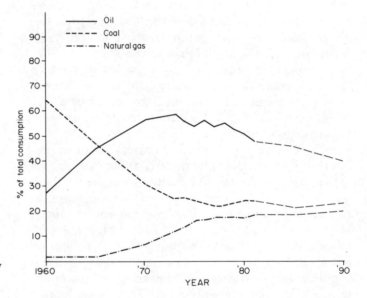

FIGURE III-1 European Economic Community primary
energy consumption, 1960–1990. (From Refs. 13 and 14.)

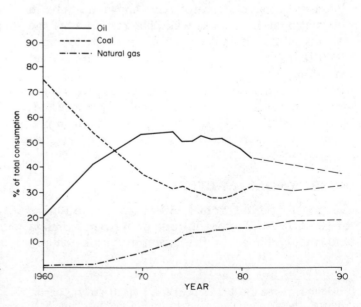

FIGURE III-2 Federal Republic of Germany primary
energy consumption, 1960–1990. (From Refs. 13 and 14.)

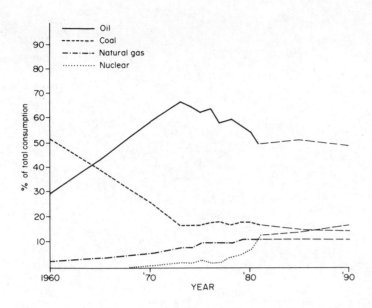

FIGURE III-3 France primary energy consumption, 1960–1990. (From Refs. 13 and 14.)

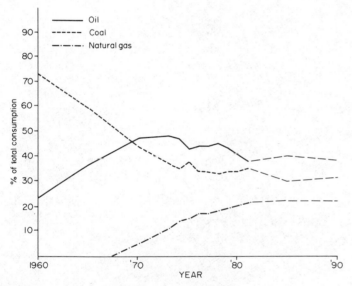

FIGURE III-4 United Kingdom primary energy consumption, 1960–1990. (From Refs. 13 and 14.)

deposits exist, there has plainly been hesitancy in attempting to explore for and develop new reserves, especially of oil and natural gas.

Domestic Energy Base of Western Europe

Coal. The UK and the FRG are by far the major coal producers in the EEC. Estimated reserves and resources are as shown in Table III-2. In 1981, coal production in the FRG increased for the third consecutive year. Bituminous output stood at 88.4 million tons, with 130.6 million tons of lignite.[3] There were 38 coal mines and 20 lignite operations. The Ruhr coalfield accounted for nearly 80% of all coal, the Saar field for another 12%. The Rheinland field provided more than 90% of West Germany's lignite.[4]

In the UK, 211 underground mines and 65 open pits produced 126.4 million tons of coal.[3] This represented a 3% decline from 1980. France, the only other EEC nation with any significant output, produced 18.6 million tons of bituminous coal and 3 million tons of lignite in 1981,[3] the first increment to French coal production in nearly 20 years. The Lorraine Basin, near the German border, is the primary producing region.

During the 1950s and 1960s, as oil became the dominant fuel, the coal industry rapidly lost its competitive edge. This fostered not only state subsidies, import quotas, guaranteed loans, and taxes on fuel oil, but even outright nationalization in France and the UK. The decline of coal is well documented—in the UK, for example, output

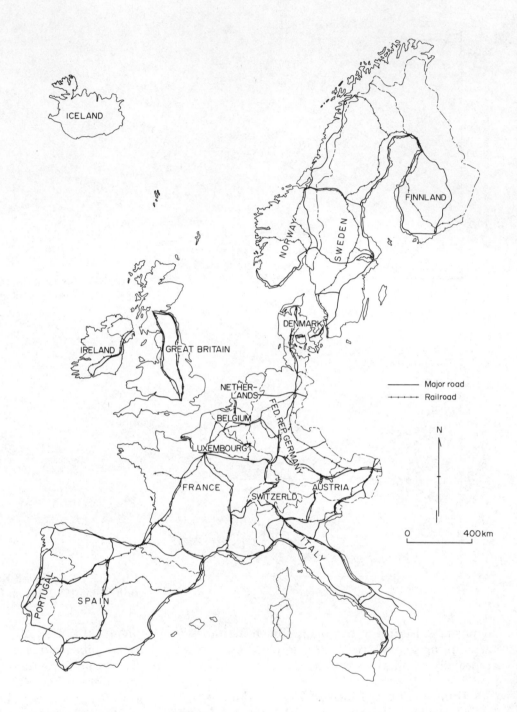

FIGURE III-5 Physical infrastructure in Western Europe.

TABLE III-2 Estimated Coal Reserves and Resources (Billion Metric Tons)

	Reserves	Resources
FRG		
Bituminous coal	24	230
Lignite	35	55
UK	45	160–220

Source: Ref. 3.

dropped nearly 45% from 1957 to 1981. Moreover, the labor force was reduced by 68%.[3,5] In the iron and steel industry, coal's use as a feed has slipped from 90% in 1960 to 62% by the late 1970s.

At present, the Western European coal industry can be characterized as labor-intensive and high-cost. Seams are increasingly deeper and geologically more complex, with many sites depleted. Robinson and Marshall[5] hold nationalization culpable for promoting high British production costs, by allowing the industry to resist reform measures (e.g., the closure of uneconomic mines) since government supports are guaranteed. In addition, the powerful National Union of Mineworkers, and the ever-present strike threat, must also bear some responsibility. This helps account for the poor operating record of many European coal sectors, such as the UK's National Coal Board, which lost almost $120 million in 1980–1981.[3]

Given these facts, as well as continued plans for further mine closures, and the long lead times inherent in bringing new capacity on line, it is unrealistic to expect sudden surges in coal production after years of neglect. Production growth in the FRG and the UK is anticipated to be slow, while output for Western Europe as a whole has been projected to stagnate at 260 million tons per year for the rest of the decade.[6] High operating costs, with slim prospects for rapid output increments, make it evident that if coal's importance in the overall energy balance is to increase, the EEC must acknowledge the necessity of importing a large share of its requirements.[1,2,6] In 1980, the EEC imported 17% of its coal, with major suppliers being the United States, South Africa, Poland, and Australia. These nations, with their large, low-cost surface mines, can provide coal more cheaply to the EEC than can the community's own coal mines. There is thus limited potential for major increases of domestically produced coal to help replace oil.

Oil. Although the British and Norwegian production of oil in the North Sea has greatly alleviated their individual oil supply problems, community-wide dependence on imported crude still exceeds 80%. Indeed, only the UK output has helped improve the position of the whole EEC from what it was in 1973 (see Table III-3). Currently, Norwegian oil reserves are listed at 924 million tons, with the UK's estimated at 1.89 billion tons of oil.[7] The UK is at present self-sufficient in oil. Output in 1982

TABLE III-3   Selected Oil Import Dependency (Percent)

	1973	1980
EEC	98	81
France	98	98
Norway[a]	75	Exporter
Italy	99	98
UK	99	—
FRG	95	96

[a]Non-EEC.

Source: Ref. 2.

was 102 million tons.[7] This ranked the UK eighth in the world, and first in production of light crude. Norway's output in 1982 stood at 24.3 million tons.[7]

It would thus seem that these two nations are capable of significant increases in production, which could greatly reduce the EEC's need to import OPEC oil. Such expectations are intensified by the fact that Norway has only recently permitted large-scale exploration north of the 62nd parallel.[8] However, the potential for marked production increments faces two constraints. First, the UK's production of North Sea oil is expected to peak in the mid-1980s, barring any major new finds.[9] Second, Norway is imposing rigid restrictions controlling the expansion of its offshore petroleum industry, in hopes of mitigating inflationary and other economic impacts on its citizens induced by large oil revenues.[8] Nor can onshore EEC production of oil be anticipated to increase markedly. Evidently, indigenous reserves of oil cannot be relied on to lower import requirements.

Natural Gas. The switch from coal to oil saw a concomitant rapid increase in the useage of natural gas. Gas presently represents 18% of total EEC energy consumption, a rise from 12% in 1973 (see Figure III-1). Western Europe is much less import dependent on gas than oil, with domestic production accounting for more than 70% of community needs. The Netherlands is the main gas producer, supplying the EEC with more than 50% of its natural gas. Dutch reserves are estimated at 1.5 trillion cubic meters.[7] In 1981, the Netherlands produced 84.6 billion cubic meters.[4] Groningen remains the most important gas field. Norway's reserves are listed at 1.6 trillion cubic meters,[7] but Norwegian gas production in 1981 was only 26 billion cubic meters.[4] British gas reserves stand at 720 billion cubic meters;[7] output in 1981 was 52 billion cubic meters.[4] Finally, the FRG produced 19 billion cubic meters of natural gas in 1981. German reserves, however, are small, listed at 289 billion cubic meters.[3]

The large reserves held by Norway, the Netherlands, and the UK would seem to indicate a favorable potential for greatly boosting natural gas output. Unfortunately, in the past all three producers have seemed firmly set against increasing output for export purposes. As indicated above, UK North Sea production will peak within the next few years. In the case of both Norway and the Netherlands, political policies have prevented rapid growth. Norway's decision is motivated primarily by economic considerations, to lessen adverse consequences of an influx of petrodollars. Even if large finds are made above 62° north latitude, production controls will be implemented, and the strategy of cautious development will continue. The Netherlands, in an attempt to preserve its domestic gas endowment, decided in the mid-1970s to decrease its domestic production.[3] This policy of conservation precluded new export contracts. As a result, exports declined as contracts expired. Given the importance of the Netherlands as a supplier of EEC natural gas, this could have presaged an actual drop, rather than increase, in gas supplies for the community from domestic sources. This possibility was underscored by the continuing dearth of a good pipeline network to distribute gas onshore from the North Sea. Now, while Norway has approved plans for an 843-kilometer pipeline from its sector to Emden, West Germany, the UK has dropped state support for its North Sea pipeline project.[3] This indicates that the problem will persist.

In July 1983, a significant policy reversal was announced by the Dutch government. Prompted by declines in energy consumption and the possibility that demand may remain stagnant, the Netherlands has decided to ease its restrictions on domestic gas sales. Although exports will probably be controlled until the 1990s,[10] this still provides one major source of potential increases of indigenous natural gas supplies. Meanwhile, there is a serious risk that Western Europe may become increasingly dependent upon natural gas from the USSR.

Nuclear Energy. Despite staunch opposition, the use of nuclear energy is assuming a vital role in the production of electricity in Western Europe. More than 100 reactors are currently utilized.[2] France remains a world leader in nuclear technology. In June 1983 nearly 44% of all French electricity was produced by nuclear plants, the highest percentage in the world.[11] For the EEC as a whole, nuclear power accounted for 16.5% of its electrical generation in 1981.[2] Western European nations with important nuclear industries include France, Sweden, Finland, Belgium, and Switzerland. In many countries, this sector is only slowly expanding (e.g., Ireland, the UK, and the FRG), while in some nuclear power is prohibited outright (e.g., Norway, Portugal, and Denmark).

One of the goals declared for 1990 by the EEC is the production of from 70 to 75% of electricity generating needs by coal and nuclear power.[12] For this to be achieved, the problems of environmental (and frequently governmental) resistance, as well as huge cost overruns and project delays, must be overcome. Nuclear energy is still projected to be increasingly used as a substitute for oil and gas in European power stations. The magnitude of such substitution, however, remains problematical. At present, the development of the nuclear industry continues below expectations. The exception is in France, where its nuclear power program must actually be scaled back, in response to a "power glut."

Hesitancy to Exploit Domestic Deposits. The perceived failure of the EEC to actively exploit its domestic energy minerals certainly engendered much of the turmoil surrounding the agreement between the USSR and several European nations to import Soviet natural gas. Particularly in the United States, it was believed that external supplies of questionable security were being pursued at the expense of ample—and certainly more secure—domestic supplies. This was more than an attempt at diversification of supply source. It represented a dilatory approach to increasing the potential from indigenous output.

As Hoffman[2] has indicated, substantial deposits remain in Western Europe of oil, natural gas, oil shale, and geothermal power. These, however, have been neglected, while natural gas imports are actually increasing. Several reasons have been cited for such neglect.[2]

1. There is a lack of comprehension concerning the potential inherent in the Western European region, especially offshore. Some experts believe unexplored districts hold vast undiscovered gas fields.
2. Insufficient motivation exists for the energy companies to explore for and develop natural gas. For example, many countries have imposed unfavorable tax structures; taxes as high as 80% over the life of a field are found in Norway. Probably most important, there are very few independent energy companies operating in Western Europe—less than 1% of the total found in the U.S.

Moreover, all underground minerals are state-owned, with exploration rights granted by individual governments. (This problem obviously overflows into the nonfuel minerals sector.) These facts imply virtually no competitive market incentive to explore dynamically for oil and gas.

Despite the probability of significant amounts of undiscovered energy, the ability of Western Europe to markedly increase production from domestic sources, as it strives to reduce its dependence on imported oil, appears limited. The Netherlands' recent reversal of its policy of restricting natural gas production does give some cause for optimism, as does the growth of nuclear power, especially in France. Generally, however, additional approaches must be applied in the drive to obtain a secure energy base.

Conservation

Lowering actual energy consumption is a second strategy that can be employed to reduce the need for importing. Clearly, the trend since 1973 is for lower total energy usage, as illustrated in Table III-4. Only in France does 1981 consumption exceed that of 1973, and this increment is minimal. Unfortunately, the cause of these reductions is not entirely clear. Although some are due to more efficient usage following the large oil price increases of 1979, unquestionably a large share of these decreases can be attributed to economic recession. The problem is to determine how large a share. If the saving in energy consumption is due mostly to poor economic times, improvement in the economy will witness concomitant large jumps in energy usage. The success of efforts to conserve thus must be questioned.

One of the few areas of EEC member cooperation in the energy arena has been achieved in the setting of goals for 1990.[12] These are intended to help reduce reliance on imported oil. The major goals are:

1. To reduce the energy consumption/GNP growth ratio to 0.7
2. To reduce oil consumption to approximately 40% of gross energy consumption
3. To generate 70 to 75% of electricity requirements by use of coal and nuclear energy, as noted previously
4. To encourage the use of renewable sources of energy
5. To pursue an energy-pricing policy aimed at achieving community objectives

TABLE III-4 Total Primary Energy Consumption (Million Tons Oil Equivalent)

	EEC	France	UK	FRG
1960	519	90	170	178
1965	654	116	193	185
1970	838	150	213	236
1973	970	182	233	270
1975	899	168	214	247
1977	952	179	212	268
1979	1019	198	221	287
1981	944	197	195	263

Sources: Refs. 13 and 14.

Unfortunately, these goals are just that—no plan of action for the EEC has been formulated to accomplish them. As a consequence, nations will have differing success rates in meeting the objectives. The UK, for example, has already cut its oil consumption to the 40% figure; France and the FRG are approaching it. Other countries, such as Italy (whose oil share of total consumption is nearly 65%), will undoubtedly fall short of this mark. The failure to substitute toward coal, and resistance to nuclear energy, will cause continued excessive reliance on oil, and thus on imports, for the foreseeable future, for Belgium, Denmark, and Ireland, as well as Italy.[2] Particularly for these countries, neglect of conservation implies a failure to minimize oil imports.

Diversification of Supply Sources

Western Europe has long maintained a much larger diversification of foreign sources of supply for nonfuel minerals than has the United States.[15] Given the restricted potential for increases in domestic energy production, since the oil crises of the 1970s the EEC has looked to augment its external sources of energy. The rationale that motivates the broadening of the import base rests on the theory of diminishing vulnerability—an interruption of the flow of imports from one supplier would have less of a negative impact on the consumer.

However, EEC attempts to diversify sources of oil away from OPEC nations to more politically stable regions have met with little success. In 1976, 64% of the community's oil imports came from the Middle East. By 1980, this hotspot supplied 66% of EEC crude.[2] Identical figures apply for the whole of Western Europe. Again, the importance of OPEC oil varies among countries, accounting for from 39% (Switzerland) to 100% (Luxembourg) of national oil imports. The roster of other current sources, such as the USSR, North Africa, and West Africa, is limited and of uncertain dependability. This indicates continued difficulty in altering significantly the structure of oil imports.

The need for increased external supplies of coal could push import requirements to nearly one-third of Western European coal consumption by 1990.[2] Increasing shipments from the United States, South Africa, and Australia, as well as from Poland and in some cases the USSR, are anticipated. Few other nations (except possibly Colombia) are prepared to exploit the potential of the European coal market, due to insufficient production surge capacity, a lack of export infrastructure, and so on. Finally, the need to import enriched uranium has been reduced greatly through construction of local enrichment plants, although most natural uranium requirements must still be imported.

It would appear, then, that natural gas supplies offer the best opportunities for diversification. The problem, however, is to avoid diversification toward unstable sources. Dependence on supplementary exporters of questionable reliability can in fact directly affect not only the security of the supply of energy, but may also have a major impact on the alternatives available to European governments in the formulation of foreign policy. The primary example here is the controversial Soviet natural gas pipeline to Western Europe. The deal presents itself as a logical choice in lowering reliance on OPEC energy, given the massive USSR natural gas reserves, their proximity to Western European markets, and the existence of extensive pipeline networks from Eastern to Western Europe. The United States has opposed the agreement, citing the security implications of dependence on the USSR for energy, as well as the favorable credit given by Western banks, and the technology transfers that aid the Soviet oil and gas industry generally. Furthermore, the pipeline will earn the Soviets billions of dollars in foreign exchange, much of which could be used to finance military operations. The pipeline is of real importance to the USSR, as the possibility looms that its oil exports are peaking. Western Europe counters opposition with the advantages it perceives accruing from the pipeline; increased jobs, at a time when they are crucial; increased business with and exports to the USSR; and the potential for further financial relations (loans) with the Soviets. Furthermore, while Soviet gas exports to the EEC are increasing, they will remain in the range of 6% of total energy consumption (see Table III-5).

Austria will be the exception. What should be emphasized, however, is that this analysis understates what in some areas is a considerable regional dependence on Soviet natural gas, such as in West Berlin.[17] Despite this and other criticisms, Hoff-

man[2] and the U.S. General Accounting Office[18] conclude that as the EEC struggles to lower imports from OPEC, the pipeline is a rational choice. Plans to import LNG from Algeria, to increase coal imports markedly, and to expand contributions from nuclear energy are all fraught with problems. Despite cutoffs of Soviet energy exports to other nations in the past in response to political events, the Western European importers declare that the USSR's need for hard currency makes such an interruption improbable in this case. In addition, a plan outlining precautionary steps has been announced by the International Energy Agency, to counter any stoppage of Soviet gas exports.[19] These include:

1. Creation of strategic natural gas reserves, similar to the one established for oil
2. Augmentation of "dual firing" in industry, to allow rapid switching from gas to oil or coal
3. Review of commercial contracts, to allow interruptions during emergencies
4. Pursuit of an interconnected European gas and oil pipeline grid, in particular to connect the continent with the UK
5. Increased attention to diversification of supplies, to avoid excessive reliance on a single supplier

Ultimately, the wisdom of the plan to import large volumes of Soviet natural gas remains a moot point.

Related to further fostering of supply diversity is a recent Spanish proposal for a North African gas pipeline, entering Western Europe at the Strait of Gibraltar, to carry Algerian, Libyan, and possibly Moroccan natural gas.[19] Importantly, the present slack demand for gas in Western Europe may presage a situation where some nations have contracted to import more gas than they actually require. This factor certainly influences future searches for more supplies. As attempts continue to maintain a reliable energy mix through diversification, especially of sources of coal, oil, and natural gas, security of supply must remain a concern of high priority for the EEC.

Restructuring of the Energy Balance

Changes in the mix of fuels consumed are limited by the economics of the alternatives. Domestic coal production faces slow growth. Suppliers of imported coal are not without major constraints themselves, as the labor problems in Australia, economic difficulties in Poland, or political unrest in South Africa affirm. Nuclear energy still must overcome outspoken opposition to substantially increase its contribution. Natural gas is anticipated to supply nearly 20% of the EEC's energy requirements by

TABLE III-5   Selected Dependence on Soviet Gas
(Percent of Total Gas and of Total Energy Consumption)

| | 1980 | | 1990 | |
	Gas	Energy	Gas	Energy
France	11	1	30	4
Italy	24	4	35	7
FRG	20	3	30	6
Austria	61	9	82	18

Sources: Refs. 2 and 16.

1990. Much of this growth must come from imported gas, including significant amounts from the USSR. The Dutch decision to rescind conservation policies implies that domestic natural gas production could supply a growing portion of the EEC's energy demands in the 1990s. Some nations, due to a failure to implement a nuclear program and hesitancy in substituting toward coal and natural gas, will remain reliant on predominantly OPEC crude. For those other countries pursuing a vigorous substitution policy, the prospects for increasing coal or natural gas imports in response to rising consumption appear inevitable.

Programs altering the structure of energy usage are greatly influenced by forecasting demand uncertainty. Such risks are inherent in planning. France is currently in the midst of an energy oversupply. This has been caused by a recession-induced drop in demand, at the same time that its ambitious coal-natural gas imports and nuclear power program were under way.[11] As a consequence, France is now forced to reassess its plans, by scaling back reactor construction, reducing its coal requirement projections, and attempting to lower its contractual obligations for Soviet gas. This scenario merely underscores the volatility of the energy economy, and the vulnerability of nations to events beyond their control.

The Need for a Comprehensive EEC Policy

The unfortunate recurring element that endures throughout the period from 1973 to the present day is the absence of a unified, coordinated approach by the members of the EEC toward this common problem of oil import dependence and supply vulnerability. The individual goals of each nation continue to shape energy policy, frequently at the expense of the common good. The failure to subdue nationalistic proclivities in favor of a doctrine of joint cooperation has historically been the major weakness of the EEC. While energy goals have been announced, a plan of action to achieve them is conspicuously missing.[2] Individual policies will probably prove insufficient to solve Western Europe's complex energy situation. Indeed, the diversity of the energy policies of individual nations virtually precludes the formation of any coherent community program. This condition may subsequently prove quite harmful, as negligence in establishing an EEC-wide energy policy could foment a return to the position that was occupied during the 1970s. Without a cessation of provincial thinking, the energy problems of Western Europe, both individually and collectively, will persist.

CONTRASTING POLICIES FOR SECURING MINERAL SUPPLIES

The initial oil crisis of 1973 that induced Western European concern over energy imports also precipitated an examination of the security of supply of other raw materials. A pioneer in industrial development, Western Europe has experienced over 200 years of high consumption. As a consequence, virtually all significant indigenous deposits have been depleted, leaving these nations highly dependent on foreign sources. The current position is depicted in Table III-6. This withering of the domestic reserve base has in no way, however, diminished the need for ever-greater mineral inputs. The region is not only a large consumer, but a major refiner as well, accounting for approximately 20% of world refined metal output. Processing is also heavily reliant on imported materials. The importance of Western Europe to the international availability of metals is thus much greater than its reserve base would indicate, due to the magnitude of its processing industries. Import requirements have produced a major geographical bias, with processing facilities normally locating on the coast to minimize transport costs.

Comparable to energy supplies, the high degree of dependence on imported nonfuel minerals has

TABLE III-6 Net EEC Import Reliance on Selected Minerals as a Percent of Consumption, 1980

Aluminum	28
Asbestos	82
Bauxite and alumina	84
Cadmium	53
Chromium	100
Cobalt	100
Copper	99
Fluorspar	18
Gold	99
Iron ore	79
Lead	70
Manganese	100
Mercury	86
Molybdenum	100
Nickel	100
Phosphate	99
Platinum-group metals	100
Potash	1
Tin	95
Titanium	100
Tungsten	77
Vanadium	100
Zinc	71

Source: U.S. Department of the Interior, Bureau of Mines.

implicit significance for the economic stability of Western Europe as well as for the defense capabilities of this region. Furthermore, the importance of Western European processing indicates that a supply interruption to the continent would have a widespread impact beyond local economies, ultimately influencing the international markets for many metals. Szuprowicz[15] suggests that the past extent of import reliance has prompted a much larger level of supply source diversification than is the case in the United States. In addition, political pragmatism has played a more important role in relations with other nations, as a component of policies designed to lessen the risks of a supply cutoff.

Domestic Reserves

With few exceptions, mineral reserves in all of Western Europe are minimal. In the case of the EEC countries, only for barite and fluorspar do cumulative reserves exceed 7% of the world's total. For all of Western Europe, Spain ranks first in reserves of mercury, accounting for nearly one-fourth of the world's supply.[20] Norway is estimated to contain the second largest reserve base of ilmenite,[20] and Sweden has significant amounts of uranium resources. For the important ferrous and nonferrous minerals, however, Western Europe is clearly a have-not region.

Domestic Production in Western Europe

The EEC is a minor producer of mined minerals. Only in the cases of barite, fluorspar, cadmium, and potash does output exceed 10% of world production. The major mineral product of the EEC continues to be steel.

Iron and Steel. France is by far the main producer of iron ore in the EEC, while Sweden is the main iron producer outside the community. Each produced approximately 20 million tons of ore in 1982.[20] Swedish recoverable iron is believed to be 2.2 billion tons, and for France 1.8 billion.[20] This production is insufficient for the needs of European steelmakers—as Table III-6 indicates, for example, the EEC itself is nearly 80% dependent on imports to meet its iron demand.

Although containing no manganese reserves, and producing less than 3% of the world's iron ore, the European Coal and Steel Community in 1982 was the third leading steel manufacturer in the world. Output in that year was 86 million tons of crude steel, accounting for 17% of the world's total. West German output, at 36 million tons, led the EEC, and was the fourth largest national producer globally.[21] Other important community steelmakers include Italy, France, and the UK.

The steel sector in the EEC is a further example of the continuing problem of member diversity. National steel industries run the gamut of extremes, from the FRG's free-enterprise approach, with a minimum of state aid, to the heavily subsidized, recently nationalized French industry and to the long-time nationalized British Steel Corporation (which may soon be at least partially "denationalized"). Such varying programs impose a formidable constraint on cooperative policies.

Community steel producers faced an especially difficult transition in the period up to 1985. The EEC's goal in that year is a profitable steel sector, which requires no state subsidies. Accordingly, major reorganization and modernization are under way, as well as plant closings, to reduce capacity to approximately 140 million tons.[22] This represents a capacity reduction of nearly 18%. The FRG has been the most vocal opponent of governmental subsidies to the steel industry, which it feels give other industries a competitive advantage.[3] The EEC program of austerity will cause a 50% cut in the labor force by 1985 as well, down from the force of 800,000 of a decade ago. The intent of this program is future viability and an end to the huge losses which presently characterize many industries. (For example, for the year to April 1983, the British Steel Corp. operated at a $591 million loss.[23]) In a time of depression in the international market for steel, and exacerbated by American import barriers to some European steel products, severe cutbacks of this kind may appear extreme, but are undoubtedly inescapable.

Aluminum. Greece is the sole producer of any significant amounts of bauxite; its output in 1982 stood at 3.4 million tons.[20] France, where bauxite was first found and mined, near the village of Baux in 1821, produced 1.8 million tons in 1981.[3] However, its reserves are nearing depletion.

Several EEC nations have notable aluminum metal production. The FRG ranked fifth in the world, producing 780,000 tons in 1982[20] from 10 smelters. France produced 420,000 tons, to rank seventh.[20] The UK production of aluminum stood at 250,000 tons.[20] Italy and the Netherlands also produced significant amounts. Non-EEC countries also had respectable output, particularly Norway, which produced 700,000 tons in 1982 to rank sixth globally, and Spain, whose production of 400,000 tons made it the eighth most important producer internationally.[20]

Tin. British reserves of tin are estimated at 260,000 tons of metal content.[20] Production in 1982 stood at 3000 tons of tin,[20] with Cornwall in the Southwest accounting for almost all output. Recent years have witnessed an increase in tin exploration and development activities. This has been accompanied, however, by a decline in smelting capacity. At present, one smelter remains in operation, at Capper Pass in Yorkshire.

Other Ore Production by the EEC. Ireland is a significant producer of cadmium, and also produces some zinc and barite. The FRG is the second-leading producer of potash. It also produces some barite, cadmium, lead, and zinc. Italy's output of asbestos is notable; it produces barite, fluorspar, and some potash as well. France is an important potash producer. It also produces some barite and fluorspar. The UK also has minor fluorspar production.

Other Significant Production outside the Community. The non-EEC nations of Western Europe, like their community counterparts, contribute, with some exceptions, relatively small amounts to the international supply of mined minerals. Spain is the second leading producer of mercury, as well as a producer of small amounts of fluorspar and zinc. Norway ranks third in ilmenite production, and also produces some vanadium. Finland is an important producer of vanadium, as well as accounting for minor amounts of chromium and cobalt. Por-

tugal and Austria are tungsten producers, and Sweden produces some zinc.

The Importance of Western European Processing. Standing in sharp contrast to Western Europe's minor mine output is the importance of its mineral processing industry. Much of the refined metals are subsequently exported, with a large volume destined for other EEC ports. Table III-7 illustrates the magnitude of Western European processing by comparing the world share of mine output by the Big 3 and by Western European producers as a whole, to the world share of refined metal output by these two groups.

The EEC is the major producer of refined cadmium. It is also the leading zinc metal and refined antimony producing center, and ranks second in production of refined lead and in cobalt metal output. Western European copper refinery capacity is nearly eight times its mine capacity. Although containing no reserves of vanadium, more than 50% of the world's trade in ferrovanadium originates from EEC member nations (Belgium-Luxembourg, France, the FRG, and Italy). If exports from Austria and Norway are included, Western Europe accounts for approximately 75% of the world's ferrovanadium exports.[24] More than one-fourth of

TABLE III-7   Comparison of Percent Share of World Mine Output to Share of World Refined Metal Output

	Big 3	Total Western Europe (Big 3 Plus:)	
Aluminum[a]			
Bauxite	2	6	(Greece)
Refined metal	10	20	(Norway, Spain, Italy, Netherlands)
Copper[b]			
Mine		2	
Refined metal		15	
Ferroalloys[b]			
Manganese, chromite	0	2.4	(Finland, Greece)
Ferroalloys	9 (excludes UK)	22	(Italy, Norway, Spain, Sweden)
Lead[b]			
Mine	2	8	(Belgium, Greece, Ireland, Spain,
Refined metal	12	20	Sweden)
Zinc[b]			
Mine	2 (excludes UK)	14	(Finland, Greenland, Ireland, Italy, Spain, Sweden)
Refined metal	9 (excludes UK)	22	(Belgium, Finland, Italy, Spain)
Petroleum[a]			
Crude	3 (UK)	4	(Norway)
Refined	10	18.5	(Italy, Netherlands, Spain, Belgium, Greece, Sweden)

[a] 1981 data.
[b] 1978 data.

Sources: Refs. 3 and 24.

the world's 38 principal ferrochrome producing companies are West European, and 18 of the 35 ferrovanadium producers are located in Europe. In addition, the UK is one of the few sources of titanium sponge, although its output is small compared to the three world leaders.

Given the heavy dependence on imported raw materials, the recent trend toward more mineral processing before export by less developed countries (LDC) producers could have a major impact on the future health of European refineries. The dual factors of limited environmental regulations and cheap energy could provide producer nations with a comparative cost advantage against which the Europeans cannot compete. The degree of success with which producers meet their goals of maximizing value added before export will largely determine the effects on processors on the continent. As Western Europe negotiates in the future for mineral supplies from LDCs, some concessions to producers in this area appear necessary. To obtain its minerals, the Europeans could well be forced to sacrifice a portion of their processing capacity. In the UK, this capacity has already eroded to a large extent, although it still is able to produce most of its own ferroalloy requirements.

The EEC and Concern over Mineral Supplies

Prior to 1973, although there was some awareness of the EEC's mineral dependency, no supply problem was envisioned. Two developments fostered concern. The first was the oil embargo of 1973. The second was an out-growth of the first. Growing numbers of LDCs, seeing the successful emergence of OPEC, became increasingly willing to exert control over their own natural resources. This was indicated not only by the nationalization of a large group of foreign-owned companies, but also by the rise of producer associations, such as CIPEC and the International Bauxite Association, as well as by the call for the New International Economic Order from the Group of 77.

Actual pessimism regarding future mineral availability was not reached, however, until the release in 1975 of the second major EEC report that dealt with nonfuel minerals. Entitled *The Community's Supplies of Raw Materials*, it concluded that there was inadequate diversification, and called for increasing use of minerals from developed nations, as well as increasing exploration and mining in LDCs.[25] In addition, the report outlined the steps necessary to reduce EEC vulnerability. These included a call for the EEC to provide the guarantee funds for investment in developing countries, where fears of expropriation frequently discouraged such activities. Other conclusions were the need for stockpiling, recycling, substitution, increasing domestic production, and utilization of seabed minerals.

The EEC report, *Relations with the Developing Countries Which Export Raw Materials*, also published in 1975, further addressed the issue of supply security. Its basic findings were the need for a coordinated approach to relations with the LDCs. Apparently, the philosophical harbinger of the Lome Convention Agreements, this paper took a distinctly conciliatory tone toward the developing nations. In this sense, it was a departure from past methods to procure supplies. The emphasis was on present-day concessions to LDCs, which would assure the long-term security of mineral imports.

The Lome Agreements. The Lome Conventions were an answer to the demands of the LDCs, as well as an attempt by the EEC to improve their political and economic relationships with the African, Carribean, and Pacific (APC) countries. The first agreement, signed in 1975, provided free access to EEC markets for ACP products (not necessarily reciprocated) and established an Export Earnings Stabilization Scheme (STABEX).[25] STABEX was an effort to counteract the harsh effects of vacillating commodity prices on producers. Under this plan, sharp decreases in export earnings would be compensated for, to be refunded later interest-free. Importantly, STABEX was limited primarily to agricultural goods, with the exception of iron ore. Mineral price fluctuations, it was felt by the EEC, were due more to cyclical demand changes.

Under the second Lome Convention Agreement (1979), the STABEX plan was expanded to include minerals. Called the Sysmin scheme, it covered cobalt, copper, tin, manganese, bauxite, and phosphate, as well as iron ore.[26] Sysmin provided similar compensation to producers should price drops cause significant harm to export earnings. Other components of the agreement included calls for increased mineral processing in ACPs, more EEC support of exploration, and other subsidization, in hopes of boosting joint mineral development in these producer nations.

This EEC position of cooperation, conciliation, and economic aid to producer nations did not, however, represent a true community consensus. Rather, the diversity inherent in the EEC hinders a joint, coherent minerals policy, just as it does an energy policy. Nations, and particularly the Big 3, follow individual, often antagonistic, paths. The Lome Agreements reflect in great measure the general mineral policies of one country—France. Much of its approach to dealings with the ACPs can be attributed to its heavy reliance on ACP markets and products—much more so than most other EEC members.[26] It is thus certainly in the French interest to avoid EEC-ACP conflict.

The Lome Agreements cannot be regarded as a true mineral policy. The consensus, the willingness by all members, to fund actively the philosophy it espouses, is absent. In addition, the ACP nations, and in fact LDCs generally, are not as important suppliers as they were in the past. Page[27] finds that a clear trend has occurred over recent years, especially for the UK, away from imports from developing nations (particularly Africa) toward other EEC nations. For example, the UK's share of metal imports supplied by LDCs dropped from 37% in 1959 to 13% in 1976, while imports from the EEC rose from 8% to 32%. There has also been a transition away from North America, toward European countries outside the community. For the EEC as a whole, in 1980 developed countries' share of mineral imports was 58.8%, while developing countries contributed 31.5%.[25] The implication here is that, due to political instability, the threat of nationalization, potential for strikes, and so on, the majority of the EEC would rather invest in "safe" projects, such as in Australia, Canada, or the United States. Since the geologic nature of some minerals, including chromium, manganese, and bauxite, limits their geographical distribution, some incentive is required to stimulate investment in LDC producer nations, and then to protect such investment. This underscores the need for a community policy. At present, however, individual national policies prevail.

The Continuing Importance of Supplies from South Africa. Much of the alarm expressed by the EEC over mineral supplies has been prompted by events in Africa. The heavy community dependence on commodities from southern Africa has been one of the primary factors instigating concern over the future flow of minerals. This is exacerbated by the fact that substantial production of many of these minerals in other parts of the world is essentially restricted to Eastern bloc nations. The political and economic volatility of southern Africa has induced a review of American reliance on this region as well. The realization that affairs in southern Africa could have serious consequences for Western Europe's mineral supply led to the French-Belgian incursion into the Shaba Province of Zaire in 1978. During the 1970s, southern Africa accounted for a major share of the community's chromite, manganese, platinum-group metals, vanadium, cobalt, asbestos, and gold. Moreover, indirect imports from

this area (i.e., African minerals processed in another nation before export to the community consumer) made southern Africa even more important. France is the sole EEC member that has minimized dependence on South Africa itself, reflecting perhaps a hesitancy to become overly reliant on Anglo-Saxon suppliers.[26] An interruption of exports from southern Africa to the EEC would affect a wide range of important commodities, and greatly increase prices, with corresponding large impacts on whole economies. For example, a German report concluded that, all things remaining static (i.e., no increases in substitution, recycling, etc.), a 30% shortage in chromite imports over a one-year period could depress GNP by as much as 25%.[28]

To reduce such serious vulnerability, a move at diversification away from the unstable southern Africa area, particularly the nation of South Africa, has been under way for several years, with mixed results. Heavy British and German dependence on South Africa for their chromite and manganese is still extant, as shown in Table III-8. For other minerals, a noticeable decrease in imports from South Africa has occurred. However, an interruption of imports from this region would still prove deleterious to steel and ferroalloy production in the FRG and the UK. This fact, combined with divergent national situations, and the absence of a willingness to surrender additional sovereignty to the EEC, accounts for the dominance of individual national minerals procurement programs.

Mineral Policies of the Big 3: Varying Approaches to Security of Supply

France. The French mineral policy appears the most extensive and consistent in all of Western Europe. In keeping with the high degree of industrial nationalization, the government is quite active in procuring mineral supplies. Concern over raw materials dates back to the late 1950s, when French mining firms were encouraged to develop the mineral potential of the former French colonies to ensure reliable supplies. The heart of policy still continues to be developing countries, particularly the former French African colonial nations.[29] The

TABLE III-8 Percent of Total Requirements for Key Minerals Supplied by South Africa, 1980

	France	FRG	UK
Chromite	31	68[a]	73
Manganese ore	40	67	49

[a]Also, 57% of ferrochrome needs.
Source: Ref. 3.

fostering of good producer-consumer relations remains preeminent. France's large reliance on LDC mineral supplies (nearly 70% of all its mineral imports) justifies this emphasis on avoidance of confrontation with producers. In addition, the bilateral producer-consumer approach taken would seem to obviate cooperation with other consumers, especially those outside the community.[26]

One of the most unique features of French policy revolves around the existence of the Bureau de Recherches Géologiques et Minières (BRGM). This is a state-owned body, functioning as two entities. It is a research and exploration institution, and it is a commercial enterprise concerned with development. Its purpose is to increase the domestic reserve base in France, as well as to extend financial and technical assistance to the development of minerals in selected foreign nations. Its joint ventures include operations involving phosphate in Senegal, bauxite in Cameroon and Guinea, copper in Morocco, iron ore in Mauritania and Guinea, and manganese in Gabon.[26] The BRGM has also been active in other countries, such as New Caledonia, French Guiana, Zaire, Peru, Australia, the Ivory Coast, and Upper Volta.[28]

Domestically, the BRGM has been involved in a detailed exploration program since 1975, with supporting funding increasing threefold from 1975 to 1979.[30] Some success has been achieved in the delineation of new deposits. When the BRGM does make a discovery, the private sector is expected to develop it. The intent is to increase domestic production, in hopes of expanding self-sufficiency for some commodities. In addition to exploration assistance, mineral development is encouraged by the government through direct investment from both French companies and the BRGM. Favorable loans are available in the cases of copper and uranium. Insurance schemes also exist to offer political risk protection.

One of the goals of French policy is privileged access to minerals from French companies. This is frequently a component of defense treaties or military cooperation agreements with the former colonies. (Although it was felt that French foreign policy in Africa would not be as militarily active under Mitterrand as in past years, the deployment of French forces in Chad again emphasizes an ability to intervene, however reticently, when such action is deemed crucial.) Another method for seeking privileged access is through the establishment of bilateral commerical agreements. BRGM activities in other nations also promote preferential treatment. The fact that the BRGM is willing to help fund projects overseas, and its readiness to participate in major exploration programs, helps create goodwill, and thus an environment encouraging stable long-term relations with producers. This helps assure mineral supply continuity.

Another aim of French minerals policy is recycling. In 1975, a broad series of incentives and subsidies aimed at allowing industry to economize on import requirements was implemented. The target by 1985 is a savings of more than 5 billion francs in imports yearly. Recycling efforts are directed toward more efficiency in product design, in processing, in mineral production, and in scrap and by-product recovery.[30]

France was the first Western Eruopean nation to establish a strategic stockpile. The decision was based on an assessment of import vulnerability, which categorized minerals according to perceived degree of risk. The findings were as shown in Table III-9. As this listing emphasizes, France does not necessarily view supplies from Western developed nations as secure.[28]

The intent of the stockpile is to provide supplies during an emergency of two months' duration. Completed stockpile costs have increased, from an original 250 million francs in 1975, to the present 4 to 5 billion. Administering body is the Caisse Française de Matières Premières (French Fund for

TABLE III-9   Import Vulnerability of Minerals

Very high risk	High risk	Medium risk	Low risk
Cobalt	Antimony	Bauxite	Iron ore
Diamonds	Copper	Chromium	Lead
Phosphate	Manganese	Tin	Nickel
Platinum metals	Molybdenum		Zinc
Silver			
Titanium			
Vanadium			
Zirconium			

Sources: Refs. 26 and 28.

Raw Materials). Composition of the stockpile remains secret.[31]

West Germany. In contrast to France's practice of active governmental intervention to assure the flow of mineral supplies, the FRG is the major proponent of a free-market approach, both domestically and overseas. German mineral policy is predicated on the belief that the private sector, rather than the state, is responsible for mineral procurement, in what can best be termed a system of laissez-faire. Warnecke[28] attributes the emphasis on market forces to the loss of German colonies and overseas mining interests, which was mandated by the Versailles Treaty. The FRG thus has continued a course, which was originally forced upon it by the Allies. Crowson[30] associated the free-market approach with the strength of the German economy. The GNP of West Germany is the largest in the EEC. This places the FRG in a superior position allowing it to outbid weaker nations for important raw materials. Hoffman[32] believes that good diversification of the sources of supply, with comparatively minor levels of direct investment overseas, allows the Germans great flexibility in responding to changing economic or political environments. In addition, their willingness to enter into long-term contracts often provides German companies with an advantage, as opposed to, for example, British firms.

These factors all contribute to the German opposition to governmental intervention in international commodity markets. In fact, it is felt that intervention promotes overproduction and inefficiency, ultimately hindering the FRG's own flexibility in procuring low-cost materials. Conequently, the state is not directly involved in contract negotiations, nor does it buy, sell, or stockpile.[32] The German government does offer some support, however. During the 1970s, state subsidization of overseas exploration increased substantially. Normally, 50% of the cost of an exploration program is funded, provided there are assurances that the FRG will be a recipient of production. The value in subsidization of this type is that exploration is encouraged during times of economic downturns, when the private sector is usually hesitant to fund it.

To stimulate overseas investment, especially in LDCs, investment guarantees, favorable loans, and minimum rate of return guarantees are available to cover a portion of foreign financing. Credits are offered by commercial banks, as well as by the Kreditanstalt für Wiederaufbau (KFW). Crowson[30] views these investment guarantee programs as particularly effective schemes for encouraging mineral

development overseas at a minimum cost. The state also provides technical assistance through the Bundesanstalt für Gewisschenschaften und Rohstoffe, or BGR.

These financial and technical supports have been useful in promoting diversification. Accordingly, the FRG government has been active in projects in Guinea, Botswana, Liberia, Niger, Brazil, Indonesia, Morocco, Thailand, and Mauritania. Domestic development of lead, zinc, fluorspar, and barite deposits has also benefited. Changes in the continental distribution of supported exploration projects during the 1970s reflect especially a shift away from Africa as diversification continues. The African share dropped from 40 to15%, while that of South America increased from 10 to 33%. Europe dropped from 25 to 15%, while North America's share went from 15 to 18%, and Asia's from 5 to 10%.[32] Further governmental assistance takes the form of R&D into conservation, substitution, and recycling.

Prior to 1980, the government also encouraged the formation of a private-sector stockpile. Key minerals to be included in the stockpile were determined on the basis of a study analyzing the risk of supply interruptions. This report found 28 commodities to be "sensitive," with the assignment of risk being greatly affected by the importance of imports from southern Africa. The most critical minerals were found to be:[32]

Chromium	Tungsten
Manganese	Cobalt
Asbestos	Vanadium
Niobium	Rutile
Tantalum	Platinum metals
Mica	Ilmenite
Fluorspar	Copper
Aluminum	Antimony
Nickel	Tin
Molybdenum	Zirconium

The aftermath of this study was the announcement of plans to establish a limited stockpile, capable of meeting requirements for one year, for chromium, manganese, cobalt, vanadium, and blue asbestos. An unusual aspect of the criteria used in defining stockpile needs was the inclusion of research and technology industries as worthy of protection.[28] Typically industrial and defense requirements take precedence, often with little regard given to the other sectors. Consistent with the government's free-market policies, private industry was primarily responsible for the collection and maintenance of the stockpile. In addition, German companies were expected to provide the funding for four months' worth of supply, with the remainder funded with governmental assistance. Despite much planning and organization, however, the stockpile project was rescinded in 1980, probably due to budgetary constraints.[31]

United Kingdom. Surprisingly, particularly considering both the magnitude of British consumption, and its heavy import dependence, the UK has until very recently shown less concern over supply security than has either France or the FRG. The fact that the London Metal Exchange, as well as many major international mining companies, have located their main operations in the British capital, fostered a belief that this confers on the UK an innate supply advantage. As important, British consumption, although large, has not exhibited the considerable growth rates seen in other EEC nations, possibly diminishing the immediacy of the need for attention.[30]

The general approach assumed by the British government has been to rely on market forces. In this feature, it is comparable to the FRG. However, the UK has not established any formal, coherent mineral policy, such as is found elsewhere. Moreover, until recently, exchange controls have prevented long-term contractual capacity for British companies. As a consequence, production from many of the mines developed with British capital has been purchased by Japan or the FRG.[30] Some of the sluggish movement toward consideration of mineral supply security has been traced to the inefficient nature of bureaucracies generally.[33] Although the government has intervened in the past to ensure acquisition of uranium, titanium, and aluminum, the aggressive procurement policies of competitors could force the UK to become more active.

There are, indeed, signs that this is occurring. Due to the strategic nature of the steel industry as a provider of many products to other important industries, heavy import dependence on the raw materials required by the steel sector has obvious security implications.[28] The vulnerability of this vital industry, coupled with existing stockpiles in at least two other nations, forced the UK in early 1983 to establish a small stockpile of its own, after years of demands for such a program.[34] Although relatively little information concerning the stockpile has been released, approximately 35 million pounds sterling have been allocated. The stockpile includes chromite, ferrochrome, manganese, ferromanganese, cobalt, and vanadium, amounting to a three- to six-month supply. Brandeis Intsel, a subsidiary of the nationalized French company Pechiney, manages the stockpile. Prior to this program, the state role was restricted to encouragement of private-sector stockpiling.

More freedom of access to mineral rights, and an easing of restrictions on direct overseas investment, have been seen as changes that could greatly improve mineral security, yet little has been accomplished in these areas.[35] Other suggested reforms include governmental guarantees of currency exchange rates on long-term contracts, insurance for high-risk investment, and increased support for domestic exploration. Although the decision to establish the strategic stockpile must be viewed as a positive step, the debate continues in the UK as to the proper function of the government in securing raw materials.

The Potential for a Comprehensive EEC Mineral Policy

Although the EEC has sponsored studies identifying problems in the security of its mineral supplies, members cannot agree on a coordinated plan of action. The lack of a common policy for the EEC concerning the dependable flow of minerals is comparable to the dearth of a community energy policy. Again, this can be attributed in large part to the great diversity, both political and economic, that is inherent in the organization.[25] National problems, the degree of state control of industries, currency values, unemployment levels, and inflation rates all vary widely from country to country. To illustrate this disparity, Table III-10 depicts differences in growth in GNP, and in the unemployment rate, for various EEC nations. The flexibility of an economically strong nation is certainly superior to that of a country experiencing high unemployment and stagnant economic growth, yet this is precisely the situation in the EEC. These gaps produce a great strain on the unity of the community.

Kilmarx[37] identifies other barriers to a common EEC mineral policy. Many arise from differences in political and economic philosophy. There is often extreme disagreement regarding state intervention in commodity markets. Thus France's active role is contrasted by the FRG's laissez-faire approach. This lack of consensus has prompted such views as "It is dangerous to seek a coherent and comprehensive minerals policy, whether at national or Community level, for such a path encourages unnecessary intervention."[38] Second, an attempt is made to avoid policies that may alienate producer countries. The willingness of some members to meet the demands of LDCs, however, will be constrained by domestic economic difficulties. Third, the sense of urgency required to address effectively

TABLE III-10 Comparisons of GNP Growth Rates and Unemployment Data for Some EEC Nations

	Percent yearly change in GNP			Percent of labor force unemployed		
	1982	1983	1984	1982	1983	1984
Belgium	-0.1	0.2	1.5	13.1	14.7	15.0
France	1.8	-0.3	0.3	8.6	8.8	9.9
FRG	-1.1	0.8	2.6	7.5	9.5	10.0
Italy	-0.2	-0.8	2.4	9.0	9.4	9.8
Netherlands	-1.4	-0.2	0.3	10.0	15.5	17.7
UK	0.9	2.2	2.6	10.7	11.8	11.8

Source: Ref. 36.

the issue of mineral security appears lacking. Such urgency, as seen during the oil crisis, is required to foster a coordinated program. Other obstacles include the inability of the minerals industry to agree on policy formulation; the decline of the political clout of this sector generally; the rise of certain special-interest groups, such as environmental factions; and finally the diminishing of confidence in the United States and in its ability to adequately defend its allies. Any variable impeding the overall unity of the EEC can thus impact the potential for a comprehensive mineral policy.

The ad hoc formulation of mineral policy as adopted by individual nations could prove to be a quite divisive system in the event of an actual supply problem. Shortages could cause tension between the industrialized nations and LDC producers, while increasingly stiff competition for minerals could inflict grievous rifts within the EEC itself, as well as with other industrialized consumers. This would particularly be the case if the flow of southern African commodities were interrupted.[37]

Stockpiling is a further area of potential discord arising from the absence of a community mineral policy. If only a few members of the EEC maintain a stockpile, the capacity of these nations to weather a supply crisis is superior. However, this may be at the expense of community solidarity. The implications for the EEC of some members accepting and others rejecting the stockpile concept are most important in the area of sharing of supplies. (That is, will there be any? How will apportionment be determined?) The fact that the large U.S. strategic stockpile has been viewed as a "threat" to the EEC[28] gives some idea of the difficulty.

A movement is now apparently underway, however, promoting the formulation of a mineral policy for the entire EEC. Chromium, manganese, phosphate, and platinum group metals have been classified as the four minerals of "prime strategic importance" to the community.[39] Other key minerals are more amenable to substitution. Thus antimony, cobalt, molybdenum, nickel, titanium, vanadium, tantalum, and niobium, although vulnerable to supply interruptions, occupy a strategically less important position.

In March 1982, analysis was begun on the feasibility of an EEC stockpile.[40] In addition, due to the increase in mining capital requirements, and the rise in risks that discourage private overseas investment, the Executive Committee of the Community began to outline proposals for an all-EEC mineral policy. It is doubtful, however, that the unyielding physical and ideological obstacles barring the realization of such a program can readily be overcome. Until a comprehensive EEC policy toward securing of raw materials is reached, individual members will continue unilaterally to seek the procurement of vital minerals, to the detriment of a strong and united community.

REFERENCES

1. Dunkerley, Joy. "The Future for Coal in Western Europe." *Resources Policy* 4, No. 3 (September 1978), pp. 151-159.
2. Hoffman, George W. *Energy Strategies of Western Europe in the 1980's: Dependable Supplies versus Security Implications.* Policy Study No. 18. Center for Energy Studies, University of Texas at Austin, July 1982.
3. U.S. Department of the Interior, Bureau of Mines. *Minerals Yearbook 1981*, Vol III: *Area Reports: International.* Washington, D.C.: U.S. Government Printing Office, 1983.
4. "Northern Europe." "Western Europe." In *Mining Annual Review, 1982.* London: Mining Journal, June 1982, pp. 485-488, 515-538.
5. Robinson, Colin, and Eileen Marshall. *What Future for British Coal?* Hobart Paper 89. London: The Institute of Economic Affairs, 1981.

6. "Europe's Energy Options." *Mining Journal*, 8 October 1982, pp. 245-246.

7. "Worldwide Oil andGas at a Glance." *Oil and Gas Journal*, 27 December 1982, pp. 78-79.

8. Earney, Fillmore C. F. "Norway's Offshore Petroleum Industry." *Resources Policy* 8, No. 2 (June 1982), pp. 133-142.

9. Brand, David. "Britain Faces Challenge As Drop in Oil Output in a Few Years Looms." *The Wall Street Journal*, 8 July 1983, p. 1.

10. Jacobs, Jan. "Fuel Glut and Recession Prompt the Dutch to Ease Natural Gas Conservation Rules." *The Wall Street Journal*, 22 July 1983, p. 18.

11. Bray, Nicholas. "France, Facing Power Glut, Scales Down Its Plans to Build More Nuclear Reactors." *The Wall Street Journal*, 28 July 1983, p. 25.

12. "Energy: Long-Term Objectives and Strategy." *Bulletin of the European Communities*, No. 5 (1980), pp. 23-25.

13. Organization for Economic Cooperation and Development. *Energy Balances of OECD Countries 1960-1974*. Paris: OECD, 1976.

14. Organization for Economic Cooperation and Development, International Energy Agency. *Energy Balances of OECD Countries 1971-1981*. Paris: OECD, 1983.

15. Szuprowicz, Bohdan. *How to Avoid Strategic Materials Shortages*. New York: John Wiley & Sons, Inc., 1981.

16. U.S. Department of State, Bureau of Public Affairs. *Soviet-West European Natural Gas Pipeline*. Current Policy No. 331. Washington, D.C.: U.S. Government Printing Office, 1981.

17. Mufson, Steve. "Anatomy of Continuing Soviet Pipeline Controversy." *The Wall Street Journal*, 31 August 1982, p. 27.

18. "GAO: Europe Has No Alternative to Soviet Gas." *Oil and Gas Journal*, 6 June 1983, pp. 44.

19. Ibrahim, Youssef M. "West Moves to Counter Gas Cutoff." *The Wall Street Journal*, 21 June 1983, p. 39.

20. U.S. Department of the Interior, Bureau of Mines. *Mineral Commodity Summaries 1983*. Washington, D.C.: U.S. Government Printing Office, 1983.

21. Thurow, Roger. "German Cabinet Approves Aid for Steel Firms." *The Wall Street Journal*, 15 June 1983, p. 34.

22. "EC Orders Cuts in Steel Capacity of Up to 19.7%." *The Wall Street Journal*, 1 July 1983, p. 16.

23. "British Steel Loss Was $591 Million in Year to April." *The Wall Street Journal*, 13 July 1983, p. 29.

24. U.S. Department of the Interior, Bureau of Mines. *Mineral Facts and Problems, 1980 Edition*. Washington, D.C.: U.S. Government Printing Office, 1980.

25. Frey-Wouters, Ellen. *The European Community and the Third World*. New York: Praeger Publishers, 1980.

26. Maull, Hans W. "French and European Community Policies for Securing the Supply of Primary Commodities." *Primary Commodities: Security of Supply Conference*, Washington, D.C., 11-12 December 1980.

27. Page, William. "UK Balance of Payments in Raw Materials." *Resources Policy* 5, No. 3 (September 1979), pp. 185-196.

28. Warnecke, Steven J. *Stockpiling of Critical Raw Materials*. London: The Royal Institute of International Affairs, 1980.

29. "Changing Times for France." *Mining Journal*, 5 February 1982, pp. 93-95.

30. Crowson, P. C. F. "The National Mineral Policies of Germany, France, and Japan." *Mining Magazine*, June 1980, pp. 537-549.

31. "Nonfuel Minerals Stockpiling by Japan, South Korea, France, and Britain." *Alert Letter on the Availability of Raw Materials*, No. 13 (June 1982), p. 16.

32. Hoffman, Lutz. "Policy of the German Federal Government and European Countries for Securing the Supply of Primary Commodities." *Primary Commodities: Security of Supply Conference*, Washington, D.C., 11-12 December 1980.

33. "Dialogue on United Kingdom Minerals Supply." *Mining Journal*, 30 May 1980, p. 435.

34. U.S. Department of the Interior, Bureau of Mines. "United Kingdom Establishes Government Stockpile of Strategic Minerals." *Minerals and Materials: A Bimonthly Survey*, February-March 1983, Washington, D.C.: U.S. Government Printing Office, 1983, p. 10.

35. Parry, Audrey. *Towards a Minerals Strategy for Britain*. Foreign Affairs Research Institute Publ. No. 5 (1981).

36. "World Economic Outlook." *US News and World Report*, 22 August 1983, pp. 122-137.

37. Kilmarx, Robert A. "Security and Strategic Considerations for Commodity Policies." *Primary Commodities: Security of Supply Conference*, Washington, D.C.: 11-12 December 1980.

38. Frame, A. G. "Role of the United Kingdom in the World Minerals Industry." Robert Pryor Memorial Lecture to the Royal School of Mines Mining and Metallurgical Society, London, 14 January 1980.

39. "Strategic Minerals—A More Precise Definition." *Mining Journal*, 19 November 1982.

40. U.S. Department of the Interior, Bureau of Mines. "European Communities and Mineral Policy." *Minerals and Materials: A Bimonthly Survey*, August-September 1982. Washington, D.C.: U.S. Government Printing Office, 1982, p. 7.

PART IV

THE USSR

CHAPTER IV-1

Introduction

In less than 70 years, the USSR evolved from a relatively backward state to its present position as the second largest industrial power in the world. In great part this was due to its colossal mineral wealth. Today, the USSR leads the world in the production of oil, asbestos, iron ore, manganese, mercury, platinum-group metals, phosphate, potash, titanium, and nickel. In addition, it ranks as the second-leading producer of aluminum, chromite, natural gas, gold, lead, zinc, tungsten, vanadium, and diamonds.

The USSR has adopted a unique approach to its policy on minerals development: a determined effort is made to obtain production solely from its domestic endowment. Cost has been relegated to secondary importance. This idea is certainly not novel. Reliable supplies influence in particular the economic health and ability of a nation to maintain a strong defense. Thus security of supply is a primary concern of all major industrial powers. However, the degree of Soviet success in pursuit of this policy is unprecedented. The sacrifices imposed on the Soviet system by the quest for minerals self-sufficiency are undoubtedly large, but the ability to sustain high production levels for virtually all important minerals from internal sources is indeed impressive.

During the past few years, however, the capacity of the USSR to preserve its mineral import independence has become suspect. Significant changes in long-established trade patterns are apparent. Soviet mineral exports to the West have declined in several key commodities, as they have to Eastern European allies. Minerals for which the USSR was considered self-sufficient have been purchased on world markets. There is evidence of exhaustion of older deposits in accessible areas and of overall ore quality declines. Newer deposits are in more remote and inhospitable Asian regions; these areas in many instances lack physical infrastructure. Consumption of minerals is growing rapidly, due to both industrial and military requirements. These factors prompt some Soviet analysts to question whether a transition to dependence on imports for certain minerals may be imminent. The pattern of slowing production evident in some energy sectors underscores this concern.

The major scenario evoked by this possible shift to Soviet mineral dependence is increased competition for the world's minerals in international arenas. Such an event could seriously disrupt present supply-demand relationships. Moreover, the need for the Soviets to obtain raw materials by military means has been suggested. To this end, even Soviet activities in minerals-rich southern Africa could be influenced by resource considerations. The latter point in particular is noteworthy, due to the importance of minerals from this troubled region to the industries and defenses of the West.

A postulated transition from Soviet mineral self-sufficiency to import dependency is therefore a pertinent topic of inquiry. Such an investigation must address certain inescapable issues. What constraints does the Soviet minerals industry actually face in its operations? What is the current status of the Soviet energy and nonfuel minerals sectors? What are the components of present Soviet minerals policy? Is mineral import dependence approaching? This study attempts answers to these questions.

At this point, a note of caution is appropriate regarding the availability of dependable Soviet data. There is practically no official information published for most minerals. No statistics on consumption are available in the West. Production figures are limited mainly to ferrous minerals. Data on reserves or trade are restricted. The predilection for secrecy in the USSR has a decided impact on the reliability of information used in research. The motivations behind such restrictions are numerous, although a major motive is probably a desire to confuse its enemies. Clearly, all data used in Western studies must be regarded as estimates and approximations at best. The typical problems of forecasting, and of formulating conclusions based on production, consumption, reserve, and trade statistics, are amplified only by this dearth of hard data. If the inputs of an investigation are only estimates, then obviously the final judgments can be no better. All analyses in the West of the Soviet mineral industry must operate under this statistical constraint. It is unfortunate, but unavoidable.

CHAPTER IV-2

Geographic and Economic Constraints on the Minerals Industry

GEOGRAPHIC CONSTRAINTS

Comprising more than 22 million square kilometers, the USSR is a nation of extreme diversity in landform, geology, and climate. Although diverse geologic features provide great potential for formation of valuable mineral deposits, formidable obstacles impede the Soviet minerals industry as it attempts to develop these resources. From the tundra of the far north to the marshes of western Siberia to the deserts of Central Asia, huge areas of virtual wilderness exist—in effect devoid of any civilizing influence. The geographical limitations of remoteness, hostile climate, and inhospitable terrain are factors that will constrain future production of Asian mineral deposits. This is a theme emphasized by most authors (e.g., Dienes and Shabad,[1] Goldman,[2] the Office of Technology Assessment (OTA),[3] the Central Intelligence Agency (CIA),[4] Hanson,[5] Hoffman,[6] and Rumer and Sternheimer.[7])

Soviet analysts[8,9] plainly have noted the depletion of easily accessible mineral reserves. This dilemma denotes a trend for mineral deposit locations increasingly distant from the population and industrial centers of the European USSR. As the transition continues, deposits in Asia will be relied on to compensate for this depletion. Figures IV-2-1 to IV-2-3 clearly portray this trend for the energy minerals. Note that for each fuel, Asian output surpassed that from areas in Soviet Europe sometime during the period from 1977 to 1979. Currently, while 75% of the people and 82% of industry are located west of the Urals, less than half the fuels are located there.[6] Similar tendencies are evident for many nonfuel minerals. What this pattern illustrates is that the supply sources are shifting eastward, away from the markets and industrial concentrations.

As this shift eastward continues, geography will increasingly hinder the availability of infrastructure, a critical key to mineral development. Although infrastructure is a real problem in all areas of the Asian USSR, it is of special concern in western

Siberia, site of the major new oil and gas fields. Here, bitter Arctic cold and adverse landforms serve to slow construction rates, create unusual technical complications, reduce the return on investment, and inflict high labor turnover rates. The consequence of these, ultimately, is a pattern of ever-rising raw material costs. "What the Arabs have done to the West, Mother Nature has done to the Soviet Union."[5] The negative impact of geographical influences on the establishment of Soviet infrastructure cannot be overemphasized.

Transportation

One of the vital components of physical infrastructure is transportation. Its importance to the Soviets was made apparent by Brezhnev, when in late 1970s he named conveyance as one of the three critical bottlenecks in the economy (the other two being energy and steel). Nationally, railroads account for some 55% of all freight carried in the USSR,[10] and 90% of all mineral freight.[11] As seen in Figure IV-2-4, almost all rail lines are in the European part of the country. The Trans-Siberian Railway, the world's longest at 9000 kilometers, was for years the only link between the Soviet Far East and the rest of the nation.

Currently, the Baikal-Amur Mainline (BAM) is under construction north of the Trans-Siberian line. Its completion date is set for 1985. The BAM, 3200 kilometers long, is crucial to the development of previously inaccessible mineral deposits, such as the huge Udokan copper deposit and the South Yakutia coal fields. It will provide a direct link to Pacific ports for potential export of these resources. The areas north of the BAM will, for the near future, however, remain a wilderness. Mineral development in these regions will be very difficult.

In western Siberia, the railway did not reach the oilfields near Surgut before 1975. Prior to this, equipment and supplies were either airlifted or delivered by river. By the latter method, deliveries could only be made during the warm months before the rivers iced over (mid-September to mid-

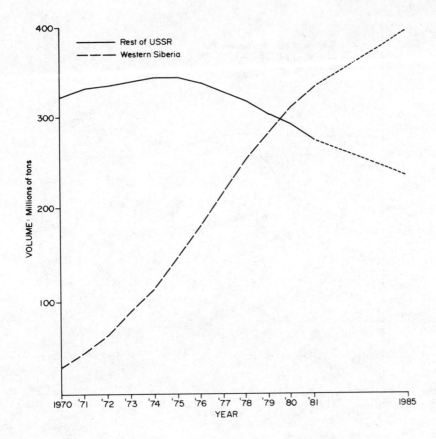

FIGURE IV-2-1 Soviet oil production: comparison of regional trends. (From Refs. 3, 10, 12, and 25.)

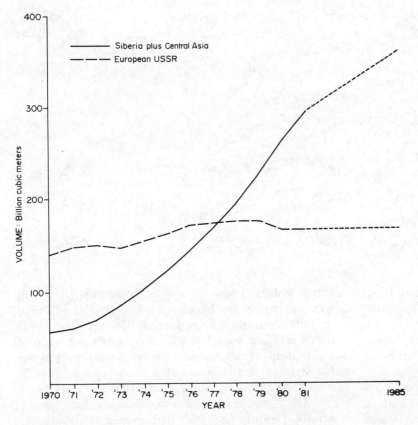

FIGURE IV-2-2 Soviet natural gas production: comparison of regional trends. (From Refs. 2, 3, 12, and 25.)

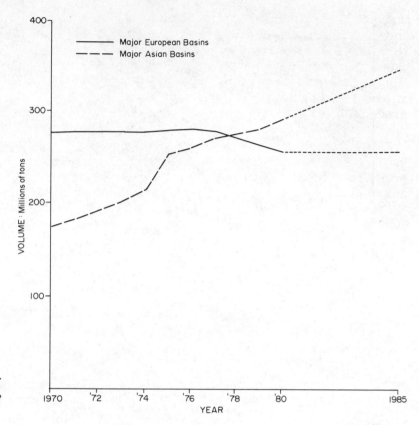

FIGURE IV-2-3 Soviet coal production: comparison of regional trends. (From Refs. 3, 25, 26, and 27.)

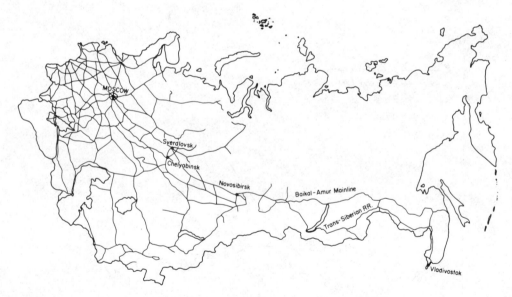

FIGURE IV-2-4 Soviet railroad network. (From Ref. 27.)

April). Storage of materials was then required until marshes froze over to permit transfer. This obviously led to significant developmental delays. Although of some regional importance, rivers cannot be relied on for all long-distance haulage, for none run east-west.

The construction of roads in Asia is time consuming, encumbered by constraints similar to those faced in rail development, owing to climate and topography. There is a particular dearth of all-weather roads in the eastern USSR. For instance, in 1980, around the important Urengoy gas fields, there was less than 100 kilometers of road, only 30 kilometers of which was paved. Construction lags far behind projections, and costs can exceed 10 times the norm in western USSR areas.[1]

The increasing use of pipelines has been a notable trend. In 1960, oil pipelines moved less than 3% of total freight turnover, and gas pipelines less than 1%. By 1980 these figures were 18% for oil and 9% for gas.[11] As might be anticipated, geographical impediments present comparable obstacles to the speed and cost of development of pipelines. For instance, in western Siberia, construction costs are 70% higher than the Soviet average.[12]

Normally, mineral production is contingent on the establishment of a reliable transportation network. This is required for movement of labor and supplies and for commodity delivery to markets. To the extent that the installation of such a network is hindered, mineral production is frustrated. As has been demonstrated, this is the case in the USSR for most regions east of the Urals. Especially in western Siberia, geographical constraints on transportation have proven a major impediment to the accelerated exploitation of energy supplies critical to the USSR.

Power

The availability of electrical power is another crucial factor influencing the minerals industry. While the energy-intensive nature of many metallurgical processes, such as the production of aluminum, is well known, power is a vital input in most mining and mineral extraction. The Soviet goal is an electrical grid based on a nationwide, unified power system. In pursuit of this goal, however, construction of both distribution stations and transmission lines lags behind schedules. This is particularly true in the rapidly developed oil and gas areas of western Siberia, where large amounts of energy are required. At present, all of Siberia is dependent upon a 3000-megawatt gas-fired plant at Surgut. In 1977, only 47 kilometers of transmission lines out of 273 planned kilometers were erected in the area. As a result, many oil and gas deposits rely on small and awkward mobile stations.[13]

> Due to the lagging development of a centralized electrical-supply system in Tyumen Province, a vast army of workers—who are in very short supply in this region—is servicing small diesel power stations. One out of every three employees . . . works in the field of power engineering.

The delay in plans to provide Asian areas with adequate amounts of energy is therefore a conspicuous constraint on the efficient use of labor and a definite barrier to attempts to reach output quotas. Although much of the impediment to the availability of power must be attributed to the structure of the Soviet economy, harsh geographic circumstances also play a major role.

Housing

As development quickens in Siberia, more than 100,000 people per year are expected to enter the region. Given the sparse population density of most areas, substantial strains are anticipated on all facets of social infrastructure. Severe housing shortages already exist, and have been called perhaps the most crucial problem facing the oil and natural gas industries.[12] This has prompted the Soviet practice of the "tour of duty" settlement to overcome housing scarcity, and the labor shortages caused by the social deprivations of a hostile environment. The tour-of-duty concept is employed particularly in western Siberia. As many as 11,000 "flying team" workers fly in to the area from the European USSR, work two weeks, then return home for two weeks. Certainly, productivity suffers from the continuous shuffling of personnel. Serious lack of proper housing is a major problem in other Asian locales as well, including the Kuzbass and Ekibastuz coal basins.[14] Labor shortages are anticipated to intensify as mineral development in Asia progresses.

The underlying theme of these infrastructural problems in housing, power, and transport is the influence of geographical factors, be they remoteness, climate, or terrain. It is true that economic variables must bear a large responsibility for some of these deficiencies. The impact of adverse geography, however, cannot be ignored; it is most responsible for making an inefficient mineral economy operate even more inefficiently.

ECONOMIC CONSTRAINTS

The economy of the USSR has been quite successful in achieving its primary purpose—the continuing dominance of the state. A concomitant ability to sustain a massive military machine has also been demonstrated. However, efficiency, as it is commonly defined in Western economies, is not a feature of the Soviet system. It can, in fact, be argued that inherent weaknesses in the economy itself impose the second major set of constraints on the optimal operation of the Soviet mineral industry. Given the magnitude of this sector, "its fate is closely linked to that of the economy as a whole."[9]

Centralization characterizes the Soviet system, which has been heavily planned since 1929. All means of production are collectivized, or state-owned. The direction and extent of economic progress is coordinated and strategically implemented through the use of sectorial output quotas (targets) in various Five-Year Plans (FYPs), with investments allocated accordingly. Typical Western market forces, such as demand or prices, have been of minor importance, and domestic consumption is permitted to increase only as far as domestic production increments exist.

Economic Stagnation

The primary effectiveness of the centrally planned economy lies in its capacity to concentrate on certain priorities, and to marshal resources for use where the leadership has decided they are required. Results are twofold: the path of economic growth can be accurately controlled, and phenomenal expansion in some areas can occur, as has been the case in the USSR. Although the obvious beneficiaries of such a system are the priority sectors (e.g., the military), other segments often face neglect (e.g., consumer goods), being allocated whatever is left, following priority funding.

In the past, the rate of economic growth in the USSR was sufficient to allow the military a large share of GNP (estimated as high as 14%), while still providing reasonable funds for other vital projects. This no longer holds true. The 6% annual growth of the 1950s fell to 5% in the 1960s, to 3.5% by the mid-1970s, and is currently less than 2%.[3,15] Such economic stagnation, especially in the face of ever-escalating military investment, can mean only increasingly stiff competition for a finite capital pool among a number of other key sectors—agriculture, energy, metallurgy, manufacturing, and transportation being among them. Obviously, some must receive inadequate funding. The ultimate outcome is the ubiquitous Soviet shortages—shortages of labor, of machinery, of construction materials, of metal, of consumer goods, and even of foodstuffs. The recent failure of the mineral industry to fulfill targets has been largely attributed to inadequate investment in all economic sectors.[16] The tightening economy will probably increase this problem, unless a clear emphasis on minerals evolves. A major strength of the Soviet system, the ability to deploy resources rapidly in priority areas, appears to have been partially negated by a malady from which the system was supposedly insulated—economic recession.

Labor Shortages

The Office of Technology Assessment[3] has cited falling rates of growth in the Soviet capital stock and labor force as underlying influences in the present economic slowdown. Labor shortages, in particular, may prove to be a problem of crisis proportions before the turn of the century. Grichar et al.,[9] Bergson,[17] Hanson,[5] the CIA,[16] the OTA,[3] Papp,[8] and Strishkov[11] emphasize this problem.

Previously, equipment or supply deficiencies could often be compensated for by using more labor input. This will no longer be possible. Population growth has slowed enormously (the USSR is actually facing an increasing infant mortality rate), while the death rate is rising. Less than 1 million new workers per year will enter the work force in this decade—far fewer than required. This slow pace of labor growth is even more ominous than it may appear, for the problem is not merely numerical:[5]

> Nearly all the growth in labor force in the 1980s will come from populations native to Central Asia and Transcaucasus. These groups are on average relatively low in skill, and the Central Asians are geographically immobile. . . . Most new entrants to the labor force will be neither in the new natural-resource development regions nor in areas where an industrial base is already built up.

Insufficient labor supply is already extant in the coal industry and in the oil and gas regions of Siberia. The situation would not be so critical if there were indications of increasing labor productivity (currently only 40% of that in the United States), but decreases are in fact reported.[11]

Inadequate Coordination among Ministries

The huge and pervasive bureaucracy that stringently supervises all aspects of Soviet business suffers from the defects common to all overly regulated systems: heavy-handedness, poor coordination, and confusion over objectives. Both horizontally and vertically, the failure to implement an integrated approach by the ministries concerned can result in some extraordinary gaps in planning, design, and implementation of important projects. For example, at Cherepovets, a major steel complex, the rolling mill was constructed well before the blast furnace, requiring the transport of metal to the mill from other areas.[18] Discord between ministries has been blamed in the Soviet press for the waste of over 1 million tons of alumina and titanium at the Khibiny apatite-nepheline complex.[19] Poor agency cooperation has been criticized in many other sectors, including the natural gas industry,[20] housing,[12] and transportation (particularly construction of the BAM).[21] Equipment and supply shortages, as well as construction delays at important projects, are frequent results, but the major consequence is waste and increased costs.

Waste and Low Quality of Industrial Output

The substantial industrial expansion achieved in the USSR has been obtained through the Five-Year Plan's emphasis on output. Poor quality and waste are, however, persistent negative outgrowths of this policy. The best illustration is the Soviet steel industry—the largest steelmaker in the world cannot make drill pipe of sufficient quality to use in its own oil wells, but must rely on imported pipe. Poor-quality steel has major implications for the effectiveness of all industrial expansion. Inadequate quality is an ever-present feature of the entire system; it must be dealt with at all levels.[11] Heavy equipment made with low quality steel is a particular concern in mining of coal and many non-fuel minerals, where safety demands reliability. Poor performance is intensified by the frequently bitter polar conditions in which several mineral industries must operate. Chronic equipment downtimes in this and other sectors are a significant problem.

Waste is "probably the single most significant Soviet mining problem."[8] Inefficiency is a result of many factors, including the planning system's emphasis on output, as well as poor pricing policy, inadequate agency cooperation, and obsolete equipment and processes. Mineral extraction rates are very low; as much as 30 to 40% of coal, 50% of gas, 70% of oil, and 20% of iron ore are left behind.[22] By-product recovery is generally poor; often only the principal metals in an ore are exploited. Losses of by-products are estimated to be greater than half the total value of the ore.[11] Mediocre recovery rates during processing waste large amounts of valuable minerals. For instance, the use of present technologies for manganese processing results in losses of up to 50%.[22] Sutulov has estimated losses during polymetallic ore concentrating and refining at as high as 25% for copper, 40% for lead, and 40 to 60% for gold and silver.[23] Much higher inputs of Soviet ore, energy, and materials are needed for the manufacture of a given item than are typical in the West. For example, consumption of metals in machine building is up to 50% higher than in the West.[11] Only 45% of all raw steel produced is used efficiently,[22] and the use of low-quality steel results in loss by corrosion of over 15 million tons of steel annually.[11] Finally, transport losses are quite large, especially for haulage of coal and iron ore, and end product mixes do not meet requirements. Soviet sources estimate that annual losses of metal in steel production, in steel consumption for machine building, and in repair of equipment and machines, amounted to 42.2 billion rubles in 1980.[11]

Some of these deficiencies could be corrected by the introduction of new technologies. However, the stifling of innovation is another impediment to economic growth. Few incentives exist for the implementation of new ideas. The introduction of innovative processes would require some interruption of normal production; such work stoppages lower output figures. Worthwhile new technologies are thus discouraged by the emphasis on the quantity of production. Furthermore, it is not merely the introduction of such innovation that is important; the key to positive impact is the success of diffusion of a technology. Such dispersal has proven to be quite slow in the USSR. Measures such as current reform attempts to change output goals from "gross" to "net production" are unlikely to make any real difference, for quantity still is the ultimate index of performance.

A major barrier to innovation diffusion is the dependence of managers on centralized leadership to suggest requisite initiatives. Generally, local managers lack the power or the incentive to promote adoption of new technologies.[24]

> Decades of centralization and subordination have instilled such a sense of powerlessness in the managerial class that decentralized initiative is now very difficult to foster, even when it is encouraged by the leadership itself.

What this has reinforced is the institutionalization of obsolete technologies. This is one of the chief causes of waste in the minerals sector. It has promoted substandard quality steel, and kept many industries (e.g., aluminum) lagging more than 10 years behind Western techniques.

DYNAMIC INTERRELATIONSHIPS OF CONSTRAINTS

It must be stressed that each economic and geographic constraint is a dynamic variable, interacting and evolving through time with other constraints. For instance, shortages of skilled labor will further retard development of needed physical infrastructure. Lack of interagency coordination prevents the planned construction of housing. Poor equipment quality leads to further waste, and so on. Without substantial reforms of the nature of the system itself, inefficiency will continue. Such reforms are unlikely. The end result of the constraints imposed on the Soviet minerals industry by its economic and geographic environment is a future of ever-swelling costs. Eventually, this will lead to constraints on the level of production. The impact of this trend will vary from one commodity to another.

REFERENCES

1. Dienes, Leslie, and Theodore Shabad. *The Soviet Energy System: Resource Use and Policies.* Washington, D.C.: V. H. Winston & Sons, 1979.

2. Goldman, Marshall I. *The Enigma of Soviet Petroleum.* London: George Allen & Unwin (Publisher) Ltd., 1980.

3. U.S. Congress, Office of Technology Assessment. *Technology and Soviet Energy Availability.* Washington, D.C.: U.S. Government Printing Office, 1981.

4. U.S. Central Intelligence Agency. *Prospects for Soviet Oil Production, A Supplemental Analysis.* Washington, D.C.: U.S. Government Printing Office, 1977.

5. Hanson, Philip. "Economic Constraints on Soviet Policies in the 1980's." *International Affairs* 57, No. 1 (Winter 1980-1981).

6. Hoffman, George W. *Energy Projections: Oil, Natural Gas, and Coal in the USSR and Eastern Europe.* Policy Study No. 3. Center for Energy Studies, University of Texas at Austin, August 1978.

7. Rumer, Boris, and Stephen Sternheimer. "The Soviet Economy: Going to Siberia?" *Harvard Business Review*, January-February 1982.

8. Papp, Daniel S. "Soviet Non-fuel Mineral Resources, Surplus or Scarcity?" *Resources Policy* 8, No. 3 (September 1982).

9. Grichar, James S., Richard Levine, and Lotfollah Nahai. *The Nonfuel Mineral Outlook for the USSR through 1990.* Washington, D.C.: U.S. Government Printing Office, 1981.

10. *USSR Facts and Figures Annual*, Vol. 6. Edited by John L. Scherer. New York: Academic International Press, 1982.

11. U.S. Department of the Interior, Bureau of Mines. "The Mineral Industry of the USSR." By V. V. Strishkov. Preprint from the *Minerals Yearbook 1980.* Washington, D.C.: U.S. Government Printing Office, 1980.

12. Wilson, David. *Soviet Oil and Gas to 1990.* Economist Intelligence Unit (EIU) Special Report 90. London: Spencer House, 1980.

13. "How Shortages Impair the Soviet Economy." *The Current Digest of the Soviet Press* 34, No. 17 (26 May 1982).

14. "Ekibastuz Complex Beset by Problems." *The Current Digest of the Soviet Press* 34, No. 30 (25 August 1982).

15. Piper, Hal. "Soviet Economic Ills May Trap Andropov." *Austin American-Statesman*, 14 November 1982.

16. U.S. Central Intelligence Agency, National Foreign Assessment Center. *The Soviet Economy in 1978-79 and Prospects for 1980.* Washington, D.C.: U.S. Government Printing Office, 1980.

17. Bergson, Abram. "Can the Soviet Slowdown Be Reversed?" *Challenge* 24, No. 5 (November-December 1981).

18. Strishkov, V. V. "Soviet Union." In *Mining Annual Review, 1981.* London: Mining Journal, June 1981.

19. "A Complex Takes Shape." *The Current Digest of the Soviet Press* 34, No. 42 (17 November 1982).

20. "The Economy." *The Current Digest of the Soviet Press* 34, No. 30 (25 August 1982).

21. Dyker, David, "Planning in Siberia on Wrong Track." *Soviet Analyst* 9, No. 2 (23 January 1980).

22. Levine, Richard M. "Soviet Union." In *Mining Annual Review, 1982.* London: Mining Journal, June 1982.

23. Sutulov, Alexander. *Mineral Resources and the Economy of the USSR.* New York: McGraw-Hill Book Company, 1973.

24. Ioffe, Olympiad S. "Law and Economy in the USSR." *Harvard Law Review* 95 (1982).

The Energy Industries of the USSR

The USSR plays a prominent role in the international energy balance. It has been the leading producer of petroleum since 1974, and the second leading exporter. It ranks second in the production of natural gas, and is the primary exporter. It is also the largest coal producer (on a tonnage basis). Finally, the USSR is one of the most advanced nations in terms of both research in and practical application of nuclear power.

Not surprisingly, production of this magnitude has internal and external ramifications. Domestically, energy consumption and economic growth are intimately linked; the ability to meet energy requirements is essential for a vigorous minerals industry. Internationally, energy exports are delivered to the West, to obtain much needed hard currency, revenue necessary for the purchase of grain and technology. Exports to the Eastern European members of the Council for Mutual Economic Assistance (CMEA) reduce their dependence on high-priced OPEC supplies. The state of the Soviet energy sector is thus a matter of great importance to the nonfuel mineral industry and to the economy as a whole.

PETROLEUM

World attention focused on the petroleum industry of the USSR as a result of the 1977 U.S. Central Intelligence Agency report, *Prospects for Soviet Oil Production*.[1] The study pessimistically predicted a peak in production by the early 1980s, sharp declines thereafter, and the need to import 175 million tons or more of oil by 1985. Such a drop would have major significance on the world availability of energy. This forecast has provoked voluminous musings on the state of Soviet oil; many studies were openly contemptuous of the CIA's findings. Ultimately, the U.S. agency attempted to make definitive projections concerning very problematical areas, ones "controlled by secrecy laws and characterized by uncertainty."[2] The report's conclusions were based on two major criteria: the reserve estimate was too low, assuming no recent large discoveries, and assessments of the Soviet use of controversial (by Western standards) production techniques were quite negative. Nonetheless, it is true that a Soviet oil production slowdown is evident.

Resources and Reserves

Attempts to present Soviet petroleum reserve statistics encounter several difficulties. The Soviets are notoriously reticent about the publication of economic information, especially so in the case of petroleum. Most of their natural resource data are secret. In 1947, the State Secret Act classified petroleum reserve figures. All production statistics for nonferrous, precious, and rare metals were declared secret in 1956, and in 1977, import and export data were restricted. Not only is the trend for more, rather than less, suppression of information, but data that are available are often thought to be intentionally distorted.[3] With this knowledge, it can be seen that Western evaluations of Soviet mineral or fuel statistics must clearly be regarded as approximations at best. When using data from this or any study written in the West, a healthy and ongoing skepticism is therefore mandatory.

The USSR uses a general mineral reserve classification system that corresponds imprecisely to Western systems.[4] The Soviet $A + B + C_1$ categories approximate our own proved plus probable reserves; again, however, this is not an exact correlation. In addition, estimation particularly of petroleum reserves is at best an uncertain science, even in the West. Given the dynamic character of a reserve base, geologic appraisals must not be regarded as definitive.

With these reservations in mind, it is not startling to learn of the wide range of published reserve estimates of Soviet oil. Lowest estimates are contained in the controversial CIA report of 1977,[1] citing reserves of only 4.1 to 4.8 billion tons* (30 to 35 billion barrels). The Swedish group Petrostudies produced the most optimistic data—16 billion tons (116 billion barrels).[5] British Petroleum estimates 8.6 billion tons (63 billion barrels),[6] and the respected scholars on the USSR, Dienes and Shabad, 10 to 11 billion tons (73–78 billion barrels).[7] Eight to thirteen billion tons (58.5 to 95 billion barrels) represents a reasonable range used by most Western authorities. The magnitude

---

*All data in this study, unless otherwise indicated, are in metric tons.

of Soviet resources is unknown, but is obviously much larger than its reserves.

Producing Regions

In 1981, the USSR produced 609 million tons of oil, an increase of only 1% over its 603-million-ton output of the year before. Western Siberia is now by far the most important petroleum-producing area, accounting for more than 50% of total national production. One supergiant field (larger than 1 billion tons) alone in the Tuymen province of Siberia—Samotlor—yields one-fourth of all Soviet oil (see Figure IV-3-1).

Historically, the USSR is one of the oldest oil-producing countries in the world; Baku in Azerbaidzahn on the west coast of the Caspian Sea was an oil region when ceded to Russia by the Persians in 1813. The Caspian and Transcaucasus districts were leading producers until the mid-1950s, when deposits in the Urals-Volga fields came on line. This area peaked in 1975, its production being surpassed by the rapid development of western Siberia.

Western Siberia. The present remarkable state of Siberian output is underscored by the fact that in 1965, the region contributed only 0.4% of the USSR's petroleum. Output has doubled here since 1975. Samotlor, with reserves exceeding 2 billion tons, produced more than 150 million tons (3.1 million barrels per day) in 1980.[8] Since oil was first found in Siberia in 1960, fast development has been recorded despite the severe geographic constraints discussed previously. This has proven quite expensive. Nonetheless, although production costs are very high, the success in Siberian extraction of both oil and natural gas is a testimony to the Soviet ability to overcome difficult barriers in pursuit of a goal. Western Siberia is expected to increase in importance; 66% of all oil is anticipated to come from this region by 1985, and 75% by 1990 (see Table IV-3-1).

Urals-Volga. The region between the Volga River and the Urals, although declining in significance, is still responsible for a third of Soviet output. It is, in fact, the third largest producing district in the world. More moderate climates and accessibility to infrastructure (and therefore markets) greatly helped in the development of this area. The Urals-Volga region is still dominated by a supergiant field, Romashkino, in the Tartar Republic. This field, found in 1948, was one of the first big finds for the Soviet oil industry. Although in decline, it is still an important reservoir.

Caspian Basin and North Caucasus. This large territory, previously the heart of the industry, contains hundreds of wells. Despite massive injection of labor and capital, yields are steadily de-

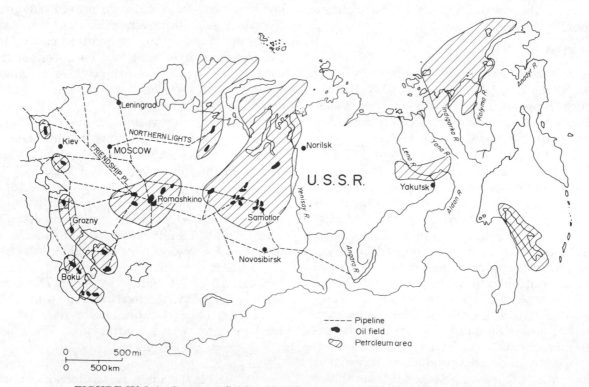

FIGURE IV-2-1   Soviet oil fields and major pipelines. (From Refs. 4, 5, 7, and 11.)

TABLE IV-3-1   Recent Soviet Petroleum Production (Million Tons)

	1975	1976	1977	1978	1979	1980	1981	1985 (plan)
Total	491	520	546	571	586	603	609	630
Siberia	148	181.7	218	254	284	312	334.3	395
Samotlor	—	—	—	—	—	150	—	—
Urals-Volga	226.7	221.7	219.2	213	201	193	—	150
Caspian	56.7	54.6	50.7	47	43	40.4	—	50
Komi	11	12	14	17	19	21	—	28
Ukraine	—	—	—	—	8.3	7.7	—	—
Sakhalin	—	—	—	—	—	2.4	—	—

Sources: Refs. 5, 11, and 31.

creasing. Production in 1980 fell short of regional quota by nearly 25 million tons; this shortfall accounted for almost the total national shortfall of 37 million for the year. The Caspian-North Caucasus is not expected to regain its important producing status, despite increasing use of advanced recovery techniques.

Ukraine. Petroleum from the Ukraine is of high quality; a further asset is its proximity to major markets and refineries. However, unfavorable geologic conditions and physical depletion now characterize the region's deposits. Beyond its local importance, the Ukraine is not a major contributor to national output; production was less than 8 million tons in 1980.

Komi. The Komi Republic lies in the Arctic, bordering the frozen Barents Sea. Although bitter climate is a disadvantage, this area is still closer to markets than western Siberia is. Infrastructure is, however, a problem. Komi is the only older producing region not now in decline. Should current exploration efforts in far North Komi prove successful, the district could be of increasing importance.

Sakhalin Island. Located off the east coast of the USSR, Sakhalin lies north of Japan's Hokkaido Island. Although present output is small (less than 3 million tons), a joint Soviet-Japanese venture is now involved in offshore development here. Sakhalin Island is viewed as a key test case for future joint projects between the two nations.[4] It is also indicative of the importance placed on energy development by the USSR. Although joint ventures are not new for the USSR, it is a decided reversal of post–World War II policy for foreign businesses to be allowed on Soviet soil.[2] It is certainly a sign that the USSR recognizes that its economy would have difficulties undertaking the move into Asia unaided.

Transport of Soviet Petroleum

Pipelines provide the cheapest means of moving raw petroleum; 85% of Soviet crude is transported in this manner.[7] Although construction costs are huge, they are largely offset by low operating and maintenance costs. Large-diameter pipe is especially cost-effective; pipelines as big as 1420 millimeters (56 inches) are now utilized.

At the end of 1980, 59,000 kilometers (36,639 miles) of oil pipelines were in place in the USSR.[9] Railroads are still used extensively to move finished petroleum products. This is due to lags in products pipeline construction and to the scattered nature of final destinations, rather than to any inherently higher efficiency in the railroad. In fact, rail transport is three times more expensive than pipeline transport. However, refineries are located in areas of established infrastructure, allowing easier movement of finished product by rail. Priority thus goes to pipeline construction in Siberia, where pipelines are the sole feasible means of moving oil from the large Tyumen and Tomsk oilfields (see Figure IV-3-1).

The impact of Siberian production is evidenced by the average distance crude is transported—this distance has tripled over the past 15 years.[10] One drawback to reliance on pipeline transportation is the Soviet dependence on the importation of high-quality steel pipe. This is of most significance in the transmission of natural gas. In addition, reliance on long pipeline networks increases the potential for serious supply interruptions due to sabotage, military action, or accidental breakage.

Petroleum Refining

There are presently 44 oil refineries in the USSR, having a total capacity of some 537 million tons.[11] As can be seen from Figure IV-3-2, there is a heavy bias toward the older producing regions in the European part of the country. Siberia, while pro-

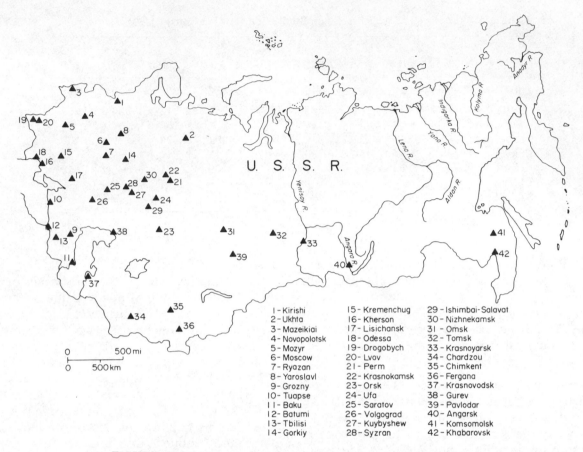

1 - Kirishi	15 - Kremenchug	29 - Ishimbai-Salavat
2 - Ukhta	16 - Kherson	30 - Nizhnekamsk
3 - Mazeikiai	17 - Lisichansk	31 - Omsk
4 - Novopolotsk	18 - Odessa	32 - Tomsk
5 - Mozyr	19 - Drogobych	33 - Krasnoyarsk
6 - Moscow	20 - Lvov	34 - Chardzou
7 - Ryazan	21 - Perm	35 - Chimkent
8 - Yaroslavl	22 - Krasnokamsk	36 - Fergana
9 - Grozny	23 - Orsk	37 - Krasnovodsk
10 - Tuapse	24 - Ufa	38 - Gurev
11 - Baku	25 - Saratov	39 - Pavlodar
12 - Batumi	26 - Volgograd	40 - Angarsk
13 - Tbilisi	27 - Kuybyshew	41 - Komsomolsk
14 - Gorkiy	28 - Syzran	42 - Khabarovsk

FIGURE IV-3-2   Soviet oil refineries. (From Refs. 5, 11, and 40.)

ducing half the petroleum, contains only one refinery, at the city of Tomsk. This limited geographic distribution strains existing rail capacity in delivering petroleum products.

Poor product quality and an inadequate end mix are named by the OTA[5] as adverse features of the Soviet refining industry. Fuel oil now represents about 50% of refinery output; this compares with 13% in the United States and 35 to 40% in Western Europe.[11] Low demand for gasoline in the USSR, due to limited private ownership of automobiles, contributes to a lack of secondary refining capacity.

Factors Contributing to Production Difficulties

Production Methods. Utilization of the turbodrill in drilling, and widespread waterflooding in well development, are the two chief Soviet production techniques that have long inspired Western disapproval. The turbodrill was invented in response to the Soviet inability to manufacture high-quality, high-stress steel pipe. Rather than use the rotary drill, standard in the West, which employs the revolution of the entire drill string, the turbodrill uses high-pressure drilling fluid forced onto turbine vanes at the well bottom to turn the bit.

Only this portion has movement, rotating at three to four times the rotary drill speed. Many Westerners claim that the turbodrill is useless at great depths, and that it drills much more slowly than its American counterpart. Others dispute these objections, claiming not only great efficiency at depth, but drilling times faster than the rotary drill can produce.[11]

Very high oil recovery rates, averaging 40 to 50% and as high as 50 to 60%, are claimed by the Soviets; this is due primarily to their practice of routine water injection (water-flooding) early in field development. In 1980, 85% of all Soviet wells used this method; in Siberia the figure was 99%.[11] The result is lower initial capital requirements due to a decreased need for the number of wells, and a higher early output. Critics say that this damages fields and reduces the ultimate yield from a deposit. The Soviets take strong exception with these views. In either case,[5]

While it is misleading to generalize about the water-cut rate for the USSR as a whole because of important variations between regions and fields, there is no doubt that poor management has led to damage in some fields, or that the fluid-lift requirements occasioned by waterflooding are burdensome.

Indeed, for 1980, the average water cut was expected to reach 57%; as early as 1976 this number was exceeded in Siberia. The true handicap induced by elevated water cuts is that a strong industry-wide need exists for high-capacity submersible electric pumps. Since Soviet-made pumps are of both low capacity and quality, importation is necessary. By 1985, the utilization of submersible pumps is anticipated to provide as much as 40% of all oil.

Imported Equipment. The low quality of Soviet equipment and spare parts extends beyond pumps, to encompass a broad spectrum of required supplies, including those needed in exploration, drilling, production, and particularly in offshore development. To rectify these deficiencies, the Soviet oil and gas sector accounted for more than 20% of Soviet nonagricultural imports in 1979, valued at $2.7 billion.[5] Interestingly, the United States has not been, or is now, the dominant supplier of oil and gas or other energy-related Soviet imports. U.S. exports in 1979 were only 3.3% of total oil and gas equipment trade from the West; Japan, West Germany, Italy, and France are by far most significant. Seismic and gas-lift equipment, as well as submersible pumps, are important trade items, but certainly the main import is large-diameter pipe.

Exploration. Exploration has been neglected in recent years in the USSR, prevailing incentives tending to work against the allocation of resources to exploration.[5] This has resulted in an emphasis on developmental drilling at the expense of exploratory work. As a consequence, the reserves to production ratio has been deteriorating since 1967.[7]

In the past, as one region declined, new large fields were available to offset depletion. For instance, all gains in net output at present are coming from Siberia. After the exhaustion of this area, however, there are no known supergiant oilfields awaiting development. Samotlor is projected to peak in this decade. While areas of great petroleum potential exist, these are for the most part in regions even more remote and less hospitable than Siberia, such as the Barents, Kara, and East Siberian Seas. Should discoveries be made in these or other offshore areas, or in Eastern Siberia, long lead times will precede production for the foreseeable future.

Future Prospects

Is the Soviet oil industry facing absolute production downturns, as the CIA has forecast? On a percentage basis, output increases show substantial declines. Figure IV-3-3 indicates yearly percent

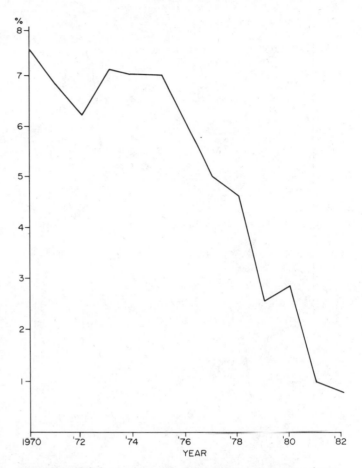

FIGURE IV-3-3 Soviet petroleum production annual percentage increase, 1970-1982. (From Refs. 4, 5, and 31.)

increases since 1970. This graph suggests that the CIA may be correct. Such an interpretation is reinforced by Figure IV-3-4, depicting yearly incremental production tonnages. Slowdowns in output are thus genuine. Note that incrementally a peak in production growth was reached in 1975. By charting total output, however, quite a different picture emerges. Figure IV-3-5 plots production since 1941; the early drop is due to the war, but beginning in 1946 there has been virtually linear increased growth.

There is ample reason to believe that the Soviets regard the petroleum industry as a crucial sector. Wilson[11] notes that the industry is more carefully monitored and planned than any other component of the economy outside the military. It is felt that any policies necessary would be implemented to ensure a reasonable supply. There is no proof of extreme measures (such as rapid resource reallocation) that might indicate clear alarm from GOSPLAN (State Planning Agency). Despite all the problems outlined above, quotas are steadily increasing, and output is generally near production targets. Huge potential still exists with enhanced recovery, with deeper or offshore drilling (still in

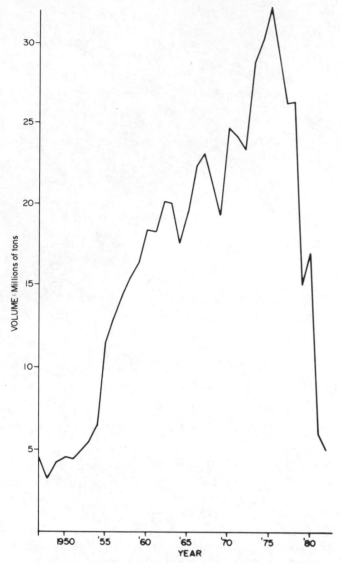

FIGURE IV-3-4 Soviet petroleum production incremental yearly output, 1947-1982. (From Refs, 4, 5, and 31.)

its infancy in the USSR), and with additional foreign technology.[4]

Finally, and perhaps most important, the potential for significant future finds must not be discounted. Thirty-seven percent of the world's total sedimentary basins lie in the USSR, compared with 11% in the Middle East (and only 2% in all of North America).[4] Although many of these basins have been explored, evaluation has been largely of a primitive nature. Again, any impact from prospective discoveries would not be significant before the next decade.

The CIA report must thus be viewed in a "worst case" perspective, of low probability. As Hoffman[12] and the OTA[5] have emphasized, a wide range of production possibility interpretations has been published for 1985, varying from 500 to 700 million tons. The Soviets themselves have targeted 614

million tons for 1982, and 630 million for 1985. A reasonable estimate seems to be, as OTA has postulated, stable or moderately increasing production. Such an appraisal could prove improper should resources be allocated in favor of the booming gas industry. Indeed, the debate over Soviet petroleum has often obscured the true latent strength of the country's energy picture. It is a strength sustained not by oil, but by natural gas.

NATURAL GAS

The Soviet natural gas industry was the only energy sector to meet 1980 production targets. In 1981, output of 465 billion cubic meters (BCM) was in excess of quota (458 BCM). Production has been increasing by 7% or more annually. In 1980 natural gas assumed second place as a percentage of total energy production (see Figure IV-3-6). Gas exports are anticipated to replace petroleum as the most important trade commodity before 1985, thus becoming the major hard-currency supplier. Finally, gas to CMEA may largely replace oil exports to this area, freeing more oil for shipment to the West. These trends all indicate that natural gas is the fuel of the future for the USSR.

Resources and Reserves

If controversy envelops Soviet oil, natural gas is marked by widespread agreement—USSR reserves are enormous. They are, in fact, so large that the typical Soviet penchant for secrecy is discarded. Listed reserves range from the Soviet's own A + B + C_1 value of 20 trillion cubic meters (TCM)[10] to a high of 34 TCM.[13] This represents approximately 40% of world reserves. The extent of resource holdings is unknown, but is assuredly mammoth.

The marked concentration of gas in western Siberia—specifically, in Tyumen province—is striking. This area, which in 1965 had listed reserves of only 400 BCM, now accounts for 27 TCM, almost 80% of all Soviet gas. Siberia contains six fields having reserves of 1 TCM or greater. These are Urengoy, the world's largest field, at 6 TCM; Yamburg, the world's second largest, at 4.4 TCM; Zapolyarny, reserves of 2.7 TCM; Medvezhe, at 1.6 TCM; Kharasavei, at 1 TCM; and Bovanenko, at 1 TCM[11] (see Figure IV-3-7).

Approximately 10% of the reserve base is found in Central Asia (mainly in the Uzbek and Turkmen Republics); the remainder is in the European USSR. From 1965 to 1976, the yearly addition to reserves exceeded 20%; while slowing in the last five years, the reserves to production ratio is today still greater than 50:1.

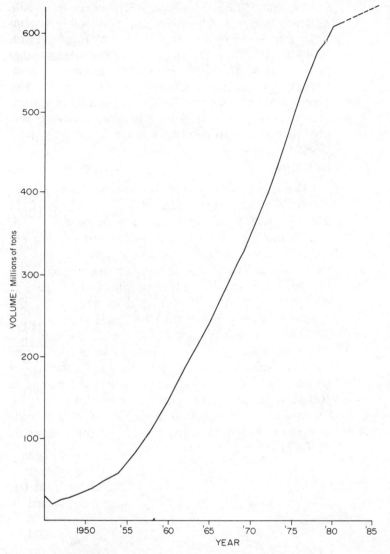

FIGURE IV-3-5 Soviet petroleum production, 1941–1985. (From Refs. 4, 5, and 31.)

Producing Regions

Natural gas production was neglected until deposits in the Ukraine were discovered in the mid-1950s. Early output was usually associated or casinghead gas, that is, gas associated with petroleum production. The leading producer area from 1960 to 1975 was the Ukraine, followed by Central Asia from 1975 to 1978, and western Siberia from 1978 to the present (see Table IV-3-2).

Western Siberia. The importance of Siberia for gas production is commensurate with its vital influence in petroleum—almost all net increase in natural gas output for the 1980s will come from there. Urengoy alone is projected to produce 275 BCM by 1985; this will be more than 40% of planned production for that year, and will be more than total Soviet gas production in 1974. The field will receive investment for the 1981–1985 period equal to all Siberian gas investment in the last (tenth) FYP.[14] Urengoy's significance is further highlighted because it is the source of the gas export pipeline to Western Europe.

In 1981, Siberian production, at 190 BCM, contributed more than 40% of Soviet natural gas (see Table IV-3-2). Output for 1985 is planned at 360 BCM, a 900% jump over 1975 Siberian gas output. This growth has occurred and is ongoing despite geographic characteristics even worse than those faced in the Siberian oil industry, since many gas deposits lie in the extreme north. Production of natural gas in Siberia began at Medvezhe in 1972. Yamburg, north of Urengoy, is projected to supply most incremental increases for the Twelfth FYP (1986–1990). Many more deposits await development, indicating that western Siberia has massive untapped potential.

Central Asia. The Central Asian region provided approximately 23% of Soviet natural gas in 1981; this desert area has been a significant producer since Uzbek's big Gazli deposit was developed

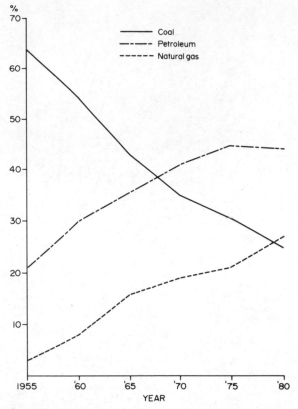

FIGURE IV-3-6 Percent share of total energy production by primary energy carrier in the USSR (From Refs. 10 and 42.)

in the mid-1960s. The high sulfur content typical of the district has necessitated processing before transport to remove impurities. Although Gazli peaked in 1971, the Turkmen Republic's huge Shatlyk field (one of the world's largest) is anticipated to maintain Central Asia's place in the production hierarchy. This is especially so since consumption here is relatively low, allowing surplus gas to be piped to the Urals or European USSR markets.

Orenburg. Located in the comparatively milder climate of the South Urals, the Orenburg gas field was discovered in 1966. Reserves exceed 1 TCM, but as in Central Asia, high sulfur and condensate levels require refining before piping to destinations. Orenburg is the source of the Soyuz Pipeline, one of the more ambitious joint CMEA projects undertaken to date. Completed in late 1978, the pipeline exports about 60% of Orenburg's output: 15.5 BCM annually to six CMEA nations, and 12.5 BCM to Western Europe. Three processing stations exist at Orenburg to upgrade the gas for transport —two treat domestic gas, one refines gas for export. Refining capacity has remained a limitation on the expansion of production here. Consequently, output is projected to remain at its current level of 48 BCM.[5]

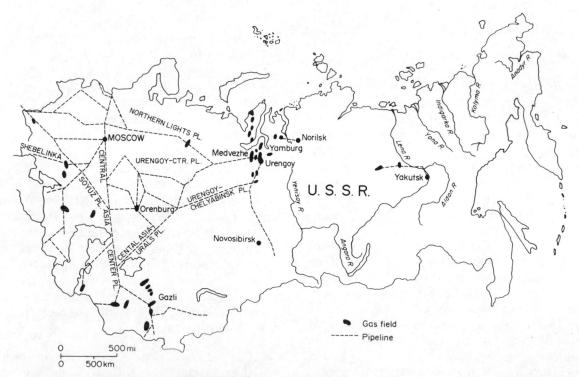

FIGURE IV-3-7 Soviet natural gas fields and major pipelines. (From Refs. 4, 5, 7, and 11.)

TABLE IV-3-2 Soviet Production of Natural Gas (Billion Cubic Meters)

	1975	1976	1977	1978	1979	1980	1981	1982	1985
Siberia	35.7	47.7	67.9	92.7	123	156	190	—	360
Central Asia	89	99	103	104	107	109.5	106	—	—
Orenburg	—	31.8	34.6	45	48	48	48	—	—
Ukraine	68.7	67.2	65.1	63	60	51	47	—	—
Komi	18.5	18.9	18.9	19	19	22	—	—	—
Total	289	321	346	374	407	435	465	492	630

Sources: Refs. 5, 11, 31, and 34.

Ukraine. The first big gas-producing district for the USSR, the Ukraine is now experiencing substantial decline. Shebelinka, the initial big field, and other large deposits, are facing rapidly increasing production costs. Nonetheless, the Ukraine's favorable proximity to industrial centers will prompt large investment, in hopes of maximizing gas recovery. In this regard it can be compared to the Caspian Basin oil region.

Komi. The far northern Komi region faces problems similar to western Siberia's climate and infrastructure deficiencies. An additional concern is the great depth at which gas occurs. Komi, specifically the Vuktyl field, has had some notable impact, as it helped initiate development of the Northern Lights Pipeline. Although Komi has undergone gradual expansion, a slow decline is anticipated for this decade.[5]

North Caucasus. Gas production here is following the declining pattern of Caucasian oil. The two major fields, Stavropol and Krasnodar, peaked as early as 1968. A huge new deposit at Astrakhan, however, should raise output above its current meager 20-BCM level. As in the Ukraine, drops in reserves and output are aggravated by rising production costs.[5]

Transport of Soviet Gas

Pipelines are indispensable in the gas industry, as they remain the single feasible medium for transport. As development occurs in increasingly remote areas, construction costs escalate exorbitantly. Pipelines presently account for 70% of total capital investment in the natural gas sector.[15] Gas is pushed through a pipeline by compressing it, often at extremely high pressures. Capacity can be increased by spacing compressor stations more closely together (usually 80 to 160 kilometers apart). A second technique to raise capacity involves cooling the gas to near 0°F. Since this method also helps prevent permafrost thawing, it is being employed on new lines from Urengoy.[16] It should be understood, however, that gas costs five times as much as oil to transport, and an oil pipeline can move three times the volume of a natural gas pipeline of equal diameter.[11]

The success attained in meeting Soviet bloc energy requirements rests primarily on the ability of the USSR to build more large-diameter gas pipelines.[9] This important feature of pipeline construction has also been emphasized by the OTA.[5] Key constraints are the climate and geography of western Siberia. Most important, however, may be the USSR's dependence on imported high-quality large-diameter pipe, and large compressor units from the West. The Soviets possess inadequate production capability for pipes and compressors. Moreover, the quality of production is again a problem. Although construction targets are rarely met, Wilson[11] finds pipelaying "fast and efficient." There are presently more than 132,000 kilometers of gas pipelines in the USSR. Despite this large network, pipeline capacity remains a prominent barrier to expansion of the gas sector.

Included in the major trunkline system today are three lines that originate from the Siberian gas fields of Urengoy, Vengapur, and Medvezhe. (See Figure IV-3-7, which maps natural gas pipeline locations.) The Northern Lights Pipeline runs through the Vuktyl field in the Komi republic, then on through Cherepovets and Minsk, one branch extending all the way to the Czech border town of Uzhgorod, allowing gas exportation. The export pipeline to Western Europe follows the Northern Lights route. The Tyumen-Moscow (Urengoy-Center) Pipeline brings gas to the capital via Nizhnaya Tura, Perm, Kazan, and Gorki. The Tyumen-South Urals (Urengoy-Chelyabinsk-Center) Pipeline runs through Chelyabinsk, Ufa, and Kuybyshev, then onward into the Ukraine, to help compensate for declining production there. Two major lines from Central Asia (particularly the Gazli and Shatlyk fields) carry gas northward: the Central Asia-Urals Pipeline brings gas from Gazli to the Chelyabinsk region; the Central Asia-Center Pipeline extends all the way to Moscow. Finally, the Soyuz Pipeline, 2750 kilometers long, exports gas through Uzhgorod from Orenburg—2.8 BCM annually go to Czechoslovakia, Bulgaria, Hungary, East Germany, and Poland, with 1.5 BCM yearly to Romania. In exchange for the gas, Romania financed the French construction of the processing unit that treats the export-bound gas (located at Orenburg); the remaining five countries supplied capital and labor, being responsible for approximately 500 kilometers of pipeline each.

Many smaller branch lines either now exist or are being constructed off the major trunklines. Most prominent of these projects are pipelines from Solenoe and Messoyakh (northeast of Urengoy) running to the important copper-nickel-cobalt-platinum metallurgical center of Norilsk (the gas to replace dependence on costly local coal); and a pipeline from the Samotlor oil field to the large Kuznetsk Basin industrial district to the southeast. This will enable associated gas formerly flared to be utilized.

In the Eleventh FYP, more than 20,000 kilometers are planned to be laid.[17] Total cost is in excess of 25 billion rubles.[18] Hewett[13] has estimated pipe requirements of 14 million tons for trunkline expansion projects until 1985, only 4 million tons of which can be domestically produced. Three million tons alone of the required pipe is earmarked for the controversial Urengoy export pipeline to Europe. Total pipe import bill is estimated greater than $6 billion.

Western technology is even more important for gas compressor stations, especially the large 16- and 25-megawatt units. These larger compressors not only increase pipeline capacity, but also shorten construction times by reducing the required number of stations.[5] Soviet production of compressors is at the incipient stage, and will prove insufficient for its needs for some time. It should again be noted that the United States has no monopoly on this gas technology, and does not even manufacture the 1420-millimeter pipe in demand by the USSR.[5]

Gas Processing

Treatment of gas is necessary before transport in both Central Asia and Orenburg to remove corrosive sulfur and condensates. Construction of processing capacity has not kept pace with natural gas production, and has been a limiting factor particularly at the latter location. Recovery of associated gas has been a noticeably weak phase of the industry, prompting major efforts to reduce wasteful flaring. As with removal of gas impurities, refining capacity remains an obstacle to this goal. All new refineries during the current FYP are to be built in Siberia; gas refining on site in Tyumen province is lagging 5 to 10 years behind requirements.[19] The goal is to totally eliminate flaring of associated gas in Siberia by 1985.

Future Prospects: The Significance of the Urengoy Export Pipeline[20]

The question of how severe a constraint pipeline construction will impose on the gas industry has been answered by the USSR-Western Europe export pipeline from Siberia. It now appears that Western participation in the project assures the USSR of the financing and supplies necessary for rapid development of its remote gas deposits. This pipeline, only one of six major trunklines planned from Urengoy during the eleventh FYP, will stretch more than 4500 kilometers to Uzhgorod, and will require more than 40 large compressors. Gas deliveries to Europe will increase from present levels of 28 BCM per year to as high as 60 BCM;[13] these were slated to begin in early 1984. Construction is greatly speeded by following the corridor of pre-existing lines.

At present, Europe relies on OPEC for more than two-thirds of its oil. To overcome this dependence on such an unstable supplier, substitution, conservation, and supply source diversification have been attempted. Opposition to nuclear power has prompted a larger role for natural gas. In the past decade consumption has grown to now account for 14% of all energy used. The potential for further major European gas production increases, however, is limited. The huge gas reserves of the USSR have thus made it a logical supply choice. Diversification of supply sources, then, is a key motivation behind Europe's role in the export pipeline.

Probably of equal importance, however, has been the boost to critically depressed sectors of the European economy (e.g., the steel industries of France and West Germany) that cooperation in the project has provided. Multimillion-dollar contracts for pipe and compressors are deals European firms badly need. After all, political leaders in Europe are just as sensitive to high unemployment statistics as are their American counterparts. European jobs, rather than supply diversification, have inspired participation with the Soviets.

The loud, belated, and ineffective opposition from the United States to the pipeline focused on several aspects. First, there is definitely some question as to the wisdom of regarding the USSR as a more secure supplier than OPEC. Historically, the USSR cut off its energy exports to Yugoslavia in

1948, Hungary and Israel in 1956, Czechoslovakia in 1968, and threatened to do so in Poland in 1981. It is true, as the Europeans point out, that gas has been imported from the USSR for a number of years with no such interruptions, and that even under planned importation increases, dependence on Soviet gas would still be about 5% of total energy consumption in all nations but Austria (where it would represent 14% by 1990)[21] (see Table IV-3-3). This does not present an altogether accurate portrayal, because certain regions will have a much greater dependence than this low average would imply. For instance, 75% of West Germany's imported gas is to be used by West Berlin.[22] Given the record of acrimony with the Soviets over this divided city, reliance on the USSR for energy supplies of this magnitude may prove disastrous. To cite a further example, Bavaria is already 80% on Soviet gas.[23]

Second, U.S. resistance has stemmed from a belief that business with the USSR may be one thing, but the extremely low interest rates on the Soviet loans (which are on a gas-for-pipe basis) amount to a subsidy for the expansion of Soviet energy. This will guarantee increased Soviet hard-currency earnings from future gas exports, at a time when such money is critically required. It has been proposed that "the large financial and technical commitment would help her [the USSR] to avoid internal reform and sectoral re-allocation of scarce resources."[24] The concern here is that subsidies allow the Soviets to free up large sums, probably to be invested in military programs. Although the United States has also been opposed to technology transfer for the project, finally leading to a temporary (and quite divisive) embargo of such technology, equipment is actually of a low technical sophistication, and is generally available from several sources. On the other hand, although the USSR declares that these imports are not required to construct the pipeline from Siberia, pipe and compressors are immensely cost-effective in terms of transport efficiency.

Two points are suggested by the Urengoy export pipeline. First, the ad hoc bilateral basis on which the deal has been made is directly attributable to the lack of a consistent U.S. policy toward the USSR. Where a dearth of Western unity of such embarrassing proportions occurs, the advantage clearly accrues to the Soviets. Second, as Karr and Robinson[23] have elucidated, "a clear distinction must be made between ordinary commercial relationships and strategic implications." Although it is important to seek out foreign trade to stimulate a depressed economy, it is crucial to avoid placing oneself in a vulnerable dependency leading to political leverage for an avowed enemy. A balance between "business" and strategic self-interests must be struck. It is precisely such a leveraging potential that makes grain purchases from the West an act of last resort for the Soviets.

Ultimately, the Urengoy export pipeline helps promise attainment of gas targets and a continuation of the expansion evident in Figure IV-3-8. Not only is gas the domestic Soviet fuel of tomorrow, it is also to be the dominant export commodity. Constant upward revisions of the 1985 quota, now at 630 BCM, are due in no small way to Europe's decision to assist in the development of Soviet natural gas.

COAL

The Soviet energy balance was once dominated by coal; industrialization under Stalin was fueled by this substance. Over time, however, coal's importance has slipped in favor of oil and gas. As Figure IV-3-6 illustrates, coal's share of production has dropped from near 70% in the 1950s to 28% in 1980. The slowdown in the oil sector, however, denotes, as in the West, a renewed emphasis on this old energy source.

Resources and Reserves

Soviet coal reserves are huge. The $A + B + C_1$ figure in 1975 was listed by the USSR at 420 billion tons, including 87 billion tons of coking coal and 190 billion tons of lignite.[10] Resource statistics are claimed to be as high as 12 trillion tons, or more than 50% of world coal resources.[25] Western sources publish a wide range of both reserve and resource data. The most conservative are derived from the World Energy Conference Survey of Energy Resources in 1978, which recorded reserves of 110 billion tons and resources of 2.4 trillion tons.[26] Highest numbers have been used by Cooper,[27] who compiled information from various sources to arrive at reserves of 500 billion tons and resources of 10 trillion tons.

TABLE IV-3-3 Soviet Gas Imports as a Percent of Total Gas and Total Energy Consumption

	1979		1990	
	Gas	Energy	Gas	Energy
West Germany	14	2	24	5
France	0	0	17–20	3
Italy	29	5	23	4
Austria	59	12	62	14

Source: Ref. 21.

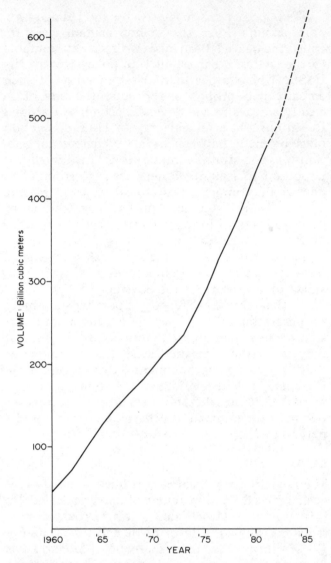

FIGURE IV-3-8 Soviet natural gas production, 1960–1985. (From Refs. 5, 31, and 41.)

Regardless of which source is adopted, the USSR plainly possesses abundant coal. The problem is, again, one of location, for due to depletion in the European USSR, from 70 to 90% of all coal now is located in Asia. It is predominately the remote character of the Tunguska (2 trillion tons) and Lena (1.5 trillion tons) Basins that prevents their inclusion in the reserve category; these areas are actually years away from development. Other basins still in the "resource" classification include Taimyr, Chukotka, and the Zyryanka (see Figure IV-3-9).

Producing Basins

Donetsk. The Donetsk Basin (Donbass) occupies 60,000 square kilometers in the Ukraine; reserves exceed 40 billion tons. This district is quite important to the USSR—it has excellent proximity to major industrial markets (especially the Krivoi Rog steel complex), and its coal is high-grade coking and anthracite. Over 40% of Soviet coking coal comes from Donetsk.

Mining conditions, however, are among the worst in the world. The Donbass is one of the Soviet's oldest underground mining areas. Serious depletion and production declines are apparent. Over one-third of mines here have been exploited to exhaustion.[7] As the highest-quality coal has become depleted, calorific value of coal produced dropped; the heating value of Donetsk steam coal declined 18% during the 1970s. Average mining depth is 566 meters, double the depth of 15 years ago, and almost one-third deeper than other Soviet basins. Seam thickness averages 0.9 meter, 25% less than 10 years ago and also one-third below the rest of the nation's mines (see Table IV-3-4). Over half of Donetsk longwall mining is on seams less than 0.7 meter thick. Excessive methane levels exist in most mines as well. In fact, on the basis of depth, methane volumes, and seam thickness, Donetsk coal would not even be classified as reserves in the United States. In 1980, the basin produced 204 million tons, a decline of 4 million tons from the preceding year (see Table IV-3-5). Despite this continuing output slowdown, Donetsk remains the leading Soviet producer.

Kuznetsk. Located in southwest Siberia, the Kuznetsk Basin (Kuzbass) covers some 26,000 square kilometers. Reserves are estimated at 60 billion tons. Kuznetsk contains good-quality coking coal in thick shallow seams, allowing production costs low enough to compete with Donetsk coal in many western areas, even after being railed 1500 miles. As Donetsk stagnates, this trend for flows of Kuzbass coal westward should intensify. Kuznetzk also figures prominently in foreign trade, representing 15% of coal exports in 1979. This basin is projected to take over as the leading metallurgical coal producer by 1985. There are, however, indications of production declines, prompted by labor shortages and the failure to add new capacity. Output was 141 million tons in 1980.

Ekibastuz. This basin covers some 160 square kilometers in north-eastern Kazakhstan and contains more than 8 billion tons of reserves. Seams are thick and amenable to strip mining. Surface operations serve to keep labor productivity high, while minimizing costs. Coal quality, however, is poor, being subbituminous, low in calorific value, and as high in ash content as 40%. Mining techniques, even by Soviet standards, are mediocre. In 1980, 46% of output was rock or waste.[28] While processing could reduce the ash content to 38%, this is still double the maximum acceptable U.S.

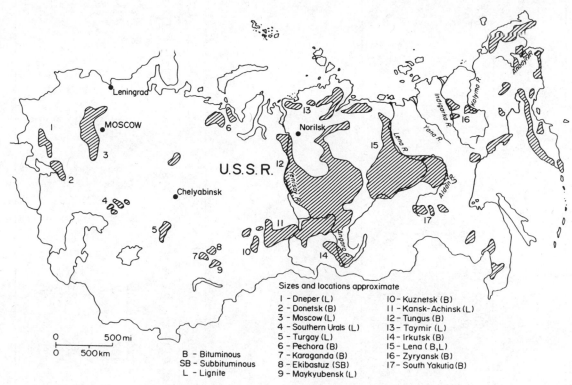

Sizes and locations approximate

1 – Dneper (L)	10 – Kuznetsk (B)
2 – Donetsk (B)	11 – Kansk-Achinsk (L)
3 – Moscow (L)	12 – Tungus (B)
4 – Southern Urals (L)	13 – Taymir (L)
5 – Turgay (L)	14 – Irkutsk (B)
6 – Pechora (B)	15 – Lena (B,L)
7 – Karaganda (B)	16 – Zyryansk (B)
8 – Ekibastuz (SB)	17 – South Yakutia (B)
9 – Maykyubensk (L)	

B – Bituminous
SB – Subbituminous
L – Lignite

FIGURE IV-3-9   Soviet coal basins. (From Ref. 5.)

standard for power plant steam coal. Four large electrical generating facilities are planned for the Ekibastuz territory, having a capacity of 4000 megawatts each. Production at the basin was 66 million tons in 1980, representing nearly one-fourth of all surface-mined Soviet coal.

Karaganda. Karaganda, southwest of Ekibastuz, covers 3000 square kilometers. Although containing some 8 billion tons of reserves, more than half are below 300 meters. Karaganda contains both coking and steam coal, much of which must be beneficiated. A shortage of treatment capacity leads to some extractable coal being left in place. Production, which began in the 1930s, has been slowing since the mid-1970s, due to serious lags in additions to mines, labor shortages, and equipment problems. This basin is an excellent example of poor planning—the city of Karaganda was built on top of 1 billion tons of coal.[5] Great effort is required to minimize surface subsidence.

Kansk-Achinsk. This basin, lying just east of Kuznetsk, contains the largest explored reserves in the USSR at 72 billion tons. Coal is surface-minable at low cost, but it is of low rank (mostly lignite); it cannot be shipped over long distances (such as to the European USSR markets) because it combusts spontaneously. Consequently, almost all output is used locally. Most important, the coal's variable

ash fusion temperatures constitute a slagging problem in boilers. A prototype boiler capable of utilizing Kansk-Achinsk coal has yet to be developed. The basin is earmarked for very ambitious projects. From eight to ten 6400-megawatt power stations are planned, with some plants having coal liquefaction capabilities. Kansk-Achinsk is scheduled for large output increases, but these are definitely contingent on suitable boiler development.[5]

Pechora. Located near the Yamal Peninsula in extreme northeastern European USSR, the Pechora Basin's bitter climate makes mining costly and difficult. The region has a high rate of methane explosions. Explored reserves are 8 billion tons, but much coal is below 300 meters. Two-thirds of production is high-ash coking coal, which helps support the Cherepovets steel center; the remainder is steam coal. Although development began here during World War II, little new capacity has been added in over 15 years. Output in 1980 was 28 million tons.

Moscow. The Moscow Basin (Podmoskovny) lies to the west of the capital; it covers some 120,000 square kilometers. Production has been declining here since 1958. While the basin contains reserves of about 5 billion tons, the coal is mostly low-ranking lignite—high in ash and sulfur, low in calorific value. Currently, the Moscow Basin is of

TABLE IV-3-4   Characteristics of Major Soviet Coal Basins

Basin	Type of mining	Explored reserves (billion tons)	Average seam thickness	Average mine depth (meters)	Average calorific value (Btu/lb)	Moisture content (%)	Ash content (%)	1980 share of production (%)
Donetsk	Underground	40	0.9	566	10,900	6.5	19.2	31
Kuznetsk	Underground and surface	60	2.5	262	9,990	10.5	19.0	19
Pechora	Underground	8	2.4	454	9,390	8.3	25.1	4
Karaganda	Underground	8	2.5	384	9,250	7.5	28.8	7
Ekibastuz	Surface	8	10–40	—	7,250	7.7	39.1	10
Kansk-Achinsk	Surface	72	—	—	6,490	33.0	10.7	4
Moscow	Underground	5	2.5	135	4,550	32.3	35.5	4

Sources: Refs. 5 and 10.

TABLE IV-3-5   Soviet Coal Production (Million Metric Tons)

	1960	1970	1971	1972	1973	1974	1975	1976	1977	1978	1979	1980	1981	1985
Total	510	624	641	655	668	685	701	712	722	724	719	716	704	770–800
Basins														
Donetsk		218	218	217	219	220	222	224	222	—	208	204	—	210
Kuznetsk		113	116	119	123	128	134	139	142	—	—	141	—	154
Ekibastuz		23	27	32	36	42	46	46	50	57	59	66	—	84
Karaganda		38	40	42	43	45	46	47	48	—	—	48	—	50
Kansk-Achinsk		—	—	—	—	—	28	29	32	—	33	35	—	48
Pechora		22	22	23	23	23	24	26	27	—	29	28	—	28
Moscow		36	37	37	36	35	34	31	30	—	27	25	—	20
South Yakutia		—	—	—	—	—	—	—	—	—	—	3	—	12

Sources: Refs. 5, 28, 31, and 42.

only local importance, providing fuel for regional industries. This decline has been caused by depletion, low coal quality, and concomitant high recovery costs.

South Yakutia. This region may prove to be quite important in the near future. Reserves, once listed at 3 billion tons, are now estimated at 40 billion tons.[25] Production of Yakutsk's thick seams is limited at present (only 3 million tons in 1980); however, this basin is the site of another major joint Soviet-Japanese project, the development including a $450 million compensation loan.[29] By 2000, South Yakutia is targeted to yield 85 million tons of coking coal for export to Japan. The new Neryungri coal complex is slated for an initial capacity of 12 million tons per year. Neryungri is quite close not only to the unfinished BAM being constructed just to the south, but also to the large Aldan Basin iron ore deposits.

Current Problems in the Soviet Coal Industry

Declining Output. Coal production in the USSR has been dwindling since 1978—the first significant decline in 20 years. Output dropped by 2% from 1980 to 1981 (see Figure IV-3-10). To aggravate the situation further, Soviet output statistics are for uncleaned coal; this may overstate tonnage by 20 to 40%.

The major cause of this decline is the depletion of high-quality, easily mined coal; it is especially serious in the older mines. Early Soviet coal mines were in the West, near market centers. As a result of many decades of mining, these deposits are nearing exhaustion. The Donbass thus typifies the problem, its remaining coal seams being deeper, thinner, and more steeply dipping. These conditions impede mechanization, affect mine safety, and promote high costs.

During the 1980s, virtually all production growth will be provided by the newer coal basins in Asia. The coal industry must thus face the climatic and infrastructural constraints confronted by its sister energy industries. As production costs steadily increase at mines in the European USSR, the huge strip-mining operations at Ekibastuz and Kansk-Achinsk, through economies of scale, are characterized by declining costs. Although these low production costs are offset somewhat by low coal quality, transport costs, and costs of providing infrastructure, the success of the Soviet coal industry hinges on the expansion of surface mining at Eastern basins. By 1985, 60% of total output is projected to be east of the Urals.[30]

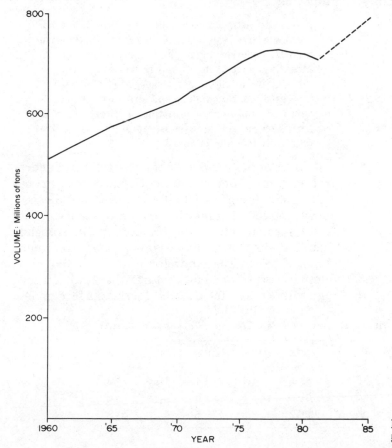

FIGURE IV-3-10 Soviet coal production. (From Refs. 5 and 7.)

Quality Problems. Not only is actual production tonnage on the decline, but calorific value (heat content) as well (see Figure IV-3-11). If past trends continue, calorific value will fall by about 1% per year until 1985, to as low as 8100 Btu per pound.[5,28] Any increment in output could be offset by this trend. In fact, production could increase, but the standard fuel equivalent could actually decline. This is due primarily to growing reliance on the use of lignite, mainly from the surface mines of the Asian USSR. Depletion of high-quality reserves in older coal districts must also be blamed.

High ash content is an additional problem. In 1980, the ash content of shipped coal was 20.3%.[10] In the Ukraine, this figure was nearly 32% in 1975. Obviously, beneficiation is required. Although most Soviet coal does require preparation, less than half is treated, due to a shortage of facilities. As a result, fully 20% of deliveries to power stations is substandard. Primarily export coal and metallurgical coal currently are treated.

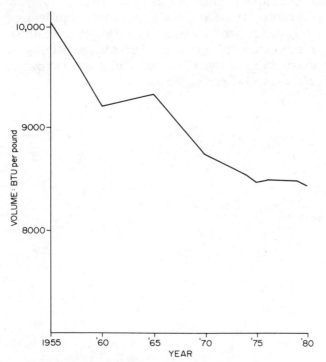

FIGURE IV-3-11   Average calorific value of Soviet coal. (From Refs. 5 and 28.)

Poor quality is an especially significant factor in the development of Siberian coal. By the mid-1980s, Ekibastuz and Kansk-Achinsk will provide one-third of total production. The slagging problems that characterize Kansk-Achinsk coal cause power plants to refuse shipments whenever possible. Ekibastuz coal fares only slightly better. Given the increasing dependence on output from these basins, the poor preparation the Soviets have exhibited for the use of Siberian coal looms as a major obstacle to any potential rapid expansion of national production.

Labor and Productivity Shortcomings. Labor shortages that plague the Soviet economy as a whole are particularly grave in the coal industry. These deficiencies are caused mainly by a low level of mechanization; more than 50% of manpower is still engaged in manual labor.[5] This accounts for the huge work force, which at more than 1 million in the early 1970s outnumbered the U.S. coal labor force more than 6 to 1. By 1981, more than 2 million workers were employed in the Soviet coal industry. Shortage problems could be intensified by a reduction of the workweek to 30 hours in many eastern sectors, this being initiated in response to high labor turnover rates. As with the oil and gas industries,[28]

> Failure to attract and retain a skilled labor force has been the largest single deterrent to Siberian industrial development. For many years special incentives have been offered to workers settling in Siberia, including higher wages, longer vacations, increased pension rights, and privileges in education and housing. Some of these indirect expenses for Siberian coal add substantially to the comparatively low, direct costs.

To accentuate these problems, declines in underground mine labor productivity have been extant since 1975 (see Table IV-3-6). These declines are compensated for somewhat by productivity in surface mines, which is more than eight times higher than in underground operations. At Donetsk, where almost 55% of the labor force is employed, productivity is not one-half that of other underground mines; it is only 15% that of Siberian surface mines.

TABLE IV-3-6   Labor Productivity in Soviet Coal Mining (Metric Tons per Worker per Month)

	1971	1972	1973	1974	1975	1976	1977	1979
Total	62.3	66.3	69.7	73.1	75.4	75.1	75.3	70.2
Underground mining	48.0	40.5	52.6	54.3	55.2	54.6	53.7	48.6
Surface mining	310.0	335.1	362.5	391.2	428.3	435.5	454.0	448.0

Source: Ref. 5.

Productivity gains at Donetsk are of obvious importance. The growing emphasis on surface mining will improve productivity averages; in 1981, however, only 39% of coal was surface mined.[31] By comparison, over 50% of U.S. output comes from open pits.

Failure to Expand Existing Capacity. Increasing competition for funding (the result of the economic slowdown and large military requirements) has prevented the investment necessary to expand present coal-mining operations. As seen in Table IV-3-7, increments in coal investment are lagging behind those in oil and natural gas. The CIA[28] has found that between 1976 and 1979, additions to existing capacity were the lowest in more than a decade. This fact, combined with rising exhaustion rates in the European USSR, causes three-fourths of new capacity to be used to offset depletion. For the Moscow and Karaganda Basins, from 1976 to 1980 all new capacity was required to compensate for depleting reserves. Considering the long lead times inherent in new additions, failure to increase capacity will have deleterious impacts on production growth for at least the 1980s. The extent of the problem is seen in the important Kuzbass, where only one new mine has opened in the last 19 years.

Inadequate Transportation Capacity. Three modes of transport can potentially move Soviet coal: rail, slurry pipeline, and high-voltage direct-current transmission from mine-mouth power plants. At present almost all coal is transported by railroad. Major transport shortages exist in Asia. Some Siberian coals cannot be transported long distances without self-igniting. For coals that are amenable to long haulage, losses during shipment are as high as 3 to 4% of national production.[29] The mismanagement and insufficient capacity that plague the transport sector directly affect coal delivery efficiency. Especially west of the Urals, railroads are overtaxed, the coal share of total shipments here being much higher than the national average. This condition will only grow worse as Siberian coal becomes more important. Until 1985, the largest increases in coal shipments are projected from the Kuznetsk and Ekibastuz Basins, primary destinations being power plants at Chelyabinsk and Sverdlovsk in the Urals. Despite some improvements, little expansion of rail capacity is anticipated.

Although mine-mouth power plants will be crucial at both Ekibastuz and Kansk-Achinsk, attempts at converting coal energy to electricity and transmitting this by wire are encountering difficulties. The problems presented by burning Kansk-Achinsk coal have been mentioned. Using current state-of-the-art 800-kilovolt (kV) lines, transmission from Siberia to the industrial centers of European USSR results in unacceptable power losses. High-voltage lines of 2400 to 2500 kV are needed to transmit over such great distances with losses less than 15%. This technology remains unavailable before 2000. The Soviets are, however, developing a 1500-kV line now that can transmit as far as the Urals.[28] The Soviet strategy of substituting coal for oil in power plants east of the Urals is a response to these problems of long distances, power losses, and high costs. The western regions will utilize nuclear energy.

Soviet slurry technology remains an anomaly. While leading the world in slurry theory (including hydraulic mining), the problems facing the oil and gas industries of insufficient domestic capabilities for the manufacture of pipe, pipeline compressors, and pumps, also hinder coal slurry production. A slurry pipeline from Kuznetsk to the European USSR is scheduled to come on-line by 1990. Many technical problems, including those induced by the severe Siberian climate, remain unsolved. In October 1982, an American corporation, Occidental Petroleum, was involved in talks with the Soviets concerning the construction of a pipeline to carry liquefied coal from Siberia to Moscow.[32] Generally, however, rail and wire transmission will convey the bulk of Soviet coal for the near future.

Low Mechanization, Poor Equipment. Despite the availability of superior foreign equipment, the USSR is not dependent on imported technology in its coal industry, as its oil and gas sectors are. In some areas, such as hydraulic mining, Soviet equipment is preeminent. This independence does produce some shortcomings, however, and these stem from the economic deficiencies of quality and poor planning:[5]

TABLE IV-3-7 Soviet Investment in Primary Energy (Millions Rubles, Constant Prices)

	Coal	Oil	Gas
1965	1426	2070	615
1970	1541	2527	1041
1975	1749	3853	1798
1976	1747	4066	1835
1977	1848	4503	2031
1978	2035	5270	2210
1979	2020	5860	2020

Source: Ref. 5.

Despite the large inventory of equipment, the level of mechanization is often low, including the main extraction operations in some basins. This is due in part to failure to produce needed quantities of equipment properly; lack of sufficient parts or repair crews; and neglect of maintenance schedules.

Beyond the limited use of mechanized mining, machinery that is used is often old, unreliable, ill-suited to increasingly complex geological conditions (especially thin seams), and frequently broken down. All these factors can certainly affect safety in a mine.

Future Prospects

The success of the coal industry in meeting production goals is dependent largely on the development of Siberian surface mines. Production costs at Ekibastuz and Kansk-Achinsk are as low as one-tenth that of Soviet underground mines. It has been demonstrated, however, that indirect costs relating to labor amenities, and especially to the construction of physical infrastructure networks, are not reflected in these low figures. Nonetheless, the cost differentials between Western and Eastern production are expected to widen, as costs rise and depletion continues in the European USSR.

The present Five-Year Plan, much less ambitious than former plans, mandates a 770- to 800-million-ton production range for 1985. This is generally regarded by Western analysts as overly optimistic; a 765-million-ton level is seen as a best-case forecast.[5] Such output would require overall annual growth of 7%. More significantly, it would entail a 21% increase in surface mine production. Such expansion is dependent on the construction of beneficiation capacity for Ekibastuz and Kansk-Achinsk coal, development of a boiler suited to coal from Kansk-Achinsk, investment adequacy, and success in meeting long-distance power transmission plans.

Until 1985, the largest increases are projected at Ekibastuz, Kuznetsk, Kansk-Achinsk, and South Yakutia. The low coal quality at Ekibastuz and Kansk-Achinsk will restrict usage. Stable or declining production is anticipated for Moscow, Pechora, and Donetsk, with small growth potential at Karaganda. In sum, the reliance placed on Siberian deposits for increments to national production must constrain any rapid substantial expansion of the Soviet coal industry for the near future if it expects to continue delivery to its present coterie of customers.

NUCLEAR ENERGY

The USSR was the initial country to utilize nuclear power commercially.[7] Since the opening of its first atomic reactor, a 5-megawatt plant at Obninsk (near Moscow) in 1954, the USSR has not only given nuclear energy high priority, it has prided itself on its leading technological role in the development of the atom. In the early 1970s, the rate of growth of electricity from nuclear plants was greater than those of either fossil fuels or hydroelectric generation. At a time when the U.S. nuclear industry was stagnating, Soviet development was leaping ahead. In 1979, nuclear energy provided 0.5% of Soviet electricity; by 1980, this had risen to more than 5%, representing some 13,460 megawatts of capacity, or about 25% of U.S. capacity. In the Eleventh FYP, this share is to increase to 14% of the USSR's total electricity generating capacity.[5]

Uranium Reserves

The magnitude of Soviet uranium reserves is surrounded by a veil of secrecy greater than for any other Soviet mineral resource. Absolutely no information on uranium production or enrichment capacity has been published, undoubtedly due to a desire to conceal its military nuclear arsenal. It is known that a wide variety of uranium deposits occur, of both sedimentary and igneous origin. Dienes and Shabad[7] have postulated several uranium mining districts on the basis of the high status given to a known mining town or province for no apparent other reason. These areas are exhibited in Figure IV-3-12. In addition, they have estimated uranium production at approximately that of Canadian output, or from 2000 to 2500 tons. Cooper[27] has used estimates of 2.1 million tons as a uranium reserve base, and 4.2 million tons of resources. Such figures must be approached skeptically. Although it can be interpreted from ambitious future plans that adequate domestic supplies and enrichment capacity exist to meet nuclear power requirements (not to mention large military needs), the OTA[5] has noted that it appears substantial uranium stockpiling is occurring.

The USSR Nuclear Program

The shift of the fuels base to Asia implies rapidly escalating costs for fossil fuels in the European sections of the USSR. This is the fundamental element underlying the increasing emphasis on the use of nuclear power west of the Urals. Of the 30 reactors on line at 14 sites in late 1981, all but one were located in Europe. The exception is at Bilibino in the remote Soviet Far East, this plant supplying electricity and heat to a gold-mining operation (see Figure IV-3-13.) As mentioned above, no more nuclear plants are projected to be built in the Asian USSR.

Soviet reactors are of two standard designs: the RBMK (large capacity channel or pressure-tube reactor), representing more than 60% of existing

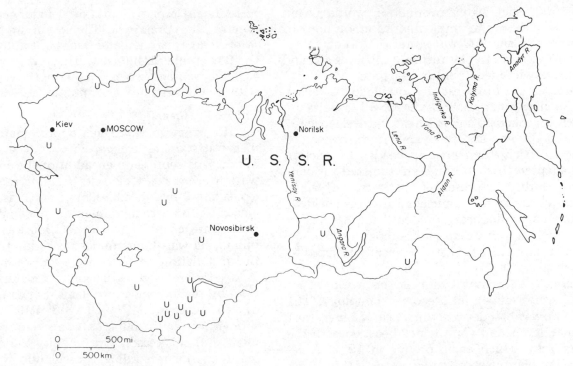

FIGURE IV-3-12   Possible Soviet uranium production regions. (From Ref. 7.)

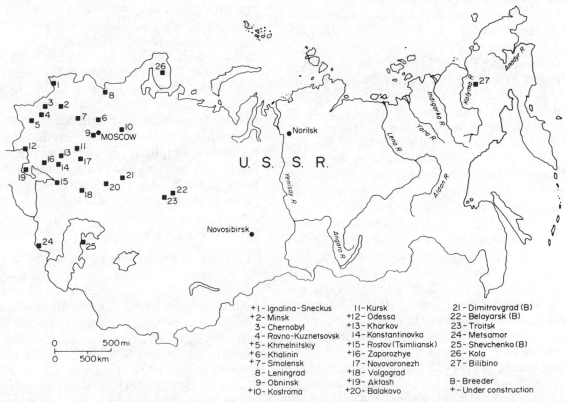

+1- Ignalina-Sneckus	11- Kursk	21- Dimitrovgrad (B)
+2- Minsk	+12- Odessa	22- Beloyarsk (B)
3- Chernobyl	+13- Kharkov	23- Troitsk
4- Rovno-Kuznetsovsk	14- Konstantinovka	24- Metsamor
+5- Khmelnitskiy	+15- Rostov (Tsimliansk)	25- Shevchenko (B)
+6- Khalinin	+16- Zaporozhye	26- Kola
+7- Smolensk	17- Novovoronezh	27- Bilibino
8- Leningrad	+18- Volgograd	
9- Obninsk	+19- Aktash	B- Breeder
+10- Kostroma	+20- Balakovo	+- Under construction

FIGURE IV-2-13   Soviet nuclear reactor sites. (From Ref. 5.)

plants, and the VVER (pressurized water reactor). All requirements are met by domestic manufacturing. No need for imported equipment or technology has been observed.[5] The Volgodonsk Heavy Machine Building Plant (Atommash) on the Don River is currently under construction. This factory is projected to produce eight 1000-megawatt VVER reactor vessels per year when it achieves full capacity in 1990. Such batch production would be an enormous aid to the implementation of nuclear development plans. Atommash raises the possibility of reactor exports in the future. Finland is already using two Soviet-made nuclear reactors.

Three areas of interest are notable in discussing Soviet nuclear energy. First, the USSR is quite interested in cogeneration of heat with electricity. Currently, Bilibino and Beloyarski both cogenerate district electricity and heating needs. Six nuclear plants adapted to such usage are targeted for construction during the present FYP. Given the boundless faith in the safety of nuclear power (see below), the Soviet's willingness to locate nuclear plants near population centers makes cogeneration an efficient treatment of what is typically referred to as "thermal waste."

Second, the Soviet Union is the world leader (along with France) in breeder technology. The original Soviet breeder reactor was a 350-megawatt plant at Shevchenko (on the northeast coast of the Caspian Sea), which came on-line in 1972. A desalination plant uses 250 megawatts of this capacity. Other breeders are located at Beloyarski and Dimitrovgrad. The Soviets have been involved in breeder research for more than 30 years; construction of more breeder plants should be anticipated.

Third, and most significantly, is the ostensibly complete acceptance of the nuclear power plant as a safe technology. The official Soviet position is that concerns over the safety of nuclear power are founded on ignorance. The implications of this philosophy are twofold: First, many of the safety systems, particularly of a redundant nature, that are standard in Western nuclear plants, are not found in the USSR. As a consequence, costs are reduced substantially, while construction times can be halved.[7] Accident-preventive measures are viewed as superfluous; therefore, there are "limited core reduction instrumentation, no fuel meltdown backup system, and no containment structures other than the steel lined reactor cavity itself."[7] Second, given the Soviet view that serious accidents are virtually impossible, safety considerations are not a primary determinant of nuclear plant siting. Cogeneration then becomes a feasible and pragmatic technique, for plants can be located quite near urban areas.

A final comment should be offered concerning the absence of public opposition to nuclear energy in the USSR. Public involvement in the development of nuclear energy in the West has tied up or prevented many projects. Goldman[4] and the OTA[5] have emphasized the strong positive effects on nuclear expansion of such a lack of environmental resistance. Despite reports of a major nuclear accident in the Urals near Khyshtym in the 1950s, safety and environmental issues are not a factor in the Soviet nuclear industry. This must be construed as an advantage when comparing Soviet progress with that desired in Western nations.

Future Prospects

If current goals are met, approximately 38,000 megawatts of nuclear capacity will be on-line by 1985. This represents an additional 25,000 megawatts during this FYP, with funding requirements exceeding 9 billion rubles, or as much as 80% of all electrical power investment.[5] Such plans may prove unreachable. Capacity in 1980 was almost 10 billion kilowatt-hours under plan, due to delays in the installation of nuclear power stations, delays induced more by construction and materials supply than by technological problems. Series production of 1000-megawatt VVER and RBMK reactors should provide the impetus for accelerated development. For this reason, the inability to bring Atommash on line according to schedule would have major implications for any Soviet ability to meet growth targets. Further limitations on future development include competition for investment capital from other sectors; possible shortages of skilled labor, materials, and equipment; poor organization; and inadequate capacity to manufacture the needed turbines and generators.[5]

Eastern European participation in Soviet nuclear projects is actively encouraged, on an electricity-for-labor and/or capital basis. To the extent that Soviet targets are not fulfilled, CMEA electrical capacity needs are also affected. Nonetheless, a continuing emphasis on nuclear power as the major mode of electrical power generation for the European USSR is expected, and the further rapid expansion of the industry is anticipated.

ENERGY POLICY

The formulation of energy policy in the USSR revolves around the rigorous pursuit of methods to free oil from domestic use, its value as a trade commodity far outweighing its worth as a fuel at home. Based on opportunity costs, the ability of petroleum to earn vital hard currency makes it far too valuable to burn. The key to Soviet policy, then, appears to be tied to the degree of success achieved in substitution for and conservation of oil.

In 1981, the value of petroleum products exported amounted to more than $29.4 billion.[33] This represented an increase of nearly 20% over 1980. Much of the large increases in export value during the last decade can be attributed to the

OPEC price hikes, from which the Soviets benefited enormously. Petroleum exports now appear to have leveled off at about 150 million tons per year. OECD countries accounted for about $14 billion in oil imports in 1981, or more than half of all Soviet hard-currency earnings. Finland, West Germany, and France were the largest Western customers.

The Eastern European members of CMEA imported Soviet oil worth $12 billion in 1981. This was purchased at below-market prices, based on an average of world prices for the preceding five years. Such subsidizing of CMEA energy has prevented larger purchases from OPEC, helping to mitigate what is still a miserable foreign debt picture. Although the Soviets obviously prefer an economically stable CMEA, the USSR must, as with domestic consumption of oil, suffer foregone opportunity costs. The Soviets are well aware of this. For this reason CMEA has been warned to anticipate a plateau of around 80 million tons of oil annually over the next several years. Moreover, the pricing mechanism for petroleum to East Europe is finally catching up with world levels—it leaped over 34% from 1980 to 1981, with another 31% increase in 1982.[33] This will bring the price to fully two-thirds of the world price. The dilemma for CMEA nations is that they are quite reliant on Soviet oil, as well as other Soviet fuels. Four of the six CMEA nations depend on the USSR for more than 90% of their oil needs.[34] Despite programs to conserve, to increase domestic production, to participate in joint ventures to develop other supply sources, or to substitute (this option being extremely limited, due to coal's already important status), the situation cannot change significantly before the late 1980s. The Soviets may thus be forced to maintain oil exports to the area—and at reduced prices—to prop up the economies there.[35]

This reasoning has been one of the central tenets prompting concerns over potential Soviet oil declines. Output drops would compel reduced exports to CMEA, a curtailment of exports to the West, or domestic shortages. Certainly, none of these are favorable prospects from the Soviet perspective. The ultimate fear in the face of severe shortages is a Soviet move toward the oil fields of the Middle East.[36]

The ability to meet energy goals impacts the nonfuel minerals sector beyond the more obvious energy availability—economic growth linkages. Should hard-currency needs fail to be met by oil and gas, other exportables would be depended upon to make up the shortfall. Currently, only arms sales and mineral exports (notably platinum-group metals, diamonds, and particularly gold) occur in sufficient volume to be of real aid. Even at present levels these could offer only marginal compensation for a drop in Western currency earnings from oil exports. Moreover, the present recession in these commodities limits their contribution.

Substitution

If the relationship between energy and the economy were not so intimate, that is, if there were a trend for lower energy/GNP ratios, then reduced energy consumption would have minimal effects on economic growth. No such trends are evident.[2,7] Energy availability thus remains crucial. Substitution and conservation efforts strive to attain the necessary energy requirements. Substitution for oil by gas, coal, and nuclear energy is now ongoing.

In recent years, the gas share of total production and consumption has shown strong increases. Oil made substantial gains until 1975, when a peak was reached (see Figures IV-3-6 and IV-3-14). This is due in part to oil production slowdowns, but was also prompted by reallocation of resources in favor of natural gas. Note the large discrepancy between U.S. and Soviet oil consumption, as indicated by Figure IV-3-14, illustrating the magnitude of Soviet petroleum exports.

The decision to emphasize natural gas as the fuel of the future followed a long, heated debate

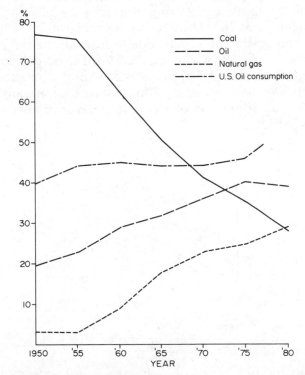

FIGURE IV-3-14 Consumption of energy in the USSR (From Refs. 4 and 10.)

in the late 1970s between advocates of coal and those supporting gas. Gas emerged the clear winner. In addition to gas deliveries to CMEA, it is planned that exported gas can help make up for any drop in oil hard-currency earnings. The impediment to domestic substitution by gas is its restriction to areas where pipeline construction is economical, that is, where use is concentrated. As seen in Figure IV-3-15, this program has been under way since 1975. The Urals-Volga region in particular has effected this replacement of oil by gas;[7] proximity to existing gas lines has speeded implementation. A second drawback to supplanting oil with gas is a lack of secondary refining capacity. It has been demonstrated above that fuel oil represents, at 50%, a comparatively large share of the refining product mix. Gas could compensate for leveling oil production, provided that secondary capacity existed to generate the increasingly-in-demand lighter and intermediate products from the freed oil. This capacity is not expected to be in evidence prior to the latter part of this decade.

Coal could, then, become the preferred substitute fuel for dispersed demand, where gas delivery would be difficult. Moreover, Goldman[4] has made an interesting point concerning the ease of readjustment from oil to coal. The longer the time lapse since the switch from coal to petroleum as dominant fuel, the more painful the change back to coal, this being chiefly a matter of disintegration of the coal infrastructure during the time of oil dominance. Given the rather recent switch from coal to oil and gas in the USSR, coal infrastructure is still healthy. (As shown in Figure IV-3-14, oil did not surpass coal until 1973, quite late when compared with Western consumption patterns.) Certainly, this is an added factor working in favor of coal as a substitute.

The problems that exist in the coal industry have been presented above. Soviet leaders apparently felt that supply constraints would limit the ability of coal to substitute for oil. Declines in overall output and coal quality, low regional quality, and the difficulties induced by the remoteness of new Asian deposits all serve to restrict the future role for coal. Although all new power plants east of the Urals are planned to be coal-fired, coal production levels could inhibit the rapid growth of electrical power in switching from oil to coal while still maintaining allocations to other sectors. This situation underscores the commitment in the eleventh FYP to natural gas and nuclear energy.

Nuclear power is replacing fossil fuels of all types in many areas of the European USSR. In the Ukraine alone the use of atomic energy frees yearly 62 million tons of oil.[37] Utilization of breeder reactors saves at least 2 million tons of oil at each breeder site.[38] However, gas, coal, and nuclear power will be unable to provide all boiler fuel needs west of the Urals for some time. Oil will of necessity remain important here.

Conservation

A major emphasis is now being placed on the more efficient use of fuel and other raw materials. The push for conservation has been aided by two recent large increases in prices (one in retail and one in wholesale petroleum products). This is a significant move, in view of Soviet reluctance to alter pricing structure, the central economy being supposedly free of such capitalistic defects as inflation. It reflects the acknowledgment that former prices did not represent the true value of oil and hence promoted waste. Conservation efforts in the USSR involve different obstacles than those facing similar efforts in the United States. For instance, in 1975 the transportation sector consumed only 10% of all Soviet energy used; this figure was almost 35% in the United States. Such statistics depict the low percentage of the population in the USSR owning cars (only 5%) compared with American consumers. Potential for conservation in the Soviet transportation sector is thus limited.[4]

Cooper,[27] however, in his examination of the electrification of the Soviet railway system, has offered compelling evidence that such modernization techniques saved more than 32 million tons of oil in 1980. This represented over 40% of the Soviet oil exported to the Eastern European members of CMEA in that year, and, at subsidized prices, a $9 billion savings. As much as 45 million tons could be conserved in 1985 by rail electrification. Obviously, the future benefits of this conservation are

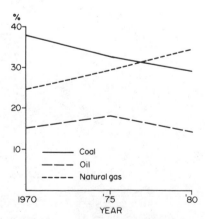

FIGURE IV-3-15 Consumption of fuel used in boilers. (From Ref. 4.)

limited, first by the fact that much of the system is now electrified, and second by the rising of oil export prices to CMEA to a level more commensurate with world prices. Nonetheless, observations stating conclusively that conservation possibilities in the transportation sector of the USSR are limited often underestimate Soviet capabilities, as Cooper has shown.

The main area where energy conservation could be effective is in industrial consumption. Usage by industry accounted for 60% of total consumption in 1975, compared with only 35% in the United States.[7] The USSR's centralized system should be advantageous here, allowing coordination on a level that is not possible in a market economy. For such a program to succeed, however, much of the waste and inefficiency inherent in the Soviet economy must be reduced; certainly, the increase in prices will be an incentive toward this end.

Many authorities (e.g., the Defense Intelligence Agency,[39] Goldman,[4] Dienes and Shabad,[7] and Wilson[11]) now predict continued growth of Soviet energy production, with a future ability to meet or near production goals. Energy planning has been called "steady, purposeful and economically rational."[7] In general, although geographic and economic variables make them more expensive and difficult to develop, the USSR does possess the fuels it needs—particularly natural gas. Indeed, the ability of the USSR to quickly make gas the dominant Soviet bloc energy form is pivotal. Western European assistance in the Urengoy export pipeline seems to ensure that the Soviet energy goals will be met. Future energy policy will embrace the diverse aims of stimulating further economic growth, supporting vulnerable CMEA countries, and earning Western hard currency, these policies being guided by a broad range of inputs from the bureaucratic views of planning agencies to the larger strategic concerns of supply security and Soviet bloc stability. Such competing interests shall continue to determine the Soviet energy balance.

REFERENCES

Petroleum

1. U.S. Central Intelligence Agency. *Prospects for Soviet Oil Production*. Washington, D.C.: U.S. Government Printing Office, 1977.
2. U.S. Congress, Joint Economic Committee, Subcommittee on International Finance and Security Economics. *Energy in Soviet Policy*. 97th Congress, 1st session. Washington, D.C.: U.S. Government Printing Office, 1981.
3. Kroncher, Allan. "Economic Progress Is Only on Paper." *Soviet Analyst* 11, No. 5 (10 March 1982).
4. Goldman, Marshall I. *The Enigma of Soviet Petroleum*. London: George Allen & Unwin (Publisher) Ltd., 1980.
5. U.S. Congress, Office of Technology Assessment. *Technology and Soviet Energy Availability*. Washington, D.C.: U.S. Government Printing Office, 1981.
6. Scanlon, A. F. G., ed. *BP Statistical Review of World Energy 1981*. London: British Petroleum Company, PLC, 1981.
7. Dienes, Leslie, and Theodore Shabad. *The Soviet Energy System: Resource Use and Policies*. Washington, D.C.: V. H. Winston & Sons, 1979.
8. "USSR's Biggest Oil Field Passes Milestone." *Oil and Gas Journal*, 31 August 1982.
9. "Pipelines Hold Key to Soviet Gas Production." *Oil and Gas Journal*, 29 June 1981.
10. U.S. Department of the Interior, Bureau of Mines. "The Mineral Industry of the USSR." By V. V. Strishkov. Preprint from the *Minerals Yearbook 1980*. Washington, D.C.: U.S. Government Printing Office, 1980.
11. Wilson, David. *Soviet Oil and Gas to 1990*. Economist Intelligence Unit (EIU) Special Report 90. London: Spencer House, 1980.
12. Hoffman, George W. *Energy Projections: Oil, Natural Gas, and Coal in the USSR and Eastern Europe*. Policy Study No. 3. Center for Energy Studies, University of Texas at Austin, August 1978.

Natural Gas

13. Hewett, Ed. A. "The Pipeline Connection: Issues for the Alliance." *The Brookings Review*, Fall 1982.
14. USSR Energy Targets: 3—Gas." *Soviet Analyst* 10, No. 13 (24 June 1981).
15. "West Siberia's Gas." *New Times*, No. 19 (May 1982).
16. "USSR Finishing Tests of Its Biggest Gas Compressors." *Oil and Gas Journal*, 21 June 1982.
17. "Soviets Press Construction of 56 in. Gas Pipelines." *Oil and Gas Journal*, 14 June 1982.
18. "Energy Projects and Plans." *The Current Digest of the Soviet Press* 34, No. 7 (17 March 1982).
19. "Industry." *The Current Digest of the Soviet Press* 34, No. 25 (21 July 1982).
20. Garn, Jake. "Questions about the Soviet Gas Pipeline." *The Wall Street Journal*, 20 May 1982.
21. U.S. Department of State, Bureau of Public Affairs. *Soviet-West European Natural Gas Pipeline. Oct. 14, 1981*. Current Policy No. 331. Washington, D.C.: U.S. Government Printing Office, November 1981.
22. Mufson, Steve. "Anatomy of Continuing Soviet Pipeline Controversy." *The Wall Street Journal*, 31 August 1982.
23. Karr, Miriam, and Roger W. Robinson, Jr. "Soviet Gas: Risk or Reward?" *The Washington Quarterly* 4, No. 4 (Autumn 1981).
24. Blau, Thomas, and Joseph Kirchheimer. "European Dependence and Soviet Leverage: The Yamal Pipeline." *Survival* 23, No. 5 (September–October 1981).

Coal

25. Ratnieks, H. "The USSR Leans on Comrade Coal." *The Geographical Magazine* 54, No. 4 (April 1982).

26. "World Energy Conference Survey of Energy Resources 1978." In Sam H. Schurr, *Energy in America's Future*. Baltimore: Johns Hopkins Press, 1979.

27. Cooper, Hal B. H., Jr. *A Comparative Analysis of Railroad Electrification between the United States and the Soviet Union*. Prepared for Senator John G. Tower, United States Senate, Armed Services Committee, 15 March 1982.

28. U.S. Central Intelligence Agency, National Foreign Assessment Center. *USSR: Coal Industry Problems and Prospects*. Washington, D.C.: U.S. Government Printing Office, 1980.

29. Dyker, David. "Planning in Siberia on Wrong Track." *Soviet Analyst* 9, No. 2 (23 January 1980).

30. "USSR Energy Targets: 4—Coal." *Soviet Analyst* 10, No. 14 (8 July 1981).

31. Levine, Richard M. "Soviet Union." In *Mining Annual Review, 1982*. London: Mining Journal, June 1982.

32. "U.S. Hopes to Bar Occidental's Hammer from Building Siberia-Moscow Pipeline." *The Wall Street Journal*, 8 October 1982.

Energy Policy

33. "Value of Soviet Oil Exports Hits Record." *Oil and Gas Journal*, 23 August 1982.

34. *USSR Facts and Figures Annual*, Vol. 6. Edited by John L. Scherer. New York: Academic International Press, 1982.

35. Hoffman, George W. *Eastern Europe's Resource Crisis, with Special Emphasis on Energy Resources: Dependence and Policy Options*. Policy Study No. 14. Center for Energy Studies, University of Texas at Austin, January 1981.

36. Meyer, Herbert E. "Why We Should Worry about the Soviet Energy Crunch." *Fortune*, 25 February 1980.

37. "Industry." *The Current Digest of the Soviet Press* 34, No. 27 (4 August 1982).

38. "How Will Urals Industry Get More Energy?" *The Current Digest of the Soviet Press* 34, No. 37 (13 October 1982).

39. "DIA: Soviets Will Meet Energy Targets." *Oil and Gas Journal*, 21 September 1981.

40. *USSR Facts and Figures Annual*, Vol. 4. Edited by John L. Scherer. New York: Academic International Press, 1980.

41. U.S. Central Intelligence Agency, National Foreign Assessment Center. *USSR: Development of the Gas Industry*. Washington, D.C.: U.S. Government Printing Office, 1978.

42. *USSR Facts and Figures Annual*, Vol. 5. Edited by John L. Scherer. New York: Academic International Press, 1981.

CHAPTER IV-4

The Ferrous Minerals

The energy wealth of the USSR is complemented by its deposits of nonfuel minerals. The USSR is fortunate to hold substantial reserves of the vital ferrous minerals required by a modern industrial state. This has fostered production levels which frequently lead the world, and has enabled the USSR to maintain its role as the leading producer of steel, a position it has held since 1971. Table IV-4-1 indicates ferrous metal production, share of world output, production ranking, reserves, and ranking of reserve holdings in 1981.

This paper uses the Western ferrous mineral classification system. In the USSR, only iron ore and manganese are considered ferrous metals. All others, including precious metals, are regarded as nonferrous ("colored") metals and are administered by the Ministry for Non-Ferrous Metallurgy.[1]

IRON ORE

Reserves

The USSR leads the world in iron ore production and in the size of its reserves. Soviet $A + B + C_1$ reserves were estimated in 1976 at 60.2 billion tons, averaging 38% Fe.[2] Of this amount, 10.3 billion tons assay more than 55% Fe and require no concentration. This is nearly a third of world reserves. More than 70% of Soviet iron ore lies in Europe. Table IV-4-2 delineates the iron content of the

TABLE IV-4-2 Distribution of Iron Ore Reserves (%)

Ukraine	31.0
European center	24.4
Urals	15.7
Kazakhstan	15.0
Siberia	7.4
Northwest	3.0
Far East	2.5
Other	1.0

Source: Ref. 4.

major regions by percent of total reserves. The major iron ore districts are shown in Figure IV-4-1.

Of all iron reserves, approximately 14% require no beneficiation, 77% need simple treatment, and 9% must undergo complex beneficiation. This is in stark contrast to 1940, when 90% of all iron was direct-shipping ore, requiring no treatment. This illustrates the major shift that occurred in developmental policy during the 1950s. As higher-grade, more accessible ores became exhausted, the Soviet steel industry was reoriented toward the use of low-grade iron ore, with concentration thus becoming more important.[3] As a consequence of this, average mine grade has declined from 50% Fe in 1950 to 38% in 1980.

One of the most significant iron areas in the USSR is the Kursk Magnetic Anomaly (KMA),

TABLE IV-4-1 Soviet Ferrous Metals Production and Reserves

Commodity	Unit[a]	1981 production	Percent share of world production	World rank	Reserves	World rank
Chromite	A	2,400	24.2	2	60,000[b]	3
Cobalt	B	2,250[c]	7.7	3	100,000[d]	
Iron ore	A	242,000	28.3	1	60,200,000	1
Manganese	A	9,400	34.8	1	250,000[d]	2
Molybdenum	B	10,900	10.0	4	200,000[d]	3
Nickel	A	158[c]	22.1	2	5,000	3
Tungsten	B	8,850	16.0	2	150,000[d]	3
Vanadium	B	10,000	27.8	2	8,000,000[d]	2

[a] A, thousand tons; B, tons.
[b] Ore.
[c] Excludes production from Cuban ores.
[d] Metal.

Sources: Refs. 2, 4, 9, and 11.

FIGURE IV-4-1   Location of Soviet iron ore and manganese deposits. (From Refs. 1, 2, 3, and 4.)

which stretches from Smolensk to Rostov. The iron formations responsible for the KMA underlie an area greater than 200,000 square kilometers. As such, it is the largest recognized iron-rich region in the world (see Figure IV-4-1).

Production and Trade

The Soviet iron industry is marked by a tendency to concentrate production in large complexes. In 1979, more than 50% of all surface-mined output came from only eight large mines, and 80% of all underground production was derived from 19 operations.[2] Output in 1981 was 242 million tons, a decline of 3 million tons from the year before.[4] Pellet production in 1981 amounted to 51 million tons. Nearly 85% of all production is from open pits. In 1980, there were 71 underground mines and 59 surface mines in operation, as well as 92 concentration facilities, 29 sinter plants, and 7 pelletizing complexes (Sokolov-Sarbaysk, Kachkanar, Lebedinsk, Kremenchug, and 3 in the important Krivoy Rog area). For 1981, however, Levine[4] has drastically revised the number of underground mines, reducing the figure to 40.

The major iron-producing regions of the USSR are:

Krivoy Rog
Kursk Magnetic Anomaly
Urals
Kazakhstan
Siberia
Kola Peninsula

The Ukraine now accounts for more than 50% of all output. The Krivoy Rog region has long been a major source of iron ore, and presently represents nearly 90% of Ukranian production. It is also one of the primary steel centers, supported by the nearby Donetsk coal basin and Nikopol manganese deposits. Krivoy Rog contains nine large surface mines and 23 underground operations, as well as the nation's largest concentrator, at Severnyy. Importantly, output is currently declining at Krivoy Rog.

The KMA, while rapidly increasing in both reserves and production, was 7 million tons under quota in 1980. Output was 40 million tons that year. The major iron ore producer in the KMA is the Lebedin facility.

The Urals began producing iron ore in the 1930s. Largely in response to defense interests, efforts were made to diversify industrialization eastward away from the huge Ukanian concentrations. This proved tactically sound, as the German capture of western Soviet territory forced reliance upon eastern industry. Currently, the Urals are experiencing iron ore shortages at concentrators, requiring shipments from other regions, especially the KMA. Export obligations have been blamed in part for this problem.

Major new iron ore facilities are projected in the Eleventh Five-Year Plan (FYP) in Hazakhstan, the KMA, and in Karelia at Kostamush, where a joint Soviet-Finn project is under way. Future growth is anticipated in the Aldan Basin of the Far East, as well as in other areas of eastern Siberia and Kazakhstan. The completion of the Baikal-Amur

Mainline (BAM) will provide potential for important exports from these Asian deposits to Pacific nations, notably Japan. The announcement of major new iron discoveries in late 1982, exceeding 1 billion tons of ore in the Angara region of the Far East, underscores this possibility.[5]

Iron ore exports in 1981 were 38 million tons. This represents less than 20% of total Soviet output. Poland, Czechoslovakia, and Hungary are the three main importers. In fact, more than 95% of all Soviet iron exported is bound for the CMEA. These nations are nearly 75% dependent on the USSR for their iron ore supplies.

Future Prospects

In addition to regional domestic shortages, Levine[4] has identified three important problems facing the Soviet iron industry: a trend for substantial increases in the depth of open pits, a 40% increase in production costs per ton since the early 1970s, and a continuing trend to lower grades. From 1975 to 1980, the average grade of mined ore dropped from 33.4% to 32.3%.

These difficulties do not portend any serious supply problems. The reserve base is adequate to meet even major consumption increases. In contrast to some other important minerals, most iron is located in accessible areas. More beneficiation may be required; a failure to invest in these facilities could prove a burden, although the real concern is whether, in providing the requirements of its own steel industry, the USSR can also support the needs of the CMEA steel plants. Exports have declined in the past five years, although not precipitously (about 7%). It may be necessary for CMEA to look for a larger percentage of its supplies on the world market, in which case its plight could be compared to that of the United States. On the other hand, if waste and inefficiency were eliminated from the Soviet steel industry, adequate supplies for both domestic consumption and export would be available.

MANGANESE

The Soviets are the world's leading producer of manganese; its reserves rank second. $A + B + C_1$ reserves are listed at 250 million tons of ore, containing 23 to 26% metal.[2] Virtually all deposits are sedimentary in origin. Soviet manganese is found chiefly in two regions: Nikopol in the Ukraine and Chiatura in Georgia (Transcaucasus) (see Figure IV-4-1). Ore from Nikopol is of low quality, though deposits are large. Manganese deposits here lie more than 80 meters deep and are barely 2 meters thick. Ore at Chiatura is the richest in the USSR and represents approximately 40% of the reserve base. Manganese was first found here in 1848.

Production in 1981 was 9.4 million tons.[4] Levine[4] has revised 1980 production estimates downward to 9.75 million tons from a previous reported level of 10.25 million tons. This indicates that 1981 was the second year of output decreases. Figure IV-4-2 charts Soviet manganese production since 1960. Note that a similar drop occurred in 1967–1969. The basis for this decline could not be ascertained. However, this dip in output was followed by a decade of expansion, and there is no reason to believe that this pattern could not be repeated.

Approximately 80% of current manganese production comes from Nikopol. This area surpassed Chiatura in the early 1960s to become the leading producer. The proximity of the Ukrainian steel centers in great part prompted this expansion, which continues today. Most incremental output increases for Soviet manganese in the Eleventh FYP are planned at Nikopol. In 1980, Nikopol contained

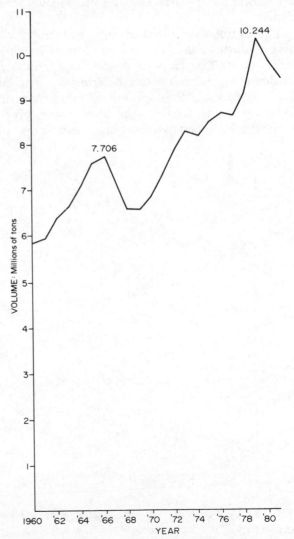

FIGURE IV-4-2 Soviet manganese production. (From Refs. 2, 4, and 14.)

19 underground mines, 10 surface mines, and 9 concentrators. Open pits provide more than 75% of output here. By contrast, 80% of production in Chiatura is derived from underground mining methods. Minor Soviet producing areas include Dzezhdy (near Dzezhazgan) and Atasuy (near Karaganda) in Kazakhstan.

Exports in 1980 were approximately 1.35 million tons. As with Soviet iron, CMEA is the primary destination, this group of nations receiving 80% of the USSR's manganese exports. CMEA is also assisting in further development at Nikopol. As seen in Figure IV-4-3, the ratio of manganese exports to total output has remained constant over the past 20 years. This is in direct conflict with the findings of such scholars as Fine,[6] who have declared that declining exports, especially to the West (to whom exports were terminated in 1977[7]), foreshadow serious problems for Soviet availability of manganese. Mere trend extrapolations based on the past two years of lowered output could well produce erroneous results. Disaster prophecies are premature.

This is not to say that problems are not evident. These include delays in construction of needed facilities, waste due to poor processing technologies, and depletion of many oxide deposits.[4] The latter feature is forcing development of manganese from carbonate ores. Despite the fact these ores are lower grade, often deeper-lying, and more diffi-

cult to concentrate, the Eleventh FYP calls for growth to occur in manganese production from carbonates. This may reverse the trend toward a stabilization of ore grade declines. In 1950, the ratio of crude to usable ore was 72%; by 1965, it had fallen to 47%.[3] In 1980, however, the ratio was only 45%.[2] Use of carbonates may initiate further quality declines. Generally, however, the reserve base is huge and in accessible areas. Even given major increases in consumption due to the emphasis on Soviet ferroalloy production, Papp's[7] judgment that the USSR must either curtail exports to CMEA, or actually import manganese, is not supported by the facts.

CHROMIUM

The USSR is ranked second behind South Africa in both production and exports of chromite. $A + B + C_1$ reserves are estimated by the Bureau of Mines at 60 million tons of 30 to 65% Cr_2O_3 content ore,[2] placing Soviet reserves third in importance internationally. The most significant region, and the only source of high-grade chromite, is at Khrom-Tau, in western Kazakhstan (see Figure IV-4-4). The only other area of important chromite concentration is at Saratov in the Urals. The low Cr_2O_3 content (20 to 40%) and high Fe/Cr ratio of Saratov chromite limits its use to refractory and chemical applications.

In 1981, production of crude ore (30 to 44% Cr_2O_3) was 3.3 million tons, a decline of 100,000 tons from the 1980 production level. Output of marketable ore (45 to 56% Cr_2O_3) declined 50,000 tons to 2.4 million tons.[4] Figure IV-4-5 illustrates Soviet production of marketable chrome ore over time. Approximately 95% of Soviet chrome comes from the Donskoye mining and concentration complex at Khrom-Tau. Much of the output from Khrom-Tau is converted into ferrochromium and other ferroalloys at Aktyubinsk, for use in the steel industry. Donskoye chrome production is projected to grow by 14% during the present FYP; the great depth at which new mines are attempting to extract ore here, as deep as 1200 meters, may prove an obstacle to this plan. Concentrate from Donskoye averaged 49.8% Cr_2O_3 in 1978. Underground mines at Saratov account for the remaining 5% of Soviet chromite production.

Seventy percent of production appears to be consumed internally, although whether stockpiling is taking place remains undetermined. In 1978, consumption by industrial sector was as follows: metal production, 45%; refractories, 35%; chemical and other uses, 20%. When compared to chromite

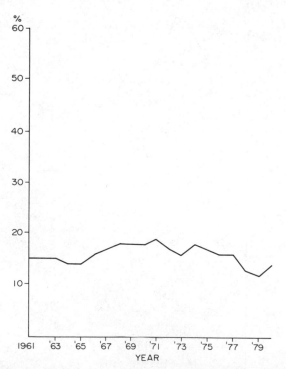

FIGURE IV-4-3 Soviet exports of manganese as a percentage of production. (From Refs. 2, 4, and 14.)

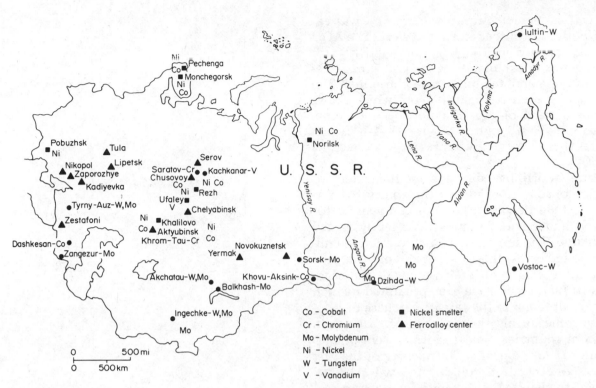

FIGURE IV-4-4   Other ferrous mineral deposits and ferroalloy centers. (From Refs. 1, 2, 3, and 4.)

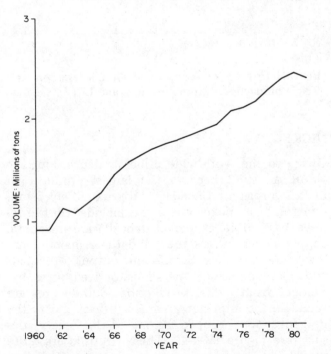

FIGURE IV-4-5   Soviet chromite production. (From Refs. 2, 4, and 14.)

end use in the United States, the Soviet refractory sector represents nearly twice the percentage of its American counterpart.

Exports totaled 550,000 tons in 1981; almost 75% of all exports are destined for CMEA. The

United States does import some Soviet raw ore (234,000 tons in 1979, 170,000 tons in 1980, and 111,000 tons in 1981), but Soviet chromite is not nearly as important now as it once was. Historically, the USSR has been a major supplier to the West of both manganese and chromite. As late as 1947, for instance, 47% of American requirements of chromite and one-third of its manganese were imported from the Soviets.[8] However, as the rift between the East and the West deepened, the Soviet Union became regarded as a supply source of increasingly dubious security. The temerity of reliance on Soviet minerals was underscored beginning in the Berlin crisis of 1948, and again during the Korean War. In both these crises, the Soviet Union embargoed shipments to the United States. Fortuitous events prevented serious U.S. supply interruptions, although during the early 1950s there were price escalations for these minerals. Ultimately, the embargoes worked to the advantage of the Americans, by hastening efforts to diversify sources of supply with minimal adverse consequences. By the early 1970s, the United States had reduced its dependence on Soviet chrome to 31%, and had eliminated imports of Soviet manganese. Presently, Soviet chrome constitutes less than 20% of U.S. consumption. This trend is characteristic of the Western steel producers generally. Before widespread application of the Argon-Oxygen Decarburization (AOD) Process, and as recently as 1975, about half of Soviet output of chromite was exported, with most intended for the West. The AOD technology opened up the vast, lower-grade South African deposits for metallurgical use, greatly reducing the demand for high-grade chromite.[9] The results can be seen in Figure IV-4-6, which shows the general trend of overall export declines, especially to the West, and in Figure IV-4-7, which illustrates that exports now represent less than 25% of output.

Future availability depends on the extent of domestic consumption and on the ability to overcome ore-grade declines. To help compensate for the latter, underground mining has recently started at Donskoye. The importance of chromium in military uses indicates that high-grade ore may be stockpiled.[2] Purchase of foreign supplies, and negotiations to buy, could have been prompted by such a strategy, although in the case of purchase of Rhodesian chrome during the embargo of that nation by Western countries, Soviet motives were primarily resale to make a profit. The opening of the first chromite concentration mill, located at Donskoye, may point to plans to develop lower-grade reserves.[7] Although this would require substantial investment in additional facilities, it would also preclude supply problems. Due to recent downturns, future production trends are difficult to project with a high measure of confidence. The magnitude of the reserve base does appear adequate to meet needs at present consumption levels.

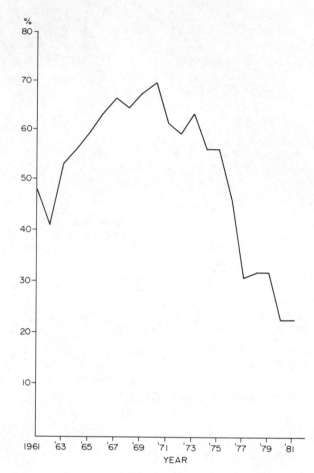

FIGURE IV-4-7 Soviet exports of chromite as a percentage of production. (From Refs. 2, 4, and 14.)

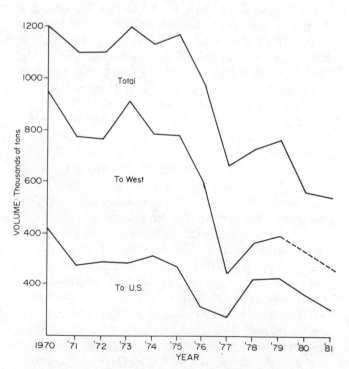

FIGURE IV-4-6 Soviet chromite exports. (From Refs. 4, 14, and 15.)

NICKEL

Owing to the worldwide minerals depression, the USSR in 1982 became the largest producer of nickel, surpassing Canada.[10] Approximately 5 million tons[2] of nickel metal are included in the reserve base; 65% of nickel deposits are magmatic copper-nickel sulfide ores, and the remainder are low-grade laterites. Significant output of cobalt and platinum-group metals is derived as a by-product from nickel production. Sulfide ores are found in the important Norilsk district and in the Pechenga-Monchegorsk regions of the Kola Peninsula (see Figure IV-4-4). These deposits are quite similar in several respects to Canada's Sudbury. Together, Norilsk and the Kola Peninsula account for 80% of Soviet nickel reserves. Major oxide deposits occur at Aktyubinsk in Kazakhstan, at Ufaley in the central Urals, and in the Ukraine. Since the USSR apparently has free access to Cuban nickel ore, reserves from that country (which total more than 3 million tons) should perhaps be included in available Soviet nickel reserves.

Nickel production has risen from 33,000 tons

in 1950, to 158,000 tons in 1981.[4] An additional 20,000 tons of metal from Cuban nickel-cobalt concentrate was also produced. Annual output since 1960 is shown in Figure IV-4-8.

The large expansion in output can be attributed largely to the development of Norilsk. Site of one of the world's greatest metallurgical complexes, Norilsk is currently the leading nickel producer. Important co-products and by-products here include more than 75% of all Soviet platinum group metals, as well as copper, cobalt, gold, silver, selenium, and other rare metals. Mineralization is associated genetically and spatially with large deep-seated mafic intrusions.[11] Norilsk development is all the more remarkable when viewed in its geographic context. Its location north of the Arctic Circle has borne the effects of climate, remoteness, and high costs. Achievements here, although slow and costly, are a reflection of the high priority placed on the district by Soviet leaders. Among recent accomplishments at Norilsk is a 33-kilometer copper-nickel slurry pipeline to the huge new Madezhda metallurgical complex.

Nickel production is also found in the Urals, and the Kola Peninsula. Figure IV-4-4 exhibits the location of nickel smelters. Norilsk is the largest, followed by three facilities in the Urals: Ufaley (opened in 1934), Rezh (1935), and Kahlilovo. The Monchegorsk and Pechenga smelters in the Kola Peninsula constitute the third major refining center. Finally, a Ukranian smelter at Pobuzhsk treats nickel laterites to produce ferronickel. The smelters at Norilsk and the Kola Peninsula are characterized by modern electrolytic refining technologies. These complexes stand in stark contrast to the antiquated facilities in the Urals.

It should be remembered that despite large increases, output growth in the Tenth FYP, at less than 9%, fell substantially short of the mandated 20 to 30% increment. Lagging construction at Norilsk is the prime cause. For the current FYP, a minimum 30% increase is planned, with Norilsk to again provide much of this. Output from Norilsk is projected to increase to the point that shipments by sea to Kola smelters for processing will be required. This is a considerable distance. As is often the case, plans are probably too ambitious. Nonetheless, Soviet production and exports are expected to only rise in importance, based on Soviet and Cuban concentrates.[4] To this end, $600 million has recently been invested in Cuba for future nickel development.

Nickel exports are presently a significant hard currency earner; exports to the United States totaled nearly $21 million in 1980, and more than $27 million for the first half of 1981. Large sales of nickel in 1981 have been seen as further depressing already low world prices.[4] There is little indication of future supply problems, given the size of deposits at Norilsk and the Kola Peninsula. Although nickel at these sites will remain difficult and costly to exploit, the priority given these regions assures continued development.

MOLYBDENUM

Molybdenum is a ferrous metal which, although not produced in the Soviet Union in sufficient quantities to allow exporting, is nonetheless adequate for domestic needs. $A + B + C_1$ reserves are believed to be 200,000 tons of contained metal,[2] ranking third in the world. Deposits are sulfides (mostly molybdenite), copper-molybdenum, and tungsten-molybdenum ores.

Production in 1981 was 10,900 tons of molybdenum metal.[4] Approximately 50% of output is a byproduct of copper operations. Armenia is the most important region in this context, although it has no processing capacity. The large Zangezur surface mine and copper-molybdenum complex not only supplies one-fourth of all Soviet molybdenum,

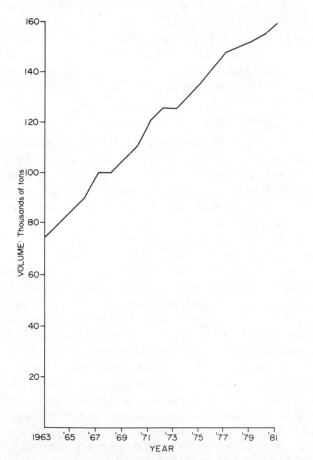

FIGURE IV-4-8 Soviet nickel production (excludes production from Cuban ores). (From Refs. 2, 4, and 14.)

but also ranks among the lowest in terms of cost. Other important producers of molybdenum from copper ores include Kazakhstan (especially the Balkhash complex), Sorsk in East Siberia (the largest single producing complex), and Siberia (particularly Tuimsk). Also of interest is a Soviet-Mongolian joint development project in Mongolia at the Erdenet open pit complex. Ores here assay 0.85% Cu and 0.012% Mo. Processing must be done in the USSR (see Figure IV-4-4).

Tungsten-molybdenum ores account for 30% of molybdenum production. Primary producing regions are the North Caucasus (especially Tyrny-Auz) and Buryat. Molybdenite is mined in Uzbek and in Siberia. These areas account for 20% of Soviet molybdenum supplies.

Both exports and imports of molybdenum are insignificant. Thus, while the USSR produces enough for its own needs, it does not produce adequate quantities for its CMEA neighbors. It appears that Soviet molybdenum output is increasing at a rate suitable to meet the expanding requirements of its steel industry. Supply availability does not appear to be a problem.

COBALT

Soviet cobalt reserves are estimated at 100,000 tons of metal.[2] Cobalt is associated with the large copper-nickel deposits found in various regions, especially Norilsk, the Kola Peninsula, and the Urals. It is also the principal mineral component at a small deposit at Dashkesan in Azerbaidzhan. This area has been worked for its cobalt since the turn of the century. In addition, a rich cobalt-arsenic deposit is found at Khovu-Aksink in the Tuva Republic north of Mongolia (see Figure IV-4-4).

Output in 1981 was 3750 tons; this included 1500 tons of production from Cuban ores. Figure IV-4-9 illustrates Soviet cobalt production over the recent past; note that production from imports of Cuban concentrate is not reflected in this graph. Virtually all cobalt produced at present is a by-product of nickel production. Norilsk is the leading producer, and has been since the Ninth FYP. Monchegorsk and Pechenga on the Kola Peninsula are also major cobalt producers. Several complexes in the Urals, including Ufaley, Rezh, and Uzhural-nikel, are also important (see Figure IV-4-4).

Cobalt production increases have been modest—16% in the 9th FYP, and 10% in the 10th. A 30% increase is hoped for from 1981 to 1985. This is largely dependent on the rate of growth at Norilsk, and is unlikely to be reached. The importance of imports from Cuba, which have been running at 40% or more of consumption, must be emphasized.

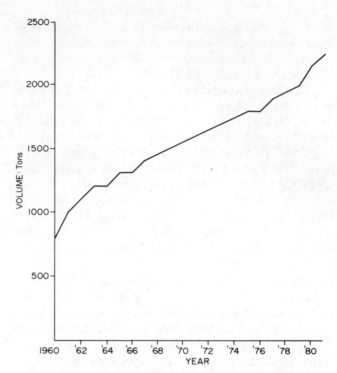

FIGURE IV-4-9 Soviet cobalt production (excludes production from Cuban ores). (From Refs. 2, 4, and 14.)

Cobalt remains a strategic mineral, especially in terms of military usage. The Soviet military machine requires steadily increasing supplies. Despite this, recovery of cobalt remains low.[2] Cuban ores thus play a critical and integral part in assuring the adequacy of Soviet cobalt supplies.

VANADIUM

The USSR ranks second in both vanadium reserves and production, behind South Africa. The primary product is vanadium-rich slag, a coproduct with iron from titaniferous magnetite deposits, located at Kachkanar in the Urals. Soviet reserves are listed at 8 million tons of vanadium content.[12]

In 1981, production was 10,000 tons of metal.[10] Nizhniy Tagil is the only modern producer. Concentrate is processed in oxygen converters, yielding steel and a 21.1% V_2O_5 slag. The other major smelter is at Scusovoy, where slag of 17.2% V_2O_5 is produced. Chusuvoy is the only source of ferrovanadium (see Figure IV-4-4). While expansion at Kachkanar is emphasized to boost vanadium availability, research is underway to recover the metal from power plant ash, which could supply up to 25% of needs,[4] and as a by-product of bauxite at alumina refineries.[2] Both production and exports are expected to become increasingly significant. Given these facts, supply appears adequate to meet even large consumption increases. Some in the South African vanadium industry even feel that major new discoveries in western Kazakhstan could propel Soviet production above that of South Africa.[13]

TUNGSTEN

According to the Bureau of Mines,[2] tungsten $A + B + C_1$ reserves of 150,000 tons of contained metal exist in the USSR, ranking the nation third in reserve holdings. Deposits are generally low grade, and are frequently associated with molybdenum. Figure IV-4-4 also locates major areas of tungsten mineralization.

Production in 1981 was 8850 tons.[4] Primary tungsten producing regions are:

1. North Caucasus, especially the Tyrny-Auz W-Mo complex. This is not only the largest single deposit, but biggest producing facility as well.
2. Kazakhstan, especially the Verkhne-Kayraktin W and Akchatau W-Mo deposits.
3. Central Asia, especially the Ingechke deposit.
4. Transbaikal, especially the Dzihda deposit in Buryat.
5. The Far East, including the extreme northern Iultin W-Sn deposit and the Vostok complex.

Before the China-Soviet rift, Chinese exports of tungsten were very important to the USSR. Those imports ceased in the late 1960s, as did Soviet exports to the West. Today, despite being the world's number two producer of tungsten, output is not sufficient to meet demands. Although exports remain insignificant, the Soviets now rely on imports for nearly 20% of consumption. China is again a supplior of some Soviet tungsten, with the remainder coming from Mongolia. Given the remote nature of many deposits, this level of import dependence, as uncharacteristic of the Soviet mineral industry as it might be, is expected to continue.

STEEL

The USSR is the largest producer of refractories, ferroalloys, crude steel, and steel pipe.[2] However, its inability to manufacture high quality rolled steel products, combined with a high internal demand for such goods due to its ambitious pipeline network construction, mandates a need to import these items. Output of crude steel in 1981 was 149 million tons,[4] almost 8 million short of target.

Eleven metallurgical centers account for more than 50% of all raw steel. These complexes, located in Figure IV-4-10, are:

Magnitogorsk
Chelyabinsk
Cherepovets
Karaganda
Krivoy Rog
Kuznetsk
Nizhniy Tagil
Novolipetsk
West Siberia (also at Kuznetsk)
Zaporozhstal
Zhdanov

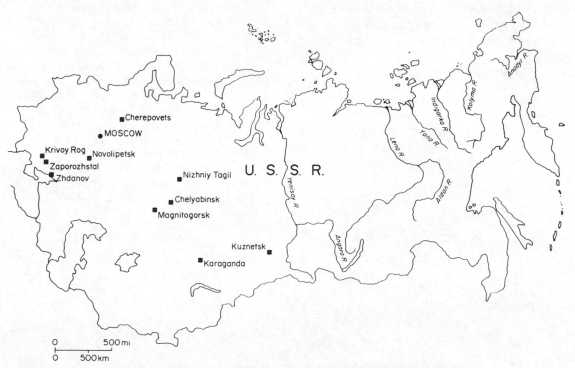

FIGURE IV-4-10 Major Soviet steel centers. (From Ref. 4.)

In virtually all of these complexes, and particularly at Karaganda, Krivoy Rog, Kuznetsk, Novolipetsk, Zaporozhstal, and Zhdanov, there are large nearby deposits of iron ore, manganese, and coking coal.

There are at present 76 total metallurgical works that produce raw steel; 42 of these are oxygen converters. In addition, 138 blast furnaces produce pig iron; these have an average capacity of 1258 cubic meters. Production is still heavily reliant on open-hearth furnaces, facilities often being more than 40 years old. Increasing emphasis is placed on more up-to-date technologies, such as oxygen conversion, electric arc furnaces, powder metallurgy, and continually cast steel production. Trends in steel production methodology are shown in Table IV-4-3.

The USSR is the largest producer of steel. It may also be the largest waster of steel. It has been estimated that only 45% of total raw steel is used efficiently in the economy; the rest is either remelted or lost.[2] In 1980, Soviet industry wasted more than 25% of steel output.[2] The making of steel itself likewise involves considerable waste, much of it due to poor metallurgical techniques.[1]

For the Eleventh FYP, investment in ferrous metallurgy is to increase 40% over the past five years to assist modernization and expansion. These improvements, however, remain slow. The priority on tonnage in the FYP induces not only low quality, but poor assortment of final products. The three factors of high waste, poor quality, and insufficient product mix all serve to not only hinder the initial consumer, but also to limit general economic development and progress throughout the network of steel utilization. Production and consumption of

TABLE IV-4-3 Methods of Steel Production (Percent)

Process	1960	1970	1976
Oxygen converter	3.8	15.3	28.0
Electric arc	8.9	10.7	7.4
Open hearth	84.4	76.6	64.0
Bessemer	2.9	1.4	0.6

Source: Ref. 14.

steel will remain high in the future—and so will these problems.

REFERENCES

1. Sutulov, Alexander. Mineral Resources and the Economy of the USSR. New York: McGraw-Hill Book Company, 1973.
2. U.S. Department of the Interior, Bureau of Mines. "The Mineral Industry of the USSR." By V. V. Strishkov. Preprint from the Minerals Yearbook 1980. Washington, D.C.: U.S. Government Printing Office, 1980.
3. Shabad, Theodore. Basic Industrial Resources of the USSR. New York: Columbia University Press, 1969.
4. Levine, Richard. "Soviet Union." In Mining Annual Review, 1982. London: Mining Journal, June 1982.
5. "Industry." The Current Digest of the Soviet Press 34, No. 45 (8 December 1982).
6. Fine, Daniel I. "Mineral Resource Dependency Crisis: Soviet Union and United States." In The Resource War in 3-D: Dependency, Diplomacy, Defense. Edited by James A. Miller, Daniel I. Fine, and R. Daniel McMichael. Pittsburgh, Pa.: World Affairs Council of Pittsburgh, 1980.
7. Papp, Daniel S. "Soviet Non-fuel Mineral Resources, Surplus or Scarcity?" Resources Policy 8, No. 3 (September 1982).
8. Eckes, Alfred E., Jr. The United States and the Global Struggle for Minerals. Austin, Tex.: University of Texas Press, 1979.
9. Shabad, Theodore. "The Soviet Mineral Potential and Environmental Constraints." Mineral Economics Symposium, Washington, D.C., 8 November 1982.
10. U.S. Department of the Interior, Bureau of Mines. Mineral Commodity Summaries 1983. Washington, D.C.: U.S. Government Printing Office, 1983.
11. Smirnov, V. I. Ore Deposits of the USSR, Vol. II. San Francisco: Pitman Publishing, Inc., 1977.
12. U.S. Department of the Interior, Bureau of Mines. Mineral Facts and Problems, 1980 Edition. Washington, D.C.: U.S. Government Printing Office, 1980.
13. U.S. Department of Commerce. Industrial Outlook Report. Minerals South Africa—1981. Washington, D.C.: U.S. Government Printing Office, 1982.
14. U.S. Department of the Interior, Bureau of Mines. Minerals Yearbook, 1960–1979. Washington, D.C.: U.S. Government Printing Office.
15. USSR Facts and Figures Annual, Vol. 6 Edited by John L. Scherer. New York: Academic International Press, 1982.

The Nonferrous Minerals

As with the ferrous minerals, Soviet output of the nonferrous metals is impressive. This is especially notable in the case of tin and aluminum, for production of these minerals did not start until long after the Bolshevik Revolution. Table IV-5-1 lists nonferrous production, share of world output, production ranking, reserves, and reserve ranking in 1981.

ALUMINUM

Aluminum was not produced in the USSR until 1932, for two main reasons: electrical power was widely unavailable, and it was believed that no domestic bauxite deposits existed. Soviet $A + B + C_1$ reserves of bauxite containing 26 to 62% alumina are now estimated at 65 million tons.[1] Bauxite is a mineral for which the shift of the supply sources to the east is particularly evident: 75% of all reserves now lie in Asia. Deposits are found in the Turgay region of Kazakhstan, the eastern slopes of the Urals, and in the Tikhvin area near Leningrad.

Bauxite is low grade and insufficient for aluminum production at desired levels. This fact has had two important consequences. First, from 40 to 50% of bauxite consumed must be imported; major countries of origin are Greece, Guinea, Hungary, and Yugoslavia. Second, the Soviets have pioneered in the production of aluminum from nonbauxitic sources. In 1980, nepheline syenite accounted for 16% of output. From 1968 to 1978 there was nearly a fivefold increase in the production of alumina from nepheline syenite. Deposits are found in the nepheline-apatite complex of the Kola Peninsula, and in Siberian nepheline-containing deposits. In addition, the Mukhalsky deposit has just been discovered in northern Transbaikal; it has been called the third largest aluminum ore deposit in the USSR.[2] Completion of the BAM will greatly speed its development. Alunite ore accounts for less than 2% of Soviet aluminum, with deposits found in Kazakhstan and the Transcaucasus. Figure IV-5-1 plots major bauxite, nepheline, and alunite locations. Recent attempts to produce alumina from kaolinite in Uzbek have apparently been abandoned.

Soviet production of aluminum in 1981 was 1.94 million tons, including 150,000 tons from scrap and 3.1 million tons of alumina.[3] Production since 1963 is illustrated in Figure IV-5-2. The Bureau of Mines[1] believes total capacity to be 2.185 million tons, distributed among 14 reduction plants. Table IV-5-2 lists these plants, and Figure IV-5-1 plots them.

TABLE IV-5-1 Soviet Nonferrous Metal Production and Reserves

Commodity	Unit[a]	1981 production	Percent share of world production	World rank	Reserves	World rank
Aluminum	A	1,790	10.7	2		
Bauxite	A	4,600	5.3		65,000	
Antimony	B	8,200	11.4	4	150[b]	4
Cadmium	B	2,900	16.7	1	50,000[b]	4
Copper	A	950	12.1	3	40,000[b]	3
Lead	A	410	12.4	2	16,000[b]	3
Mercury	C	63,000	32.4	1	1,100,000	2
Tin	B	36,000	14.6	2	600,000[b]	
Titanium	B	38,500	40.0	1	10,000,000[b]	
Zinc	A	790	13.5	2	20,000	4

[a] A, thousand tons; B, tons; C, 76-lb flasks.
[b] Metal.

Sources: Refs. 1, 3, 11, and 12.

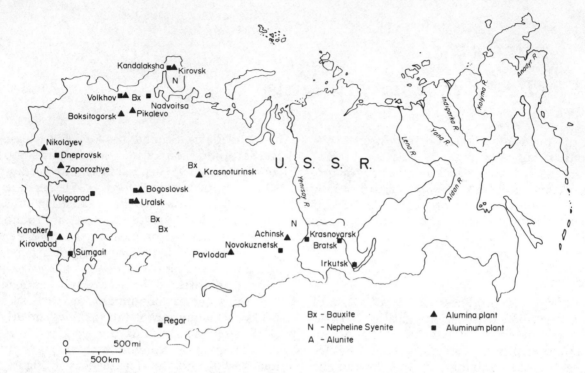

FIGURE IV-5-1 Soviet aluminum and alumina plants, deposits of bauxite, nepheline syenite, and alunite. (From Refs. 1, 3, 4, and 10.)

In addition, a smelter is under construction at Sayansk in Siberia. Alumina plants are shown below and in Table IV-5-2. (This list may be incomplete.)

Nikolayev	Achinsk
Kirovabad	Kirovsk
Pavlodar	Boksitogorsk
Volkhov	Pikalevo
Bogoslovsk	Zaporozhye
Uralsk	Krasnoturinsk

Aluminum is one of the main minerals for which a decided attempt has been undertaken to move production to the supply sources. Started during World War II, this shift continues today in response to the eastern location of reserves. In 1965, Asia accounted for only 35% of aluminum metal.[4] Smelters such as those at Krasnoyarsk, Irkutsk, and Bratsk have been placed to not only minimize raw material transport costs, but also to take advantage of the priority placed on development of massive Siberian hydroelectric projects, thus ensuring abundant cheap energy.

Four general problems appear to plague the Soviet aluminum industry. First is the ubiquitous feature of waste: it is estimated that from 8 to 20% of bauxite is lost during refining.[1] Second, as a review of Table IV-5-2 shows, many primary aluminum plants are quite old. In fact, the first two smelters in the country, Volkhov and Dne-

provsk, although small, are still in operation. Six of the 14 smelters are more than 30 years old. This is in some measure a root of the third problem, that of poor technology. The Bureau of Mines[1] feels that the Soviet industry's processes lag behind the West by more than a decade. To rectify this, France's Pechiney Ugine Kuhlmann is aiding in modernization efforts.[1] Finally, the use of non-bauxite minerals to produce alumina is blamed in at least one plant (Achinsk) for causing technical difficulties which prevent the meeting of quotas.[1] Bauxite is still the preferred material for the making of aluminum, and the USSR is keenly aware of its domestic bauxite insufficiencies. Many of the USSR's financing and technical assistance programs in foreign nations involve securing raw material supplies for its aluminum industry. Current joint ventures include a bauxite plant in Guinea and an alumina refinery in India; negotiations for another alumina plant are also underway. In addition, in 1981, the Soviets contracted to purchase 200,000 tons of bauxite from Brazil. The USSR is more dependent on bauxite and alumina imports than on any other major commodity.

On the other hand, Soviet aluminum metal exports have exceeded 500,000 tons yearly since 1973; exports are currently approximately one-third of total output, with most metal going to the CMEA. It is generally believed that aluminum is

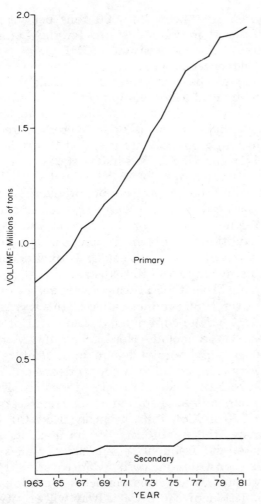

FIGURE IV-5-2 Soviet aluminum production. (From Refs. 1, 3, and 13.)

TABLE IV-5-2 Primary Aluminum Plants in the USSR

Plant	Probable capacity (thousand tons)	Began operations
	Asia	
Bratsk	540	1966
Krasnoyarsk	390	1964
Irkutsk	240	1962
Regar	160	1975
Novokuznetsk	160	1943
	Europe	
Bogoslovsk	140	1945
Uralsk	135	1939
Volgograd	125	1959
Sumgait	70	1955
Kanaker	70	1950
Dneprovsk	70	1933
Nadvoitsa	35	1954
Kandalaksha	30	1961
Volkhov	20	1932

Source: Ref. 1.

produced more for export than internal use.[1] Clearly, imports of bauxite are more than compensated for by the value of exports. Papp[5] has shown that the percent of bauxite imports has remained constant since the mid-1970s relative to aluminum exports, signifying the basis for importation. Shabad[6] sees a growing portion of the world's bauxite supplies being converted into aluminum in the USSR for re-export, the low energy costs of Siberian smelters directing this expansion. The Eleventh FYP calls for aluminum production increases of 15 to 20%, with large increments as well in alumina output.

Aluminum is thus a mineral product for which the self-sufficient aims of the USSR are not met, the domestic reserve base being a constraint to production at desired levels. However, the industry is orientated more towards export, rather than domestic consumption. Import requirements thus give an exaggerated notion of the true Soviet dependency.

COPPER

Copper production is a very old activity in the country. The metal was first mined in the Urals almost 300 years ago. The importance of the various types of Soviet copper deposits differs from that typically found in the West; this is exhibited in Table IV-5-3.

Soviet $A + B + C_1$ reserves are listed at 40 million tons of metal (3.6 billion tons of ore averaging 1.1% Cu).[1] As seen from Table IV-5-3, most are low-grade copper-nickel ores, cupriferous sandstones, massive sulfides, and porphyries. The significance of the copper-nickel ores, and relatively

TABLE IV-5-3 Types of Copper Deposits in the USSR

Genesis	Deposit type	Total percent of reserves USSR	Total percent of reserves Rest of world
Magmatic	Copper-nickel	30.6	3.1
	Vanadium-iron-copper	0.8	—
Carbonatite	Carbonatite	—	0.8
Skarn	Skarn	2.0	0.6
Hydrothermal	Copper porphyry	13.1	55.3
	Quartz-sulfide	2.0	1.2
	Native copper	—	1.0
Massive sulfide	Copper-lead-zinc	21.2	8.8
Stratiform	Cupriferous sandstones and shales	30.3	29.2

Source: Ref. 14.

minor importance of porphyry deposits, is notable. Half of all copper reserves occur in central and eastern Kazakhstan. Other important copper regions include the eastern Urals, Uzbek, eastern Siberia, the Transcaucasus, and Norilsk (see Figure IV-5-3). Minor deposits are found in the northern Caucasus, western Siberia, and the Kola Peninsula.

In 1981, 65 mines produced 128 million tons of ore,[3] 80% coming from surface mines. Output of blister copper in that year was 950,000 tons of primary and 95,000 tons of secondary metal from 13 smelters. In addition, the USSR had in operation 40 concentrators and 11 refineries. Figure IV-5-4 depicts Soviet copper production since 1960. Table IV-5-4 lists the copper smelters and refineries; these are located in Figure IV-5-3.

Central and eastern Kazakhstan account for 30% of all Soviet copper. Since the early 1960s this district has been the site of major increases in smelting and refining capacity, in response to the recognition of the Urals' decline. The porphyry deposits of Kounrad, Kalmakyr, and Bozshchakul, and the huge sandstone-hosted deposit at Dzhezkazgan, were seen as the copper supplies of the future. Kounrad is now one of the largest mines in the country, and the Dzhezkazgan metallurgical complex has likewise become very important. The Urals are still a significant producer, although depletion of mines and diminishing production continues. By 1980, as much as 80% of output was expected to come from Asian operations.[1] Copper

exports have been 240,000 tons per year since 1978; most metal is destined for the Council for Mutual Economic Assistance (CMEA).

The Soviet copper industry is experiencing many problems. The trend toward increasing use of surface mining and large equipment has helped improve efficiency.[7] Milling and smelting technologies, however, remain poor, as does byproduct recovery. Copper hydrometallurgy is still at a very incipient level in the USSR. Strishkov[8] states that two to three times more Soviet labor and capital is required to attain a given copper production increase than is needed in the West. This underscores the crucial nature of labor shortages to the industry. An additional problem is idle capacity, due to shortages of concentrate at Urals smelters and the Balkhash complex in Kazakhstan.

The Eleventh FYP calls for increases of 20 to 25% over 1980 production levels; this is greatly dependent on the continuing development of remote areas. These include Norilsk and the Mongolian Erdenet copper-molybdenum joint venture. Over the long run, the Udokan copper deposit, northeast of Lake Baikal, should begin operations. This is a huge mineralized region, containing reserves of more than 700 million tons, averaging 1.5% Cu. Unfortunately, the remoteness of the area, its hostile climate, and frequent seismic activity all serve to triple construction costs.[3] Moreover, a refining technique has not been perfected for Udokan ore. Completion of the BAM railway will rectify some

FIGURE IV-5-3   Soviet copper smelters and refineries and major copper deposits. (From Ref. 8.)

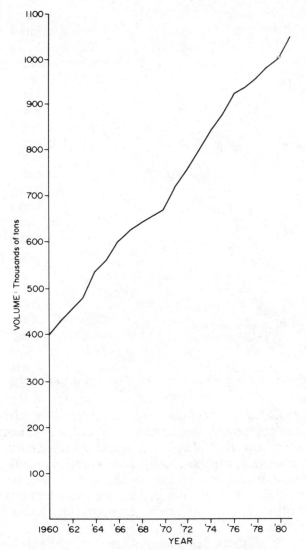

FIGURE IV-5-4 Soviet blister copper production. (From Ref. 8.)

TABLE IV-5-4 Soviet Copper Smelters and Refineries

Location	Plant	Smelter	Refinery
Kazakhstan	Balkhash	X	X
	Dzhezkazgan	X	X
	Irtysh	X	X
Uzbek	Almalyk	X	X
Urals	Kirovgrad	X	X
	Kyshtym		X
	Karabash	X	
	Krasnouralsk	X	
	Pyshma		X
	Sredneuralsk	X	
Moscow	Moscow	X	X
Transcaucasus	Alaverdy	X	X
Kola Peninsula	Severonikel	X	X
	Pechenganikel	X	X
East Siberia	Norilsk	X	X

Source: Ref. 8.

of the more obvious geographical constraints. While Udokan will undoubtedly offer substantial copper production to the USSR, this is not anticipated prior to the next decade. In the interim, although some importing of ore and concentrate may increase in importance to allow metal exports to CMEA to continue,[9] domestic copper availability should present no serious problems.

LEAD AND ZINC

The USSR ranks second in production of both refined lead and zinc. $A + B + C_1$ reserves of lead are listed at 16 million tons of contained metal and 20 million tons of zinc.[1] Approximately 85% of all ores are complex massive sulfide types, as opposed to the simpler lead or zinc ores typical in the United States. More than two-thirds of reserves are located in Kazakhstan, especially the Altai and Kara-Tau regions. Large zinc deposits also occur in the Urals. Major locations of lead and zinc are exhibited in Figure IV-5-5.

Total lead output in 1981 was 630,000 tons,[3] including 410,000 tons of primary production (unchanged since 1978) and secondary production of 220,000 tons (an increase of 5000 tons over 1980). Figure IV-5-6 graphs production of lead since 1963. Primary production estimates for 1976–1980 have recently been revised downward by 100,000 tons each year, with secondary output being increased by this amount.[3] (This revision is represented by the dashed line in Figure IV-5-6.) Zinc production totaled 870,000 tons in 1981;[3] primary output rose by 5000 tons to 790,000 tons, while secondary zinc production remained unchanged at 80,000 tons. Figure IV-5-7 illustrates zinc output since 1963. More than 10% of both lead and zinc is recovered as a by-product. Kazakhstan is by far the most important lead-zinc producer, accounting for more than 70% of all lead and approximately 50% of all zinc. Other important producers are the Urals, Uzbek, Siberia, northern Caucasus, and the Ukraine. Two-thirds of lead and zinc are mined by underground methods, although more than half of Kazakhstan's output come from open pits.

Most mining complexes, generally consisting of one or a few mines, have an associated concentrator. Lead and zinc concentrates then are transported to their respective smelters and refineries. These are listed in Table IV-5-5 and located in Figure IV-5-5. These facilities are commonly quite old, and are sporadically expanded and modernized. An unequal distribution of lead smelters and zinc refineries is extant. In contrast to lead smelters, which are generally near concentrators, zinc refinery locations are widely scattered, their energy-intensive

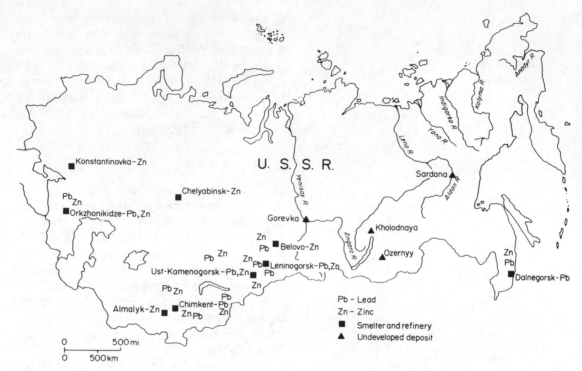

FIGURE IV-5-5   Soviet lead and zinc smelters and refineries and major deposits. (From Refs. 7 and 10.)

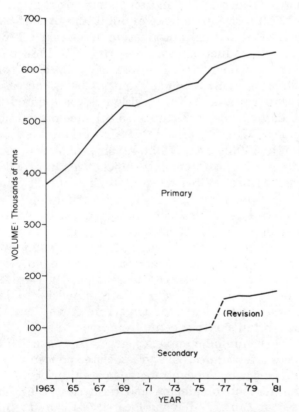

FIGURE IV-5-6   Soviet lead production. (From Refs. 1, 3, and 13.)

nature requiring positioning near coal sources or hydroelectric plants.[4] It required post–World War II development in inaccessible areas, such as eastern Kazakhstan, to establish a power system; prior

to this, zinc concentrate was transported thousands of kilometers. A similar situation is now seen in the Soviet Far East. Lead concentrates are smelted at Dalnegorsk, but no nearby zinc smelter exists. Zinc concentrates, therefore, are shipped to the opposite side of the country, to the Konstantinovka smelter in the Ukraine, a staggering transport distance.

Exorbitant raw material transport of this kind is representative of one of the chief problems in the Soviet lead-zinc industry, that of lagging construction of new mines and processing facilities. These developmental failures are blamed for supply shortages at Urals concentrators, and long-distance haulage of ore in some areas of Kazakhstan for processing.[3] Other difficulties are strongly in evidence. Metal recovery is still low—zinc recovery at concentrators from lead-zinc ores is estimated at only 60%.[1] (Recovery is generally greater than 90% in the United States.) The differential is not due so much to poor Soviet technology in this case, but rather to the complex nature of its polymetallic ores, and to the lower grades mined in the USSR.[7] Moreover, ore grades are declining as high-grade deposits become depleted. For example, in eastern Kazakhstan, metal content has declined 40% in the past 10 years.[3] These factors have all served cumulatively to make lead and zinc production costs among the highest in the entire Soviet minerals industry.[10]

Large new deposits have been discovered in the BAM zone of the Far East. These include important

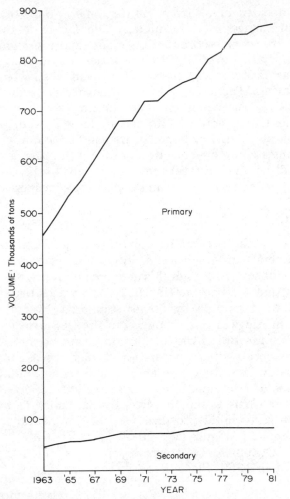

FIGURE IV-5-7 Soviet zinc production. (From Refs. 1, 3, and 14.)

TABLE IV-5-5 Lead and Zinc Smelters and Refineries

Lead	Zinc	Lead-zinc
Chimkent	Almalyk	Leninogorsk
Dalnegorsk	Belovo	Ordzhonikidze
	Chelyabinsk	Ust-Kamenogorsk
	Konstantinovka	

Source: Ref. 10.

finds at Gorevka (on the Angara River in east Siberia), Kholodnaya (northeast of Lake Baikal), Ozernyy (east of Lake Baikal in Buryat), and Sardana (on the Aldan River) (see Figure IV-5-5). However, development at these locations suffers from the same delays encountered at western operations. Certainly, the isolated nature of these deposits has done little to spur investment, especially with strong competition for funds from other (possibly higher priority) sectors. As a result, these areas are years away from production.

In contrast to the West, where a definite trend toward declining lead usage exists, Soviet domestic consumption is rapidly increasing. The CIA[10] estimated this growth at almost 15% for the years 1975-1978. Much of the increase can be attributed to the use of lead for cable sheathing, an application assumed largely by plastics in the West. Soviet zinc consumption, on the other hand, has followed the pattern of lowering use found in the Western market economies.

Importantly, exports of both lead and zinc are manifesting declining tendencies. Net exports of lead and zinc metal are depicted in Figure IV-5-8. (This graph excludes imports of ore and concentrate.) Note that for both metals decreases are apparent, since 1973 for zinc and 1976 for lead. In fact, net lead exports reached 0 in 1981, an event unparalleled in the past. The problem is clarified further by an examination of Figure IV-5-9, which portrays imports of ore and concentrate as a percent of production. These imports come predominantly from the United States, Canada, and Western Europe. The large drop in lead imports corresponds with the severe fall in net exports, suggesting exports can resume only if imports increase. Zinc ore and concentrate imports, on the other hand, show increases, yet net zinc metal exports still are dropping. In all probability, a decrease in zinc imports would result in the pattern seen for lead. Such declines will force the CMEA, the destination of most Soviet lead and zinc exports, to look elsewhere for its supplies.

The role of imports of ore and concentrate thus appears to be crucial. Papp,[5] Grichar et al.,[9] and the CIA[10] have all indicated the real possibility that the USSR will rely increasingly on such trade,

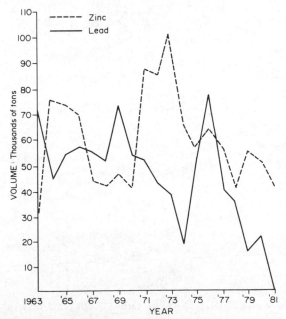

FIGURE IV-5-8 Soviet net exports of lead and zinc. (From Refs. 1, 3, and 13.)

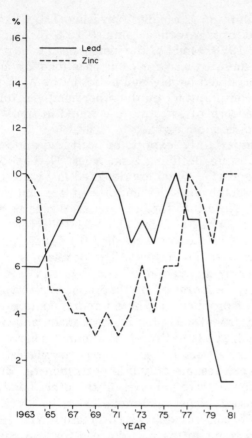

FIGURE IV-5-9 Soviet imports of lead-zinc ore and concentrates as a percentage of production. (From Refs. 1, 3, and 13.)

a decided break from the customary policy of self-sufficiency in these metals. The failure in recent years to meet production plans, and most importantly to expand existing facilities and develop new Far Eastern deposits, have proven costly to the Soviets. Particularly in the case of lead, which is still experiencing internal consumption increases, the USSR could be forced to look to the world market to satisfy domestic demand until new Asian mines come on line. It is doubtful that the Soviets will be able to meet their obligations to the CMEA unless imports of ore and concentrate are increased.

TIN

Tin is the second of the three major nonferrous minerals in which the USSR has a net import dependency. Although this dependency is not as large as that for bauxite (in 1981, 22% versus bauxite's 55%), it is high by Soviet standards. Like aluminum, tin was not produced in the USSR until well after the Bolsehvik Revolution, again because of a conviction that the nation lacked tin deposits. Today $A + B + C_1$ reserves are listed at 600,000 tons of contained metal. Ores assay 0.6 to 1% Sn.[1] Most tin is found in eastern Asia, which explains the early failure to discover tin. Tin locations are mapped in Figure IV-5-10.

Output in 1981 was 36,000 tons,[3] identical to that in 1980. Placer deposits accounted for 25% of

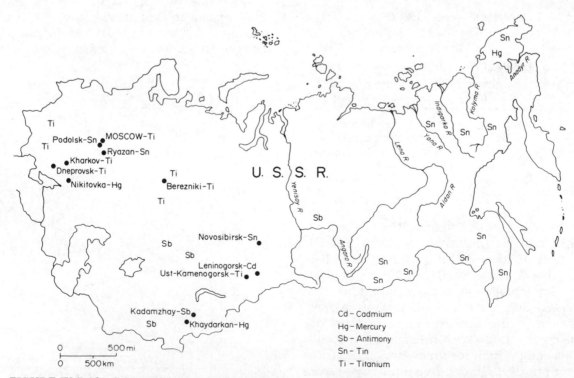

FIGURE IV-5-10 Other nonferrous mineral deposits and production centers. (From Refs. 1, 3, and 4.)

production. Main producing regions are the Soviet Far East (specifically the Maritime Territory), Yakutia, and Transbaikal. An obvious feature of these areas, particularly the first two, is the bitter Arctic climate, with eight-month winters, that must be endured in development. In this problem, tin and gold share a common geography. The largest single producer is the Khingan complex in Khabarovsk. Three tin refineries, shown in Figure IV-5-10, are found at Ryazan, Podolsk, and Novosibirsk (the main supplier). Concentrates must thus be shipped long distances for refining; most are railed to Novosibirsk.

Although ranking second in world production, tin from Malaysia and Bolivia must be imported to meet requirements. Sutulov[7] blames the remoteness of tin reserves and the low strategic importance of tin to Soviet leaders for the nation's import dependence. The latter conclusion is moot—the Soviets are exploring intensively for new deposits,[3] and continue to develop costly polar tin reserves. Moreover, the USSR and Bolivia are jointly developing a tin plant at La Placa. Increments to present output can be expected from Central Asia and the Far East. Prospects for self-sufficiency in tin appear dim, however, despite sanguine forecasts (Sutulov[7] projected tin independence by 1975).

TITANIUM

The USSR is the world's main producer of titanium sponge. $A + B + C_1$ reserves are estimated at 10 million tons of contained metal, or 70 million tons of ore averaging 10 to 20% TiO_2.[1] Both ilmenite and rutile are found in Siberia, and in placer deposits in the Ukraine, especially near the Dnieper River (see Figure IV-5-10).

Production in 1981 was 38,500 tons.[3] The expansion of the Soviet titanium industry is a fairly recent event, as shown in Figure IV-5-11. (*Note:* The dashed line represents a revision of output estimates; no changes for years before 1971 could be found.) Rapid growth is undoubtedly due to the metal's importance in military applications; huge quantities are needed for aerospace programs, and in new Soviet submarines possessing titanium hulls.[5] Major producing plants, shown in Figure IV-5-10, include the Ust-Kamenogorsk, Berrezniki, and Dneprovsk titanium-magnesium complexes, the Moscow titanium sponge facility, and the Kharkov complex. Magnesium and titanium are often produced at the same facility, magnesium being used as a reductant in the production of titanium. Although output increased by nearly 20% in the Tenth FYP, this was just half of planned growth.

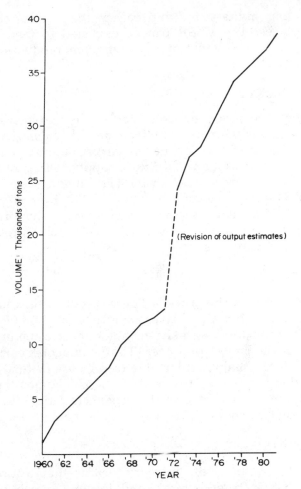

FIGURE IV-5-11 Soviet titanium sponge production. (From Refs. 1, 3, and 13.)

Changing trade patterns for this mineral were seized upon by many scholars as evidence of severe difficulties in the Soviet titanium industry. Exports of sponge of approximately 35,000 tons per year, mostly to the West, ceased abruptly in 1980, leading to widespread speculation regarding production problems, major military programs, stockpiling, reserve depletion, and so on. This was further fueled by the Soviet purchase of 44,000 tons of titanium ores from Australia in 1979. However, 1981 saw the resumption of exports, with the United States importing nearly $2 million worth during the first half of that year.[3] It also made many experts eat their own words: "there appears to be little doubt that the earlier pattern of titanium exports has been reversed."[5] These events should emphasize the danger of formulating conclusions based on Soviet trade trends, as well as the futility of attempting forecasts premised solely on extrapolations, whether Soviet or otherwise. It may be that the USSR would again halt exports of sponge—this would probably reflect its importance to the expanding military buildup rather than to supply shortages. The USSR can be expected to remain the leader in terms of both titanium production and metallurgy.

ANTIMONY

Soviet $A + B + C_1$ reserves of antimony are believed to be 150,000 tons of contained metal.[1] Kadamzhay in Kirgizia in the site of the largest deposits, and produces most of the USSR's supply. Other production occurs in Kazakhstan and Siberia (see Figure IV-5-10). Output has been 8200 tons since 1979.[3] The Soviets are 8% dependent on imports of antimony to meet demand; most supplies originate in Yugoslavia.

CADMIUM

The USSR is the world's largest producer of cadmium. $A + B + C_1$ reserves are estimated at 50,000 tons of metal.[1] Kazakhstan holds most cadmium and is the largest producer. The Leninogorsk complex is the major center here (see Figure IV-5-10). Production in 1981 was 2900 tons.[3] Almost all cadmium is produced as a by-product at various lead and zinc smelters and at some copper complexes.

MERCURY

Soviet output of mercury leads the world. $A + B + C_1$ reserves are listed by the Bureau of Mines at 1.1 billion 76-lb flasks.[1] Main deposits are in Central Asia, the Far East (especially in Magadan), and the Ukraine (see Figure IV-5-10). Mercury output in 1981 amounted to 63,000 76-lb flasks. The Khaydarkan complex in southern Kirgiz is the biggest producer. Both surface and underground mining occurs there. Number two producer is Nikitovskiy in the Ukraine; 70% of production here comes from underground mines. Exploration is underway in the Far East and other areas to locate new deposits. Additional plans call for de-

velopment of small deposits in Arctic Chukotka and other regions.

REFERENCES

1. U.S. Department of the Interior, Bureau of Mines. "The Mineral Industry of the USSR." By V. V. Strishkov. Preprint from the *Minerals Yearbook 1980.* Washington, D.C.: U.S. Government Printing Office, 1980.
2. "BAM's Aluminum." *The Current Digest of the Soviet Press* 34, No. 3 (24 February 1982).
3. Levine, Richard M. "Soviet Union." In *Mining Annual Review, 1982.* London: Mining Journal, June 1982.
4. Shabad, Theodore. *Basic Industrial Resources of the USSR.* New York: Columbia University Press, 1969.
5. Papp, Daniel S. "Soviet Non-fuel Mineral Resources, Surplus or Scarcity?" *Resources Policy* 8, No. 3 (September 1982).
6. Shabad, Theodore. "The Soviet Mineral Potential and Environmental Contraints." *Mineral Economics Symposium*, Washington, D.C., 8 November 1982.
7. Sutulov, Alexander. *Mineral Resources and the Economy of the USSR.* New York: McGraw-Hill Book Company, 1973.
8. Strishkov, V. V. "The Copper Industry of the USSR." *Mining Magazine*, March 1979 and May 1979.
9. Grichar, James S., Richard Levine, and Lotfollah Nahai. *The Nonfuel Mineral Outlook for the USSR through 1990.* Washington, D.C.: U.S. Government Printing Office, 1981.
10. U.S. Central Intelligence Agency, National Foreign Assessment Center. *The Lead and Zinc Industry in the USSR.* Washington, D.C.: U.S. Government Printing Office, 1980.
11. U.S. Department of the Interior, Bureau of Mines. *Mineral Commodity Summaries 1982/1983.* Washington, D.C.: U.S. Government Printing Office.
12. U.S. Department of the Interior, Bureau of Mines. *Mineral Facts and Problems, 1980 Edition.* Washington, D.C.: U.S. Government Printing Office, 1980.
13. U.S. Department of the Interior. Bureau of Mines. *Minerals Yearbook*, 1963–1979. Washington, D.C.: U.S. Government Printing Office.
14. Smirnov, V. I. *Ore Deposits of the USSR*, Vol. II. San Francisco: Pitman Publishing, Inc., 1977.

CHAPTER IV-6

The Precious Minerals

Production of precious minerals is crucial to the USSR. Exports of gold, platinum-group metals, and diamonds are the second major source of Western hard currency behind oil and gas. Total trade value for these commodities exceeded $4.5 billion in 1981.[1] Sales of gold and platinum group metals are large enough to be a primary determinant of world prices. The high unit value of these items makes them easy to stock, and a ready source of fast "crisis" funds. Moreover, the magnitude of the reserve base is in each instance huge, ranking first or second in the world. Clearly, the precious minerals are critical to the Soviets in terms of both economic and political value. Table IV-6-1 illustrates precious mineral production, share of world output, production ranking, reserves, and ranking of reserve holdings in 1981.

GOLD

Soviet activities in the precious minerals markets are obscured by secrecy and clouded by controversy as to their motivations. This is particularly true of gold. The USSR ranks second in terms of reserves, production, and exports of gold behind South Africa. $A + B + C_1$ reserves were estimated at 6200 tons[2] (nearly 200 million troy ounces) in 1979. Two-thirds of all gold occurs as placer deposits, with the remainder found in numerous nonferrous metal deposits. The Soviets estimate that reserves are sufficient to support 12 to 15 years of production at current levels. Gold stocks are so consequential, however, that intense exploration is continuously carried on.

Output has been listed for 1981 at 262 tons[3] (8.42 million troy ounces). Gold production, according to the U.S. Bureau of Mines, is illustrated in Figure IV-6-1. Not surprisingly, gold is a mineral for which a range of output estimates exists. The CIA[4] put Soviet gold production at 317 tons (10.2 million troy ounces) in 1980. The highly regarded Consolidated Gold Fields report places Soviet output in the range 280 to 350 tons.[5] Production estimates truly began to vary by source in 1979. Papp[6] has noted Soviet sales of gold outside major European markets as the cause.

Soviet gold operations are directed by 14 regional gold administrations ("zolotos"). More than two-thirds of all gold comes from the Soviet Far East and East Siberia. Virtually all major rivers of these areas contain important placer deposits, including the Lena, Aldan, Kolyma, Yana, Anadyr, and Indigarka. Magadan, a former zone of exile, is the main producing district. Although this area exhibits slowly declining output, having peaked in the early 1970s, 35 placer mines and 23 dredges still produce significant gold. Yakutia is the second leading producer. Both these regions suffer from severe Arctic climates and difficult developmental conditions, including eight-month winters. During the Stalin reign, most gold was mined here by forced labor.

The remaining third of Soviet gold production occurs in the Urals, Central Asia, Armenia, and western Siberia. Uzbek in Central Asia is the site of the world's largest gold mine, at Muruntau (see Figure IV-6-2). The lode is huge, but low grade.

TABLE IV-6-1 Soviet Precious Minerals Production and Reserves

Commodity	Unit[a]	1981 production	Percent share of world production	World rank	Reserves	World rank
Diamonds						
Gem	A	2,100	19.1	2	30,000	
Industrial	A	8,500	27.3	1	120,000	
Gold	B	8,420	21.4	2	200,000	2
PGM	B	3,350	48.5	1	90,000	2
Silver	B	46,500	15.0	3	1,000,000	1

[a] A, thousand carats; B, thousand troy ounces. *Note:* One ton is approximately 32,150 troy ounces.

Sources: Refs. 2, 3, 14, and 15.

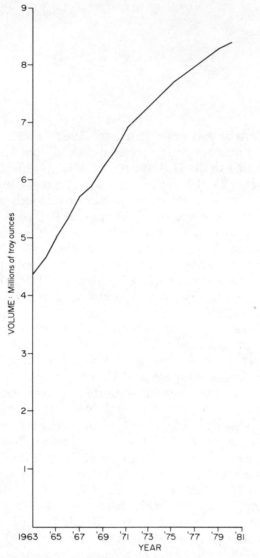

FIGURE IV-6-1   Soviet gold production. (From Refs. 2, 3, and 16.)

Both surface and underground methods are used at Muruntau. The open pit is more than 100 meters deep, with a 35:1 stripping ratio. While the Bureau of Mines[2] estimated Muruntau's annual production at 27 tons for 1980, Consolidated Gold Fields[7] placed it at 80 tons. Another major mine is located at Zod in the mountains of Armenia near the Turkish border. Zod has been worked for its gold as early as 1200 B.C. It is small, but very high grade, with native gold assaying 800 to 950 fine. As at Muruntau, mining is by both underground and surface operations. Output is estimated at 10 tons per year.

A significant amount of Soviet gold is produced as a by-product. While the Bureau of Mines[2] puts by-product gold at some 10% of total output (or some 25 tons), Consolidated Gold Fields[7] estimates

it at greater than 60 tons. The following information derives chiefly from this latter source. Copper ores provide 75% of all by-product gold. Recovery varies by location, ranging from 60 to 75%. The main source is the Kalmakyr copper complex in Uzbek, which accounts for nearly 30% of all by-product gold. Another one-third comes from copper operations in the Urals. Kazakhstan and Armenia copper production also yields gold. Lead-zinc ores provide the remaining by-product gold, centered in Central Asia, Azerbaidzhan, the Far East, and Irkutsk. Recovery is somewhat lower than that from copper ores, varying from 50 to 75%. Most unrecovered gold is lost in tailings.

Exports of gold are a vital source of hard currency for the USSR. As such, gold, more than any other mineral commodity, is relied on in times of emergency, gold being an asset with high liquidity, easily convertible to cash. Sales are not stable, but rather display sharp oscillations. During the Stalin era, gold hoarding was practiced. Upon Khruschev's ascendancy, sales began. From 1953 to 1965, 2900 tons of gold was sold.[5] These sales were larger than annual production, so gold stocks were reduced, probably to the point that some concern was registered. This is reflected in the actual purchases of gold which took place in the latter 1960s. (see Table IV-6-2). Large sales did not begin anew until 1972, when stockpiles were again large (the Muruntau deposit having come into production). High sales in 1976–1978 were also greater than annual production, and again reduced Soviet gold holdings. Figure IV-6-3 depicts gold exports since 1973. It must be emphasized that this data is a rough estimate at best, gleaned from gold traders in London, West Germany, and Switzerland.

Changing gold sales levels prompt rampant hypothesizing in the West concerning Soviet motivations. Some scholars believe that gold is sold up to a predetermined hard-currency requirement. Others feel that gold sales are closely tied to world prices; when prices are depressed, as in 1980, gold is withheld in hopes of stimulating price levels. (This "exercise in market discretion" is also practiced by the main supplier of world gold, South Africa.[8]) However, a comparison of gold sales with gold prices shows no clear relationship (Figure IV-6-3). Efforts to rebuild stockpiles depleted by the high exports of 1976–1978 could also account for low sales, such as in 1980. Fluctuating gold trade is probably determined mostly by hard-currency needs, however. Thus the Polish financial crisis, or a miserable Soviet grain crop, could cause high gold sales in periods of low prices, such as in 1981. But when exports are compared to, for exam-

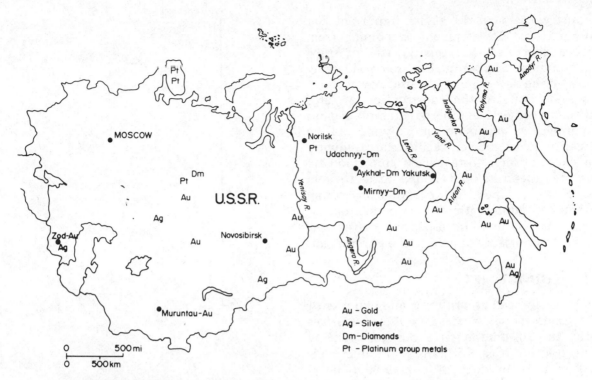

FIGURE IV-6-2 Soviet precious mineral deposits. (From Refs. 2 and 3.)

TABLE IV-6-2 Soviet Gold Sales to the West (Tons)

Year	Amount	Year	Amount
1960	177	1972	213
1961	266	1973	275
1962	178	1974	220
1963	489	1975	149
1964	400	1976	412
1965	355	1977	401
1966	-67	1978	410
1967	-5	1979	199
1968	-29	1980	80
1969	-15	1981	250
1970	-3	1982	190
1971	54		

Sources: Refs. 2, 3, 5, and 17.

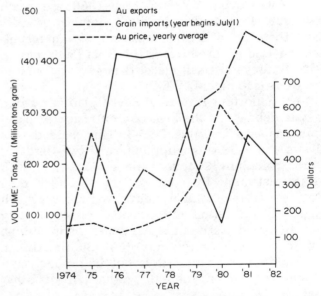

FIGURE IV-6-3 Comparisons between gold price, Soviet gold exports, and grain imports. (From Refs. 2, 3, 5, and 17.)

ple, grain imports, as in Figure IV-6-3, no direct correlation is apparent. Probably much discrepancy is due to the fact that the increases in the world price for oil have earned the USSR more foreign currency for a given volume of oil exports, thus reducing the contribution necessary from gold. In addition, Papp[6] has noted that some gold may be sold to dealers outside major gold markets, in which case sales figures are underestimated. Generally, it can be stated that the Soviets will sell into a rising market; conversely, they will avoid a falling one, or even buy large amounts to support a dropping price.[5] In either case, the USSR must now be regarded as an integral part of the international gold business. The USSR has proven itself a skillful and shrewd player of the gold market.

For the future, the USSR can be expected to maintain its position as a primary producer and exporter, at least into the next century. This is not to say that problems are nonexistent in the Soviet gold industry, particularly at its important Far Eastern districts.[9] Development is coming from ever-deeper deposits. The consequences of using mining equipment of poor quality are considerably amplified by the severe polar climate. Power shortages have precipitated the construction of large and expensive projects in otherwise remote regions (e.g., hydroelectric plants on the Kolyma, and the sole Asian nuclear power facility, at Bilibino). Massive exploration efforts have found good tin and tungsten deposits, but added little to gold reserves. However, these factors should in no way be construed as adumbrating the imminent demise of Soviet gold production, especially considering the strategic importance of gold to the government.

PLATINUM-GROUP METALS

The USSR is the leading producer and exporter of platinum-group metals. Soviet $A + B + C_1$ reserves are listed at 90 million troy ounces,[2] ranking second to South Africa. Virtually all production is achieved as a by-product of copper and nickel operations. In 1981, production was estimated at 3.35 million troy ounces[3] (104 tons), a 3% increase over 1980. Outputs since the early 1960s are shown in Figure IV-6-4. The Norilsk complex provides more than 75% of all Soviet platinum group metals. Second leading producer is the Kola Peninsula. In addition, several small placer deposits are found in the Urals (see Figure IV-6-2).

An important feature of Soviet platinum-group metals deposits is the high concentration of palladium, as seen in Table IV-6-3. This characteristic allows the USSR to supply nearly two-thirds of world palladium supplies, and almost one-fourth of its platinum. By contrast, the second leading platinum metals supplier, South Africa, has a much higher Pt/Pd ratio, and provides approximately 61% of world platinum.[10] A further distinguishing factor is the primary nature of South African platinum-group metals—it is not by-product, and hence not dependent on production of other metals.

Future production increments are clearly tied to expansion at Norilsk. In the past, construction delays here have directly resulted in failure to meet national output quotas. Admittedly, however, targets are often overambitious—the Ninth FYP called for production growth of 60%, and the Tenth for a growth of 80%. The Eleventh FYP mandates a 30% increase. Slow development has limited annual growth at Norilsk to only 4 or 5%. The ability to meet plans is therefore dependent on the success attained in achieving capacity at

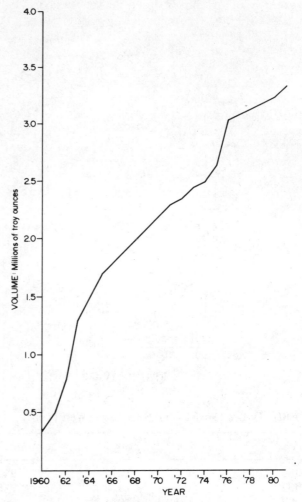

FIGURE IV-6-4 Soviet platinum-group metals production. (From Refs. 2, 3, and 16.)

TABLE IV-6-3 Concentrations of Soviet Platinum-Group Metals (Percent)

Deposit	Pt	Pd	Ir	Rh	Os	Ru
Norilsk	25	71	—	3	—	1
Monchegorsk (Kola Peninsula)	22.7	76	0.2	1	—	0.1
Uralian placers	91	1	5	0.5	2	0.5

Source: Ref. 18.

Norilsk, specifically at the new huge Nadezhda complex.

Exports of platinum-group metals are an additional major source of hard currency, bringing in more than $200 million in 1981.[1] Main markets for Soviet exports are the United States, West Germany, and Japan. There is a tendency toward declining exports, although little agreement can be found among Western experts concerning trade volumes. Thus some have noted exports falling from 800,000 troy ounces in 1975 to under 350,000 troy ounces in 1981, yet the Bureau of Mines[2] has estimated 1980 exports at 1.9 million troy ounces. Given discrepancies of this kind, prospects for the future are difficult to project. Certainly, several recent trends will benefit potential Soviet trade possibilities in the platinum metals.[10] Demand for palladium is anticipated to grow more than 4% until 1985; platinum demand is to rise by only 0.6% during this period, due to the depressed nature of the automobile, chemical, and refining industries. Japanese imports of palladium increased 24% in 1982 alone. Substitution by palladium for gold and other high-priced metals in the electronic and electrical sectors of the West is progressing; palladium is also substituting for platinum itself in many applications due to its lower price. A structural advantage thus accrues to the Soviet platinum metals industry, because of its high palladium output. In addition, the Soviets have recently proved adept at forcing higher palladium prices during annual contract negotiations with metals dealers.[11] Strong yearly increments to production, combined with features noted above, all point to a continuing dominant role by the USSR, together with South Africa, in platinum-group metals.

SILVER

Soviet silver reserves are believed to be nearly 1 billion troy ounces.[2] Practically all silver is derived as a by-product from nonferrous production, with gold treatment operations providing the remainder. Production in 1981 amounted to 46.5 million troy ounces[3] (1446 tons). Major producing areas are the Urals (where ores assay 6 to 15 grams of silver per ton), Kazakhstan, the Far East, East Siberia, and Armenia (see Figure IV-6-2). In addition, new facilities at Norilsk and the Kola Peninsula are recovering silver from copper-nickel ores.

Although the USSR is one of the world's biggest silver producers, output is insufficient for domestic needs. Moreover, import requirements are increasing, with most supplies shipped from Canada and Switzerland. In 1980, more than 31 tons (1 million troy ounces) were imported, double the level for 1978 and 1979. In October 1982 alone, more than 10 million ounces were purchased, chiefly through Swiss dealers.[12] Jumps in domestic demand could be induced by the electronics and/or photographic industries, by military programs, or (less likely) due to speculation in the market. Internal shortages have promoted research into unorthodox sources

of silver, such as recovery from the ashes of burned film.[3] Nonetheless, barring substantial consumption decreases, the USSR will continue to depend on imports, although it is doubtful that these will be allowed to become excessive.

DIAMONDS

The USSR ranks first in industrial diamond output and second in production of gem diamonds. $A + B + C_1$ reserves are 30 million carats of gem and 120 million carats of industrial stones.[2] In 1981, output of gem diamonds was 2.1 million carats, and of industrial diamonds 8.5 million carats;[3] both represent small declines from 1980 production.

The Soviet diamond industry is centered in Yakutia, especially the Mirnyy, Aykhal, and Udachnaya deposits (see Figure IV-6-2). Prior to the discovery of kimberlite pipes here in the mid-1950s, the USSR depended on imported diamonds. Diamonds are also found in the Western Urals near Perm. Gems are cut in Leningrad, Sverdlovsk, and Smolensk. Synthetic diamonds production is located at Kiev, Yerevan, Moscow, Tashkent, and Poltava.

The value of diamond exports is estimated at $175 million in 1981.[1] Gem diamonds account for most of this trade, with Belgium-Luxembourg, Japan, and West Germany principal destinations. The Soviets appear to collaborate their sales efforts with the De Beers cartel. Certainly, Soviet diamond trade has benefited from price floors established by the Central Selling Organization. In the past, diamonds have been sold to London; these items, primarily rough stones, are then passed on through a third party to De Beers.[13] Some Soviet diamonds are also marketed via Antwerp through a Belgian-Soviet export organization in which the Soviets have a controlling interest. Exploration is to be stepped up during the Eleventh FYP, and the USSR should remain a major supplier of diamonds to world markets for the foreseeable future, due to the large reserve base, and the ability of this commodity to earn hard currency.

REFERENCES

1. Prest, Michael. "Soviets Learn to Play Markets to Their Best Advantage." *The Globe and Mail*, 2 August 1982.
2. U.S. Department of the Interior, Bureau of Mines. "The Mineral Industry of the USSR." By V. V. Strishkov. Preprint from the *Minerals Yearbook 1980*. Washington, D.C.: U.S. Government Printing Office, 1980.
3. Levine, Richard M. "Soviet Union." In *Mining Annual Review, 1982*. London: Mining Journal, June 1982.
4. U.S. Central Intelligence Agency, National Foreign

Assessment Center. *Handbook of Economic Statistics 1981*. Washington, D.C.: U.S. Government Printing Office, November 1981.

5. Green, Timothy. *The New World of Gold*. New York: Walker and Company, 1981.

6. Papp, Daniel S. "Soviet Non-fuel Mineral Resources, Surplus or Scarcity?" *Resources Policy* 8, No. 3 (September 1982).

7. "Soviet Gold Production" and "By-product Gold Production in the Soviet Union." In *Gold 1980*. By David Potts. London: Consolidated Gold Fields Limited, 1980.

8. "Soviet Gold." *Mining Journal*, 3 October 1980.

9. "Gold in the Soviet Far East." *World Mining*, December 1981.

10. "Platinum's Tribulations." *Mining Journal*, 29 October 1982.

11. Behrmann, Neil, and Anne Mackay-Smith. "Russian Bear Learns New Tricks As Soviets Create a Bull Market for Their Palladium." *The Wall Street Journal*, 26 November 1982.

12. Behrmann, Neil, and Anne Mackay-Smith. "Heavy Soviet, Mideast Buying of Silver Sparked Recent Market Rise, Dealers Say." *The Wall Street Journal*, 1 November 1982.

13. U.S. Department of Commerce. *Industrial Outlook Report. Minerals. South Africa—1981*. Washington, D.C.: U.S. Government Printing Office, 1982.

14. U.S. Department of the Interior, Bureau of Mines. *Mineral Commodity Summaries 1982/1983*. Washington, D.C.: U.S. Government Printing Office.

15. U.S. Department of the Interior, Bureau of Mines. *Mineral Facts and Problems, 1980 Edition*. Washington, D.C.: U.S. Government Printing Office, 1980.

16. U.S. Department of the Interior, Bureau of Mines. *Minerals Yearbook*, 1963–1979. Washington, D.C.: Government Printing Office.

17. "Soviet Union's Gold Sales Fell 25% Last Year, Traders Say, Perhaps Signaling Economic Gains." *The Wall Street Journal*, 31 January 1982.

18. Smirnov, V. I. *Ore Deposits of the USSR*, Vol. III. San Francisco: Pitman Publishing, Inc. 1977.

The Nonmetallic Minerals

The USSR produces a large variety of nonmetallic minerals. In some, it is a major exporter, while for others it requires imports to meet demand. Included in the nonmetals are the fertilizers, especially important to the USSR because of its notoriously poor agricultural harvests. Table IV-7-1 lists major nonmetal minerals production, share of world output, production ranking, reserves, and ranking of reserve holdings in 1981.

ASBESTOS

The USSR ranks first in production and second (behind Canada) in reserves of asbestos. All Soviet asbestos is chrysotile. $A + B + C_1$ reserves are estimated at 100 million tons.[1] The USSR surpassed Canada in the 1960s to become the leading producer. In 1981 output was 2.22 million tons.[2] A persistent problem in the asbestos industry is the generally poor quality of the majority of Soviet asbestos.

The primary producing region is the Uralasbest complex near Bazhenovo in the central Urals (see Figure IV-7-1). This area accounts for more than 50% of Soviet asbestos; it was the only supplier until the 1960s, when three additional deposits were developed. Currently, more than 25% of production comes from the Dzhetygara deposit near Kustanay in north-western Kazakhstan. The Kiyembay complex near Orenburg in the south Urals provides about 6% of output; Kiyembay is a joint venture with the CMEA. Although quotas call for

170,000 tons (one-third of production) to be exported to these nations, in 1981 less than half this amount was delivered.[2] The fourth largest asbestos complex is at Tuvaasbest, north of Mongolia. Asbestos here is of higher quality than that found at other Soviet deposits. In addition, a huge deposit in East Siberia, at Sayan, containing 7 million tons of asbestos, is now being developed.

Exports of asbestos in 1981 totaled 625,000 tons. Two-thirds of this quantity was intended for CMEA nations. Given the Soviet reserve base, and the large increases in consumption apparent in the noncommunist world, the USSR should continue as the main producer and a major exporter.

FLUORSPAR

The USSR ranks as the number two producer of fluorspar. Soviet $A + B + C_1$ reserves are listed at 10 million tons,[1] with output reaching 530,000 tons in 1981.[2] Major producing regions are the Maritime Territory of the Far East (especially the Yaroslavsk deposit), Transbaikal, Uzbek, and Kazakhstan (see Figure IV-7-1). Domestic consumption, however, particularly in the steel and aluminum sectors, places demands on the fluorspar industry that it cannot meet. Consequently, the USSR is now more than 50% dependent on imports of fluorspar, with supplies coming from Mongolia, China, and Thailand. Increases in consumption are consistently outstripping production increments,[2] so this import dependency will continue.

TABLE IV-7-1 Soviet Nonmetallic Mineral Production and Reserves

Commodity	Unit[a]	1981 production	Percent share of world production	World rank	Reserves	World rank
Asbestos	A	2,220	44.6	1	100,000	2
Fluorspar	A	530	10.6	3	10,000	
Phosphate	A	75,000	21.8	2	1,500,000[b]	3
Potash	A	8,350	29.0	1	3,800,000[c]	1

[a]A, thousand tons.
[b]P_2O_5 content.
[c]K_2O content.

Sources: Refs. 1, 2, 4, and 5.

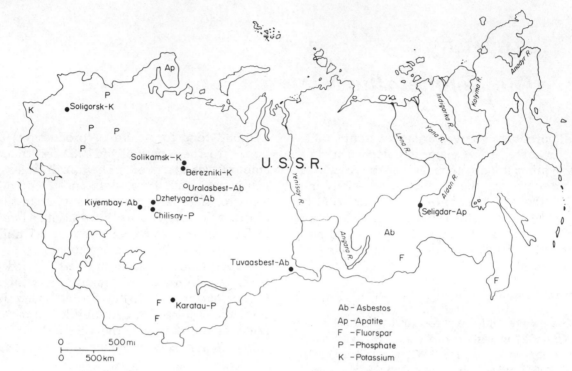

FIGURE IV-7-1 Major Soviet nonmetallic mineral deposits. (From Refs. 1 and 2.)

PHOSPHATE

Due to the depressed phosphate industry in the United States, the USSR is now the largest producer of this commodity. $A + B + C_1$ reserves are believed to be 1.5 billion tons of P_2O_5 in 14 billion tons of ore.[2] Thirty-seven percent of this amount is found in the apatite-nepheline ores of the Khibiny Mountains on the Kola Peninsula (see Figure IV-7-1). This is the largest deposit, with ore grades averaging 16.5% P_2O_5, and 39.4% after beneficiation. The phosphorite deposits at Karatau in Kazakhstan account for 26% of phosphate reserves; ores here assay 21 to 26% P_2O_5. Other small deposits are found near Moscow, Leningrad, and the upper Kama in the Urals; these are all very low grade. New large deposits are being developed in Estonia, Kazakhstan, and the Far East. The USSR shares with South Africa the distinction of having the only reserves of phosphate rock of an igneous origin.

Soviet output of phosphate in 1981 was 75 million tons,[2] including 49 million tons of apatite ore (17.7% P_2O_5) and 26 million tons of sedimentary rock (13% P_2O_5). Major producing regions are the Kola Peninsula, Karatau, the Chilisay phosphorite basin in Kazakhstan, and the Seligdar apatite deposit in Yakutia (see Figure IV-7-1). The latter complex is important in that it will remove the need to transport fertilizer to the Far East from Kazakhstan or the European USSR. Both the Kola and Karatau complexes are planned to increase production 50% in the Eleventh FYP.

Exports of phosphatic concentrate and phosphate fertilizer to Western Europe are decreasing. Trade to this area totaled some 200,000 tons in 1980. An important long-term trade agreement has been reached with Morocco, whereby Soviet assistance in developing the Meskala phosphate deposit in Morocco will be paid for by phosphate exports to the USSR. The wretched record in Soviet agriculture definitely points to consistently high demand for fertilizer. Given this fact, the Moroccan pact may indicate concern over future supply availability; some experts suggest the possibility of depletion of some high-grade Soviet deposits.[3] Alternatively, the high cost of beneficiation processes for igneous phosphate may have induced Soviet leaders to prefer cheaper imported Moroccan sedimentary rock. Such a policy would be a decided change from the usual Soviet aspirations for mineral self-sufficiency.

POTASH

The Soviet potash industry is the largest in the world. $A + B + C_1$ reserves are estimated at nearly 23 billion tons of ore assaying 16-40% K_2O, or 3.8 billion tons of K_2O;[1] reserves are chiefly carnallite and sylvite. The upper Kama Basin in the

Urals contains two-thirds of all potassium ore. Nearly 5 billion tons of sylvite assaying 16 to 20% K_2O are found in Belorussia. Other large concentrations occur in the Livov Basin of the Ukraine and in Turkmen (see Figure IV-7-1). In addition, a huge new deposit in the Irkutsk area of eastern Siberia covering some 10,000 square kilometers has begun development.

Production reached 63 million tons of ore (8.35 million tons nutrient content) in 1981.[2] This fell far short of tenth FYP goals of 85 to 90 million tons. Major producing centers are at Soligorsk in Belorussia (providing 50% of all potassic fertilizer), Solikamsk and Berezniki in the Urals, and Strebnikov and Kalush in the Ukraine. Twelve underground mines and one open pit operate in these districts. Output is planned to reach 92 million tons by 1985.

At present, Soviet potash salts are part of a barter arrangement between the USSR and the U.S.'s Occidental Petroleum. Occidental receives this potash as partial payment for construction of an $8 billion liquid ammonia plant. Generally, however, exports are destined for CMEA countries. Potash exports exceeded 6.5 million tons in 1981, with Poland and Hungary in particular dependent on imports from the USSR. Although much idle capacity exists in the Soviet industry, and output will probably continue to fall short of goals, production of potash will remain high, reflecting its importance to the farmers of both the USSR and other CMEA members.

REFERENCES

1. U.S. Department of the Interior, Bureau of Mines. "The Mineral Industry of the USSR." By V. V. Strishkov. Preprint from the *Minerals Yearbook 1980*. Washington, D.C.: U.S. Government Printing Office.
2. Levine, Richard M. "Soviet Union." In *Mining Annual Review, 1982*. London: Mining Journal, June 1982.
3. U.S. Department of Commerce. *Industrial Outlook Report. Minerals South Africa—1981*. Washington, D.C.: U.S. Government Printing Office, 1982.
4. U.S. Department of the Interior, Bureau of Mines. *Mineral Commodity Summaries 1982/1983*. Washington, D.C.: U.S. Government Printing Office.
5. U.S. Department of the Interior, Bureau of Mines. *Mineral Facts and Problems, 1980 Edition*. Washington, D.C.: U.S. Government Printing Office, 1980.

Chapter IV-8

Soviet Minerals Policy

Large production levels, supported by equally large reserves, have enabled the USSR to maintain a minimal dependence on external mineral supplies, ultimately achieving a degree of self-sufficiency unmatched by any other nation. The net import dependency of the USSR is exhibited in Table IV-8-1. There are in fact only a few minerals—bauxite, fluorspar, tungsten, tin, silver, antimony, and cobalt—for which the Soviets must rely on imports. Significantly, many of these minerals produced in great quantity in the USSR are commodities for which the United States faces sizable domestic production deficiencies. Table IV-8-2

portrays the minerals for which Soviet output ranks first or second, and for which the United States was substantially import dependent in 1982. (Note also, however, that some of these minerals are ones for which the Soviets themselves are import reliant, although not to the extent that the United States is.) This emphasis on self-sufficiency is a unique component of Soviet mineral policy. A review of the historical basis for self-sufficiency, as well as the current state of minerals policy, is in order.

HISTORICAL BACKGROUND

Following the chaotic era of the Bolshevik Revolution, the Soviet minerals industry faced stagnation. The country was fully 68% dependent on foreign supplies for its minerals consumption,[1] due in great part to the destruction of its industrial capacity. It was originally felt that the revolution would spread

TABLE IV-8-1 Soviet Net Import Dependence of Minerals as a Percent of Consumption, 1981

Metals		
Aluminum	-60	
Antimony	8	Yugoslavia
Bauxite and alumina	55	Guinea, Yugoslavia, Hungary, India
Cadmium	-14	
Chromium	-30	
Cobalt	42	Cuba
Copper	-21	
Gold	-2000	
Iron ore	-19	
Lead	0	
Manganese	-15	
Mercury	0	
Molybdenum	0	
Nickel	-19	
PGM	-100	
Silver	3	Switzerland, Canada
Steel	2	FRG, Japan, Italy, France
Tin	22	Malaysia, UK, Bolivia
Titanium		
Ilmenite	-13	
Rutile	0	
Tungsten	19	PRC, Mongolia
Vanadium	-5	
Zinc	-5	
Nonmetals		
Asbestos	-39	
Fluorspar	51	Mongolia, PRC, Thailand
Phosphate	-20	
Potassium	-58	

Source: Ref. 5.

TABLE IV-8-2 U.S. Import Dependence on Minerals for Which the USSR Is a Major Producer (Percent)

Minerals	1982 U.S. import dependence
USSR as leading producer	
Asbestos	74
Cadmium	69
Iron ore	36
Manganese	99
Mercury	43
Nickel	75
Platinum metals	85
Potash	71
USSR as No. 2 producer	
Chromium	88
Diamonds	100
Gold	43
Tin	72
Tungsten	48
Zinc	53
Other minerals for which USSR is a major producer	
Antimony	45
Cobalt	91
Fluorspar	87
Silver	59

Source: Ref. 4.

to other nations. When it became apparent this was an unrealistic hope, Stalin formulated his "socialism in one country" theory, whereby communism would first reach an advanced state in the USSR, and could then subsequently infiltrate other lands more easily. To this end, the USSR regarded itself as a nation under siege from the West ideologically, politically, and economically. Rather than depend on hostile foreign sources of raw materials, the Soviets deemed it wiser to rely on their own resources for defense.[1] The historical tradition of Russian xenophobia undoubtedly contributed to this policy. It was also felt that the USSR would provide the material base to sustain the diffusion of communism; industrial production thus assumed an essential role in the political program of the Soviet leadership, becoming in effect an ideological goal.

The policy of mineral self-sufficiency thus cannot be divorced from ideology. Marxist-Leninist theory has always placed a high premium on raw materials;[2] Stalin's "socialism in one country" only served to add further emphasis. The USSR then became the fortress state. Psychologically, the drive for self-sufficiency can be tied to a remembrance of those perilous years when the Soviet minerals industry was virtually inoperative. As a consequence, the commitment to independence at any cost, and despite the burdens imposed on the present generation, has become a fervid determination.

MINERALS POLICY

The minerals policy of the USSR is predicated chiefly upon this concept of self-sufficiency. Minerals development has been and remains a critical component of industrial expansion, and it has also frequently been used as a means of settling uninhabited territories. As such, minerals production has assumed a high strategic importance. If a commodity is believed essential, an attempt is made to produce it from internal endowments, regardless of the cost. An excellent example of this is the use of nonbauxitic sources to produce aluminum. Implicit in this policy is the exploitation of deposits that would be uneconomic, and hence undeveloped, in the West.

A distinguishing feature of the Soviet economy is the well-defined direction that minerals development can take, due to centralized planning. It is especially evident when compared to the divergent expansion in market economies.[3] Soviet goals are well formulated and are achieved through the FYP targets as set by GOSPLAN. As can be seen from

the minerals production graphs from Chapters IV-4 to IV-7, development and growth can assume almost linear proportions.

Prices and Production

World mineral prices have much less of an impact on Soviet mineral production than in a market economy. For instance, low 1982 prices slowed Canadian nickel production to almost 50% of its 1981 output.[4] This not only affected the nickel industry, however—Canadian production of platinum-group metals, a by-product of nickel, fell 51%. By contrast, even though Soviet platinum metals are also a nickel by-product, output in 1982 increased by more than 150,000 troy ounces. This seems to indicate that important by-products (in this case, a foreign currency earner) will be produced despite the world market conditions for the primary commodity.

To ensure the continual production of strategically important minerals, ventures working at a loss are kept in operation by Soviet subsidies. Although this guarantees domestic availability, it also contributes in large measure to waste at many facilities. Subsidies provide little incentive for higher efficiency. The reverse could in fact be true. Less waste may possibly cause a cutoff of government funding.

In response to this problem, an emphasis on conservation and efficiency has become an important element of present Soviet policy.[5] The wholesale price revisions of January 1982 for raw materials are an attempt to promote this strategy. It is hoped that the need to subsidize unprofitable operations will be greatly reduced by this step. Evidence of success remains to be seen, however. For example, even after the last major price increases in 1967, almost half of all lead-zinc complexes operated at a loss.[6]

Development of Arctic Minerals

The drive to self-sufficiency has led to many unusual strategies, one of the most prominent being the development of Arctic minerals. The USSR produces more commodities from polar districts than any other nation.[7] Norilsk is the most important minerals center in the Arctic, and arguably in the entire country. Norilsk and the Kola Peninsula combined contain more than 95% of Soviet platinum-group metals and 80% of nickel reserves; they produce substantial quantities of copper and cobalt as well. The Kola Peninsula is also the site of the world's largest apatite-nepheline deposit. Eastern Siberia and the Far East account for most gold, diamonds, and tin output, and significant tungsten and mercury production. The

Pechora coal basin is the second leading coal producer west of the Urals. In addition, the huge oil and gas deposits of western Siberia must be included. The main implication of large Arctic minerals extraction is high cost. Problems due to supplies and labor availability, equipment performance, construction schedules, and technological difficulties are commonplace. Nonetheless, development of polar minerals is indicative of the importance assigned to self-sufficiency, and the lengths to which Soviet leaders will go to maintain it.

Trade

Trade is a major component of minerals policy; political rather than economic objectives are frequently the chief determinant of trade volumes. Exports and imports are thus more a reflection of national goals as defined in the FYP; balance of payments factors are a key influence. Exports are not "surplus" as is usually the case in the West, but could be consumed internally. The two principal aims regulating trade are the acquisition of hard currency through exports, allowing importation of needed goods (typically grain and Western machinery and technology), and the maintenance of Soviet hegemony over the CMEA nations.

Mineral exports now account for approximately half of total exports. Raw materials trade provides 80% of hard-currency earnings,[5] especially shipments of oil (which now accounts for more than half), natural gas, gold, platinum-group metals, diamonds, chromite, and nickel. Major Western trading partners are West Germany, Finland, and Japan. Most exports to the United States are mineral commodities as well, including fuel oil, chromite, gold, nickel, platinum-group metals, aluminum scrap, and titanium. The Soviets normally will attempt to sell the minimal quantity at the highest price, although emergencies can induce sales in depressed markets, or at below world prices.[5] Recent barter agreements have also been negotiated, involving U.S. dairy products for Soviet nickel and palladium.[8]

It is the CMEA, however, that is the primary trading partner of the USSR. Approximately half of all Soviet trade, and most of its mineral exports, go to these nations. A leading objective of this exchange is to more closely integrate the East European economies with that of the USSR. One strategy to achieve this goal is the dependence of the CMEA on Soviet goods. For example, Soviet exports now furnish 90% or more of CMEA oil, natural gas, pig iron, iron ore, and electrical power consumption; two-thirds of its petroleum products, rolled ferrous metals, and phosphate fertilizer; and 60% of its

coal and manganese ores.[3] The low prices of these commodities have amounted to subsidization in the past, the hope being that reduced payment requirements would inject some needed stability into CMEA economies. Joint Soviet-CMEA ventures are also being increasingly emphasized, probably to lessen the burden imposed on the Soviets by their struggling satellites. Current projects include construction of electrical power transmission networks and two nuclear plants (at Konstantinovka and Khmelnitskiy), the Kiyembay asbestos complex, the Soyuz gas pipeline, iron ore mining in the KMA and Ukraine, petroleum refining, and ferroalloy production at Nikopol and in Kazakhstan.[9] Total cost of CMEA participation exceeds $13 billion.

Increasingly, there are signs that the economic importance of trade may be gaining in significance relative to political objectives. Despite the obvious benefit to the USSR of a stable CMEA, price hikes and export ceilings are being imposed, with the CMEA encouraged to broaden supply sources.[10] This can only mean further trade with the West, and concomitant deteriorating trade balances. As for the Soviets, in dealings with the West, compensation arrangements similar to the Urengoy gas export pipeline are becoming more frequent; joint projects of this type overcome the typical Soviet deficiencies in capital, equipment quality, and technology. Several Western nations, for example, have been approached regarding joint development of the Udokan copper deposit.[3]

Soviet imports of minerals now make up more than 10% of all imports. While mineral self-reliance is the principal aim, imports of some minerals are necessary due to insufficient reserves. Other motivations may include compensating for short-term supply bottlenecks, supplying regions far from producing areas, or as a gesture of goodwill to friendly nations. In addition, the possibility of preclusive purchasing does exist, particularly from Africa. Generally, however, imports have not yet become inordinate; self-sufficiency has remained the prevailing policy. Shall it continue to be so?

REFERENCES

1. Sutulov, Alexander. *Minerals in World Affairs*. Salt Lake City, Utah: University of Utah Press, 1972.
2. Papp, Daniel S. "Soviet Non-fuel Mineral Resources, Surplus or Scarcity?" *Resources Policy* 8, No. 3 (September 1982).
3. U.S. Department of the Interior, Bureau of Mines. "The Mineral Industry of the USSSR." By V. V. Strishkov. Preprint from the *Minerals Yearbook 1980*. Washington, D.C.: U.S. Government Printing Office, 1980.

4. U.S. Department of the Interior, Bureau of Mines. *Mineral Commodity Summaries 1983*. Washington, D.C.: U.S. Government Printing Office, 1983.

5. Levine, Richard M. "Soviet Union." In *Mining Annual Review, 1982*. London: Mining Journal: June 1982.

6. U.S. Central Intelligence Agency, National Foreign Assessment Center. *The Lead and Zinc Industry in the U.S.S.R.* Washington, D.C.: U.S. Government Printing Office, 1980.

7. U.S. Central Intelligence Agency. *Polar Regions*. Washington, D.C.: U.S. Government Printing Office, 1978.

8. "Periscope: Two Servings of Surplus Food." *Newsweek*, 24 January 1983.

9. Hoffman, George W. *Eastern Europe's Resource Crisis, with Special Emphasis on Energy Resources: Dependence and Policy Options*. Policy Study No. 14. Center for Energy Studies, University of Texas at Austin, January 1981.

10. Grichar, James S., Richard Levine, and Lotfollah Nahai. *The Nonfuel Mineral Outlook for the USSR through 1990*. Washington, D.C.: U.S. Government Printing Office, 1981.

The Feasibility of Continued Mineral Self-sufficiency

In recent years, the ability of the USSR to maintain its mineral self-sufficiency has come into question. Led by Daniel Fine[1] of MIT, a group of scholars has interpreted such variables as changing trade patterns, rapid rises in domestic consumption, ore grade declines, economic difficulties, a slowdown in the development of Siberian resources, and military stockpiling as indicating a transition to "selective resource dependency." Such a shift to outward access for certain minerals could at the least signify increased competition for supplies in world markets. Fears of a Soviet readiness to employ military force for mineral acquisition have also been expressed. In this light, the invasion of Afghanistan, ongoing Soviet military expansionism, and Soviet activities in southern Africa, have all been seen as evidence of the Soviet ability to exert its control over resource-rich areas, and its willingness to do so through the use of force.[2,3] Is this a realistic interpretation of the problems facing the Soviet mineral industry—that is, is the USSR becoming a net importer of many minerals for which it was once self-reliant?

ORE GRADE DECLINES

Decreasing assays for mined minerals are a reality. During the 1980s alone, metal content is expected to decline by 10 to 15% for most nonferrous ores and by 12 to 15% for ferrous metal ores.[4] This, combined with mine depletion in European sectors, is the primary cause of the supply centers shifting to Asia. Declining ore grade implies a need for increasing concentration, changes in mining, and certainly indicates higher costs.

Although this is a serious problem, numerous opportunities exist in the USSR that may help compensate for its consequences. To date, the Soviet minerals industry has simply not made optimum use of its available reserves. Specifically, the elements of waste, technological inadequacy, and resultant poor quality all prevent comprehensive minerals utilization. Potential for improvements abound, be they in mining, concentrating, refining, by-product recovery, transportation, and so on. Modernization and increased efficiency could in great measure counteract lowering ore grades,

definitely raising the general availability of supplies. The deficiencies caused by the Soviet planning system itself are another matter; incompetent management and poor coordination will be largely irradicable, barring basic system reform. Technological improvements and innovations, whether imported or internally derived, however, do offer considerable hope in balancing the effects of lower-quality ore.

CHANGING TRADE PATTERNS

Trends for decreasing exports of some minerals, such as chromite and titanium, and the importation of many minerals once adequately supplied domestically, such as purchases of chromite from Iran in mid-1980, and CMEA's procurement of manganese from Gabon in 1979, have been called indications of serious domestic production difficulties. This kind of analysis suffers from several shortcomings, foremost of which is a misunderstanding of the function of trade in the USSR. As has been shown, trade is used mainly to achieve political objectives. A further factor determining trade policy is balance-of-payments considerations. Exports do not reflect absolute surpluses, nor do imports necessarily indicate national shortages, although they can certainly mean regional ones. It is thus of dubious value to base any conclusions on trade in the short term. For example, low gold and platinum-group metal exports in the late 1970s to the West, and cessation of titanium sales in that period, raised many questions as to output capabilities. The resumption of significant sales of all three in the early 1980s illustrates the danger of short-term trend extrapolations of Soviet trade statistics.

Some export declines may be attributed to stockpiling; the Bureau of Mines[5] has submitted that some high-quality ore may be safeguarded from "excessive export," including chromite, titanium, nickel, and vanadium. In the case of chromite, which has exhibited strong long-term downward exporting tendencies, as seen in Figures IV-4-6 and IV-4-7, implementation of the AOD process has reduced demand for metallurgical-grade ore worldwide. The only two minerals where trade trends can be said to justifiably indicate problems are lead and zinc, where, as Figures IV-5-8 and IV-5-9 illustrate, the ability to export these commodities is apparently dependent on ore and concentrate imports. Finally, in the case of manganese, Fine's[1] declaration that exports have shown extreme declines is not supported by the facts, as proven by Figure IV-4-3.

RAPIDLY INCREASING CONSUMPTION

The continuing development of Soviet industrialization, Fine[1] affirms, is placing demands on the minerals industry that cannot be met solely through domestic production. This is a situation exacerbated by commitments to CMEA, and particularly by ever-growing military requirements. These claims may be the most justified of any of the points Fine makes. Certainly the demands of the largest steelmaker and biggest military program in the world consume substantial amounts of minerals. There may indeed by years when shortages develop due to these needs. However, the USSR is a planned economy. For all its problems because of this, such as waste and inefficiency, it has generally proven adept at matching supply and demand in sectors considered strategically vital, exclusive of the agricultural sector. In nonpriority areas, demand is simply restricted to the supplies available. Moreover, if consumption has increased, so has production—often dramatically. The production graphs from Chapters IV-4 to IV-7 speak for themselves; the USSR displays growth patterns typical of an immature industrial nation, which it could properly be classified as, according to its developmental age. The magnitudes of output, however, rival those of the most industrially advanced nations. Although it is true that production levels short of target are the norm, quotas and output continue to increase. There is little reason to believe that the USSR would allow itself to be put into a position where demand in critical sectors would have to be met for the long run by imports; it is frankly contrary to all that the Soviet efforts have been directed towards.

ECONOMIC CONSTRAINTS AND THE SUPPLY SHIFT EASTWARD

Any future major increments to output will normally be affected by the limitations of the Soviet economy itself, and by geographical constraints in Asia as new deposits here are developed. These factors have been discussed in Chapter IV-2. Of all variables hindering minerals self-sufficiency, the economic ones are the most fundamental. Although centralization may be an advantage in guiding the minerals industry and in directing growth

to meet demand, such benefits are offset by the disadvantages inherent in the system. The problems of labor shortages, poor agency planning and coordination, and stiff competition for limited capital are features that will persist.

While the transition to Asia of the mineral supply base would seem unfavorable, it is this trend, more than any other, that makes any proclamation of impending Soviet minerals dependence a questionable conclusion. It is highly unlikely that a nation so vast and industrially young could have already exhausted its reserves, even considering its large production levels. The potential for further finds in Asia is considerable, for huge areas are still not adequately explored. The difficulty is not absolute depletion, but new source location; geography, infrastructure availability, climate, and cost are all at the heart of the issue. The USSR differs from nearly all other major mineral producers in that apparently significant undeveloped mineralized regions do exist, but in remote areas. Although it may be cheaper to import some minerals, the USSR does appear to have a choice.

This leads to Fine's[1] assertion that a slowdown exists in the development of Siberian resource development, this being in effect a strategy of "in situ stockpiling," as the world (cheaper) supplies are used. Such a factor, if true, would lend much weight to a theory of imminent mineral dependence. Rumer and Sternheimer[6] have poignantly shown that investment in Siberia has not exhibited the strong increases one would expect to find (see Figure IV-9-1). This both strengthens and weakens

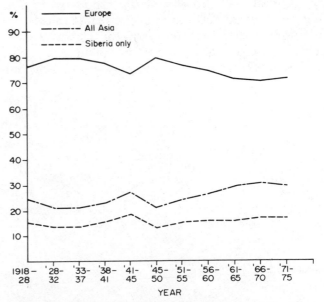

FIGURE IV-9-1 Trends in regional distribution of Soviet capital investment. (From Ref. 6.)

Fine's argument. Despite the obvious need for Siberian development, capital investment is not displaying the type of rising trends that Asian minerals production is (see Figures IV-2-1 to IV-2-3). Rather, Siberian investment as a percent of total investment is constant over a considerable period. On the other hand, development is *not* slowing. Moreover, these are cumulative investment figures; they certainly are not indicative of the huge increments in Asian energy sector funding as oil and gas development continues in western Siberia. Two additional features merit consideration. Billions have been spent on the BAM; the implications of this railroad on the future accessibility of Asian deposits, as well as the potential for subsequent mineral export, are quite manifest. Equally so is the ongoing development at Norilsk, and the continual exploration programs in Asia for new deposits. The assertion of a Siberian resources developmental slowdown tends to lead to a conclusion that Soviet leaders have dropped plans for the region. Statistics do not unalterably lead to such a finding. Variables, such as the present economic stagnation, with concomitantly higher competition for investment capital, or the massive Soviet military buildup, could affect the size and rate of Siberian investment. The fact remains that the future of Soviet minerals development lies in the East.

Self-sufficiency, then, will still be an option open to Soviet leaders, but it will cost even more than in the past, and be more difficult to maintain. The feasibility of this option shall be affected by competition for investment, by the state of detente, by the health of the economy, in short by a whole host of internal and external factors. It must again be stressed that the ability of the USSR to overcome insurmountable barriers in the quest of a goal cannot be discounted. This may not be accomplished with high efficiency or productivity, but the success rate is impressive enough in priority areas, again, excluding agriculture, to warrant notice.

The historical ideological importance of the policy of minerals self-sufficiency implies that it is unlikely to be abandoned casually. Present enormous investment and construction projects in evidence in remote districts such as Norilsk suggest it has not been. Although it is virtually impossible to authoritatively interpret the actions of a closed society such as the USSR, the Soviet minerals industry, despite its many serious problems, cannot be properly construed as a declining force. The continuation of the strategy of mineral self-sufficiency is ultimately dependent on a renewal of the commitment to this expensive policy; such a commitment remains not only possible, but probable.

REFERENCES

1. Fine, Daniel I. "Mineral Resource Dependency Crisis: Soviet Union and United States." In *The Resource War in 3-D: Dependency, Diplomacy, Defense.* Edited by James A. Miller, Daniel I. Fine, and R. Daniel McMichael. Pittsburgh, Pa.: World Affairs Council of Pittsburgh, 1980.
2. "Soviet Occupation of Afghanistan Has Resource Manifestations." *Alert Letter on the Availability of Raw Materials,* No. 13 (June 1982).
3. Rees, David. *Soviet Strategic Penetration in Africa.* London: Institute for the Study of Conflict, 1976.
4. Levine, Richard M. "Soviet Union." In *Mining Annual Review, 1982.* London: Mining Journal, June 1982.
5. U.S. Department of the Interior, Bureau of Mines. "The Mineral Industry of the USSR." By V. V. Strishkov. Preprint from the *Minerals Yearbook 1980.* Washington, D.C.: U.S. Government Printing Office, 1980.
6. Rumer, Boris, and Stephen Sternheimer. "The Soviet Economy: Going to Siberia?" *Harvard Business Review,* January–February 1982.

The Soviet Presence in Southern Africa: An Example of Resource Diplomacy

Any analysis of USSR minerals and related policy must of necessity examine Soviet activities in southern Africa, for it has been postulated that Soviet interests in this region are part of a coherent design to either acquire minerals or to deny the West access to these resources. The importance of the minerals of southern Africa to the industries of the West is well documented, if not widely known. In brief, concern has been expressed over the availability to the West of certain commodities should the entire southern region fall into the Soviet sphere of influence. The geographic concentration and production of these minerals are limited mainly to southern Africa and the USSR. Minerals of concern include chromite, manganese, platinum-group metals, gold, vanadium, and cobalt. These are commodities often having vital industrial or defense applications, for which there is little or no U.S. or other NATO production, and for which no ready substitutes exist. Spreading Soviet influence in Africa also raises the specter of cartel-like manipulations of production and prices of these materials. Finally, a strong Soviet presence threatens the security of the Cape oil sea route, a waterway of extreme supply importance. Are these feasible hypotheses? Is the Soviet presence a policy of minerals acquisition, or even of minerals denial?

It is unfortunate that this issue is debated along ideological liberal-conservative lines, for the subject is more complex than such dogma allows. A survey of selected literature is offered in Table IV-10-1. This list is not intended to be exhaustive.

Undoubtedly, U.S. minerals producers and lobbyists representing this group have used the idea of a confrontation over resources in an attempt to foster support for a stronger domestic industry. Considering the utter depression which has existed in this sector in recent times, producers are not to be blamed for using any leverage that may present itself. Equally true is the South African government's ploy of using the Soviet African presence to garnish Western backing in their struggle with the black nationalist movements. Again, given the odds facing South Africans, this should not be unexpected. Liberal arguments generally point to these

facts, claim such distortions thus prove resources are not an issue, and call for severance of all economic and political ties with the apartheid South African government. In fact, as with most complex problems, the truth lies somewhere in between.

Chapter IV-8 attempted to show that the Soviets are not in any immediate danger of becoming import dependent, so Soviet activities are probably not an excursion for new supplies on a far-off continent. As for "minerals denial to the West," the typical arguments used to negate the concept are:

1. It is doubtful that the Soviets would engage in such an expensive policy of unclear outcome. Southern Africa is an area of great diversity geographically and tribally; control over such a region would be difficult to achieve, much less maintain. It would also be very costly, given the distance from the motherland.

2. The Soviet record in Africa is certainly not one of consistent success. USSR strategies have generally centered on military, rather than economic development. This may attract independence-minded nationalists in the short term, but in the long run does little to improve life in these countries. Soviet failures in Egypt, Guinea, and the Sudan can be attributed in large measure to this fact. Thus, even if there were a clear aim to tie up African minerals, success is not guaranteed. This is hardly comforting.

3. The Cape sea route argument is countered by claims that if the Soviets desired to cut oil shipments, they could do so most easily closer to the source, namely in the Middle East itself or in the Straits of Hormuz. This analysis is most persuasive; it does seem reasonable that any attempted disruption would occur as close to the USSR as possible.

4. The main contention used to refute the potential of a cutoff of Western supplies centers around regional economies. Minerals production is an element vital to the economic health of this area. Any disruption of trade would have serious consequences. Conspiring with the

TABLE IV-10-1 Opinions of Various Authors Regarding Soviet Intentions in Southern Africa

Author	Resources-related	Not resources-related
Strauss[a]	X	
Szuprowicz (21st Century Research)[b]	X	
Fine[c]	X	
Shabad[d]		X
Vogely[e]		X
Strishkov (BOM)[f,g]		X
Bowman[h]		X
Shafer[i]		X
Rees (Institute for the Study of Conflict)[j]	X	
Council of Economics and National Security[k,l]	X	
Resources for the Future[m]		X
Soviet Analyst[n]	X	
Meyer[o]	X	
Tilton and Landsberg[p]		X
Foreign Affairs Research Institute[q]	X	
ALARM (Miller)[r]	X	
South African Embassy[s]	X	
Slay[t]	X	

[a]Simon D. Strauss, "Mineral Self-Sufficiency—The Contrast between the Soviet Union and the United States." *Mining Congress Journal*, November 1979.

[b]Bohdan Szuprowicz, *How to Avoid Strategic Materials Shortages.* New York: John Wiley & Sons, Inc., 1981.

[c]Daniel I. Fine, "Mineral Resource Dependency Crisis: Soviet Union and United States." In *The Resource War in 3-D: Dependency, Diplomacy, Defense.* Edited by James A. Miller, Daniel I. Fine, and R. Daniel McMichael. Pittsburgh, Pa.: World Affairs Council of Pittsburgh, 1980.

[d]Theodore Shabad, "The Soviet Mineral Potential and Environmental Constraints." *Mineral Economics Symposium*, Washington, D.C., 8 November 1982.

[e]William A. Vogely, "Resource War?" *Materials and Society* 6, No. 1 (1982).

[f]U.S. Department of the Interior, Bureau of Mines, "The Mineral Industry of the USSR." by V. V. Strishkov. Preprint from the *1980 Minerals Yearbook.* Washington, D.C.: U.S. Government Printing Office, 1980.

[g]Bohdan Szuprowicz, "Critical Materials Cut-off Feared." *High Technology*, November–December 1981.

[h]Larry W. Bowman, "The Strategic Importance of South Africa to the United States: An Appraisal and Policy Analysis." *African Affairs* 81, No. 323 (April 1982).

[i]Michael Shafer, "Mineral Myths." *Foreign Policy*, No. 47 (Summer 1982).

[j]David Rees, *Soviet Strategic Penetration in Africa.* London: Institute for the Study of Conflict, 1976.

[k]Council on Economics and National Security, The "Resource War" and the U.S. Business Community: The Case

for a Council on Economics and National Security. Washington, D.C.: Council on Economics and National Security, 1980.

[l]Council on Economics and National Security, *Strategic Minerals: A Resource Crisis.* Washington, D.C.: CENS, 1981.

[m]"What Next for U.S. Minerals Policy?" *Resources*, No. 71 (October 1982).

[n]Brian Crozier, "Strategic Decisions for the West." *Soviet Analyst* 10, No. 5 (4 March 1981).

[o]Herbert E. Meyer, "Russia's Sudden Reach for Raw Materials." Reprinted from *Fortune*, 28 July 1980.

[p]John E. Tilton and Hans H. Landsberg, "Nonfuel Minerals: The Fear of Shortages and the Search for Policies." *RFF Forum*, Washington, D.C., 21 October 1982.

[q]Foreign Affairs Research Institute, *The Need to Safeguard NATO's Strategic Raw Materials from Africa.* No. 13 (1977).

[r]"'Resource War' Targeted as a Major National Problem." *Alert Letter on the Availability of Raw Materials*, No. 1 (May 1981).

[s]"South Africa: Persian Gulf of Minerals. Strategic Minerals Threat." Backgrounder issued by the Information Counselor, South African Embassy. No. 9 (December 1980).

[t]General Alton D. Slay, "Minerals and National Defense." *The Mines Magazine*, December 1980.

Source: Philip R. Ballinger, "An Analysis of the Energy and Nonfuel Minerals of the USSR." Master's thesis. University of Texas at Austin, 1983.

Soviets to halt mineral flow to the West would certainly mean a loss of Western investment. Moreover, it would be tantamount to "Finlandization," implying a decided abandonment of national identity and autonomy.

There are, however, many points that argue strongly for the inclusion of resource considerations in the formulation of Soviet foreign policy.

1. Arguments based on the importance of southern African mineral economies assume that the only markets for African commodities are Western. Such thinking ignores the possibility of preemptive buying. This is a crucial factor. The USSR could, through coercion, or political manuevering with new and friendly socialist governments, buy up African minerals. It has been suggested that the Soviets do not possess the currency to implement such a plan. Rest assured that the Soviets could accumulate the required capital (gold?)—if there was the promise of very large profits from resale to the West at much higher prices. There is a recent precedent for this scenario—the U.S. embargo of Rhodesian chrome forced purchases of the mineral from the USSR. Evidence indicates that the Soviets merely bought the African chrome, then passed it on to the United States at inflated

prices.[1] Preemptive purchases would allow the African producers continued sale of their minerals, would provide the USSR with a large additional source of hard currency, and would certainly foment economic and political tensions in the West. A further consideration is the role of gold in the nation of South Africa. Earnings from gold sales are so significant, due to high prices, that exports of other minerals are not truly critical from a survival standpoint. Excluding gold, then, there does not appear to be an acute requirement for the primary minerals producer in southern Africa to export its commodities.

2. Although geopolitical concerns should not be overestimated, neither should the probability of their inclusion in overall Soviet policy be underestimated. It is undoubtedly overly simplistic to view Soviet actions in the south of Africa as driven solely or even primarily by raw materials, for Soviet policy is multi-faceted, with formulation of goals more intricate than this. Still oriented to eventual world domination, it is characterized by opportunism, flexibility, and the ability to hold available several options.

Conversely, the USSR is positively cognizant of the Western dependence on these minerals. To think that the Soviets would not leap to take advantage of easy opportunities is to display a disregard for past history. Indeed, a strategy of denying the West resources would be quite attractive from the Soviet viewpoint. It would be very low risk, involving minimal visibility and confrontation, particularly if proxy (Cuban, East German) forces were employed.

3. It is most probably, however, the sheer political volatility of southern Africa that has brought the Soviet presence. The national liberation movements provide decided avenues for Soviet gain, the issues of race and colonialism being used to promote Marxist-Leninist doctrine. It is the Soviet support of guerrilla operations that is the real danger. Guerrilla tactics by their very nature are intentionally designed to disrupt normal activities in an economy. The importance of both the transportation network and mineral production in southern Africa makes them natural targets. Soviet support of nationalist movements could thus lead to a disruption of production for those minerals critical to Western industries and defenses. This would require no grand Soviet purpose, yet the outcome could be the same.

4. It has become the vogue in recent months for many authorities (e.g., Shabad,[2] Vogely,[3] Strishkov[4] to dismiss offhandedly and self-assuredly (and often pompously) the feasibility of a Soviet resource diplomacy. Most usually point to the desire of the USSR to be regarded as a good and reliable business partner, and to signs of growing Soviet dependence on the international marketplace for some goods, as factors that must mitigate adventurism. This is spurious reasoning—the Soviet invasion of Afghanistan, as well as its recent martial law stranglehold in Poland, occurred during the decade which saw the largest volumes of Soviet trade turnover in history. Such a view of the USSR as a "good trading partner" could prove a costly optimism, as the European natural gas consumers may discover.

Ultimately the extent to which resources play a role in the formulation and implementation of Soviet policy in southern Africa remains problematical, despite the definitive conclusions of some experts to the contrary. A reasonable position should deem it prudent not to reject the possibility of a Soviet design to manipulate raw materials. Shortsightedness may be philosophically self-satisfying, but is a disservice to the needs of a mineral-dependent West.

REFERENCES

1. Szuprowicz, Bohdan. "Critical Materials Cut-off Feared." *High Technology*, November–December 1981.
2. Shabad, Theodore. "The Soviet Mineral Potential and Environmental Constraints." *Mineral Economics Symposium*, Washington, D.C., 8 November 1982.
3. Vogely, William A. "Resource War?" *Materials and Society* 6, No. 1 (1982).
4. U.S. Department of the Interior, Bureau of Mines. "The Mineral Industry of the USSR." By V. V. Strishkov. Preprint from the *Minerals Yearbook 1980*. Washington, D.C.: U.S. Government Printing Office, 1980.

CHAPTER IV-11

Conclusions

1. The most fundamental element seen in this study is the transition from the European to Asian regions of the USSR for most sources of mineral supplies. This shift must have a tremendous impact on the costs of production in the Soviet minerals industry. These eastern areas are characterized by remoteness, bitter climates, unfavorable terrains, and a dearth of infrastructure. Construction of the Baikal-Amur Mainline will provide access to some of these remote deposits. An additional consideration is the lengthening supply lines implied by the move to Asia. Reliance on long transport networks to bring mineral commodities to western centers of Soviet industry increases the potential for supply disruptions.

2. Economic factors will serve to further increase costs. Many economic problems can be traced to the very nature of the Soviet planned economy. Coordination among supervisory agencies is inadequate. The quality of industrial output is poor. Waste is pervasive, yet few incentives exist to promote increased efficiency. Innovation is stifled. Serious labor shortages are imminent. These all serve to further reduce the effectual operation of the minerals industry.

3. Oil is essential to the USSR to fuel domestic economic expansion, to earn hard currency from the West, and to help stabilize vulnerable Eastern European economies. World concern over the possibility of a Soviet requirement to import oil appears unfounded. Although USSR oil production growth is slowing, potential in several areas, such as enhanced recovery, offshore development, or future discovery, indicates that a renewed production surge could be evident in the 1990s.

4. The ability of the USSR to rapidly develop its huge natural gas reserves in western Siberia is the key to meeting energy needs for the entire Soviet bloc. Western European participation in the Urengoy natural gas export pipeline removes capital and technological obstacles to this development.

5. Problems apparent in the Soviet coal industry hinder its capacity for expansion. These include depletion of older mines in Soviet Europe, low coal quality in the major eastern deposits, low labor productivity, labor shortages, and lagging invest-

ment. The low rank of most Asian coals may prove particularly troublesome—quality constraints on transport distances, as well as limited transport availability in Asia, restrict coal's potential as an oil substitute to those power plants east of the Urals.

6. An emphasis on nuclear power, unimpeded by environmental opposition, and aided by CMEA participation, suggests continued growth. Nuclear energy is earmarked as the mode of electrical power generation in the European USSR.

7. Soviet energy policy centers around efforts to substitute for and conserve oil, due to its value as a hard-currency earner. Gas will bypass oil in terms of export value sometime in this decade. The speed with which this transition occurs will determine whether ceilings must be imposed on either domestic consumption or on oil exports to Eastern Europe.

8. The USSR will remain a major producer of virtually all important ferrous, nonferrous, precious, and nonmetallic minerals. Only manganese and chromite show recent production downturns. These declines could prove short-lived. Imports of lead and zinc ores and concentrates may be required to allow future exports of these metals.

9. Soviet mineral policy is based on self-sufficiency. The historical and ideological importance of this strategy, combined with the crucial role of minerals in industrial development, suggests further pursuit of mineral independence. Unique tactics designed to further this goal include the development of Arctic minerals and the relegation of prices to secondary importance.

10. Trade is often used as a tool for attaining political as well as economic objectives. Current trade goals include the acquisition of hard currency and the increasing integration of CMEA economies with that of the USSR. Gas and oil remain the primary hard-currency earners. Should the Europeans be able to procure other supply sources for their natural gas needs, such as from northern Norwegian fields, even gas could prove an unreliable long-term provider of Soviet hard currency. Other mineral exports, notably gold, diamonds, and platinum-group metals, can contribute only to a limited degree. Moreover, demand for these commodities

is much more price-elastic than that for either oil or natural gas exports. This suggests a continuation of the Soviet currency squeeze. Most mineral exports remain primarily directed toward Eastern Europe. The trend for a decreasing importance of Soviet minerals in Western consumption will continue.

11. The majority of the evidence does not support the contention of some authors, such as Fine, that the USSR is moving from a policy of minerals self-sufficiency to one of outward access. However, the increasing trend toward expansion of Soviet mineral production in the remote and inhospitable eastern and northeastern parts of the country will require huge expenditures for mine development and associated physical infrastructure, and is likely to be constrained by labor problems. For this reason, the Soviets may find it more economic to import some minerals in the short to medium term. This does not mean that the USSR is physically running out of minerals. In all likelihood, pursuit of mineral self-sufficiency will prevail for most commodities, after the mines and physical infrastructure are in place in Soviet Asia.

12. No definitive judgment can be made regarding the extent to which resource considerations, specifically a strategy of denying the West important minerals, make up Soviet foreign policy in southern Africa. The recent pattern for many analysts is to dismiss the possibility of a Soviet resource diplomacy. This reasoning ignores particularly the potential for preemptive purchases by the USSR. It also fails to consider possible Soviet requirements to import selective minerals, until development is finalized in the eastern parts of the country.

Bibliography

PART I: THE UNITED STATES

Abbott, C. C. "Economic Penetration and Power Politics." *Harvard Business Review* 26 (1948), pp. 410-424.

Alert Letter on the Availability of Raw Materials, No. 64 (January 1984), p. 8.

American Mining Congress Journal 70, No. 4 (23 February, 1984), p. 2.

Anstett, T. F., et al. *Platinum Availability—Market Economy Countries*. U.S. Bureau of Mines Information Circular 8897. Washington, D.C.: U.S. Department of the Interior, 1982.

Bateman, Alan. "Wartime Dependence on Foreign Minerals." *Economic Geology*, June-July 1946, pp. 308-327.

Bayless, Alan. "Natural-Gas Pressure Builds to a Head," *The Wall Street Journal*, 9 January, 1984.

Bennethum, Gary, and L. Courtland Lee. "Is Our Account Overdrawn?" *Mining Congress Journal*, September 1975, pp. 33-48.

Brown, Stuart. "Fiber Optics to Replace 4M lbs. of Copper at AT&T." *American Metal Market/Metalworking News*, 23 August 1982, p. 24.

Calaway, Lee, and W. C. J. van Rensburg. "U.S. Strategic Minerals Policy Options." *Resources Policy*, June 1982.

Cameron, Eugene N. "The Contribution of the United States to National and World Mineral Supplies." In *The Mineral Position of the United States*. Edited by Eugene Cameron. Madison, Wis.: University of Wisconsin Press, 1973.

Cameron, Eugene N. "Changes in the Political and Social Framework of United States Mineral Resource Development, 1905-1980." *Economic Geology, Seventy-fifth Anniversary Volume, 1905-1980*. El Paso, Tex.: The Economic Geology Publishing Co., 1981.

Carroll, B. A. *Design for Total War: Arms and Economics in the Third Reich*. The Hague: Mouton Publishers, 1968.

The Changing Relationship: The Australian Government and the Mining Industry. As cited by W. C. J. van Rensburg and S. Bambrick in *The Economics of the World's Mineral Industries*. New York: McGraw-Hill Book Company, 1978.

Cockran, C. N. "Energy Balance of Aluminum from Production to Application." *Journal of Metals* July 1981, pp. 45-48.

Dresher, W. H. *Technological Innovation and Forces for Change in the American Industry*. Washington, D.C.: National Research Council, 1978.

Drury, Orcott, U.S. Department of Commerce, Washington, D.C. Interview, 27 May 1982.

Dupree, Walter, et al. *Energy Perspectives 2*. U.S. Department of Energy. Washington, D.C.: U.S. Government Printing Office, June 1976.

Eckes, Alfred E., Jr. *The United States and the Global Struggle for Minerals*. Austin, Tex.: University of Texas Press, 1979.

Engineering Mining Journal, February 1982. "Managing Political Vulnerability." An interview with Dr. Warnuck Davies.

Everest Consulting Association, Inc., and CRV Consultants, Inc. *The International Competitiveness of the U.S. Non-ferrous Smelting Industry and the Clean Air Act*. Princeton Junction, N.J., and New York, April 1982.

Fine, D. I. "Mineral Resource Dependency Crisis: Soviet Union and United States." In *The Resource War in 3-D: Dependency, Diplomacy, Defense*, 18th World Affairs Forum. Edited by James A. Miller, Daniel I. Fine, and R. Daniel McMichael. Pittsburgh, Pa.: World Affairs Council of Pittsburgh, 1980.

Freeman, Alan. "Alumax Likely to Join Venture with Pechiney." *The Wall Street Journal*, 6 February, 1984.

Gagnon, James, U.S. Department of State, Washington, D.C. Interview, 4 June 1982.

"The Gas Glut Has Alaska and Canada Hustling." *Business Week*, 25 October 1982.

Gates, Paul W. "The Federal Lands: Why We Retained Them." In *Rethinking the Federal Lands*. Edited by Sterling Brubaker. Washington, D.C.: Resources for the Future, Inc., p. 35.

U.S. General Accounting Office. *The U.S. Mining and Mineral Processing Industry: An Analysis of Trends and Implications*. Washington, D.C.: U.S. Government Printing Office, 1979.

"Government Measures." *Mining Journal*, 7 January 1983.

Government official, U.S. Department of Commerce, Washington, D.C. Interview, 7 January 1983.

Grace, Richard P. "Metals Recycling: A Comparative National Analysis." *Resources Policy* 4, No. 4 (December 1978). p. 254.

Murray, Hadyn H. "Nonbauxite Alumina Resources." In *Cameron Volume on Unconventional Mineral Deposits*. Edited by Wayne C. Shanks III. New York: Society of Mining Engineers, 1983, pp. 111-120.

Hagenstein, Perry R. "The Federal Lands Today: Uses and Limits." In *Rethinking the Federal Lands*. Edited by Sterling Brubaker. Washington, D.C.: Resources for the Future, Inc., p. 78.

Honkala, R. A., and K. R. Knoblock. "A Cartographic Look at Constraints to Mineral Exploration and Development." *Mining Congress Journal*, February 1980.

H.R. 33. 98th Congress, 1st session, 3 January 1983.

Huddle, Frank P. "The Evolving National Policy of Minerals." *Science* 191 (20 February 1976), pp. 654-659.

Hughes, Kathleen A. "Eleven U.S. Copper Producers to Request Trade Commission Put Limits on Imports." *The Wall Street Journal*, 26 January, 1984, p. 45.

Ickes, H. "The War and Our Vanishing Resources." *American Magazine* 140 (1945), pp. 20-22.

Irving, Robert R. "How Can We Stave Off an OPEC in Metals?" *Iron Age*, 25 May 1981, pp. 79-85.

Kahn, Herman, and Anthony Weimer. *The Year 2000: A Framework for Speculation on the Next Thirty-Three Years.* New York: MacMillan Publishing Co., Inc., 1967.

Kroft, D. J. "The Geopolitics of Non-energy Minerals." *Air Force Magazine*, June 1979, p. 76.

Landsberg, Hans. "Is a U.S. Materials Policy Really the Answer?" *Professional Engineer*, September 1982.

Leith, C. K. *World Minerals and World Politics.* London: Kennikat Press, 1931. Reissue 1970.

Leith, L. K. "Principles of Foreign Mineral Policy of the United States." *Mining and Metallurgy* 25 (1946), p. 14.

Lemons, Jim F., Jr., et al. *Chromium Availability—Domestic.* U.S. Bureau of Mines Information Circular 8895. Washington, D.C.: U.S. Department of the Interior, 1982.

Lesemann, Robert H. "U.S. Smelters Losing Access to Copper Concentrates." *American Mining Congress Journal* 69, No. 19 (19 October 1983), p. 18.

MacAvoy, Paul W. "A policy That Closes U.S. Copper Mines." *The New York Times*, 19 December 1982.

McGovern, G. Imports of Minerals from South Africa by the United States and the OECD Countries. U.S. Senate Subcommittee on African Affairs, September 1980.

Malthus, Thomas R. *An Essay on Population.* London: J. M. Dent & Sons Ltd., 1803.

Manes, Allan S. *U.S. Merchant Marine, Sealift Acquisition Policy and National Security.* Congressional Research Service, Liberty of Congress, 1981.

May, E. R. *Lessons of the Past: The Use and Misuse of History in American Foreign Policy.* New York: Oxford University Press, Inc., 1973.

Meadows, Donella H. *The Limits to Growth: A Report to the Club of Rome on the Predicament of Mankind.* New York: Universe Books, 1974.

Miller, Betty. *The Petroleum Potential of Wilderness Lands in the Western United States.* Geological Survey, U.S. Department of the Interior, December 1983.

Miller, James. "Cal Nickel Targets Gasquet Mountain Strategic Minerals." *Alert Letter on the Availability of Raw Materials*, No. 52 (September 1983).

Miller, James. "Experts Make Cobalt Policy Recommendations." *Alert Letter on the Availability of Raw Materials*, No. 68 (February 1984).

Miller, James Arnold. "Cobalt Bought from Zaire for the U.S. Stockpile." *Alert Letter on the Availability of Raw Materials*, No. 4 (August 1981).

Miller, James Arnold. "Defense Production Act: Relief for Domestic Mining Industry." *Alert Letter on the Availability of Raw Materials*, No. 6 (August 1981).

Miller, James Arnold. "Cobalt Availability: Thought-Provoking Commentary by M.I.T.'s Clark." *Alert Letter on the Availability of Raw Materials*, No. 19 (October 1982), p. 9.

Miller, James Arnold. "DOD Management of the Stockpile." *Alert Letter on the Availability of Raw Materials*, No. 34 (March 1983).

Miller, James Arnold. "AMC Supports Transfer of Stockpile to Defense Department." *Alert Letter on the Availability of Raw Materials*, No. 37 (April 1983).

Miller, Richard E. "Export Opportunities for Western Coal." *Mining Congress Journal*, February 1982.

Mining Journal 300, No. 7712 (10 June 1983), p. 393.

Mobray, Jo. *Geography of the United States Uranium Supply: Resources, Production, and Institutions.* Public Information Report 6. Center for Energy Studies, University of Texas at Austin, 1981.

Moss, R. A White Paper, Council on Economic and National Security, August 1980, p. 42.

"Most Stockpile Cobalt Not Good for Defense." *American Mining Congress Journal*, 69, No. 22 (7 December 1983), p. 3.

National Materials Advisory Board, Commission on Socio-technical Systems. *Considerations in Choice of Form for Materials for the National Stockpile.* Washington, D.C.: National Academy Press, 1982.

National Materials and Minerals Program Plan and Report to Congress. 5 April 1982.

Nichols, Mike. "Stockpiling for Strategic and Economic Purposes." Term paper. University of Texas at Austin, Spring Semester 1980.

Nixon, R. M. *The Real War.* New York: Warner Books, 1980.

Nulty, Peter. "How to Pay a Lot for Cobalt." *Fortune*, 4 April 1983, pp. 151-155.

O'Boyle, Thomas F. "Steelmaker's Excess Ore Capacity Hindering the Industry's Recovery." *The Wall Street Journal*, 29 August 1983, p. 15.

Oil and Gas Journal, 26 December 1983.

Oil and Gas Journal. "Forecast/Review." 30 January 1984, and earlier annual review issues.

"115th Annual Review and Outlook." *Engineering and Mining Journal* 185, No. 3 (March 1984).

Park, Charles. *Earthbound: Minerals, Energy and Man's Future.* San Francisco: Freeman, Cooper & Company, 1975.

Pasztov, Andy. "U.S.-Backed Synfuels Program Is Likely to End in 1984; $5 Billion Seen Returned." *The Wall Street Journal*, 19 December 1983.

Pasztov, Andy, and Raymond A. Joseph. "EPA to Propose, in Next Few Months, Ban or Phase-out of Remaining Asbestos Uses." *The Wall Street Journal*, 4 October 1983, p. 20.

Paul, John H. "Offshore Imports Threaten Potash Industry." *Mining Congress Journal*, 7 March 1984, p. 16.

Peterson, G. R., and S. J. Arbelbide. *Aluminum Availability*

—*Market Economy Countries.* Bureau of Mines Information Circular 8917. Washington, D.C.: U.S. Department of the Interior, 1983.

"Phosphate Production Plans at New North Carolina Mine." *Mining Magazine* 150, No. 3 (March 1984), p. 195.

"Phosroc: U.S. Industry Disadvantaged." *Mining Journal,* 301, No. 7729 (7 October 1983), p. 257.

"Precious Metal Possibilities in the U.S. Rocky Mountains." *Mining Magazine* 149, No. 5 (November 1983), pp. 286–287.

Redfield, W. C. *Dependent America: A Study of the Economic Bases of Our International Relations.* Boston: Houghton Mifflin Company, 1926.

Santini, J. D. Response to McGovern's Report on "Imports of Minerals from South Africa by the United States and the OECD Countries." 22 October 1980.

Schmidt, Helmut, and Manfred Kruszona. *Regional Distribution of Mining Production and Reserves of Mineral Commodities in the World.* Hanover, West Germany: Federal Institute for Geosciences and Natural Resources, January 1982.

Scott, Douglass P. "The Macroeconomic Environment for 1979 Stockpile Goals." Internal report to FEMA.

Shroder, John F., Jr. "The USSR and Afghanistan Mineral Resources." In *International Minerals: A National Perspective.* Edited by Allen F. Agnew. Boulder, Colo.: Westview Press, Inc., 1983.

Solow, Robert M. "The Economics of Resources or the Resources of Economics." *Proceedings of the American Economic Association* 64 (May 1974).

Spector, Stewart R. "Price and Availability of Energy in the Aluminum Industry." *Journal of Metals,* June 1981, pp. 138–139.

The Stockpile Story. Washington, D.C.: American Mining Congress, 1963.

Strategic and Critical Materials Stock Piling Act of 1946, P.L. 520.

Strategic and Critical Materials Stock Piling Revision Act of 1979.

"Strategic Stockpile of Key Raw Materials 'Set Up by UK.'" *Financial Times,* 14 February 1983, p. 1.

The Economist. "Strategic Stockpiles: Who's Hoarding What?" 24 May 1980.

Szuprowicz, Bohdan O. *How to Avoid Strategic Materials Shortages.* New York: John Wiley & Sons, Inc., 1981.

Technical Information Center, Institute of Gas Technology. *Energy Statistics 1st Quarter 1984,* Vol. 7, No. 1. Chicago: IGT, 1984.

"Temporary Respite for Uranium?" *Mining Journal,* 4 February 1983.

Tilton, John E. "U.S. Policy for Securing the Supply of Mineral Commodities." *Primary Commodities: Security of Supply Conference,* Second Round Table, Washington, D.C., 11–12 December 1980.

"Titanium Producers at Odds." *Mining Journal* 302, No. 7744 (20 January 1984), pp. 33–35.

"U.S. Coal Exports Derailed." *Mining Journal,* 2 December 1983.

U.S. Congress. House. Committee on Armed Services. *National Defense Stockpile Hearings on H.R. 2603, H.R. 2784, H.R. 2912, and H.R. 3364 before the Seapower and Strategic and Critical Materials Subcommittee of the Committee on Armed Services House of Representatives.* 97th Congress, 1st session, 2 and 4 June 1981.

U.S. Congress. House. Committee on Interior and Insular Affairs. *U.S. Minerals Vulnerability: National Policy Implications. A Report Prepared by the Subcommittee on Mines and Mining of the Committee on Interior and Insular Affairs of the U.S. House of Representatives.* 96th Congress, 2nd session, 1980.

U.S. Congress, Office of Technology Assessment. *Management of Fuel and Nonfuel Minerals in Federal Land: Current Status and Issues.* Washington, D.C., April 1979.

U.S. Congress. Senate. Committee on Armed Services. *Strategic and Critical Materials Stock Piling Act Revision Hearing before the Subcommittee on Military Construction and Stockpiles of the Committee on Armed Services of the U.S. Senate.* 96th Congress, 1st session, 1979.

U.S. Congress. Senate. Committee on the Judiciary. *Soviet, East German and Cuban Involvement in Fomenting Terrorism in Southern Africa. A Report of the Chairman of the Subcommittee on Security and Terrorism to the Committee on the Judiciary United States Senate.* 97th Congress, 2nd session, 1982.

U.S. Department of Commerce, Bureau of Census. *Statistical Abstract of the United States, 1982–1983,* 103rd ed. Washingtn, D.C.: U.S. Government Printing Office, 1983.

U.S. Department of Energy, Energy Information Administration. *1982 Survey of U.S. Uranium Exploration Activity.* Washington, D.C.: U.S. Government Printing Office, August 1983.

U.S. Department of Energy, Energy Information Administration. *Historical Overview of U.S. Coal Exports, 1973-1982,* Washington, D.C.: U.S. Government Printing Office, November 1983.

U.S. Department of Energy, Energy Information Administration. *Weekly Coal Production: November 5, 1983.* Washington, D.C.: U.S. Government Printing Office, November 1983.

U.S. Department of Energy, Energy Information Administration. *Monthly Energy Review, December 1983.* Washington, D.C.: U.S. Government Printing Office, March 1984.

U.S. Department of the Interior. *Energy Resources of Federally Administered Lands.* Washington, D.C.: U.S. Government Printing Office, November 1981.

U.S. Department of the Interior, Bureau of Land Management. *Public Land Statistics 1982.* Washington, D.C.: U.S. Government Printing Office, April 1983, pp. 3–4.

U.S. Department of the Interior, Bureau of Mines. *Alumina Availability—Domestic.* Circular 8861. By G. R. Peterson, R. L. Davidoff, et al., Undated.

U.S. Department of the Interior, Bureau of Mines. *Mineral Commodity Summaries.* Washington, D.C.: U.S. Government Printing Office, various years.

U.S. Department of the Interior, Bureau of Mines. *Mineral Facts and Problems, 1980 Edition*. Washington, D.C.: U.S. Government Printing Office, 1980.

U.S. Department of the Interior, Bureau of Mines. *The Domestic Supply of Critical Minerals*. Washington, D.C.: U.S. Government Printing Office, 1983.

U.S. Department of the Interior, Bureau of Mines. *Mineral Commodity Summaries 1983*. Washington, D.C.: U.S. Government Printing Office, 1984, and back issues.

U.S. Department of the Interior, Bureau of Mines. *Minerals Yearbook, Centennial Edition 1981*. Washington, D.C.: U.S. Government Printing Office, 1983, and other back issues.

U.S. Department of the Interior, Geological Survey. Map of Curtis Bay Quadrangle. Baltimore County, Maryland. 7.5 Minute Series (Topographic), 1957.

U.S. Department of the Interior, Geological Survey. Map of Somerville Quadrangle. New Jersey. 7.5 Minutes Series (Topographic), 1969.

U.S. Department of the Interior, Geological Survey. Map of Maples Quadrangle. Allen County, Indiana. 7.5 Minute Series (Topographic), 1971.

U.S. Department of the Interior, Geological Survey. Map of West Point Quandrangle. Orange County, New York. 7.5 Minute Series (Topographic), 1971.

U.S. Department of the Interior, Geological Survey. *Principal Federal Lands Where Exploration and Development of Mineral Resources Are Restricted*, NAS-R-0401-75MO1. Washington, D.C., January 1981.

U.S. Federal Emergency Management Agency. *Stockpile Report to Congress October 1980–March 1981*.

U.S. Federal Emergency Management Agency. *Stockpile Report to the Congress October 1981–March 1982*.

U.S. Federal Emergency Management Agency, National De-*Report to Congress October 1980–March 1981*.

U.S. Federal Emergency Management Agency, National Defense Stockpile Policy Division. *Cobalt Project*. By Marilyn B. Biviano. 26 November 1980.

U.S. General Accounting Office. Report by the Comptroller General. *National Defense-Related Silver Needs Should Be Reevaluated and Alternative Disposal Methods Explored*. 11 January 1982.

U.S. General Services Administration. "Budget Includes $245 Million for Stockpile Acquisitions." *GSA News Release*, 23 January 1978.

U.S. General Services Administration, Federal Preparedness Agency. *A Study of the Effect of Lead-Times, Substitutability, and Civilian Austerity on the Determination of Stockpile Objectives*, 15 July 1975.

U.S. General Services Administration, Office of Stockpile Transaction, Program Report and Accounting Staff. *Depot Totals per Commodity Fiscal 1981*.

U.S. National Foreign Assessment Center, Central Intelligence Agency. *The World Factbook—1981*. Washington, D.C.: U.S. Government Printing Office, April 1981.

U.S. Office of the Federal Register. *The United States Government Manual 1982-83*. Washington, D.C.: U.S. Government Printing Office, July 1982, p. 299.

U.S. President's Materials Policy Committee. *Paley Report*, June 1952, Vol. 1-5.

U.S. Senate Republican Policy Committee. "National Defense Stockpile Inventory of Strategic and Critical Materials 1962-80 Comparison." *Republican Report*, 1980.

"U.S. Tentatively Rules 3 Countries Subsidize Various Steel Exports." *The Wall Street Journal*, 8 February 1984, p. 3.

Warneke, Steven J. *Stockpiling of Critical Raw Materials*. London: The Royal Institute of International Affairs, 1980.

Weinberg, G. L. *The Foreign Policy of Hitler's Germany: Diplomatic Revolution in Europe, 1933-1936*. Chicago: The University of Chicago Press, 1970.

White, Lane. "Custom Copper Concentrates." *Engineering and Mining Journal*, May 1982, pp. 72-75.

Wilson, Carroll. *Coal—Bridge to the Future: Report of the World Coal Study*, Cambridge, Mass.: Ballinger Publishing Co., 1980.

PART II: JAPAN

Books and Articles

Agress, Philip. "U.S. Calls on Japan to Eliminate Import Barriers." *Business America* 5, No. 6 (22 March 1982).

Alexander, W. O., and E. G. West. "Metallurgical Operations on Islands." *The Metallurgist and Materials Technologist* 14 (April 1982).

"Amax to Slash 1982 Moly Output." *Mining Journal* 298, No. 7652 (16 April 1982).

American Bureau of Metal Statistics. *Non-ferrous Metal Data*, various years.

". . . and Rare Metal Plans." *Mining Journal* 299, No. 7670 (20 August 1982).

"Anguin Agreement." *Mining Journal* 298, No. 7658 (28 May 1982).

"Anschutz Seeks Federal Aid in Developing Cobalt Mine." *Mining Engineering* 34 (November 1982).

"Arco Unit to End Copper Mining in Butte, Mont." *The Wall Street Journal*, 10 January 1983.

Asano, J. "Nippon Kokan Makes Major Strategy Change: From Defensive to Preemptive Attack." *The Oriental Economist* 50 (November 1982).

"Asia Oil Will Cut Purchases from Mobil and Exxon to Half." *Japan Economic Journal*, 11 January 1983.

Atlas of the World, 5th ed. Washington, D.C.: National Geographic Society, 1981.

Atsumi, Keiko. "Foreign Aid: New Departure in Policy in 1981 Was to Place Emphasis on Doubling ODA." *Industrial Review of Japan, 1982*. Tokyo: Japan Economic Journal, 1982.

"Australia/Japan Coal Liquefaction Plant." *Mining Magazine*, May 1982.

"Australia: Japan Drives Down Coal Prices." *Mining Journal* 300, No. 7697 (25 February 1983).

"Australia/Japan Nuclear Agreement." *Energy World*, No. 95 (August-September 1982).

"Australia Looks to Labor." *Mining Journal* 300, No. 7689 (11 March 1983).

"Australia Losing Some Japanese Iron Orders to Brazilian

Producers." *The Wall Street Journal*, 30 September 1982.

"Australian Firm to Get Loan of $633 Million for Coal Production." *The Wall Street Journal*, 1 April 1983.

"Australian/Japan U Trade." *Mining Magazine*, March 1982.

Baldridge, Malcolm. "Limited R&D Partnerships Can Spur High Technology." *American Metal Market/Metalworking News* 91, No. 50 (14 March 1983).

Bambrick, Susan, National Energy Research and Development Council, Canberra, Australia. Interview, 29 November 1982.

"Basic Key Materials, Technology Top Priority for MITI Budget Plan." *Tokyo Petroleum News*, 25 August 1982.

Bayless, Alan. "Japan Wants Coal Producers to Cut Prices and Deliveries As Steel Demand Declines." *The Wall Street Journal*, 21 January 1983.

"B.C. Coal Signs Pact to Sell from New Mine to Japanese Mills." *The Wall Street Journal*, 9 February 1983.

Beal, Tim. "Economics: Japan: Japanese Bids to Outwit Protectionists." *South*, No. 17 (March 1982).

Borsuk, Richard. "Union Oil Faces Production Problems at Gulf of Thailand Natural Gas Field." *The Wall Street Journal*, 28 July 1982.

Boyer, E. "How Japan Manages Declining Industries." *Fortune*, 10 January 1983.

Boyle, Thomas. "Forging Ahead: Laid Low by Recession Big Steel Companies Consider Major Change." *The Wall Street Journal*, 27 May 1983.

"Brazil: An Opening Ceremony Was Held August 5. . . ." *Engineering and Mining Journal* 183 (September 1982).

Brooks, Rosanne. "Anti-trust Law Changes Could Ease Cooperative R&D Efforts." *American Metal Market/Metalworking News* 91, No. 65 (4 April 1983).

Brown, Stuart. "Staving Off Glass Inroads: Gains in Electronics Enable Copper Wires to Transmit More Data." *American Metal Market/Metalworking News*, Copper Supplement, 22 November 1982.

"Business Week/Harris Poll: Executives Split on Saving Smokestack Industries." *Business Week*, 4 April 1983.

Calaway, Lee, and W. C. J. van Rensburg. "U.S. Strategic Minerals: Policy Options." *Resources Policy* 8 (June 1982).

"Canada Agency Clears Exports of Crude Oil to Japan, South Korea." *The Wall Street Journal*, 27 April 1983.

"Canadian Colliery Development Talks Progress; Coal Imports Up." *Japan Economic Journal*, 15 August 1982.

Chegwidden, Judith. "Rare Earths." In *Mining Annual Review, 1981*. London: Mining Journal, 1981.

Chiba, Atsuko. "Japan's Aluminum Woes." *Forbes*, 21 July 1980.

"China Launches Steel Complex's Second Portion." *The Wall Street Journal*, 9 June 1983.

"China Seems Due to Purchase Three Million Tons of Steel." *Japan Economic Journal* 21, No. 1046 (1 March 1983).

"Chinese Envisage Big Project for Production of Aluminum." *Japan Economic Journal*, 25 January 1983.

Cobbe, James. *Governments and Mining Companies in Developing Countries*. Boulder, Colo.: Westview Press, Inc., 1979.

"Colombian Profile." *Mining Journal* 298, No. 7638 (8 January 1982).

"Comalco Assesses Aluminum's Future at Smelter Opening." *Engineering and Mining Journal* 183 (October 1982).

"CRA Ltd. to Acquire 50% of Japan Firm; RTZ Earnings Drop." *The Wall Street Journal*, 16 April 1982.

Crowson, P. C. F. "The National Mineral Policies of Germany, France and Japan." *Mining Magazine*, June 1980.

"CVRD Obtains Mining Concession for Carajas Manganese Reserves." *Engineering and Mining Journal* 183 (November 1982).

"Delay in Uranium Group Formation." *Japan Economic Journal*, 18 January 1983.

"Developing Technopolises for the 21st Century." *Asia Research Bulletin* 12, No. 8 (31 January 1983).

"Direct Foreign Investment Hit Record $8.9 Billion." *Japan Economic Journal*, 15 February 1983.

"Dome Petroleum Asks Canada Permission to Continue LNG Plan." *The Wall Street Journal*, 24 February 1983.

"Drilling Operations Completed." *Asia Research Bulletin* 12, No. 2, Report 5 (31 July 1982).

Eckes, Alfred E. *The United States and the Global Struggle for Minerals*. Austin, Tex.: University of Texas Press, 1979.

"Eight Japanese Companies. . . ." *Mining Engineering* 35 (August 1983).

Ellison, T. D. "Manganese." In *Mining Annual Review, 1982*. London: Mining Journal, 1982.

Emmerson, J. K. *Arms, Yen and Power: The Japanese Dilemma*. New York: Dunellen, 1977.

Environmental Policies in Japan. Paris: Organization for Economic Cooperation and Development, 1977.

"Esmark Unit, Japanese Firm to Set Up Mining Venture." *The Wall Street Journal*, 29 April 1983.

Exporter's Encyclopedia, 77th ed. New York: Dun & Bradstreet International Ltd., March 1982.

Farr, P. J. "Rare Earths." In *Mining Annual Review, 1982*. London: Mining Journal, 1982.

"Ferrochrome Supply and Demand." *Mining Journal* 298, No. 7639 (15 January 1982).

Financial Times. *International Mining Company Yearbook 1981*. Harlow, Essex, England: Longman Group Ltd., 1981.

"First Al Exports from Indonesia." *Mining Journal* 299, No. 7680 (29 October 1982).

"A First Long-Term Deal to Buy North Sea Oil." *Asia Research Bulletin* 12, No. 4, Report 5 (30 September 1982).

Fitzgerald, M. D., and G. Pollio. "Aluminum: The Next Twenty Years." *Journal of Metals* 34 (December 1982).

"Four Japanese Companies Appear Set to Build NASCO Steel Plant." *Japan Economic Journal*, 15 February 1983.

Franklin, J. M., D. M. Sangster, and J. W. Lydon. "Volcanic-Associated Massive Sulfide Deposits." *Economic Geology, Seventy-Fifth Anniversary Volume, 1905–*

1980. El Paso, Tex.: Economic Geology Publishing Co., 1981.

"Fruita Delayed." *Mining Journal* 300, No. 7691 (14 January 1983).

Fuller, Ronald. "Zirconium and Hafnium." In *Mining Annual Review, 1981*. London: Mining Journal, 1981.

Furukawa, Tsukasa. "Japanese Tend to Bigness." *American Metal Market/Metalworking News*, Specialty Steels Supplement, 2 August 1982.

Gadsen, P. H. "Titanium." In *Mining Annual Review, 1975*. London: Mining Journal, 1975.

"Giant Traders Taking a High-Tech Stance." *The Oriental Economist* 50 (July 1982).

"Go Ahead for Coal Project." *Mining Journal* 298, No. 7638 (8 January 1982).

"Government Okays Jabiluka U Mine." *World Mining* 35 (October 1982).

Gresser, Julian, Koichiro Fujikura, and Akio Morishima. *Environmental Law in Japan*. Cambridge, Mass.: The MIT Press, 1981.

Hanks, R. J. *The Pacific Far East: Endangered American Strategic Position*, Special Report. Cambridge, Mass.: Institute for Foreign Policy Analysis, Inc., November 1981.

"The Hatachi Copper Mine. . . ." *Engineering and Mining Journal* 182 (September 1981).

Horio, Koichi, Nobuyuki Kitamura, and Yutaka Ariake. "Waste Energy Recovery at Kashima Steel Works." *Iron and Steel Engineer* 59 (July 1982).

Howard, Al. "Dry Quench Needs Lower Cost." *American Metal Market/Metalworking News*, Steelmaking Today Supplement, 27 September 1982.

Ibrahim, Youssef. "Iranians Propel Petroleum Sales with Cut Rates." *The Wall Street Journal*, 16 April 1982.

Imai, Hideki. *Geological Studies of the Mineral Deposits in Japan and East Asia*. Tokyo: University of Tokyo Press, 1978.

Inaba, M. "Japan Shuns Metal Substitution." *American Metal Market/Metalworking News*, 20 September 1982.

Inaba, Minoru. "In Japan, Computer Development Continues." *American Metal Market/Metalworking News* 91, No. 70 (11 April 1983).

Independent Commission on Development Issues. *North-South: A Programme for Survival*. Cambridge, Mass.: The MIT Press, 1980.

"Indonesia Needs Japanese Assistance." *Ironmaking and Steelmaking* 9, No. 2 (1982).

"Indonesia: Negotiations Are under Way. . . ." *Engineering and Mining Journal* 182 (July 1981).

Ishizuka, M. "Industrialists Are Not Convinced Whether It Will Become a Panacea: Law for 'Structural Improvement' of Industries." *Japan Economic Journal*, 22 February 1983.

Itoh, F., and T. Hirota. "On Development and Researches of Marine Mineral Resources in Japan." In *World Mining and Metals Technology*, Proceedings of the Joint MMIJ-AIME Meeting, Denver, Colo., September 1976, Vol. 1. Edited by A. Weiss, New York: American Institute of Mining, Metallurgical, and Petroleum Engineers, 1976.

Jameson, Sam. "MITI Policy Irks Japan Smelters: Trade Rift over Aluminum." *Los Angeles Times*, 21 March 1983.

"Japan & Australia Launch Joint Brown Coal Liquefaction Project." *The Oriental Economist* 50 (April, 1982).

"Japan: Antimony Import Problems. . . ." *Mining Journal* 299, No. 7670 (20 August 1982).

"Japan Approves Deep Sea Law." *Mining Journal* 299, No. 7665 (16 July 1982).

"Japan: Coal Liquefaction." *Energy World*, Bulletin of the Institute of Energy of the United Kingdom, No. 97 (November 1982).

"Japan: Coal Mine Reopened in Northern Japan." *World Coal* 7 (July–August 1981).

"Japan: Declining Oil Stockpile." *Asia Research Bulletin* 12, No. 7, Report 5 (31 December 1982).

"Japan Demand for Gold Likely to Recover." *Asia Research Bulletin* 12, No. 11, Report 5 (30 April 1983).

"Japan Drives Down Coal Prices." *Mining Journal* 300, No. 7697 (25 February 1983).

"Japan: Energy Plans Swing Back to Coal." *World Business Weekly*, 15 June 1981.

"Japan Eyes U.S. West Coast Coal." *Japan Economic Review*, 15 February 1982.

"Japan: Gold Imports Totalled 167.3 Tons in 1981." *Asia Research Bulletin* 11, No. 12, Report 5 (30 April 1982).

"Japan: Government Signs Sea Law Code." *Asia Research Bulletin* 12, No. 10, Report 5 (31 March 1983).

"Japan: International." *Bulletin of Anglo-Japan Economic Institute*, The Economic and Trade Picture, No. 234 (November–December 1982).

"Japan: Joint Venture with Russia Resumed." *Asia Research Bulletin* 12, No. 4, Report 5 (30 September 1982).

"Japan: Loan for Coal Projects." *Asia Research Bulletin* 12, No. 10, Report 5 (31 March 1983).

"Japan May Open Market Further to U.S., but Steps Might Not Satisfy White House." *The Wall Street Journal*, 6 December 1982.

"Japan's Nuclear Industry at a Turning Point," Part I. *Japan Finance and Industry*, Quarterly Survey, No. 50 (January–June 1982).

"Japan: Occidental Talks about Pungshuo in China." *World Coal* 8 (November–December 1982).

"Japan Pulls Out of Argentinian Steel Project." *Mining Journal* 298, No. 7677 (8 October 1982).

"Japan: Record Copper Output." *Mining Journal* 300, No. 7694 (4 February 1983).

"Japan Seeks Price Reduction." *Mining Journal* 299, No. 7688 (24 December 1982).

"Japan to Continue Soviet Energy Project despite Ban by U.S." *The Wall Street Journal*, 7 July 1982.

"Japan to Stockpile Aluminum." *Mining Journal* 298, No. 7651 (9 April 1982).

"Japan Today." *Mining Magazine* 125 (November 1971).

"Japan: U.S. $27b. Nuclear Energy Scheme." *Asia Research Bulletin* 12, No. 3, Report 5 (31 August 1982).

"Japan Was Scheduled to Begin Stockpiling." *Engineering and Mining Journal* 183 (May 1982).

"Japan: World's Largest Importer Is Now Facing Problems." *World Coal* 8 (November–December 1982).

"Japan's Steel Industry: Trouble Hits the East." *Iron and Steel International* 55 (October 1982).

"Japanese Al Predictions." *Mining Journal* 300, No. 7691 (14 January 1983)

Japanese Aluminum Federation News, No. 4 (1982).

"Japanese Banks to Extend Credits to Colombia Project." *The Wall Street Journal*, 27 December 1982.

"Japanese Interest in Coal Canyon Mine." *Mining Journal* 298, No. 7641 (29 January 1982).

"Japanese Interested in Developing Petaquilla." *World Mining* 35 (June 1982).

"Japanese Loan for Peruvian Development." *Mining Journal* 299, No. 7679 (22 October 1982).

"Japanese Mineral Demand May Boost Pacific Northwest and Alaska." *Mining Engineering* 35 (February 1983).

"Japanese Mission to Discuss Albras." *Mining Journal* 299, No. 7682 (12 November 1982).

"Japanese Prime Minister Aims at Easing of Monetary Policy as Way to Aid Growth." *The Wall Street Journal*, 3 January 1983.

"Japanese Reach Accord with Egyptians on Steel Plant." *Japan Economic Journal*, 6 October 1981.

"Japanese Stockpiling." *Mining Journal* 300, No. 7690 (6 January 1983).

"Japanese to Withdraw from Mt. Arthur." *Mining Journal* 299, No. 7680 (29 October 1982).

"Japanese Tungsten Discovery." *Mining Journal* 299, No. 7688 (24 December 1982).

"Japanese Utility Quits Coal-Mining Project in New South Wales." *The Wall Street Journal*, 27 October 1982.

"JISEA Figures Exports of Steel Will Drop in 1983." *Japan Economic Journal*, 8 February 1983.

Johnson, Chalmers. *MITI and the Japanese Miracle*. Stanford, Calif.: Stanford University Press, 1982.

"Joint Oil Exploration with China." *Asia Research Bulletin* 12, No. 2, Report 5 (31 July 1982).

Jones, S. L. "Samarium-Cobalt Magnets Spur Electronic Advances." *American Metal Market/Metalworking News*, 28 June 1982.

Jones, S. L. "Rare Earth's Development Said to Be in Its Infancy." *American Metal Market/Metalworking News*, 9 August 1982.

Kanabayashi, Masayoshi. "Japan's Recession-Hit Companies Make Complex Arrangements to Avoid Layoffs." *The Wall Street Journal*, 17 February 1983.

"Kawasaki Heavy Industries: Concludes Tieup with a German Firm in Coal-Fired DR Plants." *The Oriental Economist* 50 (January 1982).

"Kawasaki Steel Corp: Will Offer Reconstruction Aids to Spanish Steel Industry." *The Oriental Economist* 50 (January 1982).

Kawata, Sukeyuki, ed. *Japan's Iron and Steel Industry—1982*. Tokyo: Kawata Publicity, Inc., 1982.

Keatley, Robert. "The Far East Worries As U.S. Pressures Japan to Rearm." *The Wall Street Journal*, 29 November 1982.

Kingston, John. "RMI/Kobe Steel Reach Titanium Accord." *American Metal Market/Metalworking News* 91, No. 40 (28 February 1983).

Kohno, H., H. Asao, and T. Amano. "Use of Alternative Fuel at Onahama." *Journal of Metals* 34 (July 1982).

Kohno, Teruyuki. "Science and Technology: Priority Will Go to New Materials, Information Processing and Biotechnology." In *Industrial Review of Japan, 1982*. Tokyo: Japan Economic Journal, 1982.

Korchynsky, M. "Vanadium—Present and Future." *CIM Bulletin* 75, No. 843 (July 1982).

"La Caridad Financing." *Mining Journal* 299, No. 7686 (10 December 1982).

Lambert, Ian, and Takeo Sato. "The Kuroko and Associated Ore Deposits of Japan: A Review of Their Features and Metallogenesis." *Economic Geology* 69 (December 1974).

Lehner, U. C., and Masayoshi Kanabayashi. "Japan Minimizes New U.S. Curbs on Soviet Trade." *The Wall Street Journal*, 23 June 1982.

Lehner, Urban. "Japan's Kyushu Promoted as 'Sun Belt' in Hopes of Luring Factories to the Sticks." *The Wall Street Journal*, 16 June 1982.

Lehner, Urban. "The Outlook: Japan's High Savings: Boon or Burden?" *The Wall Street Journal*, 25 October 1982.

"Less SA Ferrochrome for Japan." *Mining Journal* 298, No. 7648 (19 March 1982).

Lin, Kuang-Ming, and W. R. Hoskins. "Understanding Japan's International Trading Companies." *Business* 31 (September–October 1981).

Lohr, S. "Japanese Premier Urges an Increase in Armed Strength." *The New York Times*, 28 November 1982.

Mahler, Walter. "Japan's Adjustment to the Increased Cost of Energy." *Finance and Development* 18 (December 1981).

"Main Points of Mining Policies for FY '83." *Japan Metal Journal* 12, No. 31 (2 August 1982).

"Major Growth in Oil and Gas for Dennison." *Mining Journal* 298, No. 7642 (5 February 1982).

Makiuchi, Iwao. "Non-ferrous Metals: Aluminum Smelters Had Worst Year in 1981; Copper, Lead and Zinc Makers Also Fared Poorly." In *Industrial Review of Japan, 1982*. Tokyo: Japan Economic Journal, 1982.

"Malaysia Agrees to Make Shipments of LNG to Japanese." *The Wall Street Journal*, 28 March 1983.

"Malaysia and Japan's Nippon Steel Will Build DR Steel Plant." *Engineering and Mining Journal* 183 (January 1982).

"Marubeni Awarded Mexican Copper Refinery Contract." *Mining Journal* 299, No. 7677 (8 October 1982).

Matsumoto, Toshio. "Coal: Demand Is Indicating Increasing Trend, but Domestic Miners Suffer from Deficits." In *Industrial Review of Japan, 1982*. Tokyo: Japan Economic Journal, 1982.

Matsumoto, Toshio. "New Basic Materials: Development of New Types Becomes Matter of Increasing Interest in Industry." In *Industrial Review of Japan, 1982*. Tokyo: Japan Economic Journal, 1982.

Matsumoto, Yoshio. "Coal: Some Miners Appear to Have Broken Even from Industries Shifting Away from Oil." In *Industrial Review of Japan, 1981*. Tokyo: Japan Economic Journal, 1981.

"McIntyre Coal for Japan." *Mining Journal*, 299, No. 7680 (29 October 1982).

McLeroy, Priscilla, Thomas Braschayko, and W. C. J. van

Rensburg. "U.S. Resources Policy: The 1982 Report to Congress." *Resources Policy* 8 (December 1982).

McManus, George. "A New Role for Ladle Metallurgy: High Tonnage Output." *Iron Age*, 7 June 1982.

"Military Spending Still Less than 1% GNP." *Asia Research Bulletin* 12, No. 8 (31 January 1983).

Miller, J. A. "Reagan Administration's Materials and Minerals Policy: A Tremendous Step in the Right Direction." *Alert Letter on the Availability of Raw Materials*, No. 12 (May 1982).

Miller, R. E. "Export Opportunities for Western Coal." *Mining Congress Journal* 68 (February 1982).

Mining Annual Review, various years. London: Mining Journal, 1970-82.

"Minor Precious Metals." *Metal Bulletin*, No. 6527 (30 September 1980). Cited by U.S. Department of the Interior, Bureau of Mines. "Zirconium and Hafnium." By William Kirk. *Minerals Yearbook 1980*, Vol. I. Washington, D.C.: U.S. Government Printing Office, 1981.

"MITI Decides to Make National Stockpile of 11 Rare Metals from Next July." *Japan Metal Journal*, 30 August 1982.

"MITI Studies Ways to Aid Basic Materials Industries: Bogged Down in Serious Situation." *Japan Economic Journal*, 1 September 1981.

"Mitsui Postpones SRC Plant." *Mining Journal* 298, No. 7661 (18 June 1982).

"Mixed Fortunes for Minor Metals." *Mining Journal* 298, No. 7650 (2 April 1982).

"Mobil and Nissho Iwai to Join in Developing Indonesian Coal Tract." *The Wall Street Journal*, 18 November 1982.

Murray, Alan. "'In Japan, FTC Is Challenged by Not Only Business but Bureaucrats': Hashiguchi Notes Growth of FTC since 1977." *Japan Economic Journal*, 15 June 1982.

Nag, Amal. "Japanese Steel Concerns, Fearing Quotas, Weigh Buying Stakes in U.S. Counterparts." *The Wall Street Journal*, 4 June 1982.

Nagano, Takeshi. "The History of Copper Smelting in Japan." *Journal of Metals* 34 (June 1982).

Nakamura, N., Yl Togino, and T. Adachi. "Philosophy of Blending Coals and Coke-Making Technology in Japan." *Ironmaking and Steelmaking* 5, No. 2 (1978).

"Nakasone Tells Cabinet to Map Out Steps to Further Open Japan to Foreign Goods." *The Wall Street Journal*, 8 December 1982.

"New Japanese Process to Smelt Aluminum from Clay Tested." *Engineering and Mining Journal* 182 (August 1981).

"New Process Is Devised for Gallium Arsenide Crystals." *Japan Economic Journal*, 1 February 1983.

"New Technology Boosts Demand for Strontium." *Mining Journal* 299, No. 7688 (24 December 1982).

"New Technology for Zirconium Sponge Production under Test by Japanese." *Engineering and Mining Journal* 181 (November 1980).

"News from Japanese Industry." *Bulletin of American-Japanese Economic Institute*, No. 234 (November-December 1982).

"News from NKK." *Ironmaking and Steelmaking* 9, No. 4 (1982).

"Nickel Imports Thinned." *Mining Journal* 298, No. 7662 (25 June 1982).

1981-1982 World Mines Register. San Francisco: World Mining, 1981.

Ninose, Tomoko. "Chemicals: Companies in 1981 Made Rush to Advance into Biotechnology as New Field of Growth." In *Industrial Review of Japan, 1982*. Tokyo: Japan Economic Journal, 1982.

"Niobec to Idle Columbium Mines." *American Metal Market/Metalworking News* 91, No. 55 (21 March 1983).

"Nippon Amazon Aluminum Co. . . ." *Mining Engineering* 35 (March 1983).

"Nippon to Increase Copper Output." *Mining Journal* 299, No. 7678 (15 October 1982).

"N-Power Plants Operate at 70.2% Capacity." *Japan Economic Journal*, 1 February 1983.

"NZ Coal Consortium." *Mining Journal* 299, No. 7663 (2 July 1982).

Ogino, Junichi. "New Basic Materials: Development of New Types Become a Matter of Increasing Interest in Industry: Metals." In *Industrial Review of Japan, 1982*. Tokyo: Japan Economic Journal, 1982.

"Oil and Gas are Produced Successfully from Fifth Test Well in Gulf of Bohai." *Japan Economic Journal*, 22 February 1983.

"Oil from Brown Coal Planned for Victoria, Australia." *Mining Engineering* 31 (June 1982).

"Oil Imports Drop to Lowest Level in 12 Years." *Japan Economic Journal*, 1 February 1983.

Okabe, Naoki, and Hisao Saida. "Rescue Plan for Troubled Industries Draws Suspicion at Home and Abroad: Special Law for Structurally Depressed Industries." *Japan Economic Journal*, 25 January 1983.

Okita, Saburo. "Natural Resource Dependency and Japanese Foreign Policy." *Foreign Affairs* 52 (July 1974).

Padan, John, U.S. Department of Commerce, National Oceanic and Atmospheric Administration, Ocean Minerals and Energy Division, Washington, D.C. Personal correspondence, 27 April 1983.

"PASAR Copper Smelter Heads to 1983 Opening." *World Mining* 35 (June 1982).

Pehlke, Robert. "An Overview of Contemporary Steelmaking Processes." *Journal of Metals* 34 (May 1982).

Pempel, T. *Policy and Politics in Japan: Creative Conservation*. Philadelphia: Temple University Press, 1982.

Peters, Hans-Jurgen. "Japan Steel to U.S., EEC: Market Conditions Dictate Export Flow." *American Metal Market/Metalworking News*, 24 January 1983.

"Philippines Subsidizes Its Copper Production to Keep Mines Running." *The Wall Street Journal*, 6 July 1982.

"Philippines: The Philippine Government Has Increased. . . ." *Engineering and Mining Journal* 183 (December 1982).

"Philippines: Under a $2 Million Joint Exploration Agreement. . . ." *Engineering and Mining Journal* 182 (November 1981).

Portney, Paul, ed. *Current Issues in Natural Resource Policy*. Baltimore, Md: The Johns Hopkins University Press, 1982.

Post, Charles. "Gallium Arsenide: Silicon's New Competition." *Iron Age*, 1 November 1982.

"Power Reactor Corp. Enters Talks for Saskatchewan U Mining." *Japan Economic Journal*, 25 January 1983.

"Production Adjustments in Aluminum." *The Oriental Economist* 50 (June 1982).

"Queensland Has Integrated Aluminum Industry." *World Mining* 35 (November 1982).

Ramsey, Douglas, Kim Willenson, Ayako Doi, and Frank Gibney. "Japan's High-Tech Challenge." *Newsweek*, 9 August 1982.

"Recycling Key in Japanese Steel Success: Recovering Energy." *The Northern Miner* (Canada), 16 September 1982.

Ridge, John D. *Annotated Bibliographies of Mineral Deposits in Africa, Asia (Exclusive of the USSR) and Australasia*. Oxford: Pergamon Press Ltd., 1976.

Ridley, R. S., and A. F. Kaba. "Brazil Battles the Jungle to Mine Carajas Minerals." *World Mining* 36 (January 1983).

"Rise in Japanese Gold Imports." *Mining Journal* 298, No. 7646 (5 March 1982).

Rokrik, Dani. "Managing Resource Dependency: The United States and Japan in the Markets for Copper, Iron Ore and Bauxite." *World Development* 10 (July 1982).

Sage, A. M. "Vanadium." In *Mining Annual Review, 1982*. London: Mining Journal, 1982.

Saida, Hisao. "High-Level Overseas Investment Becomes Common among Japanese Enterprises." In *Industrial Review of Japan, 1982*. Tokyo: Japan Economic Journal, 1982.

"Sakhalin Project Uses Soviet Drilling Rig." *The Wall Street Journal*, 29 July 1982.

"Sanctions: And When They Are Only Halfway Up. . . ." *The Economist*, 22 January 1983.

Scanlon, A. F. G., ed. *BP Statistical Review of World Energy 1981*. London: The British Petroleum Company, PLC, 1981.

Schwartz, Lloyd. "Japan Shifts Stress in Industrial Policy." *American Metal Market/Metalworking News*, 20 September 1982.

"Science and Technology Expenditure in FY 1981 Reached Y5.9 Trillion." *Japan Economic Journal*, 18 January 1983.

"Sea-Bed Mining: Japan Proposes Legislation." *Mining Journal* 296, No. 7591 (13 February 1981).

"Sea-Law Pact Is Signed by 118 Nations." *The Wall Street Journal*, 13 December 1982.

"Seagram Unit Fights Thailand Government to Control Gas Project." *The Wall Street Journal*, 1 November 1982.

"Second Japanese Stockpile Plan." *Metal Bulletin*, No. 6700 (29 June 1982).

Shikazono, Naotatsu. "Mineralization and Chemical Environment of the Toyoha Lead–Zinc Vein-Type Deposits, Hokkaido, Japan." *Economic Geology* 70 (June–July 1975).

"Shimokawa Mine to Close." *Mining Journal* 299, No. 7682 (12 November 1982).

Shiota, S. "Balancing Ferro-Silicon Supply and Demand." *Metal Bulletin*, No. 107 (November 1979). Cited in U.S. Department of the Interior, Bureau of Mines.

"Ferroalloys." By Frederick Schottman. *Minerals Yearbook 1978–79*, Vol. I. Washington, D.C.: U.S. Government Printing Office, 1979.

Shishido, Toshio. "Japanese Industrial Development and Policies for Science and Technology." *Science* 219, No. 4582 (21 January 1983).

"Showa Halts Nickel Output." *Mining Journal* 299, No. 7683 (19 November 1982).

Siddiqi, T. A., and D. James. "Coal Use in Asia and the Pacific: Some Environmental Considerations." *Energy* 7 (March 1982).

"Six Aluminum Refiners Announce Production Curtailment Plans." *The Oriental Economist* 50 (January 1982).

Smith, B. "Scarcity and Stability in Mineral Markets: The Role of Long Term Contracts." *The World Economy*, January 1979. Cited by James Walsh, "The Growth of Develop-for-Import Projects." *Resources Policy* 8 (December 1982).

Solidi, C. G. "Peru." In *Mining Annual Review, 1982*. London: Mining Journal, 1982.

Starrels, John. "Steel's Stiff Competition." *The Wall Street Journal*, 9 July 1982.

"Steel Contract with China." *Asia Research Bulletin* 12 No. 8, Report 5 (31 January 1983).

"Steel Industry's Case against Japan Is Rejected by U.S." *The Wall Street Journal*, 28 February 1983.

"Steel Output Dips below 100 Mil. Tons." *Japan Economic Journal*, 1 February 1983.

"Stockpile Program." *Mining Journal* 298, No. 7637 (1 January 1982).

Stubbs, R. L. "Lead and Zinc." In *Mining Annual Review, 1981*. London: Mining Journal, 1981.

Stubbs, R. L. "Lead and Zinc." In *Mining Annual Review, 1982*. London: Mining Journal, 1982.

Suda, Satoru. "Japan." In *Mining Annual Review*. London: Mining Journal, 1979–1982.

"Sumikei Aluminum Formally Decides on Ending Business." *Japan Economic Journal*, 8 June 1982.

"Sumitomo Finalizes Chinese Copper Refinery Agreement." *Mining Journal* 299, No. 7677 (8 October 1982).

"Sumitomo Firms Will Tie-Up with Alleghany Ludlum Steel Corp." *Japan Economic Journal*, 18 January 1983.

"Sumitomo Metal Mining. . . ." *Engineering and Mining Journal* 183 (October 1982).

"Sumitomo Metal Mining Reveals a New Gold Vein." *The Oriental Economist* 50 (May 1982).

"Sumitomo Metal Will Take Over U.S. Pipe Joint Firm, Tube Turns." *Japan Economic Journal*, 18 January 1983.

"Survey: Japan: New International Strategies." *World Business Weekly*, 6 October 1980.

Suzuki, Kenji. "Iron and Steel: Slump of Demand Compelled Curtailment of Production in Last Half of 1980." In *Industrial Review of Japan, 1981*. Tokyo: Japan Economic Journal, 1981.

Suzuki, Kenji. "Iron and Steel: Booming Seamless Pipe Exports Greatly Contributed to Earning Gain in 1981." In *Industrial Review of Japan, 1982*. Tokyo: Japan Economic Journal, 1982.

Tamura, Hideo. "Foreign Aid: Japan's ODA Ratio to GNP in 1980 Reached Figure of 0.31 Per Cent." In *Industrial Review of Japan, 1981*. Tokyo: Japan Economic Journal, 1981.

"Thai Shale to Be Studied." *Mining Journal* 299, No. 7687 (17 December 1982).

Thompson, A. G. "Chromite." In *Mining Annual Review, 1979*. London: Mining Journal, 1979.

Thompson, A. G. "Manganese." In *Mining Annual Review, 1981*. London: Mining Journal, 1981.

"Time of Trial for General Traders: Vaunted Versatility Will Stand Them in Good Stead." *The Oriental Economist* 47 (January 1979).

"Titanium: No More Boom-or-Bust Cycles?" *Chemical Week*, 26 August 1981.

Toyoda, Shigeru. "Changes in the Use of Energy in Japanese Steel Industry—With an Emphasis on the Countermeasures Taken after the Oil Crisis," special lecture. *Transactions of the Iron and Steel Institute of Japan* 23 (January 1983).

"Trade Showdown with Japan." *Newsweek*, 15 May 1982.

Tsurumi, Yoshi. *The Japanese Are Coming*. Cambridge, Mass.: Ballinger Publishing Co., 1976.

Tsurutani, Taketsugu. *Japanese Policy and East Asian Security*. New York: Praeger Publishers, 1981.

"Twenty-Six Japanese Banks Join Loan for Canada's Quintette Mine." *Japan Economic Journal*, 25 January 1983.

Ueda, Katsumi. "Overseas Strategy: 'Country Risk' Becomes Serious Problem with International Tensions and Troubles." In *Industrial Review of Japan, 1981*. Tokyo: Japan Economic Journal, 1981.

"Ulan Sales Contract Signed." *Mining Journal* 300, No. 7700 (18 March 1983).

Ulman, N. "Brazil's Japanese Population Thrives, Helps Tokyo Forge Ties with Brasilia." *The Wall Street Journal*, 24 September 1982.

Ulman, N. "Brazil Pursues Amazon's Riches with a Project on a Huge Scale Befitting the Region's Vastness." *The Wall Street Journal*, 28 October 1982.

Ulman, N. "U.S. and Japan: Defense Tensions." *The Wall Street Journal*, 1 December 1982.

Ulman, N. "U.S. Runs Risk Pressuring Japanese." *The Wall Street Journal*, 8 December 1982.

"U.N. Approves Sea Law Treaty." *Austin-American Statesman*, 1 May 1982.

"United States: Joint Venture with Japan for Colorado Mine." *World Coal* 7 (November–December 1981).

"U.S. Electronics Firms Form Venture to Stem Challenge by Japanese." *The Wall Street Journal*, 26 August 1982.

"U.S. Gov't Suggests New Formula for Accepting Uranium Enriching." *Japan Economic Journal*, 11 January 1983.

van Rensburg, W. C. J., and Susan Bambrick. *The Economics of the World's Minerals Industries*. New York: McGraw-Hill Book Company, 1978.

Vernyi, Bruce. "Cutting Down on Cobalt." *American Metal Market/Metalworking News*, Aerospace Metals and Machines Supplement, 1983.

Walsh, James. "The Growth of Develop-for-Import Projects." *Resources Policy* 8 (December 1982).

Ward, Peter. "The Onward March of Japan That Threatens to Trample Australia." *The Australian*, 11 March 1983.

Warsh, David. "The MCC Gamble: Consortium Bets Research Effort Will Triumph over Knot of Hazards." *Austin-American Statesman*, 17 April 1983.

Way, H. J. R. "Antimony." In *Mining Annual Review, 1981*. London: Mining Journal, 1981.

White, Lane. "Custom Copper Concentrates: Why Are Copper Concentrates Produced in Arizona and Montana Being Smelted in Japan?" *Engineering and Mining Journal* 183 (May 1982).

"Work on Thailand's Zinc Smelter Moves toward 1984 Completion." *Engineering and Mining Journal* 183 (December 1982).

Yoshii, Hideo. "Petroleum: Japan Widens Its Purchases of Crude Oil to Such New Places as Mexico and Africa." In *Industrial Review of Japan, 1981*. Tokyo: Japan Economic Journal, 1981.

Yoshinara, Kunio. *Sogo Shosha: The Vanguard of the Japanese Economy*. Oxford: Oxford University Press, 1982.

Yoshino, M. Y. *Japan's Multinational Enterprises*. Cambridge, Mass.: Harvard University Press, 1976.

Young, Alexander. *The Sogo Shosha: Japan's Multinational Trading Companies*. Boulder, Colo.: Westview Press, Inc., 1979.

Government Documents

Australia

Australian Trade Development Council. *Minerals Processing*. Canberra: Australian Government Publishing Service, 1980.

Canada

Canada. Department of External Affairs. *Canada's Export Development Plan for Japan*, August 1982.

Japan

Japan. Economic Planning Agency. *New Economic and Social Seven-Year Plan*, August 1979.

Japan. Environment Agency. "Environmental Pollution." *The White Papers of Japan 1979–80*, annual abstracts of official reports and statistics of the Japanese government. Tokyo: Japan Institute of International Affairs Editorial Section of English Annual, 1981.

Japan. *Environmental Agency Establishment Law*, Law No. 88, 1971. Cited in Julian Gresser, Koichiro Fujikura, and Akio Morishima, *Environmental Law in Japan*. Cambridge, Mass.: The MIT Press, 1981.

Japan. Geological Survey. *Geology and Mineral Resources of Japan*, 2nd ed. Edited by M. Saito, K. Hashimoto, H. Sawata, and Y. Shimazaki. Japan: Geological Survey of Japan, 1960.

Japan. Ministry of Finance, Minister's Secretarial, Research and Planning Division. *Quarterly Bulletin of Financial Statistics*, September 1982.

Japan: Ministry of Finance, Study Group on Structural Adjustment. *An Outline of Japanese Taxes*, 1979, Appendix 7.6.

Japan. Prime Minister's Office, Statistics Bureau. *Japan Statistical Yearbook*. Tokyo: Japan Statistical Association and the Mainichi Newspapers, 1960–82.

Japan. Science and Technology Agency. "Atomic Energy." *The White Papers of Japan 1979–80*, annual abstracts of the Japanese government. Tokyo: Japan Institute of International Affairs Editorial Section of English Annual, 1981.

Republic of South Africa

Republic of South Africa, Department of Mining and the Environment. *Atlas of Marine Resources*, 1974.

United States

Export Trading Company Act of 1982: Export Trading Companies. U.S. Code, No. 9 (November 1982).

U.S. Congress, Office of Technology Assessment. *Management of Fuel and Non-fuel Minerals in Federal Land*. Washington, D.C.: U.S. Government Printing Office, April 1979.

U.S. Congress, Office of Technology Assessment. *Technology and Soviet Energy Availability*. Washington, D.C.: U.S. Government Printing Office, 1981.

U.S. Department of State. Telegram 18,018, Tokyo, October 1982.

U.S. Department of the Interior, Bureau of Mines. "The Mineral Industry of Japan." *Minerals Yearbook*, various years, Vol. III. Washington, D.C.: U.S. Government Printing Office, 1963–82.

U.S. Department of the Interior, Bureau of Mines. "Silicon." *Minerals Yearbook*, Vol. I, various years, 1978–81. Washington, D.C.: U.S. Government Printing Office, 1979–82.

U.S. Department of the Interior, Bureau of Mines. "Cobalt." *Minerals Yearbook*, various years 1978–81. Washington, D.C.: U.S. Government Printing Office, 1980–82.

U.S. Department of the Interior, Bureau of Mines. *Mineral Facts and Problems, 1980 Edition*. Washington, D.C.: U.S. Government Printing Office, 1980.

U.S. Department of the Interior, Bureau of Mines. "Chromium." By E. C. Peterson. *Minerals Yearbook 1980*, Vol. I. Washington, D.C.: U.S. Government Printing Office, 1981.

U.S. Department of the Interior, Bureau of Mines. "The Mineral Industry of Mexico." By O. Martinez. *Minerals Yearbook 1980*, Vol. III. Washington, D.C.: U.S. Government Printing Office, 1981.

U.S. Department of the Interior, Bureau of Mines. "Chromium." By John Papp. *Minerals Yearbook 1981*, Vol. I, preprint. Washington, D.C.: U.S. Government Printing Office, 1982.

U.S. Department of the Interior, Bureau of Mines. "Ferroalloys." By Raymond Brown. *Minerals Yearbook 1981*, Vol. I. Washington, D.C.: U.S. Government Printing Office, 1982.

U.S. Department of the Interior, Bureau of Mines. "Japan: Metals Stockpile." *Metals and Materials: A Bimonthly Survey*, August–September 1982. Washington, D.C.: U.S. Government Printing Office, 1982.

U.S. Department of the Interior, Bureau of Mines. "Japan: Steelmakers Hurt by World Recession." *Minerals and Materials: A Bimonthly Survey*, October–November 1982. Washington, D.C.: U.S. Government Printing Office, 1982.

U.S. Department of the Interior, Bureau of Mines. "The Mineral Industry of Australia." By Charlie Wyche. *Minerals Yearbook 1981*, Vol. III. Washington, D.C.: U.S. Government Printing Office, 1982.

U.S. Department of the Interior, Bureau of Mines. "The Mineral Industry of the Philippines." By John Wu. *Minerals Yearbook 1981*, Vol. III. Washington, D.C.: U.S. Government Printing Office, 1982.

U.S. Department of the Interior, Bureau of Mines. "Nickel." By Scott Sibley. *Minerals Yearbook 1981*, Vol. I. Washington, D.C.: U.S. Government Printing Office, 1982.

U.S. Department of the Interior, Bureau of Mines. "Rare Earth Minerals and Metals." By James Hendrick. *Minerals Yearbook 1981*, Vol. I. Washington, D.C.: U.S. Government Printing Office, 1982.

U.S. Department of the Interior, Bureau of Mines. "Zirconium and Hafnium." By William Kirk. *Minerals Yearbook 1981*, Vol. I. Washington, D.C.: U.S. Government Printing Office, 1982.

U.S. Department of the Interior, Bureau of Mines. "Japan: Japanese Private Metal Stockpile." *Minerals and Materials: A Bimonthly Survey*, December 1982–January 1983. Washington, D.C.: U.S. Government Printing Office, 1983.

U.S. Department of the Interior, Bureau of Mines. *Mineral Commodity Summaries 1983*. Washington, D.C.: U.S. Government Printing Office, 1983.

U.S. White House Report. *National Materials and Minerals Program Plan and Report to Congress*, No. 1982-505-002/60. Washington, D.C.: U.S. Government Printing Office, April 1982.

PART III: WESTERN EUROPE

Brand, David. "Britain Faces Challenge As Drop in Oil Output in a Few Years Looms." *The Wall Street Journal*, 8 July 1983, p. 1.

Bray, Nicholas. "France, Facing Power Glut, Scales Down Its Plans to Build More Nuclear Reactors." *The Wall Street Journal*, 28 July 1983, p. 25.

"British Steel Loss Was $591 Million in Year to April." *The Wall Street Journal*, 13 July 1983, p. 29.

"Changing Times for France." *Mining Journal*, 5 February 1982, pp. 93–95.

Crowson, P. C. F. "The National Mineral Policies of Germany, France, and Japan." *Mining Magazine*, June 1980, pp. 537–549.

"Dialogue on United Kingdom Minerals Supply." *Mining Journal*, 30 May 1980, p. 435.

Dunkerley, Joy. "The Future for Coal in Western Europe."

Resources Policy 4, No. 3 (September 1978), pp. 151–159.

Earney, Fillmore C. F. "Norway's Offshore Petroleum Industry." *Resources Policy* 8, No. 2 (June 1982), pp. 133–142.

"EC Orders Cuts in Steel Capacity of Up to 19.7%." *The Wall Street Journal*, 1 July 1983, p. 16.

"Energy: Long-term Objectives and Strategy." *Bulletin of the European Communities*, No. 5 (1980), pp. 23–25.

"Europe's Energy Options." *Mining Journal*, 8 October 1982, pp. 245–246.

Frame, A. G. "Role of the United Kingdom in the World Minerals Industry." Robert Pryor Memorial Lecture to the Royal School of Mines Mining and Metallurgical Society, London, 14 January 1980.

Frey-Wouters, Ellen. *The European Community and the Third World*. New York: Praeger Publishers, 1980.

"GAO: Europe Has No Alternative to Soviet Gas." *Oil and Gas Journal*, 6 June 1983, p. 44.

Hoffman, George W. *Energy Strategies of Western Europe in the 1980's: Dependable Supplies versus Security Implications*. Policy Study No. 18. Center for Energy Studies, University of Texas at Austin, July 1982.

Hoffman, Lutz. "Policy of the German Federal Government and European Countries for Securing the Supply of Primary Commodities." In *Primary Commodities: Security of Supply Conference*, Washington, D.C., 11-12 December 1980.

Ibrahim, Youssef M. "West Moves to Counter Gas Cutoff." *The Wall Street Journal*, 21 June 1983, p. 39.

Jacobs, Jan. "Fuel Glut and Recession Prompt the Dutch to Ease Natural Gas Conservation Rules." *The Wall Street Journal*, 22 July 1983, p. 18.

Kilmarx, Robert A. "Security and Strategic Considerations for Commodity Policies." In *Primary Commodities: Security of Supply Conference*, Washington, D.C., 11-12 December 1980.

Maull, Hans W. "French and European Community Policies for Securing the Supply of Primary Commodities." In *Primary Commodities: Security of Supply Conference*, Washington, D.C., 11-12 December 1980.

Mufson, Steve. "Anatomy of Continuing Soviet Pipeline Controversy." *The Wall Street Journal*, 31 August 1982, p. 27.

"Nonfuel Minerals Stockpiling by Japan, South Korea, France, and Britain." *Alert Letter on the Availability of Raw Materials*, No. 13 (June 1982), p. 16.

"Northern Europe." "Western Europe." In *Mining Annual Review, 1982*. London: Mining Journal, June 1982. pp. 485–488, 515–538.

Organization for Economic Cooperation and Development. *Energy Balances of OECD Countries 1960–1974*. Paris: OECD, 1976.

Organization for Economic Cooperation and Development, International Energy Agency. *Energy Balances of OECD Countries 1971–1981*. Paris: OECD, 1983.

Page, William. "UK Balance of Payments in Raw Materials." *Resources Policy* 5, No. 3 (September 1979), pp. 185–196.

Parry, Audrey. *Towards a Minerals Strategy for Britain."* Foreign Affairs Research Institute Publ. No. 5 (1981).

Robinson, Colin, and Eileen Marshall. *What Future for British Coal?* Hobart Paper No. 89. London: The Institute of Economic Affairs, 1981.

"Strategic Minerals—A More Precise Definition." *Mining Journal*, 19 November 1982.

Szuprowicz, Bohdan. *How to Avoid Strategic Materials Shortages*. New York: John Wiley & Sons, Inc., 1981.

Thurow, Roger. "German Cabinet Approves Aid for Steel Firms." *The Wall Street Journal*, 15 June 1983, p. 34.

U.S. Department of State, Bureau of Public Affairs. *Soviet–West European Natural Gas Pipeline*. Current Policy No. 331. Washington, D.C.: U.S. Government Printing Office, 1981.

U.S. Department of the Interior, Bureau of Mines. *Mineral Facts and Problems, 1980 Edition*. Washington, D.C.: U.S. Government Printing Office, 1980.

U.S. Department of the Interior, Bureau of Mines. "European Communities and Mineral Policy." *Minerals and Materials: A Bimonthly Survey*, August–September 1982. Washington, D.C.: U.S. Government Printing Office, 1982, p. 7.

U.S. Department of the Interior, Bureau of Mines. *Mineral Commodity Summaries, 1983*. Washington, D.C.: U.S. Government Printing Office, 1983.

U.S. Department of the Interior, Bureau of Mines. *Minerals Yearbook 1981*, Vol. III: *Area Reports: International*. Washington, D.C.: U.S. Government Printing Office, 1983.

U.S. Department of the Interior, Bureau of Mines. "United Kingdom Establishes Government Stockpile of Strategic Minerals." *Minerals and Materials: A Bimonthly Survey*, February–March 1983. Washington, D.C.: U.S. Government Printing Office, 1983. p. 10.

Warnecke, Steven J. *Stockpiling of Critical Raw Materials*. London: The Royal Institute of International Affairs, 1980.

"World Economic Outlook." *US News and World Report*, 22 August 1983, pp. 122–137.

"Worldwide Oil and Gas at a Glance." *Oil and Gas Journal*, 27 December 1982, pp. 78–79.

PART IV: THE SOVIET UNION

"Action and Reaction." *Mining Journal*, 14 March 1980, pp. 201–203.

"BAM's Aluminum." *The Current Digest of the Soviet Press* 34, No. 3 (24 February 1982), p. 21.

"Bankrolling the Slave Trade?" *The Wall Street Journal*, 3 May 1981, p. 30.

Barnea, Joseph, and William Epstein. "Europe's Gas: The Pipeline and Economies." *The Wall Street Journal*, 3 August 1982, p. 30.

Behrman, Neil. "Soviet's Large Foreign-Exchange Needs Force Export of Surplus Raw Materials." *The Wall Street Journal*, 13 July 1982, p. 40.

Behrman, Neil, and Anne Mackay-Smith. "Heavy Soviet, Mideast Buying of Silver Sparked Recent Market Rise,

Dealers Say." *The Wall Street Journal*, 1 November 1982, p. 38.

Behrman, Neil, and Anne Mackay-Smith. "Russian Bear Learns New Tricks As Soviets Create a Bull Market for Their Palladium." *The Wall Street Journal*, 26 November 1982, p. 14.

"Belgium Delays Setting Contract for Soviet Gas." *The Wall Street Journal*, 27 July 1982, p. 33.

Bergson, Abram. "Can the Soviet Slowdown Be Reversed?" *Challenge* 24, No. 5 (November-December 1981), pp. 33-42.

Bienen, Henry, Soviet Political Relations with Africa." *International Security* 6, No. 4 (Spring 1982), pp. 153-173.

Blau, Thomas, and Joseph Kirchheimer. "European Dependence and Soviet Leverage: The Yamal Pipeline." *Survival* 23, No. 5 (September-October 1981), pp. 209-214.

Bowman, Larry W. "The Strategic Importance of South Africa to the United States: An Appraisal and Policy Analysis." *African Affairs* 81, No. 323 (April 1981), pp. 159-191.

Brand, David. "Add the Soviet Union to the List of Those with Cash Problems." *The Wall Street Journal* 18 March 1982, p. 1.

Brand, David. "Europeans Subsidized Soviet Pipeline Work Mainly to Save Jobs." *The Wall Street Journal*, 2 November 1982, p. 1.

Bratchenko, B. F. "Basis for the Development of the Coal Industry of the USSR." *Natural Resources Forum* 1, No. 1 (October 1976), pp. 47-54.

Carter, Hodding, III. "Reagan's Two-Track Approach to Soviet Trade." *The Wall Street Journal*, 29 July 1982, p. 19.

Chesshire, J. H., and C. Huggett. "Primary Energy Production in the Soviet Union." *Energy Policy*, September 1975, pp. 223-244.

"Chromium in the Eastern Bloc." *Mining Journal*, 3 December 1982, pp. 389-390.

"COMECON in the 1980's." *Mining Journal*, 30 April 1982, pp. 315-317.

"A Complex Takes Shape." *The Current Digest of the Soviet Press* 34, No. 42 (17 November 1982), pp. 27-28.

"Construction Lags at Urengoi Gas Fields." *The Current Digest of the Soviet Press* 34, No. 23 (7 July 1982), pp. 14-15.

Cooper, Hal B. H., Jr., Dannenbaum Engineering Corporation. *A Comparative Analysis of Railroad Electrification between the United States and the Soviet Union.* Prepared for Senator John G. Tower, United States Senate, Armed Services Committee, 15 March 1982.

Council on Economics and National Security. *The "Resource War" and the U.S. Business Community: The Case for a Council on Economics and National Security.* Washington, D.C.: Council on Economics and National Security, 1980.

Council on Economics and National Security. *Strategic Minerals: A Resource Crisis.* Washington, D.C.: Council on Economics and National Security, 1981.

Cressey, George B. *Soviet Potentials: A Geographic Appraisal.* Syracuse, N.Y.: Syracuse University Press, 1962.

Crozier, Brian. "Strategic Decisions for the West." *Soviet Analyst* 10, No. 5 (4 March 1981), pp. 1-3.

"Cutting Oil Sales to the West." *World Business Weekly*, 30 March 1981, pp. 16-17.

"Deflating Pipeline Buster." *The Wall Street Journal*, 30 July 1982, p. 18.

"Delayed Automation, Low Productivity." *Soviet Analyst* 11, No. 22 (10 November 1982), pp. 4-5.

"DIA: Soviets Will Meet Energy Targets." *Oil and Gas Journal*, 21 September 1981, pp. 94-95.

Dienes, Leslie, and Theodore Shabad. *The Soviet Energy System: Resource Use and Policies.* Washington, D.C.: V. H. Winston & Sons, 1979.

Dyker, David. "Planning in Siberia on Wrong Track." *Soviet Analyst* 9, No. 2 (23 January 1980), pp. 4-5.

Eckes, Alfred E., Jr. *The United States and the Global Struggle for Minerals.* Austin, Tex.: University of Texas Press, 1979.

"The Economy." *The Current Digest of the Soviet Press* 34, No. 30 (25 August 1982), pp. 20-21.

"Ekibastuz Complex Beset by Problems." *The Current Digest of the Soviet Press* 34, No. 30 (25 August 1982), pp. 1-5.

"Energy Crisis Looms in the Soviet Bloc." *Soviet Analyst* 7, No. 25 (21 December 1978), pp. 2-5.

"Energy Projects and Plans." *The Current Digest of the Soviet Press* 34, No. 7 (17 March 1982), pp. 1-4.

"Europe Gets Ready to Strike Back." *Business Week*, 19 July 1982, pp. 50-51.

Fine, Daniel I. "Fresh Fears That the Soviets Will Cut Off Critical Minerals." *Business Week*, 28 January 1980, pp. 62-63.

Fine, Daniel I. "Mineral Resource Dependency Crisis: Soviet Union and United States." In *The Resource War in 3-D: Dependency, Diplomacy, Defense.* Edited by James A. Miller, Daniel, I. Fine and R. Daniel McMichael. Pittsburgh, Pa.: World Affairs Council of Pittsburgh, 1980, pp. 37-56.

Foreign Affairs Research Institute. *The Need to Safeguard NATO's Strategic Raw Materials from Africa.* No. 13 (1977).

"The Four-Letter Pipeline." *The Wall Street Journal*, 6 October 1982, p. 26.

Frantzell, Lennart. "Soviet Economic Problems Forecast." *Soviet Analyst* 9, No. 1 (9 January 1980), pp. 5-6.

Garn, Jake. "Questions about the Soviet Gas Pipeline." *The Wall Street Journal*, 20 May 1982, p. 30.

"Gas Pipeline to West Makes Progress." *The Current Digest of the Soviet Press* 34, No. 33 (15 September 1982), pp. 1-5.

Geddes, John M. "Ruhrgas Says US Ban, Other Problems Won't Halt Soviet Gas Pipeline to West." *The Wall Street Journal*, 8 June 1982, p. 38.

Geddes, John M. "Effect of Reagan Ban on Pipeline Gear Is Disputed." *The Wall Street Journal*, 28 June 1982, p. 16.

"Gold in the Soviet Far East." *World Mining*, December 1981, pp. 47–49.

Goldman, Marshall I. *The Enigma of Soviet Petroleum*. London: George Allen & Unwin (Publisher) Ltd., 1980.

Green, Timothy. *The New World of Gold*. New York: Walker and Company, 1981.

Grichar, James S., Richard Levine, and Lotfallah Nahai. *The Nonfuel Mineral Outlook for the USSR through 1990*. Washington, D.C.: U.S. Government Printing Office, 1981.

Gromyko, Anatoly. "Soviet Foreign Policy and Africa." *International Affairs (Moscow)*, February 1982, pp. 30–35.

Gustafson, Thane. "Energy and the Soviet Bloc." *International Security* 6, No. 3 (Winter 1981-82), pp. 65–89.

Gwertzman, Bernard. "CIA Revises Estimate, Sees Soviet as Oil-Independent through 80's." *The New York Times*, 19 May 1981, p. A1.

Hanson, Philip. "Economic Constraints on Soviet Policies in the 1980's." *International Affairs* 57, No. 1 (Winter 1980-81), pp. 21–42.

Hegge, Per Egil. "Norway's Gas Reserves and the Soviet Pipeline." *The Wall Street Journal*, 14 July 1982, p. 27.

Hewett, Ed A. "The Pipeline Connection: Issues for the Alliance." *The Brookings Review*, Fall 1982, pp. 15–20.

Hoffman, George W. *Energy Projections: Oil, Natural Gas, and Coal in the USSR and Eastern Europe*. Policy Study No. 3. Center for Energy Studies, University of Texas at Austin, August 1978.

Hoffman, George W. *Eastern Europe's Resource Crisis, with Special Emphasis on Energy Resources: Dependence and Policy Options*. Policy Study No. 14. Center for Energy Studies, University of Texas at Austin, January 1981.

Hoffman, George W. *Energy Strategies of Western Europe in the 1980's: Dependable Supplies versus Security Implications*. Policy Study No. 18. Center for Energy Studies, University of Texas at Austin, July 1982.

"How Shortages Impair the Soviet Economy." *The Current Digest of the Soviet Press* 34, No. 17 (26 May 1982), pp. 4–5.

"How Will Urals Get More Energy?" *The Current Digest of the Soviet Press* 34, No. 37 (13 October 1982), pp. 11–12.

"Impasse on the Soviet Pipeline Deal." *World Business Weekly*, 30 March 1981, p. 16.

"Industrial, Agricultural Figures Give Added Grounds for Gloom about Soviet Economy." *Soviet World Outlook* 7, No. 3 (15 March 1982), p. 6–7.

"Industry." *The Current Digest of the Soviet Press* 34, No. 25 (28 July 1982), p. 21.

"Industry." *The Current Digest of the Soviet Press* 34, No. 27 (4 August 1982), p. 19.

"Industry." *The Current Digest of the Soviet Press* 34, No. 37 (13 October 1982), pp. 20–21.

Ioffe, Olympiad S. "Law and Economy in the USSR." *Harvard Law Review* 95 (1982), pp. 1591-1625.

"Italy Likely to Honor Soviet Pipeline Pacts, despite US Embargo." *The Wall Street Journal*, 9 July 1982, p. 16.

Joffe, Josef. "Pipeline Embargo Aims at Russians, Hits Europeans." *The Wall Street Journal*, 28 July 1982, p. 19.

Karr, Miriam, and Roger W. Robinson, Jr. "Soviet Gas: Risk or Reward?" *The Washington Quarterly* 4, No. 4 (Autumn 1981), pp. 3–11.

Kessler, Felix. "Italy Follows French Stance on Pipeline." *The Wall Street Journal*, 26 July 1982, p. 21.

Kilmarx, Robert A. "Security and Strategic Considerations for Commodity Policies." In *Primary Commodities: Security of Supply Conference*, Washington, D.C., 11-12 December 1980.

Kramer, Barry. "Soviet Union's Major Propaganda Blitz Calls US Pipeline Embargo Ineffective." *The Wall Street Journal*, 3 August 1982, p. 34.

Krauss, Melvyn B. "The Siberian Pipeline and Europe's Welfare State." *The Wall Street Journal*, 19 July 1982, p. 16.

"Kremlin Concern: Not Enough Russians." *US News and World Report*, 13 September 1982, p. 12.

Kroncher, Allan. "Economic Progress Is Only on Paper." *Soviet Analyst* 11, No. 5 (10 March 1982), pp. 3–5.

"Kuzbass Coal Mining Is Developing Slowly. *The Current Digest of the Soviet Press* 34, No. 26 (4 August 1982), pp. 5–6.

Levine, Richard M. "Soviet Union." In *Mining Annual Review, 1982*. London: Mining Journal, June 1982, pp. 489–503.

"Life with an Andropov." *The Economist*, 9 October 1982, pp. 11–12.

"Looking into Industrial Bottlenecks." *The Current Digest of the Soviet Press* 34, No. 17 (26 May 1982), pp. 6–8.

Mathieson, R. S. *The Soviet Union: An Economic Geography*. London: Heinemann Educational Books Ltd., 1975.

Maxwell, Kenneth. "A New Scramble for Africa?" *The Conduct of Soviet Foreign Policy*. Edited by Erik P. Hoffmann and Frederic J. Fleron, Jr. Chicago: Aldine Publishing Company, 1980, pp. 515–534.

"Metal Signals." *Mining Journal*, 14 August 1981, pp. 105–106.

Meyer, Herbert E. "Why We Should Worry about the Soviet Energy Crunch." *Fortune*, 25 February 1980, pp. 82–88.

Meyer, Herbert E. "Russia's Sudden Reach for Raw Materials." Reprinted from *Fortune*, 28 July 1980.

Morgan, Michael. "Minerals: Key to Soviet Current Strategies?" *Defense and Foreign Affairs Digest*, December 1979, pp. 17–20.

"Moscow Combats Sanctions with Bluff." *Soviet Analyst* 11, No. 14 (14 July 1982), pp. 1–3.

"Moscow Presses Total Energy Program as Top Priority." *Soviet World Outlook* 4, No. 4 (16 April 1979), pp. 3–4.

Mufson, Steve. "US Efforts to Block Soviet Gas Pipeline Recalls Failed Embargo of 20 Years Ago." *The Wall Street Journal*, 14 July 1982, p. 30.

Mufson, Steve. "Allegations Soviets Using 'Slave Labor' Heat Up Debate over Pipeline to Europe." *The Wall Street Journal*, 17 August 1982, p. 32.

Mufson, Steve. "Anatomy of Continuing Soviet Pipeline

Controversy." *The Wall Street Journal*, 31 August 1982, p. 27.

Muller, Robert L., and David Brand. "Britain Orders 4 Firms to Defy U.S. Pipeline Ban." *The Wall Street Journal*, 3 August 1982, p. 35.

"The Nasty Pipeline Mess: Any Way Out?" *US News and World Report*, 13 September 1982, pp. 27-29.

"Oil Resource Debate Hots Up." *Soviet Analyst* 8, No. 21 (25 October 1979), pp. 7-8.

"Organizing Tuymen Oil, Gas Development." *The Current Digest of the Soviet Press* 34, No. 9 (31 March 1982), pp. 4-6.

Papp, Daniel S. "Soviet Non-fuel Mineral Resources, Surplus or Scarcity?" *Resources Policy* 8, No. 3 (September 1982), pp. 155-176.

"Periscope: Two Servings of Surplus Food." *Newsweek*, 24 January 1983, p. 15.

"Pipeline Crisis: The Feud Heats Up." *US News and World Report*, 2 August 1982, pp. 25-26.

"Pipeline Dispute: Who Benefits?" *Soviet Analyst* 11, No. 17 (1 September 1982), pp. 1-3.

"Pipelines Hold Key to Soviet Gas Production." *Oil and Gas Journal*, 29 June 1981, pp. 39-43.

Piper, Hal. "Soviet Economic Ills May Trap Andropov." *Austin American-Statesman*, 14 November 1982, p. C1.

"Platinum: A New Market Catalyst?" *Mining Journal*, 21 January 1983, pp. 33-35.

"Platinum's Tribulations." *Mining Journal*, 29 October 1982, pp. 301-302.

"Politicizing Politics." *The Wall Street Journal*, 12 July 1982, p. 16.

Prest, Michael. "Soviets Learn to Play Markets to Their Best Advantage." *The Globe and Mail*, 2 August 1982, p. B6.

Prewo, Wilfried. "The Pipeline: White Elephant or Trojan Horse?" *The Wall Street Journal*, 28 September 1982, p. 32.

"Progress Resumes on the Soviet Gas Pipeline to Western Europe." *World Business Weekly*, 29 June 1981, pp. 15-16.

Ratneiks, H. "The USSR Leans on Comrade Coal." *The Geographical Magazine* 54, No. 4 (April 1982), pp. 207-211.

Rees, David. *Soviet Strategic Penetration in Africa*. London: Institute for the Study of Conflict, 1976.

" 'Resource War' Targeted as a Major National Problem." *Alert Letter on the Availability of Raw Materials*, No. 1 (May 1981), pp. 1-2.

"Rubles on the Barrelhead." *The Wall Street Journal*, 19 July 1982, p. 16.

Rumer, Boris, and Stephen Sternheimer. "The Soviet Economy: Going to Siberia?" *Harvard Business Review*, January-February 1982, pp. 16-38.

Russett, Bruce. "Security and the Resources Scramble: Will 1984 Be Like 1914?" *International Affairs* 58, No. 1 (Winter 1981-82), pp. 42-58.

Scanlon, Tony. "Outlook for Soviet Oil." *Science* 217, No. 4557 (23 July 1982), pp. 325-330.

Seib, Gerald F. "U.S. Policy and Russia's Gas Pipeline." *The Wall Street Journal*, 25 May 1982, p. 27.

Seib, Gerald F. "Reagan Move Widening Soviet Sanctions Is Generating New Friction among Aides." *The Wall Street Journal*, 12 July 1982, p. 6.

Shabad, Theodore. *Basic Industrial Resources of the USSR*. New York: Columbia University Press, 1969.

Shabad, Theodore. "News Notes—Arctic Aspects of the Soviet Five-Year Plan (1981-85)." *Polar Geography and Geology* 5, No. 2 (April-June 1981), pp. 122-124.

Shabad, Theodore. "The Soviet Mineral Potential and Environmental Constraints." *Mineral Economics Symposium*, Washington, D.C., 8 November 1982.

Shabad, Theodore, and Victor L. Mote. *Gateway to Siberian Resources (The BAM)*. New York: John Wiley & Sons, Inc., 1977.

Shafer, Michael. "Mineral Myths." *Foreign Policy*, No. 47 (Summer 1982), pp. 154-171.

Shimkin, Demitri B. *Minerals—A Key to Soviet Power*. Cambridge, Mass.: Harvard University Press, 1953.

"Showdown Time over the Gas Pipeline." *US News and World Report*, 6 September 1982, p. 7.

"Siberian Pipeline Problems." *Soviet Analyst* 10, No. 21 (21 October 1981), pp. 7-8.

Slay, General Alton D. "Minerals and National Defense." *The Mines Magazine*, December 1980, pp. 4-7.

Smirnov, V. I. *Ore Deposits of the USSR*. Vols. I, II, and III. San Francisco: Pitman Publishing, Inc., 1977.

"South Africa: Persian Gulf of Minerals. Strategic Minerals Threat." Backgrounder issued by the Information Counselor, South African Embassy. No. 9 (December 1980).

"Soviet Deficit Foreseen." *Daily Texan*, 13 September 1982, p. 14.

"Soviet Gas and Western Hot Air." *Soviet Analyst* 11, No. 2 (27 January 1982), pp. 5-6.

"Soviet Gold." *Mining Journal*, 3 October 1980, pp. 265-267.

"Soviet Gold Production" and "By-Product Gold Production in the Soviet Union." In *Gold 1980*. By David Potts. London: Consolidated Gold Fields Limited, 1980, pp. 46-58.

"Soviet Industry Had Slower Growth Rate in First Half of '82." *The Wall Street Journal*, 22 July 1982, p. 22.

"Soviet Occupation of Afghanistan Has Resource Manifestations." *Alert Letter on the Availability of Raw Materials*, June 1982, pp. 8-9.

"Soviet Oil: Rise and Fall?" *Soviet Analyst* 6, No. 9 (5 May 1977), pp. 4-6.

"Soviets Boost Oil Production Goal in W. Siberia." *Oil and Gas Journal*, 18 January 1982, p. 62.

"Soviets Eye Output from Big Gas Field." *Oil and Gas Journal*, 12 April 1982, p. 184.

"Soviets Press Construction of 56 in. Gas Pipelines." *Oil and Gas Journal*, 14 June 1982, pp. 27-30.

"Soviets Say Pipeline Won't Be Delayed by U.S. Sanctions." *The Wall Street Journal*, 2 August 1982, p. 16.

"Soviets Seen Cutting Palladium Supplies by 25% to the West." *The Wall Street Journal*, 20 December 1982, p. 26.

"Soviet Union's Gold Sales Fell 25% Last Year, Traders Say, Perhaps Signaling Economic Gains." *The Wall Street Journal*, 31 January 1982, p. 29.

"Soviet Union Starts Its First Drilling Project in Arctic Waters." *Oil and Gas Journal*, 22 February 1982, p. 41.

Strauss, Michael J. "Soviets to Boost Gas Production at Urengoi Field." *The Wall Street Journal*, 17 June 1982, p. 29.

Strauss, Simon D. "Mineral Self-Sufficiency—The Contrast between the Soviet Union and the United States. *Mining Congress Journal*, November 1979, pp. 49-59.

Strishkov, V. V. "Soviet Union." In *Mining Annual Review*. London: Mining Journal, 1976-1980.

Strishkov, V. V. "The Copper Industry of the USSR." *Mining Magazine*, "Part One, 1640-1945," March 1979, pp. 242-253; "Part Two, 1946-1980," May 1979, pp. 429-441.

Sutulov, Alexander. *Mineral Resources and the Economy of the USSR*. New York: McGraw-Hill Book Company, 1973.

Sutulov, Alexander. *The Soviet Challenge in Base Metals*. Salt Lake City, Utah: University of Utah Press, 1971.

Szuprowicz, Bohdan. "Fear Soviet Supercartel for Critical Minerals." *Purchasing*, 8 November 1978, pp. 42-49.

Szuprowicz, Bohdan. "Russian Drive to Control Africa's Strategic Minerals." *The Bulletin*, 15 May 1979, pp. 78-84.

Szuprowicz, Bohdan. "Critical Materials Cut-Off Feared." *High Technology*, November-December 1981, pp. 44-50.

Szuprowicz, Bohdan. *How to Avoid Strategic Materials Shortages*. New York: John Wiley & Sons, Inc., 1981.

"Tell-Tale Gaps in USSR Statistics." *Soviet Analyst* 10, No. 6 (18 March 1981), pp. 6-8.

Tilton, John E., and Hans H. Landsberg. "Nonfuel Minerals: The Fear of Shortages and the Search for Policies." *RFF Forum*, Washington, D.C., 21 October 1982.

"Trying to Outbluff Reagan in the Pipeline Poker Game." *Business Week*, 16 August 1982, pp. 42-44.

"Uneven Start for Large-Scale Project." *Soviet Analyst* 7, No. 13 (29 June 1978), pp. 6-7.

U.S. Central Intelligence Agency. *Prospects for Soviet Oil Production*. Washington, D.C.: U.S. Government Printing Office, 1977.

U.S. Central Intelligence Agency. *Prospects for Soviet Oil Production, A Supplemental Analysis*. Washington, D.C.: U.S. Government Printing Office, 1977.

U.S. Central Intelligence Agency. *Polar Regions*. Washington, D.C.: U.S. Government Printing Office, 1978, pp. 22-23.

U.S. Central Intelligence Agency, National Foreign Assessment Center. *USSR: Development of the Gas Industry*. Washington, D.C.: U.S. Government Printing Office, 1978.

U.S. Central Intelligence Agency, National Foreign Assessment Center. *The Lead and Zinc Industry in the USSR*. Washington, D.C.: U.S. Government Printing Office, 1980.

U.S. Central Intelligence Agency, National Foreign Assessment Center. *The Soviet Economy in 1978-1979 and Prospects for 1980*. Washington, D.C.: U.S. Government Printing Office, 1980.

U.S. Central Intelligence Agency, National Foreign Assessment Center. *USSR: Coal Industry Problems and Prospects*. Washington, D.C.: U.S. Government Printing Office, 1980.

U.S. Central Intelligence Agency, National Foreign Assessment Center. *Handbook of Economic Statistics 1981*. Washington, D.C.: U.S. Government Printing Office, 1981.

U.S. Congress. Committee on Interior and Insular Affairs, Minerals, Materials, and Fuels Subcommittee. *Mineral Resources of and Background Information on the Eastern Hemisphere Including the Soviet Union and Satellite Countries*. 85th Congress, 2nd session. Washington, D.C.: U.S. Government Printing Office, 1958.

U.S. Congress. House of Representatives. Committee on Banking, Finance, and Urban Affairs, Subcommittee on Economic Stabilization. *A Congressional Handbook on US Materials Import Dependency/Vulnerability*. By Congressional Research Service, Library of Congress. Washington, D.C.: U.S. Government Printing Office, 1981.

U.S. Congress. Joint Economic Committee, Subcommittee on International Trade, Finance, and Security Economics. *Energy in Soviet Policy*. 97th Congress, 1st session. Washington, D.C.: U.S. Government Printing Office, 1981.

U.S. Congress, Office of Technology Assessment. *Technology and Soviet Energy Availability*. Washington, D.C.: U.S. Government Printing Office, 1981.

U.S. Department of Commerce. *Industrial Outlook Report. Minerals. South Africa—1981*. Washington, D.C.: U.S. Government Printing Office, 1982.

U.S. Department of State, Bureau of Public Affairs. *Soviet-West European Natural Gas Pipeline, Oct. 14, 1981*. Current Policy No. 331. Washington, D.C.: U.S. Government Printing Office, 1981.

U.S. Department of the Interior, Bureau of Mines. *Minerals Yearbook*, 1963-1979. Washington, D.C.: U.S. Government Printing Office.

U.S. Department of the Interior, Bureau of Mines. *Mineral Industries of Eastern Europe and the USSR*. Washington, D.C.: U.S. Government Printing Office, 1978.

U.S. Department of the Interior, Bureau of Mines. *Mineral Industries of the USSR*. By V.V. Strishkov. Washington, D.C.: U.S. Government Printing Office, 1979.

U.S. Department of the Interior, Bureau of Mines. *Mineral Facts and Problems, 1980 Edition*. Washington, D.C.: U.S. Government Printing Office, 1980.

U.S. Department of the Interior, Bureau of Mines. "The Mineral Industry of the USSR." By V. V. Strishkov. Preprint from the *Minerals Yearbook 1980*. Washington, D.C.: U.S. Government Printing Office, 1980.

U.S. Department of the Interior, Bureau of Mines. *Mineral Commodity Summaries 1982/1983*. Washington, D.C.: U.S. Government Printing Office, 1982 and 1983.

"US Hopes to Bar Occidental's Hammer from Building Siberia-Moscow Pipeline." *The Wall Street Journal*, 8 October 1982, p. 5.

"USSR Energy Targets: 1—Electricity." *Soviet Analyst* 10, No. 11 (27 May 1981), pp. 7-8.

"USSR Energy Targets: 2—Oil." *Soviet Analyst* 10, No. 12 (10 June 1981), pp. 4-6.

"USSR Energy Targets: 3—Gas." *Soviet Analyst* 10, No. 13 (24 June 1981), pp. 5-7.

"USSR Energy Targets: 4—Coal." *Soviet Analyst* 10, No. 14 (8 July 1981), pp. 5-6.

"USSR Finishing Tests of Its Biggest Gas Compressors." *Oil and Gas Journal*, 21 June 1982, pp. 112-113.

"USSR—New Plan, Old Leaders." *Mining Journal*, 13 March 1981, pp. 189-190.

"USSR's Biggest Oil Field Passes Milestone." *Oil and Gas Journal*, 31 August 1982, p. 30.

"USSR's Pipeline Bluff." *Soviet Analyst* 11, No. 2 (27 January 1982), pp. 7-8.

"Value of Soviet Oil Exports Hits Record." *Oil and Gas Journal*, 23 August 1982, pp. 66-67.

Vanous, Jan. "East European Economic Slowdown." *Problems of Communism* 31, No. 4 (July-August 1982), pp. 1-19.

Vogely, William A. "Resource War?" *Materials and Society* 6, No. 1 (1982), pp. 1-3.

Von der Heydt, Peter. "The Pipeline Dispute: Both Sides Have Erred." *The Wall Street Journal*, 4 August 1982, p. 21.

Walker, General Sir Walter. *The Next Domino?* Sandton City, South Africa: Valiant Publishers (Pty.) Ltd., 1980.

"West Siberia's Gas." *New Times*, No. 19 (May 1982), pp. 18-21.

"What Can Russia Afford?" *Foreign Report*, 11 November 1982, pp. 2-3.

"What Next for U.S. Minerals Policy?" *Resources*, No. 71 (October 1982), pp. 9-10.

"Why the Soviet Economy Slights Innovation." *The Current Digest of the Soviet Press* 34, No. 18 (2 June 1982), pp. 1-4.

Wilson, David. *Soviet Oil and Gas to 1990*. Economist Intelligence Unit (EIU) Special Report 90. London: Spencer House, 1980.

"The Worries of Soviet Oilmen." *Soviet Analyst* 9, No. 10 (14 May 1980), pp. 5-8.

"Yamal Gas Pipeline: Growing Risks." *Soviet Analyst* 10, No. 4 (18 February 1981), pp. 6-7.

Index